Hybrid Water Electrolysis

Hybrid Water Electrolysis

Non-Oxide Electrocatalysts in Small Molecule Oxidation

Edited by

Myong Yong Choi

Jayaraman Theerthagiri

M. L. Aruna Kumari

Gilberto Maia

Soorathep Kheawhom

WILEY-VCH

Editors

Myong Yong Choi
Department of Chemistry
Gyeongsang National University
Jinju, Republic of Korea

Jayaraman Theerthagiri
Department of Chemistry
Gyeongsang National University
Jinju, Republic of Korea

M. L. Aruna Kumari
Department of Chemistry
The Oxford College of Science
Bengaluru, India

Gilberto Maia
Institute of Chemistry
Federal University of Mato Grosso do Sul
Campo Grande, Brazil

Soorathep Kheawhom
Department of Chemical Engineering
Chulalongkorn University
Bangkok, Thailand

Cover Design: Wiley
Cover Image: BOY ANTHONY/Shutterstock

Typesetting: Lumina Datamatics, Madurai, India

Printing and Binding: CPI Group (UK) Ltd, Croydon, CR0 4YY

Library of Congress Card No.: applied for

British Library Cataloguing-in-Publication Data
A catalogue record for this book is available from the British Library.

Bibliographic information published by the Deutsche Nationalbibliothek
The Deutsche Nationalbibliothek lists this publication in the Deutsche Nationalbibliografie; detailed bibliographic data are available on the Internet at <http://dnb.d-nb.de>.

Print ISBN: 9783527355822
ePDF ISBN: 9783527854462
ePub ISBN: 9783527854455
oBook ISBN: 9783527854479

C9783527355822_250626

Contents

List of Contributors

Akram Alfantazi
Department of Chemical and Petroleum Engineering
Khalifa University of Science and Technology
Abu Dhabi, United Arab Emirates

Rafael M. P. Araujo
Laboratory of Electrochemistry and Advanced Materials (LEMA), Institute of Chemistry
Federal University of Mato Grosso do Sul
Campo Grande, Brazil

M. L. Aruna Kumari
Department of Chemistry
The Oxford College of Science
Bengaluru, India

Vipada Aupama
Department of Chemical Engineering and Department of Chemical Technology
Chulalongkorn University
Bangkok, Thailand

Talshyn Begildayeva
Institute for Integrated Catalysis
Pacific Northwest National Laboratory
Richland, WA, USA

Giuseppe A. Camara
Laboratory of Electrochemistry and Advanced Materials (LEMA), Institute of Chemistry
Federal University of Mato Grosso do Sul
Campo Grande, Brazil

Eduardo dos Santos Freitas Cardoso
São Carlos Institute of Chemistry
University of São Paulo
São Carlos, Brazil

Paulo N. S. Casciano
Department of Analytical and Physical Chemistry
Federal University of Ceará,
Pici Campus
Fortaleza, Brazil

K. Chandana
Department of Chemistry,
Adichunchanagiri School of Natural Sciences
Adichunchanagiri University
Mandya, India

Vishwajit Chavda
Department of Chemistry, Institute of Science, Humanities and Liberal Studies
Indus University
Ahmedabad, India

Adriana N. Correia
Department of Analytical and Physical Chemistry
Federal University of Ceará,
Pici Campus
Fortaleza, Brazil

Sidney C. Z. Costa
Laboratory of Electrochemistry and Advanced Materials (LEMA), Institute of Chemistry
Federal University of Mato Grosso do Sul
Campo Grande, Brazil

and

Federal Institute of Mato Grosso do Sul
Corumbá, Brazil

Vijayakumar Elayappan
Advanced Energy Materials R&D Division
Dongkwang Co. Ltd.,
Anyang-si, Republic of Korea

Guilherme V. Fortunato
Sustainable Energy Materials
Technical University of Munich, TUM Campus Straubing for Biotechnology and Sustainability
Straubing, Germany

and

São Carlos Institute of Chemistry
University of São Paulo
São Carlos, Brazil

P. H. Gangadharagowda
Department of Chemistry, BGS Institute of Technology
Adichunchanagiri University
Mandya, India

M. Janete Giz
Laboratory of Electrochemistry and Advanced Materials (LEMA), Institute of Chemistry
Federal University of Mato Grosso do Sul
Campo Grande, Brazil

K. Karuppasamy
Department of Chemical and Petroleum Engineering
Khalifa University of Science and Technology
Abu Dhabi, United Arab Emirates

P. Keerthana
Department of Science
Alliance University
Bengaluru, India

Rahul D. Kerkar
Government College of Arts, Science and Commerce
Khandola-Marcel, India

Soorathep Kheawhom
Center of Excellence on Advanced Materials for Energy Storage; Department of Chemical Engineering; and Bio-Circular-Green Economy Technology and Engineering Center, Department of Chemical Technology
Chulalongkorn University
Bangkok, Thailand

Hyun-Seok Kim
Division of Electronics and Electrical Engineering
Dongguk University
Seoul, Republic of Korea

K. Yogesh Kumar
Department of Sciences
Alliance School of Sciences,
Alliance University, Central Campus
Bengaluru, India

Marcos Roberto de Vasconcelos Lanza
São Carlos Institute of Chemistry
University of São Paulo
São Carlos, Brazil

Marc Ledendecker
Sustainable Energy Materials
Technical University of Munich, TUM
Campus Straubing for Biotechnology
and Sustainability
Straubing, Germany

and

Forschungszentrum Jülich GmbH
Helmholtz Institute Erlangen-Nürnberg
for Renewable Energy (IET-2)
Erlangen, Germany

Pedro de Lima-Neto
Department of Analytical and Physical
Chemistry
Federal University of Ceará, Pici Campus
Fortaleza, Brazil

Juan A. Lopez-Ruiz
Institute for Integrated Catalysis
Pacific Northwest National
Laboratory
Richland, WA, USA

Julio C. Lourenço
São Carlos Institute of Chemistry
University of São Paulo
São Carlos, Brazil

Gilberto Maia
Institute of Chemistry
Federal University of Mato Grosso
do Sul
Campo Grande, Brazil

Soumya Ranjan Mishra
Centre for Nano and Material
Sciences
Jain University, Jain Global Campus
Bengaluru, India

Ahmad Azmin Mohamad
School of Materials and Mineral
Resources Engineering
Malaysia Science University
Nibong Tebal, Malaysia

and

Department of Chemical Engineering
Chulalongkorn University
Bangkok, Thailand

B. M. Nagaraja
Centre for Nano and Material
Sciences
Jain University, Jain Global Campus
Bengaluru, India

Shreyanka Shankar Naik
National Institute for Scientific
Research—Energy, Materials, and
Telecommunications Center
Varennes, QC, Canada

Senthilkumar Nangan
Centre for Research and Development,
Department of Chemistry, Vinayaka
Mission's Kirupananda Variyar Arts &
Science College
Vinayaka Mission's Research
Foundation
Salem, India

Jessica J. P. Nascimento
Department of Analytical and Physical
Chemistry
Federal University of Ceará, Pici
Campus
Fortaleza, Brazil

and

Postgraduate Program in Engineering and Materials Science
Federal University of Ceará, Pici Campus
Fortaleza, Brazil

Thiago M. B. F. Oliveira
Center for Science and Technology
Federal University of Cariri
Juazeiro do Norte, Brazil

Dhananjai Pangotra
Department of Bio-Electro and Chemocatalysis (BioCat)
Fraunhofer Institute for Interfacial Engineering and Biotechnology IGB
Straubing, Germany

Kiruthika Paramasivam
Materials Research and Product Laboratory, Department of Chemistry
Coimbatore Institute of Technology
Coimbatore, India

Hemalatha Parangusan
Department of Physics
Mannar Thirumalai Naicker College
Madurai, India

Tulio P. Porto
Lorena School of Engineering
University of São Paulo
Lorena, Brazil

Soorya S. Raj
Department of Chemistry
The Oxford College of Science
Bengaluru, India

Shalu Rawat
Centre for Nano and Material Sciences
Jain University, Jain Global Campus
Bengaluru, India

Francisco W. P. Ribeiro
Institute of Exact and Natural Sciences
University of International Integration of Afro-Brazilian Lusophony, Auroras Campus
Redenção, Brazil

Robson S. Rocha
Lorena School of Engineering
University of São Paulo
Lorena, Brazil

Marciélli Karoline Rodrigues de Souza
São Carlos Institute of Chemistry
University of São Paulo
São Carlos, Brazil

Pooja Sahoo
Department of Physics
Veer Surendra Sai University of Technology
Sambalpur, India

Binaya Kumar Sahu
Department of Physics, Panchayat Degree College
Sambalpur University
Sambalpur, India

Alexsandro J. dos Santos
Department of Chemistry and Environmental Sciences
São Paulo State University
São José do Rio Preto, Brazil

Aravindhan Selvaraj
Department of Chemistry
Dr. B. S. Abdur Rahman Crescent Institute of Science and Technology
Chennai, India

Yogapriya Selvaraj
Department of Chemical Engineering
Chulalongkorn University
Bangkok, Thailand

Phonnapha Tangthuam
Department of Chemical Engineering and Department of Chemical Technology
Chulalongkorn University
Bangkok, Thailand

Arun Varghese
Department of Sciences
Alliance School of Sciences,
Alliance University, Central Campus
Bengaluru, India

Anbazhagan Venkattappan
Centre for Research and Development
Department of Chemistry, Vinayaka Mission's Kirupananda Variyar Arts & Science College
Vinayaka Mission's Research Foundation
Salem, India

Luciana Vieira
Department of Bio-Electro and Chemocatalysis (BioCat)
Fraunhofer Institute for Interfacial Engineering and Biotechnology IGB
Straubing, Germany

Dhanasekaran Vikraman
Division of Electronics and Electrical Engineering
Dongguk University
Seoul, Republic of Korea

S. K. Vinoth
Department of Chemistry
Jain University
Bengaluru, India

Huize Wang
Forschungszentrum Jülich GmbH
Helmholtz Institute Erlangen-Nürnberg for Renewable Energy (IET-2)
Erlangen, Germany

Xiangyu You
Sustainable Energy Materials
Technical University of Munich, TUM Campus Straubing for Biotechnology and Sustainability
Straubing, Germany

About the Editors

Professor Myong Yong Choi is a distinguished physical chemist with expertise in spectroscopy, laser photochemistry, and nanomaterials. He is a senior faculty member in the Department of Chemistry at Gyeongsang National University (GNU), Republic of Korea, where he also serves as the Director of the Core-Facility Center for Photochemistry and Nanomaterials and the Director of the Graduate School for Molecular Material Chemistry. Prof. Choi earned his MS in Chemistry from the University of Idaho and his PhD from the University of North Carolina at Chapel Hill. He subsequently conducted postdoctoral research in the Faculty of Science at the University of Southern California. His research group has pioneered a variety of laser-assisted techniques for the synthesis and characterization of metal nanoparticles, alloys, nanocomposites, and coordination polymers. These efforts have significantly advanced applications in spectroscopy, chemical sensing, and catalysis. More recently, his team has expanded into the laser-mediated fabrication of functional nanomaterials and biomaterials aimed at energy conversion and environmental remediation. Prof. Choi has authored over 250 peer-reviewed publications in high-impact international journals, holds 15 patents, and has contributed to two edited books and 12 book chapters. He has also been invited to deliver numerous plenary, keynote, and invited lectures at leading international conferences and research institutions. In addition to his research accomplishments, Prof. Choi has served on key administrative and advisory roles at GNU, including the Center for Laboratory Safety, the Center for Research Facilities, and as Chair of the Department of Chemistry. He has also contributed to scientific advisory boards of external institutions. His contributions to science and education have been recognized through multiple awards and honors, including the Commendation Award, KCS/Sigma-Aldrich Excellent Chemistry Award, the Ministry of Education's Top 50 Excellent Research Achievements, Lifetime Membership in the Gaechuck Hanlim Academy, the Gaechuck Award for Excellence in Teaching, the KCS Kim Myung Soo Award, the KCS Young Physical Chemist Award, and the Ministry of Environment's Top 50 Environmental Technology Award.

Dr. Jayaraman Theerthagiri is currently working as Research Professor in the Department of Chemistry, GNU, Republic of Korea. Dr. Theerthagiri always expresses deep gratitude to his beloved children, Yokith and Ranika, for their unconditional love.

He graduated from Sacred Heart College, Tirupathur (affiliated to Thiruvalluvar University, India) with a BSc Chemistry degree in 2009. He received his MSc Chemistry degree from Ramakrishna Mission Vivekananda College, Chennai (affiliated to the University of Madras, India) in 2011, and his PhD degree from Thiruvalluvar University in 2017 and later, he worked as an Assistant Professor/Scientist-C in the Centre for Nanoscience and Nanotechnology, Sathyabama Institute of Science and Technology (Deemed to be University), Chennai, India from 2017 to 2019. His current research is focused on the development of electrocatalysts for the production of energy-saving fuels and value-added chemicals.

Dr. M. L. Aruna Kumari is presently working as an Assistant Professor and the Head of the Department of Chemistry, The Oxford College of Science, Bengaluru, India. She completed her Bachelor's degree in Chemistry from Tumkur University, India, in 2009, and Master's degree in Organic Chemistry from Kuvempu University, India, in 2011. She received her Doctoral degree from Bangalore University, India, in 2017. Her current research focuses on photocatalytic organic transformations, synthesis of metal oxides and their organic/inorganic hybrids for energy and environmental applications, and developing advanced oxidation processes (AOPs) for the removal of antibacterial- and antimicrobial-resistant organisms from water. She has published several research articles and book chapters of scientific repute.

Professor Gilberto Maia is a senior academic staff member at the Institute of Chemistry, Federal University of Mato Grosso do Sul, Brazil. He received his BSc degree in Chemistry from the Federal University of Ceará, Brazil, in 1987, and MSc degree in Physical-Chemistry from São Carlos Institute of Physics and Chemistry, University of São Paulo, Brazil, in 1990. He completed his PhD degree in Physical-Chemistry from São Carlos Institute of Physics and Chemistry in 1994. He has been conducting research particularly on nanostructured metal electrocatalysts (supported on carbon-based materials or otherwise) for use in the oxygen reduction reaction (ORR), yielding water or hydrogen peroxide, hydrogen evolution reaction (HER), oxygen evolution reaction (OER), CO_2 reduction reaction (CO_2RR), and nitrate reduction reaction (NO_3RR); and density functional theory for metallic surfaces (supported on carbon-based materials or otherwise) to be used in electrocatalysis.

Professor Soorathep Kheawhom received his PhD degree in Chemical System Engineering from The University of Tokyo, Japan. He is an Associate Professor at the Department of Chemical Engineering, Chulalongkorn University, Thailand. His main research focuses on sustainable energy storage technologies, i.e., zinc–air batteries and zinc-ion batteries. Currently, he is leading the research cluster on energy storage in Chulalongkorn University.

Preface

Hybrid water electrolysis has emerged as a powerful strategy to advance sustainable hydrogen production and value-added chemical synthesis. By combining conventional electrolysis with the oxidation of small organic molecules, this approach offers a promising route to overcome one of the most significant challenges in water splitting: the slow oxygen evolution reaction (OER). Over the past decade, growing interest in advanced electrocatalysts and nanostructured materials has reshaped our understanding of small-molecule oxidation and opened new pathways for efficient energy conversion. This book is driven by the need to consolidate these rapid developments into a single, coherent resource for the scientific and industrial communities.

Researchers in electrocatalysis, nanotechnology, materials science, and renewable energy will find in this book a comprehensive overview of recent advancements in hybrid water electrolysis. The book explores the fundamental mechanisms governing small-molecule oxidation and highlights how functional electrocatalysts such as metal sulfides, carbides, nitrides, phosphides, and single-atom catalysts can significantly improve reaction rates, selectivity, and stability. Special attention is given to the oxidation of important organic molecules including benzyl alcohol, methanol, ethanol, urea, hydrazine, furfural, and formic acid, all of which offer pathways to more energy-efficient hydrogen production and the cogeneration of valuable chemical products.

The rapid expansion of nanomaterials research and its increasing industrial importance provide further motivation for this work. Advances in pulsed laser techniques, surface engineering, and structure–property optimization have made it possible to design electrocatalysts that are abundant, cost effective, environmentally friendly, and scalable. Understanding how nanostructured materials influence electron transfer, active-site distribution, and catalytic durability is essential for developing next-generation hybrid electrolyzers capable of integrating renewable energy sources.

This book aims not only to summarize technological progress but also to give readers the conceptual tools needed to guide future innovations. It addresses key scientific challenges in hybrid water electrolysis, outlines emerging trends in catalyst design and electrochemical engineering, and offers perspectives on how these technologies may shape the future of sustainable energy systems. By bringing together insights from materials synthesis, electrochemical characterization, theoretical modeling, and industrial application, this book encourages interdisciplinary collaboration and stimulates new directions in research.

Ultimately, this book serves as a comprehensive yet accessible resource for postgraduate students, early career researchers, experienced scientists, and industry technologists. Whether one is designing new electrocatalysts, developing hybrid electrolyzer systems, or exploring pathways for cleaner chemical production, the knowledge presented here will support informed decision-making and inspire continued innovation. It is our hope that this work will accelerate the advancement of hybrid water electrolysis technologies and contribute meaningfully to global efforts toward clean energy, environmental sustainability, and a more flexible technological future.

Part I

Fundamental

1

Fundamentals and Advantages of Hybrid Water Electrolysis

Vipada Aupama[1,2], *Phonnapha Tangthuam*[1,2], *Ahmad Azmin Mohamad*[1,3], *and Soorathep Kheawhom*[1,2,4,*]

[1]*Department of Chemical Engineering, Chulalongkorn University, Bangkok, Thailand*
[2]*Department of Chemical Technology, Chulalongkorn University, Bangkok, Thailand*
[3]*School of Materials and Mineral Resources Engineering, Malaysia Science University, Nibong Tebal, Malaysia*
[4]*Center of Excellence on Advanced Materials for Energy Storage, Chulalongkorn University, Bangkok, Thailand*

*Corresponding author: soorathep.k@chula.ac.th

1.1 Concept and Scope

Conventional low-temperature water electrolysis relies on two coupled half-reactions: hydrogen evolution reaction (HER) at the cathode and oxygen evolution reaction (OER) at the anode. The corresponding reactions in alkaline and acidic media are shown in Eqs. (1.1)–(1.6) [1]:

alkaline media

$$\text{cathode (HER)} : 2H_2O + 2e^- \rightarrow H_2 + 2OH^- \tag{1.1}$$

$$\text{anode(OER)} : 4OH^- \rightarrow O_2 + 2H_2O + 4e^- \tag{1.2}$$

$$\text{overall} : 2H_2O \rightarrow 2H_2 + O_2 \tag{1.3}$$

acidic media

$$\text{cathode (HER)} : 2H^+ + 2e^- \rightarrow H_2 \tag{1.4}$$

$$\text{anode (OER)} : 2H_2O \rightarrow O_2 + 4H^+ + 4e^- \tag{1.5}$$

$$\text{overall} : 2H_2O \rightarrow 2H_2 + O_2 \tag{1.6}$$

HER can be accelerated across a wide range of catalyst classes [2, 3]. In contrast, OER is the dominant efficiency bottleneck in most practical electrolyzers because it involves multi-electron transfer, multiple adsorbed intermediates, and intrinsically sluggish kinetics, which together result in large overpotentials and substantial energy losses [4, 5]. In Figure 1.1a, a conceptual layout of a hybrid electrolyzer that retains HER while modifying the anodic reaction is shown.

Hybrid water electrolysis is based on a straightforward yet far-reaching modification. The anode is redesigned to partially or fully replace OER with an alternative anodic oxidation reaction (AOR), while HER remains the cathodic reaction responsible for hydrogen production [6, 7]. Electrons generated by anodic oxidation flow through the external circuit to drive HER. Charge-compensating ionic species, i.e., OH^- in alkaline or anion-exchange membrane systems, move through the electrolyte or membrane between compartments (Figure 1.1b). When AOR is chosen over OER due to its lower thermodynamic potential and more favorable kinetics, the anode can operate at a lower potential. This approach lowers the full cell voltage required to reach a given current density, thus enhancing the overall electricity-to-hydrogen efficiency [8].

In the literature, several terms are used to describe closely related concepts, including hybrid water electrolysis [9], assisted water electrolysis [10], paired electrolysis [11], and electrochemical reforming [12]. These notions overlap but are not identical. In this chapter, hybrid water electrolysis is defined as an electrolysis

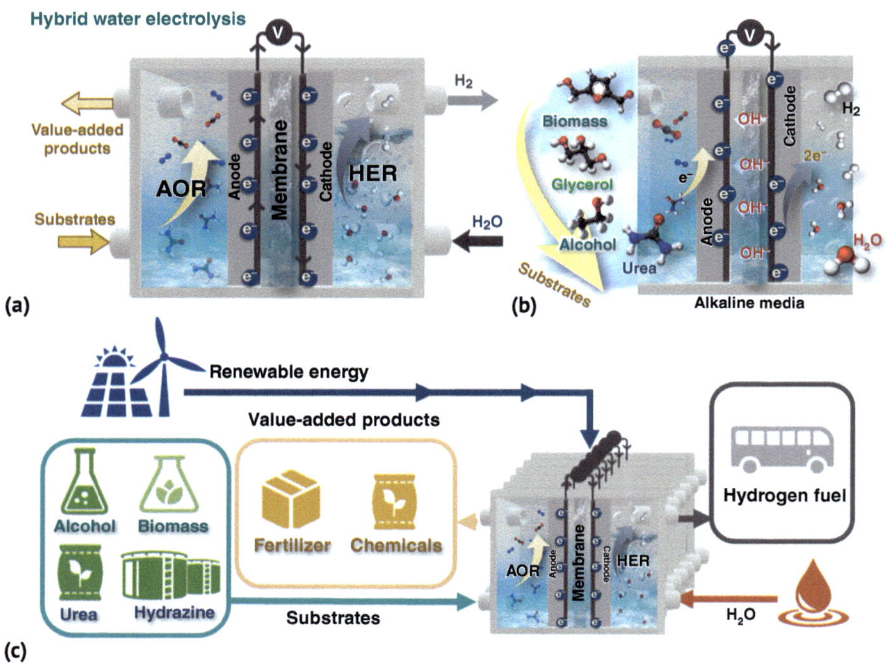

Figure 1.1 Schematic illustration of hybrid water electrolysis: (a) hybrid electrolyzer in which HER is coupled with AOR that replaces OER; (b) charge and species transport showing electron flow through the external circuit and ion transport through the electrolyte or membrane, e.g., OH^- in alkaline or anion-exchange membrane systems, with separated product streams; and (c) process outputs emphasizing the dual-function anode: hydrogen generation at the cathode and selective formation of value-added chemicals and/or treated effluent at the anode, depending on the chosen substrate.

configuration that couples HER with AOR. To reduce the cell voltage relative to conventional OER-driven water splitting, an AOR is selected, enabling the formation of functional anodic products such as value-added chemicals or detoxified waste-derived species [13]. This definition underscores the anode as an active chemical conversion module rather than merely an oxygen generator. It further emphasizes that both hydrogen production and the fate of the anodic substrate must be taken into account when assessing performance. In Figure 1.1c, the dual-output nature of the process is illustrated, with hydrogen produced at the cathode and value-added products or treated effluent generated at the anode.

Hybrid electrolysis can be viewed at three nested levels: reaction pairing, electrode function, and system design. At the reaction level, the core feature is a coupled pair in which cathodic HER consumes electrons supplied via an external circuit, and anodic oxidation supplies electrons by oxidizing a chosen substrate. At the electrode-function level, the anode is no longer optimized solely for OER stability. It must regulate adsorption, dehydrogenation, oxygen insertion, and often C—C or C—O bond transformations, with selectivity serving as a central metric. At the system level, the electrolyte, membrane (if present), electrode architecture, and operating mode (batch, flow-through, fed-batch, or continuous) must accommodate organic feed and product streams, including their transport, crossover tendencies, and downstream recovery requirements. Given the strong interactions among these layers, hybrid electrolysis is best regarded as a coupled electrochemical–chemical process rather than a minor variant of conventional water splitting.

The practical motivation for hybrid electrolysis extends beyond voltage reduction, although this is often the most visible advantage. A defining attribute is the ability to coproduce chemicals with a value greater than that of oxygen. Anodic oxidation of alcohols can yield aldehydes, carboxylic acids, or other oxygenates, while oxidation of primary amines can be directed toward nitriles under suitable catalytic environments [14, 15]. Certain substrates are also attractive because they are abundant in waste streams, such as urea-containing wastewater, which permits simultaneous hydrogen production and wastewater treatment or pollutant abatement [16].

The anodic reaction space is broad and includes urea [17], alcohols and polyols [18, 19], amines [20], and other organics, as well as biomass-derived molecules [21] and complex waste streams [22]. Within this space, thermodynamic advantages are often shown using representative potentials, as shown in Figure 1.2a. Urea oxidation has a markedly lower theoretical oxidation potential (about 0.37 V vs. RHE) than OER (1.23 V vs. RHE), and carbon oxidation has been proposed to yield even lower values in some conceptual schemes [23]. Reports of very low operating cell voltages in small-molecule systems demonstrate the magnitude of voltage reduction achievable when the anodic reaction proceeds efficiently, and competing OER is suppressed [24, 25]. At the same time, these values should be treated as indicators of feasibility rather than guarantees of device performance, because kinetic barriers, mass transport, and parasitic reactions often govern the observed voltage at practical current densities.

Such considerations motivate the scope boundaries adopted here. First, hybrid electrolysis is not defined by the mere presence of an organic molecule in

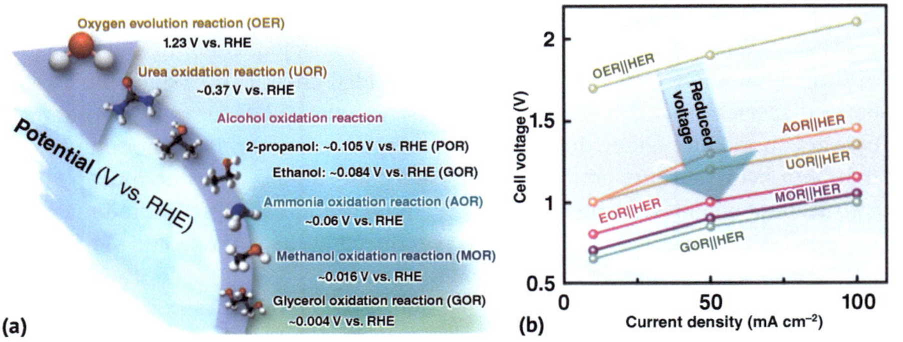

Figure 1.2 Thermodynamic and kinetic basis for energy reduction in hybrid electrolysis: (a) qualitative comparison of equilibrium potentials vs. RHE, highlighting that many AORs have lower thermodynamic potentials than OER, thereby reducing the minimum reversible cell voltage when paired with HER; and (b) schematic polarization curves comparing conventional OER-coupled electrolysis with AOR-coupled operation, illustrating how reduced anodic overpotential and suppressed OER shift the operating voltage downward at a given current density.

the electrolyte. Instead, it is defined by whether the organic oxidation reaction controls anodic charge transfer in the operating regime of interest. Competing OER can reemerge at elevated potentials, particularly if anodic oxidation pathways become transport-limited, surface poisoning reduces turnover, or the catalyst preferentially stabilizes oxygen-evolution intermediates. In these cases, the apparent advantage can diminish when the current is high, and the anode output may consist of a mixture of partial-oxidation products, deep oxidation products, and oxygen. Hybrid electrolysis must be discussed in terms of faradaic partitioning. This concept shows how current divides among competing anodic pathways and why overall current density alone is not sufficient.

Second, the anode typically produces liquid-phase products, making selectivity and carbon balance key design considerations. Organic oxidation proceeds through multistep pathways involving shared intermediates. Consequently, product distributions are sensitive to potential, local pH, substrate concentration, and mass transport. A catalyst that lowers the anode potential but generates a complex product mixture may be scientifically useful yet economically unattractive if separation costs are high [26]. In contrast, systems that produce few products at high faradaic efficiency (FE) can remain attractive even when the voltage advantage is small, due to product value and ease of separation.

Third, hybrid electrolysis should be regarded as a class of electrochemical technologies. It can be applied across multiple electrolyzer configurations rather than being limited to a single type of device. Many demonstrations employ alkaline electrolytes because alkaline conditions support a wide range of non-precious catalysts and can be compatible with anion exchange membranes. However, membranes introduce additional constraints, including organic crossover, chemical compatibility, and the management of local pH and water [27]. These factors influence achievable current density, durability, and the practicality of collecting anodic products at usable concentrations.

Finally, hybrid electrolysis carries distinct safety and scalability implications. Replacing OER can reduce oxygen generation at the anode. Some schemes do not generate gaseous products, e.g., CO_2 or N_2, thereby mitigating risks associated with H_2/O_2 coexistence and gas crossover. At the same time, organic feeds introduce new safety and handling considerations, including the flammability of particular substrates and the need for proper containment and separation [28]. Successful scale-up requires maintaining selectivity, catalyst stability, and system performance at industrially relevant operating conditions.

In summary, hybrid water electrolysis is defined here as a coupled electrochemical platform that retains cathodic HER while replacing anodic OER with a strategically chosen oxidation reaction, thereby reducing energy input and generating valuable anodic products or remediation outcomes. Its scope encompasses reaction thermodynamics and kinetics, selectivity and carbon balance, and device-level integration, because the motivating advantages are realized only when these elements are optimized together.

1.2 Thermodynamic and Kinetic Basis for Energy Reduction

The central energy argument for hybrid water electrolysis is most clearly expressed by distinguishing between what is thermodynamically possible and what is kinetically achievable at practical current densities. In conventional water electrolysis, the minimum reversible voltage is set by the equilibrium potentials of HER at the cathode and OER at the anode. The operating voltage is always higher because activation, ohmic, and mass-transport losses must be overcome. Hybrid water electrolysis modifies the equilibrium requirement at the anode by replacing OER with an alternative AOR. This approach mitigates kinetic losses, as many AORs proceed through pathways that are, under appropriate conditions, less kinetically demanding than OER.

Thermodynamically, the full-cell reversible voltage is the difference between the anode and cathode equilibrium potentials with additional terms accounting for the activities of reactants and products through the Nernst equation. OER is intrinsically energy-intensive because it is linked to the oxygen or water redox couple and sets the anode equilibrium potential around 1.23 V vs. RHE under standard conditions. In contrast, many AORs exhibit much lower theoretical oxidation potentials, thus reducing the reversible voltage required to couple the anodic reaction to HER. Urea oxidation, with a theoretical oxidation potential of ~0.37 V vs. RHE, is a frequently cited example. Carbon oxidation, with a representative potential near 0.21 V vs. RHE, exemplifies the thermodynamic advantage of choosing a suitable anodic reaction.

A lower equilibrium potential does not automatically guarantee lower energy consumption in practice. At a given current density, the operating cell voltage is best understood as the sum of the reversible voltage and the losses from overpotentials and resistance. Hybrid electrolysis primarily targets the reversible anode potential and, in favorable cases, the anode activation overpotential. At high current densities, ohmic losses and mass-transport overpotentials can dominate. Device-level demonstrations should be evaluated under practical operating conditions, rather than at low current densities where transport limitations are minimal. In Figure 1.2b, the conceptual differences between OER-coupled and AOR-coupled polarization behaviors and their impacts on operating voltage are illustrated.

The kinetic advantage of replacing OER arises from fundamental differences in reaction complexity. OER is a multistep, four-electron process that typically involves the formation and transformation of oxygenated intermediates and an O—O bond formation step, which together contribute to sluggish kinetics and large overpotential requirements. Consequently, OER typically sets the anodic overpotential and limits performance in conventional low-temperature water electrolyzers. Organic oxidation reactions involve complex, multistep pathways. When catalysts provide suitable adsorption energetics for dehydrogenation and oxygen-insertion steps, oxidation reactions can proceed at lower anode potentials. This effect is particularly evident in alkaline media, where surface hydroxyl species play a central role. At a given current density, the anodic polarization curve shifts to more negative potentials, thereby reducing the full-cell voltage required to sustain HER at the cathode.

Voltage reduction translates directly into energy savings. Under ideal conditions with 100% FE toward H_2, the production of 1 mol of hydrogen requires a charge of 2 F. The electrical energy input approximates the operating cell voltage multiplied by 2 F. Even a modest decrease in operating voltage, sustained at high current density over long periods, can substantially reduce electricity consumption.

Hybrid electrolysis shifts part of the energy demand from electricity to chemical feedstock. Organic substrates are oxidized at the anode, providing electrons and chemical energy that reduce the electrical input. This shift is intrinsic to electrochemical reforming and paired electrolysis concepts. A fair assessment distinguishes between electrical energy savings and total energy consumption. It considers the chemical energy content and cost of the feedstock as well as the economic value of the anodic products.

In particularly favorable cases, anodic oxidation is sufficiently advantageous that the required external voltage becomes very small. Reports of extremely low operating voltages in selected small-molecule systems highlight this upper bound. Some aldehyde-based systems, for example, have been reported to operate at cell voltages as low as 0.3 V. This finding illustrates how dramatically electrical input can be reduced relative to OER-driven operation when the anodic chemistry is appropriately selected and efficiently catalyzed [29].

In practice, the kinetics are more complex. Organic oxidation rarely follows a single pathway, and the observed rate depends strongly on active surface, local pH, and mass transport. Many AORs proceed through multiple dehydrogenation steps and oxygenated intermediates, so higher activity can come at the expense of selectivity [30]. Conditions that maximize current can lead to deeper oxidation or C—C cleavage. In contrast, conditions that favor selective partial oxidation may become transport-limited at high current density. The kinetic advantage over OER is not automatic. It is achieved only when catalysts sustain fast turnover rates at low anode potentials, avoid persistent poisoning, and suppress OER.

Competition between AOR and OER determines voltage reduction. Even if AOR has a lower thermodynamic potential than OER, the anode potential can still shift into a region where OER becomes significant. This shift can occur when the substrate is locally depleted, products accumulate and block active sites, or the surface reconstructs into a more OER-active state. In such a scenario, some of the anodic current produces oxygen instead of driving organic oxidation, thereby reducing both the voltage benefit and FE toward the desired products. Practical energy gain depends not only on a lower equilibrium potential but also on maintaining selectivity for organic oxidation over OER across the intended operating range.

Device configuration further influences the realization of thermodynamic and kinetic advantages. In membrane electrolyzers, such as those based on anion exchange membranes, both membrane and ionomer contribute additional ohmic resistance. They also impose constraints on water management and ion transport, which can reduce the voltage benefit of replacing OER. On the other hand, membrane separation can improve practical performance by isolating products and reducing crossover. It also enables continuous operation that stabilizes substrate supply and removes products [21]. Recent demonstrations using membrane-based or flow-through designs emphasize that coupling reaction thermodynamics with transport-aware engineering is essential for sustaining low-voltage operation at meaningful current densities.

In summary, hybrid water electrolysis reduces electrical energy input through a combination of thermodynamic and kinetic effects. Replacing OER with an AOR lowers the reversible anode potential and, under favorable catalytic conditions, can reduce activation overpotentials relative to OER. The extent of the realized benefit is determined by the total voltage loss and by competition between organic oxidation and residual OER. The most credible path to durable energy reduction is to identify a low-potential anodic reaction. This must be coupled with the design of catalysts and electrolyzer architectures that keep the anode in a regime where organic oxidation dominates charge transfer. Product formation must remain controlled, and transport limitations should not drive the system back toward OER-like operation.

1.3 Anodic Reaction Space and Substrate Selection

The anodic reaction space in hybrid water electrolysis is considerably broader than that of conventional water splitting. The anode is no longer confined to OER. Instead, it can be coupled to the oxidation of nitrogen-containing wastes, oxygenated organics, biomass-derived feeds, and other substrates. These reactions can occur at potentials lower than OER and yield products with either economic value or environmental benefit. This breadth is a core strength, but it also creates a practical challenge. Hybrid electrolysis is not a single chemistry. The most suitable substrate depends on which constraint dominates in the targeted deployment, whether electricity cost, feedstock availability, product value, separation energy, safety, or wastewater treatment needs. In Table 1.1, a structured overview of representative substrate classes, their primary value propositions, and key constraints is provided.

Table 1.1 Summary of representative substrate classes, including urea, alcohols and polyols, aldehydes, amines, and biomass-derived streams, and their primary value propositions, such as voltage reduction, chemical coproduction, or remediation.

Representative feed	Primary advantage(s)	Selectivity challenge	Main risks	Best-fit configuration(s)	Source
N-containing feeds					
Urea	Remediation/voltage	Low–med	Carbonate crossover	Alkaline AEM, flow	[31, 32]
Amines/ammonia/ammonium	Remediation/voltage	Med	N-speciation crossover	Alkaline AEM, flow	[33, 34]
Hydrazine	Voltage	Low–med	Safety/handling crossover	Alkaline, flow (controlled)	[19, 35]
O-containing small molecules					
Alcohols (C_1–C_3)	Voltage reduction, coproduct	Med–high	Over-oxidation, OER	Alkaline, batch/flow	[21]
Polyols (e.g., glycerol)	Coproduct	High	Product mixtures, transport limits	Alkaline, flow	[27]
Aldehydes	Coproduct	High	Over-oxidation, catalyst poisoning	Alkaline (controlled), batch/flow	[36, 37]
Biomass/waste streams					
Crude glycerol	Remediation/coproduct	High	Impurities, fouling	Flow + pretreatment	[38]
Wastewater organics	Remediation	High	Variability, poisoning	Flow, robust electrodes	[21]
Biorefinery-derived feeds	Coproduct	High	Mixtures instability	Flow-controlled conditions	[39]

A useful framework for discussion is to classify anodic substrates according to their primary value proposition. The first category comprises energy-driven substrates, which are mainly employed to reduce the overall cell voltage and, consequently, the electrical energy input required for hydrogen production. Urea represents a canonical example, as its theoretical oxidation potential is significantly lower than that of OER. This thermodynamic advantage can translate into substantial voltage reductions when urea oxidation dominates the anodic current. However, translating these thermodynamic advantages into practical systems is constrained by electrode consumption and coproduct management. Carbon-based oxidation strategies illustrate extreme cases of voltage minimization but further highlight the trade-offs between energetic gains, material durability, and downstream handling requirements.

The second group comprises coproduction substrates, in which the anode is designed to generate market-relevant chemicals instead of low-value oxygen. In this category, alcohols and polyols are prominent because they can be converted into acids, aldehydes, or other oxygenates. In contrast, primary amines can be converted into nitriles under appropriate catalytic conditions. In this configuration, the hybrid electrolyzer operates as a paired system that produces hydrogen and a second product stream with potential economic value.

The third group comprises remediation-oriented substrates in which the anodic reaction addresses an environmental liability, such as wastewater constituents or pollutants, while still enabling hydrogen generation. In such cases, the product may be treated as effluent or a less harmful chemical form rather than a commodity chemical.

Alcohols and polyols constitute the most extensively explored substrate family for coproduction. Their oxidation networks offer multiple controllable branching points. Under selective reaction pathways, ethanol oxidation can yield acetic acid. Similarly, benzyl alcohol oxidation can produce benzoic acid, while various biomass-derived polyols can be converted into formate, aldehydes, or carboxylic acids, depending on catalyst identity, electrolyte environment, and applied potential. These systems illustrate a broader principle of substrate selection [40]. The best substrate is not necessarily the one with the lowest theoretical anode potential, but rather the one that can deliver a stable, high-FE pathway to a concentrated, separable product slate at the intended current density. This consideration is crucial for polyols and biomass-derived oxygenates, where deep oxidation and C—C cleavage can quickly broaden the product distribution unless the catalyst surface and operating window are tuned to favor a defined subset of intermediates and products.

Aldehydes constitute a distinct region of the anodic reaction space, enabling very-low-voltage operation under selected conditions. When adsorption and electron-transfer steps are well aligned, oxidation can be both thermodynamically favorable and kinetically rapid. In this context, small-molecule aldehyde systems have been reported to approach cell voltages as low as ~0.3 V vs. RHE in specific demonstrations. At the same time, aldehydes are highly reactive and can introduce competing side reactions [41]. To avoid loss of selectivity, volatility issues, or the need for complex downstream separations, their oxidation products may require careful management. As a result, aldehydes are most compelling when they are available within controlled feedstock streams and when product capture and purification can be integrated into the process design.

Nitrogen-containing substrates occupy another strategically important segment of the anodic reaction space. Urea oxidation is particularly attractive, as it can be integrated with wastewater remediation by converting a prevalent waste constituent into benign products while simultaneously lowering the anode potential relative to OER. Beyond urea, oxidation of amines to nitriles has been discussed as both a chemical upgrading route and, in some contexts, an electro-dehydrogenation strategy relevant to liquid organic hydrogen carriers. For these nitrogen-bearing feeds, substrate selection must explicitly account for safety, toxicity, and membrane compatibility, because volatility, odor, and crossover can impose operational constraints even when electrochemical performance is favorable. Furthermore, nitriles and related products can be valuable, but they require stringent control of selectivity and purification [20].

Biomass- and waste-derived substrates extend hybrid electrolysis beyond a laboratory concept to a circular-economy platform, but they impose stringent selection criteria. Raw sugarcane juice, for example, has been reported to support the coproduction of formic acid and hydrogen, demonstrating that minimally processed agricultural streams can serve as anodic feeds rather than solely purified model compounds [42]. Likewise, lignin-derived molecules and other complex streams are promising feedstock sources that support sustainability narratives. However, these feeds are chemically complex and often contain impurities that can poison catalysts, foul porous electrodes, or accelerate membrane and ionomer degradation [43]. When wastewater or low-quality water is targeted, substrate selection must therefore account for impurity tolerance, resistance to poisoning and fouling, and stable operation under variable composition.

Across these substrate families, selection can be framed as a multi-constraint optimization problem instead of a single-criterion choice. At a minimum, five coupled criteria should be evaluated. First, within the intended electrolyte and concentration range, thermodynamic advantage must be meaningful since real feeds rarely operate under standard conditions. Second, kinetics must remain favorable at the target current density, which requires an intrinsically active catalyst, sufficient substrate mass transport to the anode, and effective removal of products that may inhibit turnover. Third, anodic reaction must maintain high faradaic selectivity against competing OER, particularly as the anode potential increases during high-rate operation. Fourth, alongside separability, product value must be considered since a low-value or diluted product mixture can negate the benefit of voltage reduction. Fifth, feedstock logistics and robustness must be plausible at scale, including availability, cost, safety, storage, and compatibility with membranes and balance-of-plant components.

Whether a substrate is practically viable, two additional considerations need to be determined. One is carbon balance, and the other is the desired oxidation depth. For certain applications, deep oxidation into small molecules is acceptable or preferred. For chemical coproduction, partial-oxidation selectivity is essential and must remain stable over time. Integration strategy is equally critical, as batch, flow-through, and fed-batch configurations impose different concentration profiles and product accumulation behaviors that strongly affect reaction kinetics and selectivity. Reports emphasizing flow-based operation for alcohol-assisted systems indicate that substrate selection should be aligned with the operating mode that maintains the surface in a selective regime [15].

In summary, in hybrid water electrolysis, the anodic reaction space spans energy-saving, coproduction, and remediation chemistries. Currently, alcohols and polyols, aldehydes, urea, amines, and biomass-derived feeds are among the most prominent substrate classes. Practical selection of an anodic substrate should be driven by an explicit match between thermodynamic and kinetic feasibility at the target current density. In addition, the ability to suppress competing OER while maintaining high selectivity, together with product value and feedstock robustness, ultimately governs scalability.

1.4 Advantages Beyond Voltage: Coproduction, Safety, and Sustainability

The most visible benefit of hybrid water electrolysis is voltage reduction. However, voltage reduction alone is insufficient to reflect the overall value of hybrid water electrolysis. In practice, hybrid systems can offer three additional advantages that are often decisive for deployment. These are the coproduction of value-added chemicals at the anode, improved safety and operational flexibility through suppression of OER, and sustainability gains enabled by coupling hydrogen production with waste conversion and the use of circular feedstocks. In Table 1.2, these advantages

Table 1.2 Summary of advantages beyond voltage and associated practical constraints. Comparative overview of the principal non-voltage benefits of hybrid electrolysis: namely, coproduction of value-added chemicals, improved safety through reduced OER, and sustainability gains via waste and biomass utilization.

Advantage	What you gain	What must be true	If not addressed
Chemical coproduction	Value-added liquids	High FE to target, stable selectivity, feasible separation	Broad mixtures, separation penalty, low net value
Product quality/ separability	Simplified processing	Narrow product slate, minimal crossover, electrolyte pH control	Cross-contamination, neutralization burden, low-purity streams
Safety (gas management)	Reduced O_2 handling	Suppressed OER, separation reliability, low gas crossover	O_2 formation, mixed-gas risk, pressure/venting limits
Durability (materials stability)	Extended lifetime, voltage stability	Corrosion-tolerant, membrane stability, fouling control	Activity drift, membrane decay, performance decay
Sustainability/ waste valorization	Waste removal, H_2 production	Impurity tolerance, pretreatment, stable operation	Poisoning/fouling, feed instability, frequent shutdowns

are summarized. Such factors are interconnected and become practically significant only when selectivity, durability, and separations are engineered alongside electrochemical performance.

1.4.1 Coproduction: Transforming the Anode into a Chemical Manufacturing Module

In conventional water electrolysis, the anode produces oxygen, which is frequently treated as a low-value by-product, particularly in distributed hydrogen production scenarios where oxygen capture, compression, and utilization are not economically attractive. Hybrid electrolysis reframes the role of the anode by enabling the selective generation of targeted chemicals. This shift goes beyond producing a single product stream and fundamentally alters the economics of the electrolyzer, as the anode acts as a chemical conversion module whose outputs can carry significant market value. Alcohols, polyols, aldehydes, and amines are frequently examined as anodic substrates, as their oxidation pathways can yield commodity or specialty oxygenates, acids, or nitriles.

When the anodic product slate is narrow, stable over time, and sufficiently concentrated for economical separation, coproduction advantage is strongest. In selective oxidation systems, small changes in potential, local pH, or surface coverage can shift reaction pathways, meaning that high instantaneous FE alone is insufficient if product distributions drift during continuous operation. For practical deployment, performance must approach that of electrosynthesis, with stable selectivity and carbon balance maintained at relevant current densities and timescales. When these criteria are met, hybrid electrolysis can operate as a paired process that produces hydrogen while upgrading low-value feedstocks into higher-value chemicals.

Coproduction also permits application-specific optimization. The most suitable anodic reaction depends on local feedstock availability and the product value chain. In regions with abundant bio-derived polyols or agricultural streams, oxidation pathways that yield organic acids or formate may be attractive. In chemical manufacturing contexts, selective oxidation to aldehydes, ketones, or nitriles may be prioritized. Hybrid electrolysis can thus complement conventional electrolyzers rather than compete directly with them. In applications where oxygen is valuable and oxygen-handling infrastructure is available, OER-based systems remain compelling. For settings where oxygen is a burden, hybrid coproduction becomes more attractive.

1.4.2 Safety and Operational Flexibility: Reducing Oxygen-related Constraints

Safety considerations are integral to electrolyzer design. The concurrent generation of hydrogen and oxygen poses inherent risks associated with gas crossover and mixing, particularly in compact systems where membrane crossover, sealing failures, or off-nominal operating conditions can lead to hazardous compositions. By suppressing OER under operating regimes in which organic oxidation predominates, hybrid electrolysis offers an intrinsic safety lever. Reduced or eliminated oxygen production mitigates hazards related to H_2/O_2 coexistence, simplifies gas-handling requirements, and alleviates demands on oxygen-separation and venting infrastructure.

The practical impact of this safety advantage depends strongly on the anodic chemistry. In many alcohol-assisted systems, the anodic product stream is primarily liquid-phase and generates minimal gaseous oxygen when OER is effectively suppressed. In other hybrid chemistries, gaseous products such as CO_2 or N_2 may be produced, depending on the substrate and pathway. These gases can be less reactive than oxygen with respect to hydrogen safety, but still require appropriate venting and management. The safety benefit should therefore be stated precisely. Hybrid electrolysis can reduce OER and thus oxygen-specific hazards, but it does not eliminate the need for gas-management strategies.

Hybrid electrolysis can provide operational flexibility. When anodic kinetics are more favorable than OER, lower cell voltages at a given current density may be achieved, expanding the operating envelope under fluctuating renewable power inputs. However, this flexibility is conditional. Depletion of the anodic substrate, catalyst poisoning, or mass-transport limitations can increase anode potential and promote the reemergence of OER, diminishing both voltage and safety advantages. Sustained operation requires maintaining the anode in a regime where the intended oxidation reaction remains dominant. This requirement depends on catalyst selection, electrode architecture, and operating strategy, often favoring continuous-flow or fed-batch configurations.

1.4.3 Sustainability and Circularity: Coupling Hydrogen with Waste Conversion

For hybrid electrolysis, the sustainability argument is most compelling. It enables coupling renewable electricity with the conversion of waste streams into hydrogen and other valuable products or benign effluents. Urea-containing wastewater is a frequently discussed case. In domestic and agricultural contexts, it is abundant and can serve as a low-potential anodic reaction while contributing to nitrogen waste mitigation. Within this framework, hybrid electrolysis functions as a combined energy and environmental technology that produces hydrogen while addressing wastewater treatment needs.

More generally, biomass-derived substrates, including lignin-derived aromatics, agricultural juices, and polyol-rich streams, allow hybrid electrolysis to be integrated into biorefineries and agro-industrial value chains. The use of sugarcane juice as an anodic feed for formic acid and hydrogen coproduction demonstrates that minimally processed streams can be used instead of purified model compounds. When coupled with biomass upgrading, hybrid electrolysis supports circular carbon use by partially converting organic streams into valuable oxygenates while simultaneously producing hydrogen. This approach is particularly attractive for distributed systems in regions with abundant biomass resources and growing renewable electricity capacity.

Sustainability benefits must be assessed realistically, especially for waste-derived feeds. Such feeds are chemically complex and can contain impurities that cause catalyst poisoning, membrane and ionomer degradation, and electrode fouling, increasing maintenance needs and reducing component lifetimes. To achieve genuine environmental benefits, hybrid electrolysis systems must use impurity-tolerant catalysts and materials, apply minimal but effective pretreatment, and avoid shifting environmental burdens elsewhere in the process.

1.4.4 Interdependence of Advantages and the Role of System-level Metrics

Coproduction, safety, and sustainability are not independent attributes. Coproduction is meaningful only if selectivity is maintained and products can be recovered at acceptable energy and cost. Moreover, safety benefits associated with oxygen suppression are realized only when operating conditions maintain OER suppression under realistic operation, including transient events and long-term degradation. Sustainability benefits derived from waste conversion depend on tolerance to real feeds and on a favorable overall environmental balance once separations, pretreatment, and component replacement are taken into account.

This interdependence implies that hybrid electrolysis should be evaluated using system-level metrics rather than isolated half-cell performance indicators. In addition to cell voltage and hydrogen FE, key metrics include product-resolved FE, carbon balance closure, stability of product distribution, and energy consumption per unit of hydrogen and per unit of target chemical. Where remediation is a primary goal, metrics such as removal efficiency and effluent quality are also essential. These measurements provide a quantitative basis for assessing whether the broader advantages of hybrid electrolysis translate into practical impact and for identifying configurations that merit further development and scale-up.

In summary, hybrid water electrolysis offers advantages that extend well beyond voltage reduction. By enabling the coproduction of value-added chemicals, it can improve process economics and create new opportunities for integration with chemical and biomass value chains. By suppressing OER under relevant operating regimes, it can mitigate oxygen-related hazards and simplify gas-handling requirements, provided that OER suppression is maintained under realistic and long-term conditions. By utilizing waste-derived or biomass-derived substrates, hybrid electrolysis can couple hydrogen production with environmental remediation and circular resource utilization. Realizing these advantages at scale will require precise control over selectivity and stability, ensuring compatibility with membranes and balance-of-plant components, and the development of operating modes and separation strategies that translate electrochemical performance into robust, deployable systems.

1.5 Remaining Challenges and Outlook

Hybrid water electrolysis is attractive. It can reduce electrical energy input by replacing OER with an alternative AOR. This approach enables chemical coproduction and, in some cases, the valorization of wastewater or biomass-derived substrates.

However, the use of complex anodic chemistries introduces practical challenges. These challenges extend beyond achieving low voltage and high current in short-term tests. Instead, they relate to maintaining selectivity, durability, and economic viability under realistic feed conditions, device designs, and long-term operation.

In hybrid water electrolysis, a primary bottleneck is product selectivity and faradaic partitioning at the anode. Organic oxidation reactions proceed through multiple surface-bound intermediates, and their pathways depend on the applied potential, local pH, substrate concentration, and changes in the catalyst surface. As a result, a catalyst can sustain high anodic currents while producing an undesired or unstable product distribution, thereby undermining the rationale for coproduction and complicating downstream separation. Selectivity should therefore be considered a time-dependent property rather than a single-point metric. As the catalyst reconstructs and active sites become blocked, or the local reaction environment within porous electrodes evolves, product distributions can change. These effects are more severe at industrially relevant current densities, where concentration gradients, mass-transport limitations, and shifts in anode potential promote deeper oxidation and side reactions.

Closely related to the challenge of maintaining selectivity is competition from residual OER at the anode. Even when an alternative AOR is thermodynamically favored, OER can reappear if the desired organic pathway is impeded by substrate depletion, catalyst poisoning, or mass-transport limitations. This competition leads to a dual penalty: part of the current is diverted to OER, reducing the voltage advantage, while the charge contributing to target products decreases, lowering FE and product yield. Suppressing OER is both a catalytic challenge and a challenge in reactor design, requiring a stable substrate supply and effective removal of inhibitory products to keep the anode potential out of the OER-dominated regime.

Catalyst robustness under realistic operating conditions represents a second major challenge for hybrid water electrolysis. Hybrid water electrolysis exposes the anode catalyst to coupled electrochemical and chemical stresses. Reactive organic intermediates, strongly oxidizing potentials during transients, local pH shifts, and continuous adsorption–desorption cycles can restructure the surface. Even when the bulk phase is nominally stable, the catalytically active surface often evolves into oxygenated or hydroxylated states. This evolution is beneficial only if it remains controlled and electronically connected. Uncontrolled reconstruction can reduce active-site density, alter adsorption energetics in ways that degrade selectivity, or accelerate dissolution and detachment from supports. Durability must be evaluated in terms of both activity and selectivity. Retention of both is essential, because a stable current density can otherwise be sustained by a gradual shift toward less selective pathways or increased OER contributions.

System design constraints constitute a third barrier that often determines whether promising catalysts can be translated into practical devices. Membranes and ionomers play a critical role in many electrolyzer configurations, particularly when product separation and gas management are required. However, organic feedstocks introduce additional compatibility challenges, including swelling, chemical degradation, and changes in ionic conductivity. Crossover of organic substrates and products can reduce coulombic efficiency, contaminate the cathode, and complicate downstream processing. At the same time, membranes can provide important benefits by enabling compartmentalization, stabilizing selectivity, and supporting continuous operation, provided that crossover is effectively controlled. The associated engineering challenge is therefore a trade-off: transport of the desired ionic species must be maximized, while transport of neutral organics and reactive products should be minimized.

A fourth major challenge concerns separations and process intensification. When the anode produces a usable chemical stream, hybrid water electrolysis offers clear value. However, electrosynthetic product concentrations are often low, and product mixtures can be complex. If product recovery requires energy-intensive separations, the system-level benefit may be diminished even at low electrolyzer voltages. Substrate selection, catalyst selectivity, and reactor configuration must therefore be evaluated in an integrated manner. Flow operation can increase effective product concentration and limit over-oxidation by controlling residence time. In contrast, batch operation may lead to progressive product degradation or secondary reactions as electrolysis proceeds. In system-level design, the anode should be treated as an integrated conversion and pre-separation unit rather than merely as an oxygen-evolving electrode.

Feedstock realism is a fifth and increasingly important barrier. Many hybrid systems are demonstrated using purified model compounds, but scale-relevant deployment is likely to involve complex feeds such as biomass-derived streams, industrial effluents, or wastewater constituents. These feeds contain inorganic ions and diverse organics that can poison catalysts, foul porous electrodes, and degrade membranes. Addressing this gap requires catalysts with impurity tolerance, electrode architectures that resist fouling while maintaining mass transport, and pretreatment strategies that are minimal but effective. The appeal of using waste-derived substrates and even waste-derived catalysts as a route toward circular green hydrogen has been emphasized. Still, it also poses challenges for long-term stability and reproducibility under variable feed compositions.

Benchmarking and reporting practices constitute a sixth challenge that shapes the field's evolution. Because hybrid electrolysis simultaneously targets hydrogen production and selective oxidation, performance claims must account for both aspects of the system. Meaningful benchmarking requires, at a minimum, cell voltage at specified current densities, hydrogen FE, product-resolved FE, carbon balance closure, and stability over time under controlled operating conditions. When suppression of OER is claimed, quantifying O_2 evolution or demonstrating its absence within the detection limit is necessary. Without consistent reporting, comparisons across catalyst families and substrates remain ambiguous, and conclusions about deployment potential can be fragile.

In summary, the remaining challenges for hybrid water electrolysis center on selectivity control and suppression of competing OER, catalyst and component durability, membrane and crossover management, separations feasibility, and validation under realistic feeds and operating conditions. The chapter sequence is designed to address these issues systematically: first by establishing shared performance metrics, second by mapping catalyst families to substrate classes, and finally by connecting material advances to engineering and deployment constraints.

References

1 Shoaib, M., Naz, M. Y., Wu, T., Rehman, H. U., Sun, E., Li, A., Zhu, Q., & Wang, N. (2025). Review of single-atom electrocatalysts for hydrogen and oxygen evolution reactions from water-splitting. *Fuel 390*, 134704. https://doi.org/10.1016/j.fuel.2025.134704

2 Luo, X., Li, S., Du, B., Lai, W., Xiao, H., Luo, J., Li, J., Chen, Y., Yu, F., & Li, D. (2024). Tuning intrinsic defects of Pt-based carbon supports by high-energy ball-milling for enhanced alkaline hydrogen evolution. *Int. J. Hydrog. Energy 51*, 1616–1629. https://doi.org/10.1016/j.ijhydene.2023.09.258

3 Palaniyandy, N., Govindarajan, D., Devaraj, L., Khezri, R., Chinnakutti, K., Kempahanumakkagari, S., Thippeswamy, R., Amer, M. S., Arunachalam, P., Al-Mayouf, A. M., & Kheawhom, S. (2024). An overview of recent advances in Pt and Pd-based materials: From design strategies to reaction mechanisms. *J. Ind. Eng. Chem. 146*, 213–237. https://doi.org/10.1016/j.jiec.2024.11.057

4 Etesami, M., Mano, P., Namuangruk, S., Khezri, R., Gopalakrishnan, M., Limphirat, W., Yonezawa, T., Motlagh, S. R., Somwangthanaroj, A., & Kheawhom, S. (2025). Solvent-free synthesis of FeCo alloy nanoparticle-embedded nitrogen-doped carbon nanotubes for oxygen reduction in zinc-air batteries. *Int. J. Hydrog. Energy 153*, 150289. https://doi.org/10.1016/j.ijhydene.2025.150289

5 Jiménez-Arévalo, V. M., Martin, P., Sepúlveda, M. F., Azocar, M. I., Zhou, X., Galleguillos Madrid, F. M., Ramirez, C. G., Zagal, J. H., & Páez, M. (2025). Recent advances in high-entropy alloys for electrocatalysis: From rational design to functional performance. *Mater. Des. 258*, 114633. https://doi.org/10.1016/j.matdes.2025.114633

6 Li, D., Tu, J., Lu, Y., & Zhang, B. (2023). Recent advances in hybrid water electrolysis for energy-saving hydrogen production. *Green Chem. Eng. 4*, 17–29. https://doi.org/10.1016/j.gce.2022.11.001

7 Roh, H., Oh, S., & Yong, K. (2025). Multiscale hierarchical electrocatalyst-driven low-energy hybrid water electrolysis in an AEM electrolyzer. *Chem. Eng. J. 519*, 165197. https://doi.org/10.1016/j.cej.2025.165197

8 Singh, B. & Verma, N. (2024). Exploring anodic oxidation reactions in hybrid water electrolysis: Special emphasis on substrate choice. *Catal. Sci. Technol. 14*, 6732–6760. https://doi.org/10.1039/d4cy00907j

9 Singh, B. & Gupta, H. (2024). Metal-organic frameworks (MOFs) for hybrid water electrolysis: Structure-property-performance correlation. *Chem. Commun. 60*, 8020–8038. https://doi.org/10.1039/d4cc02729a

10 Lee, J., Lee, S. A., Lee, T. H., & Jang, H. W. (2025). Unlocking the potential of chemical-assisted water electrolysis for green hydrogen production. *Ind. Chem. Mater. 3*, 277–310. https://doi.org/10.1039/d4im00163j

11 Zhang, Z., Li, J., Yang, S., Tao, E., & Peng, C. (2025). Advances in paired electrolysis for furfural conversion: From design principle, mechanisms to perspectives. *Green Chem. 27*, 6967–6988. https://doi.org/10.1039/d5gc01306b
12 Jiang, Y., Li, J., Guo, X., Chen, Y., Sun, W., & Peng, C. (2025). Electrocatalytic reforming of polyethylene terephthalate waste plastics into high-value-added chemicals with green hydrogen generation. *J. Colloid Interface Sci. 685*, 29–37. https://doi.org/10.1016/j.jcis.2025.01.109
13 Zhang, X., Wang, X., Sun, T., Wang, M., Zhu, J., & Wang, J. (2025). Design of alternative oxidation processes for hybrid water electrolysis systems: Recent progress and perspective. *Chem. Commun. 61*, 11353–11363. https://doi.org/10.1039/d5cc02862k
14 Ingavale, S., Marbaniang, P., Somwangthanaroj, A., Thamyongkit, P., Kidkhunthod, P., & Kheawhom, S. (2025). Electrocatalysts for ammonia production and nitrogen cycle management in zinc-NO_x batteries: Progress, challenges, and future perspectives. *Mater. Today Sustain. 29*, 101066. https://doi.org/10.1016/j.mtsust.2024.101066
15 Wang, G., Chen, A., Chen, Y., Qiao, F., Wang, J., Yang, N., Zhang, H., & Wen, Z. (2025). Advancements in electrochemical synthesis: Expanding from water electrolysis to dual-value-added products. *eScience 5*, 100333. https://doi.org/10.1016/j.esci.2024.100333
16 Zhang, Y., Zhang, W., Huang, J., Cai, W., & Lai, Y. (2024). Ce-induced NiS bifunctional catalyst transformation: enhancing urea oxidation coupled with hydrogen electrolysis. *EES Catal. 2*, 1306–1313. https://doi.org/10.1039/d4ey00119b
17 Ghosh, D., Banerjee, R., Bhadu, G. R., Bhaduri, S. N., Mondal, A., Srivastava, D. N., & Biswas, P. (2022). A nano-structured nickel trithiocarbonate complex supported on g-C_3N_4 as an efficient electrocatalyst for urea electro-oxidation. *Mater. Adv. 3*, 6831–6841. https://doi.org/10.1039/d2ma00308b
18 Begildayeva, T., Theerthagiri, J., Min, A., Moon, C. J., & Choi, M. Y. (2024). Density-controlled metalloporphyrin with mutated surface via pulsed laser for oxidative refining of alcohols to benzoic acid and H_2 production using linear tandem electrolysis. *Appl. Catal. B: Environ. 350*, 123907. https://doi.org/10.1016/j.apcatb.2024.123907
19 Liu, J., Li, T., Wang, Q., Liu, H., Wu, J., Sui, Y., Li, H., Tang, P., & Wang, Y. (2024). Bifunctional PdMoPt trimetallene boosts alcohol-water electrolysis. *Chem. Sci. 15*, 16660–16668. https://doi.org/10.1039/d4sc04764h
20 Guenani, N., Solera-Rojas, J., Carvajal, D., Mejuto, C., Mollar-Cuni, A., Guerrero, A., Fabregat-Santiago, F., Mata, J. A., & Mas-Marzá, E. (2024). Room temperature hydrogen production *via* electro-dehydrogenation of amines into nitriles: advancements in liquid organic hydrogen carriers. *Green Chem. 26*, 8768–8776. https://doi.org/10.1039/d4gc01275e
21 MariaJoseph, A., Nangan, S., Verma, D., Gnanasekaran, L., Rajendran, S., Natesan, T., Pattananuwat, P., & Okhawilai, M. (2024). Rational coupling of selective electrochemical oxidation and reduction reactions for in-situ

value-added chemical generation. *Fuel 367*, 131408. https://doi.org/10.1016/j.fuel.2024.131408

22 Wang, Y., Shan, G., Ma, K., Yang, L., Gao, L., Zhang, M., Huo, X., Li, X., Zhang, J., & Li, W. (2024). Alkaline electrolyzer-improving electrocatalytic oxidation of landfill leachate coupling with hydrogen production. *Int. J. Hydrog. Energy 64*, 926–934. https://doi.org/10.1016/j.ijhydene.2024.03.046

23 El Ouardi, M., El Idrissi, A., Arab, M., Zbair, M., Haspel, H., Saadi, M., & Ait Ahsaine, H. (2024). Review of photoelectrochemical water splitting: From quantitative approaches to effect of sacrificial agents, oxygen vacancies, thermal and magnetic field on (photo)electrolysis. *Int. J. Hydrog. Energy 51*, 1044–1067. https://doi.org/10.1016/j.ijhydene.2023.09.111

24 Sheng, W., Zhou, X., Ajmal, S., Chen, X., Ma, Y., Chen, P., Zhu, M., & Li, P. (2025). Dual-doped medium-entropy phosphides for complete urea electrolysis. *J. Colloid Interface Sci. 678*, 1192–1202. https://doi.org/10.1016/j.jcis.2024.09.142

25 Yu, H., Jiang, S., Zhan, W., Liang, Y., Deng, K., Wang, Z., Xu, Y., Wang, H., & Wang, L. (2023). Formaldehyde oxidation boosts ultra-low cell voltage industrial current density water electrolysis for dual hydrogen production. *Chem. Eng. J. 475*, 146210. https://doi.org/10.1016/j.cej.2023.146210

26 Wei, Z., Huang, X., Duan, H., Shao, M., Li, R., Zhang, J., Li, C., & Duan, X. (2024). Electrochemical synthesis in company with hydrogen production via renewable energy: Opportunities and challenges. *Chin. J. Catal. 58*, 1–6. https://doi.org/10.1016/s1872-2067(23)64625-x

27 Wang, G., Chen, J., Qiao, F., Wang, J., & Wen, Z. (2025). Hybrid dual-electrolyte electrochemical cells for glycerol oxidation upgradation. *Chem. Sci. 16*, 12651–12678. https://doi.org/10.1039/d5sc02411k

28 Guo, Y., Qi, P., Zhang, Q., Li, M., Liu, J., & Sun, H. (2025). Control strategy for hydrogen production system using HTO-based hybrid electrolyzers. *Energy Rep. 13*, 2354–2364. https://doi.org/10.1016/j.egyr.2025.01.012

29 Gao, G., Sun, Z., Chen, X., Zhu, G., Sun, B., Huang, X. L., Liu, H. K., & Dou, S. X. (2024). Recent advances in hydrogen production coupled with alternative oxidation reactions. *Coord. Chem. Rev. 509*, 215777. https://doi.org/10.1016/j.ccr.2024.215777

30 Yang, Y., Jiang, M., Wang, Y., Zhang, Y., Wu, Y., & Wu, D. (2025). Solvent-engineered NiMo-based electrocatalysts for simultaneous hydrogen evolution and PET plastic waste upcycling. *Mater. Today Phys. 56*, 101775. https://doi.org/10.1016/j.mtphys.2025.101775

31 Liu, D., Xing, S., & Chen, X. (2025). Unlocking efficient urea oxidation via built-in field modulation in $NiFe_2O_4/CeO_2$ Mott-Schottky heterojunction. *Electrochim. Acta 538*, 147007. https://doi.org/10.1016/j.electacta.2025.147007

32 Xu, X., Dong, Y., Liu, F., Wang, H., Wang, X., Li, X., Ren, J., & Wang, R. (2024). The Ni-P dual vacancies crystalline/amorphous bifunctional electrode with ZnO "electron pump" for urea-assisted electrolysis of water to produce hydrogen. *Int. J. Hydrog. Energy 49*, 984–994. https://doi.org/10.1016/j.ijhydene.2023.10.234

33 Gecgel, O., Alvarez-Pugliese, C. E., Solis, J., & Botte, G. G. (2025). Toward net energy-positive ammonia electrolysis: A 3,000 cm^2 bi-polar electrolyzer for hydrogen production and ammonia remediation. *Int. J. Hydrog. Energy 141*, 1349–1358. https://doi.org/10.1016/j.ijhydene.2025.03.191

34 Yang, X., Sun, H., Liu, C., Yu, L., & Chen, H. (2022). Stable electrolysis of ammonia on platinum enhanced by methanol in non-aqueous electrolyte for an in-situ hydrogen production. *Chem. Eng. J. 442*, 136167. https://doi.org/10.1016/j.cej.2022.136167

35 Darband, Gh. B., Lotfi, N., Aliabadi, A., Hyun, S., & Shanmugam, S. (2021). Hydrazine-assisted electrochemical hydrogen production by efficient and self-supported electrodeposited Ni-Cu-P@Ni-Cu nano-micro dendrite catalyst. *Electrochim. Acta 382*, 138335. https://doi.org/10.1016/j.electacta.2021.138335

36 Ashoka, S., Akshaya, H. M., Shirisha, M., Venkataramanan, N. S., Yogesh, K., Sanaga, N., & Yogeeshwari, R. T. (2025). Insights on the electrooxidation of formaldehyde over bimetallic $Co_2V_2O_7$ nanorod and its implication towards water electrolysis. *Solid State Sci. 163*, 107903. https://doi.org/10.1016/j.solidstatesciences.2025.107903

37 Ji, H., Wu, Z., Li, T., Ju, G., Li, H., Wang, Y., Xu, H., & Liu, J. (2025). Bipolar hydrogen generation via furfural-water co-electrolysis over bifunctional CuPt catalysts. *Int. J. Hydrog. Energy 102*, 856–865. https://doi.org/10.1016/j.ijhydene.2025.01.096

38 Srinivaas, M., Lee, Y.-Y., Huang, B.-W., Li, I.-C., & Chang-Chien, G.-P. (2025). High-entropy alloys as advanced electrocatalysts for biomass conversion and sustainable hydrogen production. *Int. J. Hydrog. Energy 137*, 1322–1354. https://doi.org/10.1016/j.ijhydene.2024.12.430

39 Ma, Y., Zhang, Y., Yuan, W., Du, M., Kang, S., & Qiu, B. (2023). Electroreforming injects a new life into solid waste. *EES Catal. 1*, 892–920. https://doi.org/10.1039/d3ey00147d

40 Zhou, J., Yu, Y., Han, J., Lai, J., & Wang, L. (2025). A review of electrocatalyst design for the selective oxidation of alcohols to carboxylic acids. *Chem. Eng. J. 525*, 170054. https://doi.org/10.1016/j.cej.2025.170054

41 Pan, Y., Li, Y., Dong, C.-L., Huang, Y.-C., Wu, J., Shi, J., Lu, Y., Yang, M., Wang, S., & Zou, Y. (2023). Unveiling the synergistic effect of multi-valence Cu species to promote formaldehyde oxidation for anodic hydrogen production. *Chem 9*, 963–977. https://doi.org/10.1016/j.chempr.2022.12.008

42 Gunerhan, A., Altuntas, O., & Açıkkalp, E. (2025). A comprehensive analysis of the production of H_2 and value-added chemicals from the electrolysis of biomass and derived feedstocks. *Biomass Bioenergy 192*, 107510. https://doi.org/10.1016/j.biombioe.2024.107510

43 Li, J. & Duan, H. (2024). Recent progress in energy-saving hydrogen production by coupling with value-added anodic reactions. *Chem 10*, 3008–3039. https://doi.org/10.1016/j.chempr.2024.08.009

Part II

Alcohol-assisted Water Electrolysis

2

Metal Carbides and Nitrides Electrocatalysts for Alcohol-assisted Water Electrolysis

Phonnapha Tangthuam[1,2], *Vipada Aupama*[1,2], *Ahmad Azmin Mohamad*[1,3], *and Soorathep Kheawhom*[1,2,4,*]

[1]*Department of Chemical Engineering, Chulalongkorn University, Bangkok, Thailand*
[2]*Department of Chemical Technology, Chulalongkorn University, Bangkok, Thailand*
[3]*School of Materials and Mineral Resources Engineering, Malaysia Science University, Nibong Tebal, Malaysia*
[4]*Center of Excellence on Advanced Materials for Energy Storage, Chulalongkorn University, Bangkok, Thailand*

*Corresponding author: soorathep.k@chula.ac.th

2.1 Background and Scope

For applications where direct electrification is challenging, such as high-temperature industrial processes, e-fuels for long-distance transport, and long-duration energy storage, hydrogen from water electrolysis is widely regarded as an important energy carrier [1]. The practical benefits of electrolytic hydrogen depend strongly on the electrical energy consumed per kilogram of H_2 and on the durability of the system under sustained high-current operation. In conventional alkaline or proton-exchange-membrane electrolyzers, oxygen evolution reaction (OER) at the anode is both thermodynamically demanding and kinetically sluggish. OER typically requires high anodic overpotentials and corrosion-resistant components. As a result, it contributes substantially to the operating cell voltage and imposes stringent stability requirements on catalysts and electrode architectures [2]. In Figure 2.1, a conceptual comparison between OER-based and alcohol-assisted operations is summarized.

Alcohol-assisted (hybrid) water electrolysis addresses the main limitation of conventional electrolysis by replacing OER with anodic (alcohol) oxidation reaction (AOR), while keeping hydrogen evolution reaction (HER) at the cathode. Many alcohol oxidations require a lower thermodynamic potential than OER and reach a given rate at lower overpotential on appropriate catalysts. Thus, the anode can operate at a lower potential, which reduces the full-cell voltage at the same current density. In practice, the operating voltage accounts for ohmic and mass-transport losses, in addition to thermodynamic and kinetic terms. Voltage benefit is realized

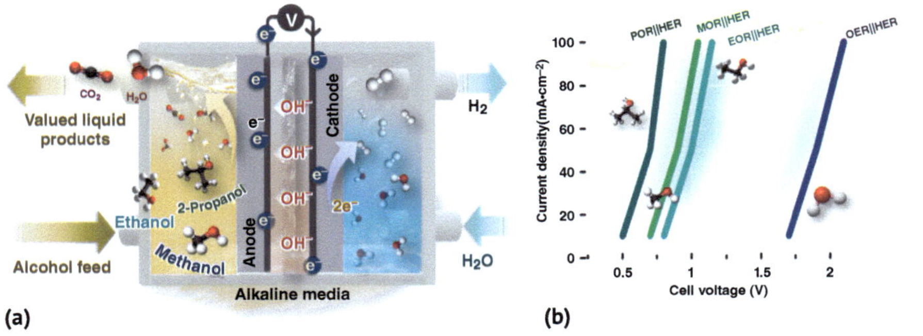

Figure 2.1 Schematic illustration of alcohol-assisted (hybrid) water electrolysis: (a) two-electrode cell in which anodic AOR replaces OER, coupled with cathodic HER; and (b) qualitative energy/polarization comparison showing how substituting AOR for OER can reduce the required anodic potential and overall cell voltage at a given current density.

only if the electrode and cell design avoid introducing new resistive or transport penalties.

Hybrid electrolysis differs from conventional water splitting by offering a tunable second product stream. Instead of evolving oxygen, the anode can generate aldehydes, ketones, carboxylic acids, or other oxygenates, depending on the alcohol substrate and catalyst-controlled pathway. When the anodic products are market-relevant chemicals or intermediates that integrate with existing value chains, this coproduction can be valuable. For example, alcohol-assisted electrolysis using biomass-derived substrates such as glycerol naturally connects to broader efforts in biomass upgrading and circular carbon utilization. At the same time, the presence of organics introduces new constraints. Selectivity, carbon balance, separation requirements, and feedstock availability become central when defining optimal performance.

The complexity of alcohol oxidation implies that hybrid electrolysis is not inherently advantageous. Alcohol oxidation is a multi-electron, multistep network involving C—H and O—H bond activation, oxygen insertion, and, in some cases, C—C bond cleavage [3, 4]. These steps compete with one another and with OER. Hence, they are strongly influenced by adsorption energies and surface coverage. A catalyst that generates high current may be undesirable if it drives deep oxidation to low-value products, accumulates strongly bound poisons that degrade activity, or promotes side reactions that compromise stability and carbon balance [5].

Under alkaline conditions and at high potentials, the anode environment is chemically aggressive, where surface oxidation, dissolution, and reconstruction are common. The net benefit of hybrid electrolysis emerges from coupled optimization: low cell voltage and high current density must be maintained together with stable, selective product formation that is compatible with practical downstream handling.

Within this landscape, catalyst design is critical. Alkaline hydrogen evolution involves water dissociation and hydrogen adsorption–desorption and is sensitive to surface electronic structure and interfacial water. At the anode, alcohol oxidation requires sites that activate alcohol molecules, handle oxygenated intermediates without excessive blocking, and remain stable under oxidizing bias. These functions can be separated across different catalysts or combined in bifunctional materials. Even then, hydrogen evolution and alcohol oxidation depend on different descriptors, making simultaneous optimization difficult. Electrode-level factors such as porosity, wettability, ionic pathways, and bubble management further influence the observed performance.

Transition-metal carbides and nitrides have emerged as promising non-oxide catalysts. Noble metals and their alloys remain important benchmarks, but their cost, resource constraints, and susceptibility to poisoning motivate the search for earth-abundant alternatives, particularly for large-scale hydrogen production [6]. In alkaline media, nickel-based catalysts and transition-metal (oxy)hydroxides are widely studied. Their alcohol-oxidation selectivity and stability can vary strongly with composition, potential window, and electrolyte. Carbides and nitrides offer a distinct alternative [7, 8]. They combine high electrical conductivity with robust metal–nonmetal bonding and span a broad compositional and defect space, enabling systematic tuning of electronic structure. In certain regimes, their surfaces can exhibit noble-metal-like adsorption characteristics while retaining lower cost and good structural integrity, making them attractive platforms for both HER and AOR.

The performance of carbides and nitrides is inseparable from their dynamic surface behavior. Under electrochemical operation, especially at anodic potentials, their surfaces rarely remain in the pristine state inferred from ex-situ characterization. Partial oxidation, oxycarbide or oxynitride layer formation, defect generation, and reconstruction are commonly observed. Depending on their extent and reversibility, these changes can generate active motifs or drive catalyst deactivation. Alcohol-assisted electrolysis must be understood as the interplay between a conductive bulk phase and an evolving surface chemistry, shaped by electrolyte composition, applied potential, and adsorbed intermediates.

In this framework, the chapter focuses on carbides and nitrides as electrocatalysts for alcohol-assisted hybrid water electrolysis. It relates their intrinsic properties to catalytic kinetics, selectivity, and operational stability. First, it considers which fundamental properties of carbides and nitrides make them promising for HER and AOR in hybrid systems, and how their composition and structure tune these properties. Second, it discusses how these materials can be engineered at the atomic, nanoscale, and electrode levels to stabilize active states and steer alcohol oxidation toward desirable products at low cell voltages and high current densities.

2.2 Intrinsic Properties Governing Electrocatalytic Activity

Transition-metal carbides and nitrides occupy a distinct position among non-oxide electrocatalysts. Their behavior arises from three tightly coupled intrinsic attributes. A metal-rich electronic structure provides high electrical conductivity. Strong metal–nonmetal bonding stabilizes the lattice and reshapes adsorption energetics, while composition, stoichiometry, and defects enable extensive tunability.

These properties are central to alcohol-assisted electrolysis, in which both half-reactions are governed by interfacial adsorption and coupled charge transfer.

High electrical conductivity is a defining feature governing electrocatalytic activity. Many carbides and nitrides exhibit metal-like conductivities because metal d-states remain partially filled and contribute strongly to the density of states near Fermi level. This conductivity reduces internal ohmic losses within catalyst particles and helps maintain a uniform potential distribution across porous electrodes [9]. At high current densities, conductivity directly affects apparent activity. If two catalysts have similar intrinsic turnover frequencies, the more conductive phase often delivers higher geometric current because charge-transfer and contact resistances are lower, allowing fuller utilization of the catalyst mass. Fast electron transport supports rapid anodic oxidation, in which the surface coverage of reaction intermediates varies with potential and reactant concentration. Local charge-transfer limitations can give rise to nonuniform reaction zones that promote selectivity drift and accelerate catalyst degradation.

A second core attribute is metal–carbon and metal–nitrogen bonding. In many relevant phases, carbon or nitrogen occupies interstitial sites in a close-packed metal lattice, leading to mixed metallic and covalent bonding and strong orbital hybridization between metal d-states and C/N p-states. This hybridization broadens the d-band, redistributes electron density, and can shift the effective d-band center. Adsorption strengths of H*, OH*, and alcohol-derived fragments are then directly influenced by the overlap between these surface states and adsorbate orbitals [10, 11]. In this context, the often-cited Pt-like behavior of certain carbides should be understood. Some compositions fall within a regime in which hydrogen-related intermediates experience near-optimal binding while maintaining high conductivity and structural integrity. Nitrides often occupy a distinct region of the catalytic landscape. Nitrogen's higher electronegativity and distinct bonding can produce surface electron densities and binding preferences that are particularly favorable for oxygenated intermediates relevant to alcohol oxidation.

Catalytic activity is frequently described using adsorption-based descriptors. For HER, a widely used descriptor is the Gibbs free energy of hydrogen adsorption, ΔG_{H*}, since catalysts that bind hydrogen too weakly or too strongly perform poorly. In alkaline media, HER kinetics depend strongly on water dissociation. Therefore, adsorption and activation of water or hydroxyl species are equally important. Carbides and nitrides are attractive because their composition and defect chemistry can be adjusted to achieve intermediate hydrogen binding. Selected sites can also retain sufficient oxophilicity to promote water activation. In practice, multi-component carbides and nitrides or heterostructured surfaces often outperform single-component phases. Distinct active sites facilitate both water activation and near-optimal hydrogen adsorption/desorption, thereby accelerating overall kinetics.

For alcohol oxidation, the reaction landscape is more complex. Alcohols adsorb and undergo sequential dehydrogenation and oxygen insertion steps, generating a series of oxygenated intermediates that compete for surface sites with OH* and other co-adsorbates. No single adsorption energy captures performance; rather, activity and selectivity depend on a balance of adsorption strengths across key intermediates and oxygenated species. If the surface binds oxygenated intermediates too weakly, oxidation is sluggish and may require higher potentials, eroding the voltage advantage. If binding is too strong, the surface can become blocked or shift toward deeper oxidation pathways. Carbides and nitrides provide valuable leverage because the same electronic tuning that adjusts ΔG_{H*} also modulates binding of oxygenates and carbonaceous fragments.

Tunability through composition, stoichiometry, and defects is the third intrinsic dimension [12]. Vacancies, non-stoichiometry, and metal substitution can adjust local coordination and electron density, creating active motifs with distinct adsorption behavior. Vacancies, for instance, can increase the density of unsaturated sites and alter the work function locally, thereby enhancing adsorption and promoting charge transfer. However, defects often increase chemical reactivity toward oxidation. The optimal defect concentration is a compromise, being high enough to create active sites and enhance kinetics, yet low enough to avoid rapid surface corrosion or uncontrolled reconstruction under anodic bias.

Multimetallic carbides and nitrides offer an additional degree of control. Bimetallic systems can tune electronic structure through ligand and ensemble effects, shifting adsorption energies and stabilizing phases that would otherwise be less favorable. In hybrid electrolysis, multimetallic designs are particularly valuable because they can decouple bulk conductivity from surface chemistry. One metal can provide a conductive backbone and structural stability, whereas another tunes adsorption and reconstruction behavior at the catalytic interface.

Many carbides and nitrides form oxygenated surface terminations after air exposure or electrochemical conditioning and can develop thin oxycarbide or oxynitride layers under bias. These surface terminations affect hydrophilicity, interfacial water structure, and the adsorption of OH and alcohol molecules. In alkaline hydrogen evolution, surface hydroxyl species can facilitate water activation. In alcohol oxidation, oxygenated surface motifs can aid dehydrogenation but may also cause over-oxidation or surface blocking if not controlled. Therefore, intrinsic behavior is best discussed in terms of the surface state that forms under relevant potentials rather than an idealized pristine surface.

In summary, the intrinsic value of transition-metal carbides and nitrides for alcohol-assisted electrolysis stems from their conductive, metal-rich electronic structures, strong and tunable metal–nonmetal bonding that reshapes adsorption energetics, and compositional and defect flexibility. These attributes define a design space in which hydrogen binding, hydroxyl affinity, and adsorption of alcohol-derived intermediates can be systematically adjusted to support reduced cell voltage, high current density, and controlled anodic product distributions.

2.3 Catalytic Roles in Alcohol-assisted Electrolysis Systems

In alcohol-assisted electrolysis, the full-cell voltage, current density, and stability arise from the coupled behavior of HER and AOR. Since both half-reactions carry the same current, limitations at either electrode manifest as increased cell voltage or accelerated degradation. Anodic chemistry differs markedly from conventional OER-driven systems, as it involves a complex organic reaction network whose selectivity and surface coverage strongly influence both kinetics and durability. Carbides and nitrides must be evaluated in a system context. A useful material is one that provides high intrinsic activity, maintains a stable functional surface, and integrates into electrodes that support transport and long-term operation at practical current densities.

2.3.1 Cathode: HER

At the cathode, the goal is efficient hydrogen generation with minimal kinetic and resistive losses. For alkaline HER, carbides and nitrides have been extensively studied because their metal-rich electronic structures provide high conductivity, and their adsorption properties can approach favorable hydrogen-binding regimes. When integrated into porous electrodes or deposited on conductive scaffolds, these materials can reduce charge-transfer resistance and improve potential uniformity across the catalyst layer. This effect enhances catalyst utilization and mitigates localized regions of high overpotential that often trigger degradation.

Mechanistically, alkaline HER proceeds through water dissociation (Volmer step, Eq. 2.1), followed by either electrochemical desorption (Heyrovsky step, Eq. 2.2) or chemical recombination (Tafel step, Eq. 2.3). Effective catalysts must promote water activation and maintain a balance between H* adsorption and desorption. They must also limit the buildup of interfacial OH^-/OH^* species that can block active sites and impede charge transfer. Carbides and nitrides can satisfy these requirements by tuning the surface electronic structure toward an intermediate hydrogen-binding state and by providing oxophilic motifs that facilitate water dissociation. In many cases, the highest performance is achieved using engineered compositions, in which defects, dopants, or heterostructured interfaces create complementary sites that jointly accelerate water activation and hydrogen release:

$$\text{Volmer}: \ H_2O + e^- \rightarrow H^* + OH^- \tag{2.1}$$

$$\text{Heyrovsky}: \ H_2O + e^- + H^* \rightarrow H_2 + OH^- \tag{2.2}$$

$$\text{Tafel}: \ H^* + H^* \rightarrow H_2 \tag{2.3}$$

In addition, the cathode must withstand sustained gas evolution. To avoid transport-induced voltage losses that could be mistaken for intrinsic deactivation, electrode architectures must be carefully designed. Such architectures should promote efficient bubble detachment while maintaining effective wetting. Carbides and nitrides integrated into robust, hierarchical structures can provide the mechanical strength and conductive pathways needed to sustain high-rate HER.

2.3.2 Anode: AOR

At the anode, alcohol oxidation is inherently more complex than OER. The process involves alcohol adsorption, stepwise dehydrogenation and oxygen insertion, and, in some cases, C—C bond cleavage, all of which compete with OER and catalyst surface oxidation. Activity alone is not sufficient; selectivity and its temporal stability are central, because the hybrid concept relies on generating usable anodic products as well as hydrogen. Surfaces that bind intermediates too strongly can suffer from poisoning and rapid performance loss, whereas surfaces that require high potentials to sustain oxidation may negate the intended voltage reduction and increase the contribution of OER.

Carbides and nitrides offer several advantages as anode catalysts [13]. Their tunable electronic structures enable control over the adsorption energies of OH* and alcohol-derived intermediates, which is a primary lever for steering reaction pathways. A catalyst must stabilize key intermediates long enough to direct them toward desired products, yet avoid persistent site blocking. Nitrides are particularly promising, as nitrogen incorporation modifies the charge distribution at metal centers and can alter the adsorption configurations of oxygenated intermediates, potentially improving selectivity. Although carbides can exhibit high activity and conductivity, care is needed to prevent excessively strong binding of carbonaceous fragments, which can lead to poisoning or nonselective over-oxidation.

Under anodic bias, the functional surface state is again critical. Many carbides and nitrides develop oxygenated terminations or reconstructed surface layers that participate during catalysis. Thin, conductive oxygenated layers can provide reactive sites for alcohol dehydrogenation and oxygen insertion, while the underlying carbides or nitrides sustain fast electron transport. Excessive oxidation, by contrast, can diminish conductivity, alter adsorption behavior, and shift selectivity. Effective anode design aims to stabilize a controlled, self-limiting surface state that remains electronically coupled to the conductive core and preserves both activity and selectivity.

Multicomponent carbides and nitrides provide enhanced tunability. Bimetallic systems can adjust adsorption energies and oxidation resistance through ligand and ensemble effects, enabling more precise tuning of intermediate stabilization and durability [14]. In practice, one component may be selected for conductivity and structural stability. The other component shapes surface chemistry and governs pathway selection.

Overall, the performance of carbides and nitrides in alcohol-assisted electrolysis emerges from the interplay of electronic structure, operando surface states, and electrode architecture. At the cathode, these materials support efficient alkaline HER. At the anode, they govern a multistep oxidation network in which selectivity, surface coverage, and stability are tightly linked.

2.4 Surface Dynamics and Stability Under Operating Conditions

The performance of carbides and nitrides in alcohol-assisted electrolysis is determined as much by surface evolution under bias as by their initial crystal structures. Potential, pH, reactant concentration, and adsorbed intermediates continuously reshape near-surface coordination, oxidation state, and termination chemistry.

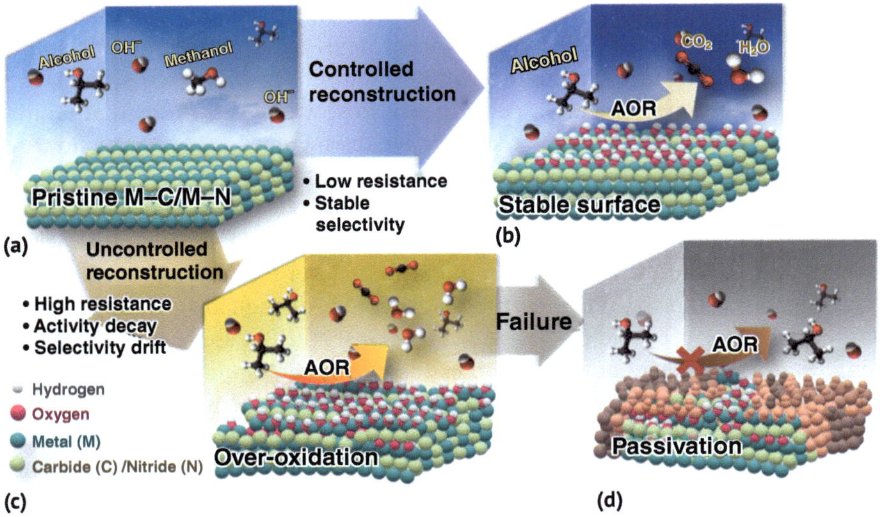

Figure 2.2 Surface evolution of carbides and nitrides under hybrid electrolysis conditions: (a) pristine carbide/nitride surface before polarization; (b) controlled reconstruction forming a thin oxygenated (oxycarbide/oxynitride-like) active shell supported by a conductive core, preserving electronic connectivity and stable activity; (c) uncontrolled oxidation or passivation leads to a thicker, less conductive surface layer, increased charge-transfer resistance, and activity decay; and (d) adsorbate-driven effects, such as intermediate poisoning and carbonaceous deposition, alter surface coverage and can drive selectivity drift during alcohol oxidation.

Consequently, the active catalyst during operation often differs from the pristine phase identified by ex-situ techniques. Stability must therefore be understood as the ability to maintain a beneficial functional surface state. It also requires preserving conductivity, accessible sites, and product selectivity. In Figure 2.2, key modes of surface evolution are summarized.

A useful distinction exists between bulk and interfacial stability. Bulk stability characterizes the retention of long-range structure and mechanical integrity in carbides and nitrides under electrolyte exposure and during cycling. Interfacial stability requires the near-surface region to remain electronically connected to the current collector, maintain a controlled population of active motifs, and avoid irreversible changes that block sites or alter adsorption energetics. In hybrid electrolysis, interfacial stability is often a limiting factor. The anode experiences high anodic potentials and complex organic film formation. The cathode, in contrast, undergoes vigorous gas evolution and repeated wetting–dewetting cycles that can strain the catalyst layer.

2.4.1 Surface Reconstruction as Activation and Degradation

Carbides and nitrides are generally robust materials. Their surfaces are rarely static under operating conditions. Under cathodic HER conditions, catalyst surfaces may undergo mild restructuring due to the adsorption of H* and OH*, changes in hydration, and local variations in proton activity. Under anodic alcohol oxidation, the driving forces for surface transformation are substantially stronger. Partial oxidation, the formation of oxygenated terminations, and the development of mixed oxycarbide or oxynitride phases are commonly observed. These transformations can be beneficial when they generate active motifs that promote dehydrogenation and oxygen insertion or improve hydrophilicity and reactant accessibility. When oxidation extends beyond a thin conductive layer, however, it reduces electronic conductivity, induces passivation, and shifts product selectivity. The central stability question is not whether reconstruction occurs, but whether it remains controlled and electronically connected to the conductive core. A material that appears to activate during conditioning may do so because a thin reconstructed layer forms, exposing catalytically competent sites. A material that degrades may experience further transformation into insulating oxide domains or lose mechanical integrity at the surface.

2.4.2 Oxidation, Corrosion, and the Conductive-core/Active-shell Picture

Under anodic bias, oxidation is the dominant chemical stress. In carbides, near-surface metal–carbon environments can convert to oxygenated metal sites, with carbon removed, reorganized, or transformed into disordered species. In nitrides, metal–nitrogen coordination can similarly convert to oxygenated coordination, with nitrogen displaced or redistributed. A common functional picture is that of a conductive carbide or nitride core supporting an oxygenated surface shell. When this shell remains thin and porous and is well electronically coupled, it can provide active sites. The core ensures rapid charge transport. If the shell thickens or becomes compact and poorly conductive, charge transfer is hindered, activity declines, and selectivity often drifts.

Electrolyte composition influences these processes. High alkalinity can promote the formation of surface (oxy)hydroxides and, under certain potentials, the dissolution of specific metal species. Organic intermediates may transiently shield surfaces by occupying reactive sites, but strongly bound fragments can also act as poisons if they accumulate. The net outcome depends on the balance between reversible adsorption and surface oxidation kinetics.

2.4.3 Adsorbate-driven Effects and Selectivity Drift

Alcohol oxidation generates dense populations of carbon-containing intermediates and oxygenated fragments that compete for surface sites. High coverage of alkoxy or dehydrogenated species can restrict access of OH^- or water to the surface, shifting pathways and, in some cases, forcing operation at higher potentials that accelerate catalyst oxidation. Strongly bound intermediates can act as poisons, reducing active-site availability and altering apparent kinetics. Over time, such coverage effects can change selectivity even if the total current remains approximately constant.

For selectivity control, nitrides are often explored. Their nitrogen-modified electronic structures influence adsorption preferences and stabilize key intermediates. This advantage is realized only if nitride-derived surface characteristics persist under operation. If the surface becomes extensively oxidized, the adsorption landscape can approach that of an oxide, and selectivity benefits may diminish. Despite their high activity, carbides can suffer from carbon poisoning and carbon deposition if they bind certain fragments too strongly. In both families, selectivity control and stability are inseparable.

2.4.4 Defects, Vacancies, and Stability–Activity Trade-offs

Defect engineering is a powerful strategy to enhance catalytic activity. Vacancies and unsaturated coordination sites can improve adsorption, facilitate water activation, and promote alcohol dehydrogenation. Nevertheless, defects also increase chemical reactivity and can exacerbate oxidation, dissolution, or phase transformations at the anode. The optimal defect concentration is therefore moderate, boosting activity while retaining a self-limiting reconstruction regime. This balance becomes increasingly difficult to sustain at high current densities. Under such conditions, pH gradients, intermediate fluxes, and local heating intensify degradation mechanisms that may remain negligible during low-current screening.

2.4.5 Electrode-level Stability

Stability is influenced by electrode architecture. During HER, bubble generation exerts mechanical stress and can locally dry out regions of the catalyst layer, leading to cracking or detachment when adhesion is weak. During anodic alcohol oxidation, even with limited gas evolution, changes in viscosity and product accumulation can cause transport limitations and uneven reaction rates. In both cases, localized high overpotentials can develop, promoting corrosion and uncontrolled reconstruction, thereby degrading transport and selectivity. Strong integration of carbides and nitrides with conductive supports and mechanically robust architectures is essential. Continuous electron pathways, stable wetting, and controlled porosity help prevent local voltage excursions and ensure uniform reactant delivery and product removal.

2.4.6 Assessing Stability in Hybrid Electrolysis

Durable performance is best evaluated using sustained operation at representative current densities, combined with time-resolved product analysis. Chronopotentiometric or chronoamperometric tests under relevant conditions, coupled with periodic analysis of product distributions and carbon balance, can reveal selectivity drift that may not be evident from current alone. Posttest characterization should then be linked to changes in overpotential, charge-transfer resistance, and faradaic efficiency (FE), rather than reported in isolation.

2.4.7 Design Implications

The design objectives for carbides and nitrides can be summarized. They consist of a conductive core with an active, electronically connected surface that remains thin and permeable. Adsorption energies must be maintained within a range that enables fast kinetics without persistent poisoning. Achieving these objectives requires coordinated control of material composition and electrode architecture. Surface reconstruction must remain controlled rather than destructive, and mass-transport limitations must be avoided to prevent operation in regimes dominated by OER and deep oxidation.

2.5 Reaction Pathways and Product Selectivity in Alcohol Oxidation

Alcohol oxidation at the anode is the defining chemical feature of hybrid electrolysis. Unlike OER, which is typically treated as a single target reaction, alcohol oxidation is a network of competing pathways that share intermediates, whose economic value varies widely. Anode performance must be judged by the combined outcome of activity, selectivity, and stability of the product distribution over time.

2.5.1 Elementary Steps and Pathway Competition

In alkaline media, most alcohol oxidation begins with adsorption of the alcohol, followed by stepwise dehydrogenation of the O—H and C—H bonds to generate alkoxy species and progressively dehydrogenated intermediates. Oxygenated surface species, typically OH^* formed from OH^- (or from water under bias), are pivotal because they facilitate dehydrogenation and enable oxygen insertion. OH^* also competes with organic intermediates for surface sites and can accelerate catalyst surface oxidation at elevated anodic potentials. The anodic potential and local pH, therefore, regulate OH^* coverage and reactivity. This control largely determines which pathways dominate.

A compact, generic elementary-step set that captures this interplay is:

$$\text{OH* formation: } * + OH^- \rightarrow OH^* + e^- \tag{2.4}$$

$$\text{Alcohol adsorption: } RCH_2OH + {}^* \rightarrow RCH_2OH^* \tag{2.5}$$

$$\begin{gathered}\text{O—H activation/alkoxy formation:}\\ RCH_2OH^* + OH^- \rightarrow RCH_2O^* + H_2O + e^-\end{gathered} \tag{2.6}$$

$$\begin{gathered}\text{Further dehydrogeneration to carbonyl:}\\ RCH_2O^* + OH^- \rightarrow RCHO^* + H_2O + e^-\end{gathered} \tag{2.7}$$

$$\text{Further oxidation (generic)}: RCHO + 3OH^- \rightarrow RCOO^- + 2H_2O + 2e^- \tag{2.8}$$

These steps are intentionally simplified, but they make the point explicit: OH^*-mediated chemistry can accelerate alcohol oxidation. Yet the same OH^* coverage can impose site competition and stability constraints as both potential and pH shift.

Primary alcohols typically oxidize first to aldehydes and, with further oxidation, to carboxylic acids. Secondary alcohols instead preferentially yield ketones. Polyols such as glycerol exhibit richer product networks because multiple carbon sites can be oxidized and C—C bond cleavage can compete, generating smaller oxygenates. Whether the reaction remains at partial oxidation or proceeds toward deeper oxidation and fragmentation is largely set by how the catalyst stabilizes key intermediates and by its tendency to promote or inhibit C—C scission.

2.5.2 Adsorption Energetics as a Selectivity Lever

For carbides and nitrides, control over surface electron density and adsorption energies is the principal means of steering pathways. Selectivity can be viewed as a balance among the following three requirements. The alcohol must adsorb strongly enough to be activated without saturating the surface with unreactive species. Key intermediates must be stabilized sufficiently to direct them toward desired products, though not so strongly that they persist as poisons. Oxygenated co-adsorbates must be present at coverages that support dehydrogenation and oxygen insertion without blocking sites or driving extensive over-oxidation.

Carbides and nitrides can be tuned to occupy intermediate adsorption regimes that are difficult to reach with simpler oxides or pure metals. Carbides often exhibit strong adsorption and excellent electrical conductivity, but may require modification to avoid excessively strong binding of carbonaceous species. Nitrides introduce nitrogen-modified electronic structures that can shift adsorption preferences and stabilize intermediates, offering additional degrees of freedom for pathway control when nitride-derived surface characteristics persist during operation.

2.5.3 Surface Oxygenated Species and Reconstructed States

Under anodic bias, carbides and nitrides acquire oxygenated surface terminations or thin reconstructed layers. These surface motifs can be catalytically beneficial because alcohol oxidation proceeds via oxygen insertion steps facilitated by OH* or lattice-like oxygenated species. A thin, conductive reconstructed layer can thus provide the appropriate environment for selective oxidation. The underlying bulk then maintains charge transport. If oxidation progresses and generates thick, less conductive surface layers, adsorption energetics can shift, activity declines, and product distributions can change. Selectivity engineering must consider both the initial composition and the tendency to form specific surface states under bias. Strategies that stabilize beneficial reconstructed layers or limit oxidation to a self-limiting depth often enhance both activity and selectivity stability.

2.5.4 Poisoning, Carbon Balance, and Long-term Selectivity

Carbon-containing intermediates in AOR can promote catalyst deactivation by causing surface poisoning or by the accumulation of carbonaceous deposits. These changes can force operation at higher potentials, increasing OER and deep oxidation and altering product distributions [15]. In hybrid electrolysis, this shift may lead to a stable current but a gradual decline in product value.

Carbon balance is a primary metric of selectivity. High FE toward the desired product is meaningful only if most of the carbon is accounted for in the measured products, with minimal loss to unquantified species such as volatile compounds or carbonate. Rigorous analysis of both gas and liquid phases, and carbonate in alkaline media when relevant, is essential for meaningful comparisons.

2.5.5 Multicomponent Carbides and Nitrides in Selectivity Tuning

To tune selectivity, multicomponent carbides and nitrides, including bimetallic and doped phases and heterostructures, provide practical routes. Ligand effects can shift electron density at active sites. Ensemble effects can create adjacent motifs that support cooperative steps. For instance, one site stabilizes an alkoxy intermediate and a neighbor provides reactive OH*. These configurations can improve overall kinetics and guide pathways toward specific products. In application-focused design, the target product defines which intermediates should be stabilized and which transformations should be suppressed. Catalyst design then aims to match adsorption and reconstruction behavior to the desired pathway under the intended potential and current density.

2.5.6 Selectivity Metrics and Benchmarking

Meaningful selectivity benchmarking requires more than single-point FE. At a minimum, reporting should include FE for major products, closure of the carbon balance, and selectivity stability over time at defined current densities and operating conditions. Because hybrid electrolysis is pursued partly to reduce electrical energy input, selectivity metrics should be interpreted alongside cell voltage. Energy consumption per unit of hydrogen produced and per unit of the target product should also be considered. These metrics are electrochemical performance indicators rather than full economic measures.

In summary, alcohol-oxidation selectivity in hybrid electrolysis is governed by competition among adsorption-controlled reaction pathways. These pathways are determined by surface oxygenated species, surface reconstruction, poisoning effects, and carbon balance. Carbides and nitrides constitute a tunable materials platform for selectivity engineering. Their composition, defect chemistry, and multicomponent design can be tailored to regulate the functional surface state under operating conditions.

2.6 Engineering Strategies to Enhance Performance

For alcohol-assisted electrolysis, intrinsic conductivity and tunable adsorption properties make carbides and nitrides attractive candidates. However, high device-level performance rarely follows from composition alone. Hybrid systems are

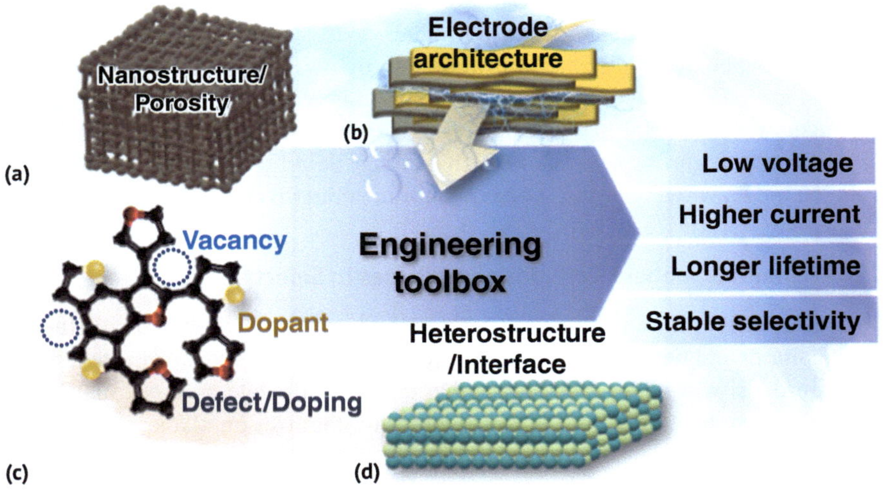

Figure 2.3 Engineering toolbox for performance enhancement in alcohol-assisted electrolysis: (a) nanostructuring and hierarchical porosity to increase accessible active sites and improve reactant and product transport; (b) electrode-architecture design, including wetting control, ionic pathways, bubble management, and mechanical adhesion, to enable stable operation at high current densities; (c) defect, vacancy, and heteroatom-doping strategies to tune adsorption energetics and stabilize beneficial surface states; and (d) heterostructures and engineered interfaces, including strong electronic coupling to conductive scaffolds, to promote charge redistribution and reduce contact resistance.

constrained by kinetics, transport, and stability. Performance improvements require coordinated engineering of local atomic environments, nano- and microstructures, and electrode architectures. In Figure 2.3, the main elements of this engineering toolbox are outlined.

2.6.1 Nanostructuring and Hierarchical Porosity

To increase accessible active-site density, reducing feature sizes and introducing hierarchical porosity are straightforward strategies. Nanostructuring increases the fraction of undercoordinated surface sites, thereby accelerating both HER and alcohol oxidation. Porosity, however, must be engineered across multiple length scales. Micropores add surface area but can trap products and hinder convective replenishment of alcohol. By contrast, mesopores and macropores provide lower-tortuosity pathways for electrolyte penetration, facilitate the removal of soluble products, and support stable operation at high current densities by reducing concentration gradients. At the cathode, gas evolution necessitates a porous structure and adequate wettability to enable bubble detachment and maintain electrolyte access. At the anode, porosity and connectivity govern alcohol delivery and product removal and help prevent local depletion or accumulation that can shift selectivity. These structural features must be balanced against mechanical robustness, as fragile, high-surface-area materials can degrade under gas-evolution stresses.

2.6.2 Morphology and Surface Termination Control

Morphology governs the exposure of specific facets and terminations, thereby influencing adsorption and reactivity. Different carbide and nitride facets can exhibit distinct metal-site densities, local coordination environments, and affinities for oxygenated species, making facet exposure a potential design variable for both HER and AOR. Although facet–selectivity relationships remain less developed than for noble metals, controlling morphology is a practical route to adjusting the adsorption landscape.

Equally important is near-surface termination chemistry. Many carbides and nitrides develop oxygenated terminations or thin oxycarbide/oxynitride layers after synthesis or during electrochemical conditioning. Surface treatments and synthesis approaches that yield stable, catalytically productive terminations can improve durability, provided the underlying bulk remains conductive.

2.6.3 Defect and Vacancy Engineering

Defects and vacancies create undercoordinated sites and modify the local electron density, often enhancing adsorption and lowering activation barriers for water activation and alcohol dehydrogenation. However, they also increase susceptibility to oxidation and dissolution. Effective defect engineering in hybrid electrolysis aims for an intermediate defect density and often confines highly defective regions to near-surface layers over a more ordered, conductive core.

2.6.4 Heteroatom Doping

Doping with nonmetals or secondary metals is commonly used to tune adsorption, adjust the work function, and influence surface reconstruction under bias. For HER, doping can shift hydrogen binding toward optimal values and introduce sites that promote water dissociation. For alcohol oxidation, doping can modify the binding of oxygenated intermediates and alter pathway preferences. Gradients or core–shell configurations, in which dopants are enriched near the surface, enable targeted modification of the catalytic interface while leaving bulk properties largely unchanged. Such doping strategies can also stabilize self-limiting reconstructed layers and suppress the formation of thick, insulating oxides.

2.6.5 Multicomponent Carbides and Nitrides

Bimetallic and multicomponent carbides and nitrides expand the design space by enabling ligand and ensemble effects that can modulate adsorption energies, reaction barriers, and oxidation resistance. In hybrid electrolysis, this flexibility can decouple conductivity from selectivity. Distinct roles can be assigned to different components: one provides a conductive framework, while the other tunes adsorption and stabilizes active surface states.

2.6.6 Heterostructures and Interfaces

Heterostructures, such as carbide–oxide hybrids, nitride–phosphide junctions, or carbide/nitride–carbon composites, can exhibit properties beyond those of the individual components. Through charge redistribution and band alignment, interfaces introduce new electronic environments and can tune adsorption and reconstruction behavior. Strongly coupled interfaces reduce contact resistance, promote a more uniform distribution of current, and can stabilize active surface motifs even under partial oxidation.

2.6.7 Conductive Scaffolds and Self-supported Electrodes

Self-supported electrodes grown directly on conductive substrates, such as metal foams or carbon textiles, can improve stability and simplify current collection. Eliminating weak particle–binder interfaces and ensuring continuous electron pathways help maintain performance under gas-evolution and oxidizing conditions. When binders or ionomers are used, their chemical compatibility with carbides, nitrides, and organic reactants must be considered to avoid swelling or adsorption phenomena that impede transport or shift selectivity.

2.6.8 Electrode Architecture and Transport

Ionic pathways, wetting, and product removal are governed by electrode architecture. At the cathode, a balance between hydrophilic and hydrophobic behaviors influences bubble dynamics and water supply. At the anode, architecture influences alcohol delivery, product desorption and removal, and the accumulation of viscous or strongly adsorbing species. Strategically designed architectures reduce ohmic and transport losses and help maintain local potentials within regimes where AOR remains dominant over OER.

2.6.9 Benchmarking During Engineering Optimization

Engineering strategies should be evaluated using performance metrics that reflect both electrochemical and chemical outcomes. For HER, these include overpotential at specified current densities, Tafel slopes, impedance-derived charge-transfer resistance, and durability under continuous gas evolution. For AOR, they include current density, onset and operating potentials, product-resolved FE, carbon balance, and selectivity stability. Full-cell tests are particularly important for hybrid systems, as three-electrode measurements alone may not capture interactions among electrodes, transport processes, and product management.

2.7 Comparative Summary of Carbides vs. Nitrides

Carbides and nitrides share several attractive features, including high electrical conductivity, robust metal–nonmetal bonding, and broad compositional flexibility. Still, they often occupy different regions of the activity–selectivity–stability.

A comparative analysis helps clarify which intrinsic tendencies and engineering levers are most suitable for a given operating window, alcohol substrate, and target product set. In Table 2.1, key contrasts and design implications are summarized.

Both families typically exhibit high conductivity, which supports the use of thick, high-loading electrodes and good catalyst utilization. Both can be mechanically robust and chemically resilient relative to many non-oxide alternatives, particularly when integrated into supported or self-supporting architectures. A common constraint is that their functional surface state under operation may differ from the bulk phase label, because partial oxidation and reconstructed surface layers can become catalytically dominant. Performance, therefore, depends on whether these transformations remain controlled and self-limiting or evolve into thick, poorly conductive oxides.

Carbides are frequently valued for their metallic electronic character and can exhibit strong apparent activity for HER and, in some cases, for oxidation reactions where fast charge transfer is advantageous. They can be particularly beneficial on the cathode, where low resistance and robust electrode integrity under gas evolution are essential. At the same time, strong adsorption of carbonaceous fragments during AOR can increase the risk of catalyst poisoning or selectivity drift if not moderated. Surface oxidation under anodic bias can gradually alter conductivity and adsorption behavior. Carbides often benefit from multimetal designs that moderate adsorption, introduce defects in a controlled manner, and form strong couplings with conductive scaffolds to maintain electron pathways despite partial surface oxidation.

Nitrides, in contrast, often exhibit nitrogen-modified adsorption landscapes that can be exploited to steer pathways in alcohol oxidation. They are frequently used as anode catalysts when the selective formation of specific oxygenates is a priority. Their practical advantage hinges on whether nitride-derived surface characteristics persist under long-term operation. Engineering strategies that stabilize nitrogen-influenced near-surface environments (through compositional tuning, heterostructures, or strong interfacial coupling) are thus crucial.

In practice, carbides are favored when the primary constraint is high-rate cathodic operation, whereas nitrides are preferred when anodic selectivity and pathway control are paramount. When durability under oxidizing bias and complex coverage are the dominant concerns, neither family is inherently superior. Instead, the decisive factors are the quality of electrode integration and the ability to maintain a controlled, conductive functional surface state.

2.8 Outlook and Remaining Challenges

Alcohol-assisted electrolysis is attractive because it can reduce the electrical energy required for hydrogen production while enabling anodic coproduction of oxygenates. Transition-metal carbides and nitrides are promising materials within this concept because they combine high conductivity, structural robustness, and electronic tunability. Their successful deployment, however, depends on addressing several linked challenges that span active-state identification, selectivity control, durability, electrode integration, and process considerations.

Table 2.1 Summary of transition-metal carbides vs. nitrides for alcohol-assisted electrolysis.

Dimension	Metal carbides	Metal nitrides	Best engineering levers
Conductivity/ electron transport	Metal-like high electrical conductivity (efficient electron transport)	Generally high electrical conductivity, arising from metallic bonding within a ceramic-like matrix	Nanostructuring and conductive supports to maximize electron percolation and contact
Typical adsorption tendency	Strong chemisorption of H and oxygenates; high carbophilicity can lead to deep deoxygenation (Pt-like binding strength)	Moderate adsorbate binding at surface N sites, imparting basicity and facilitating activation of polar molecules	Electronic structure tuning, e.g., doping or alloying, to modulate adsorption strength and selectivity of intermediates
Operando surface evolution risk	Prone to oxidative surface transformation under anodic bias (carbide → oxide + metal ion dissolution, with carbon residue)	Susceptible to surface oxides (oxynitride formation) and N loss under harsh conditions, limiting long-term stability	Protective coatings, e.g., carbon shells and phase stabilization (alloying), to mitigate oxidation and preserve active phase
HER suitability	Excellent HER catalysts; certain TMCs, e.g., Mo_2C and WC, exhibit high activity rivaling Pt	High HER activity, especially in alkaline media, approaching Pt-level performance	Optimize H-binding via surface modification, e.g., monolayer Pt or transition-metal overlays, to reach ideal HER kinetics
AOR selectivity tendency	Tend to promote complete oxidation (C—C/C—H bond breakage), risking over-oxidation to CO_2 and mixed products	Often enable more selective partial-oxidation pathways (tunable bond scission), favoring value-added intermediates over complete combustion	Bifunctional catalyst design, e.g., coupling carbide/nitride with oxyhydroxide phases, to direct desired products and avoid excessive oxidation
Main deactivation modes	Oxidation and leaching under anodic conditions: corrosion produces dissolved metal oxides and deactivating carbon deposits on the surface	Surface passivation and structural degradation via oxide formation and N de-intercalation; moderate chemical stability relative to PGMs, with gradual activity loss at high current densities	Preventive measures like in situ passivation layers, corrosion-resistant alloys, and operation within stable potential windows to prolong catalyst life
Most effective architectures	Nanoporous or supported carbide nanostructures; PGM-decorated carbides (core–shell) for enhanced activity and poison resistance	Carbon-protected nitride nanoparticles and bimetallic nitride phases can improve stability and catalytic activity	Core–shell and hybrid architectures, e.g., carbide–nitride composites, that synergistically combine high intrinsic activity with structural durability

A first challenge is to establish the true active surface state under operating conditions. For both carbides and nitrides, the catalytically relevant surface often differs from the pristine bulk phase due to reconstruction, oxygenated terminations, defect evolution, and partial oxidation. In alcohol-oxidation environments, sustained coverage by organic intermediates and oxygenated species can further reshape the surface relative to OER-only conditions. Reliable structure–activity relationships require operando and in situ characterization combined with electrochemical protocols and quantitative product analytics that are designed to distinguish stable active motifs from transient states and irreversible degradation pathways.

A second challenge is to develop selectivity frameworks that remain valid across alcohol substrates, electrolytes, and current densities. Alcohol oxidation proceeds through competing pathways, and the steps that control selectivity are highly sensitive to catalyst identity and interfacial conditions. This is because the value proposition of hybrid electrolysis relies on selective co-production rather than indiscriminate oxidation. Accordingly, mechanistically guided selectivity maps that link adsorption energetics and oxygenated-species coverage to product distributions will be vital. Rigorous carbon accounting should accompany these efforts to ensure complete carbon-balance closure across liquid and gas products, with carbonate formation included where selectivity is relevant.

Durability under combined chemical and mechanical stresses is a third challenge. At the cathode, bubble evolution and wetting–drying cycles can damage catalyst layers and disrupt electrical percolation. At the anode, oxidizing potentials, intermediate coverage, and potential excursions can accelerate surface oxidation, dissolution, and poisoning. These effects often reinforce each other. Transport limitations can increase local overpotentials, which then accelerate reconstruction or corrosion. The resulting surface changes can further degrade transport by altering wetting and porosity. Therefore, durable operation will require catalysts and electrode architectures that tolerate dynamic surface states while preserving conductivity and a stable adsorption landscape.

A fourth challenge constitutes device-level translation and benchmarking under realistic conditions. Many carbide and nitride catalysts are still evaluated primarily in three-electrode cells under conditions that do not reflect practical electrolyzers, particularly in terms of current density, catalyst loading, hydrodynamics, and temperature. Comparisons across studies are further complicated by inconsistent normalization and incomplete product accounting. Progress will be accelerated if the community converges on a minimal reporting set for hybrid electrolysis. This set should include cell voltage at relevant current densities, time-resolved FE for individual products, carbon-balance closure, durability tests beyond short operation, and posttest characterization explicitly linked to the observed performance evolution. Full-cell measurements, including membrane-based configurations where appropriate, are necessary to confirm that materials strategies translate into system-level voltage reduction and stable product distributions.

A fifth challenge concerns coupling catalytic selectivity with separations and process constraints. The value of anodic coproducts depends on their concentration, purity, and separation complexity. Catalysts that deliver narrow, stable product slates at meaningful concentrations are more likely to support practical deployment than those that generate complex mixtures, even when the latter exhibit high apparent activity. Early consideration of product streams, substrate availability and quality, and compatibility with existing infrastructure can help focus materials development toward configurations with stronger implementation potential.

Looking ahead, several directions appear particularly promising. Design approaches can prioritize stabilization of functional surface states over idealized phase purity by targeting self-limiting reconstruction that creates productive oxygenated motifs while maintaining electronic connectivity. Multicomponent carbides and nitrides, engineered interfaces, and conductive scaffolds can be developed with explicit selectivity and durability targets, supported by combined experimental and computational analysis of adsorption landscapes and reaction pathways. At the electrode and system levels, durability gains will likely come from coupling material chemistry with architectures that govern transport, bubble dynamics, and potential distribution under high-current operation.

In summary, carbides and nitrides provide a credible pathway toward efficient alcohol-assisted electrolysis. Realizing this potential will require control of operando surface evolution, transferable selectivity principles supported by rigorous carbon accounting, and durable electrode designs validated under realistic operating conditions, with product distributions aligned to practical process requirements. The chapter closes by identifying priorities for translation: robust identification of the true active surface, transferable selectivity frameworks across alcohol families, and electrode-level integration that preserves low-voltage operation and stable product slates at industrially relevant rates.

References

1 SaberiKamarposhti, M., Kamyab, H., Krishnan, S., Yusuf, M., Rezania, S., Chelliapan, S., & Khorami, M. (2024). A comprehensive review of AI-enhanced smart grid integration for hydrogen energy: Advances, challenges, and future prospects. *Int. J. Hydrog. Energy 67*, 1009–1025. https://doi.org/10.1016/j.ijhydene.2024.01.129

2 Zhang, J., Fang, Y., Chen, Y., Zhang, X., Xiao, H., Zhao, M., Zhao, C., Ma, X., Hu, T., Luo, E., Jia, J., & Wu, H. (2024). In-situ fabrication of bimetallic $FeCo_2O_4$-$FeCo_2S_4$ heterostruçture for high-efficient alkaline freshwater/seawater electrolysis. *J. Colloid Interface Sci. 653*, 821–832. https://doi.org/10.1016/j.jcis.2023.09.126

3 Tang, C., Li, B., Yan, S., & Yi, Y. (2026). 3D honeycomb-like PdAg nanoframework with optimized geometric and interface electronic structure enable efficient ethanol oxidation via dual-pathway mechanism. *J. Colloid Interface Sci. 707*, 139644. https://doi.org/10.1016/j.jcis.2025.139644

4 Wang, X., Wei, L., Wu, J., Zhu, A., Zhang, Q., & Liu, Q. (2025). Supported core-shell catalysts for enhancing ethanol electrooxidation by C1 pathway. *J. Colloid Interface Sci. 694*, 137719. https://doi.org/10.1016/j.jcis.2025.137719

5 Fu, X., Wan, C., Huyan, H., Wang, S., Zhang, A., Zhou, J., Zhang, H., Zhao, X., Chen, J., Pan, X., Huang, Y., & Duan, X. (2024). Medium entropy alloy wavy nanowires as highly effective and selective alcohol oxidation reaction catalysts for energy-saving hydrogen production and alcohol upgrade. *EES Catal. 2*, 1285–1292. https://doi.org/10.1039/d4ey00090k

6 Chen, X., Xie, S., Zhou, F., Gao, B., Zhang, X., Wu, S., Huang, L., Li, Y., Chu, P. K., & Peng, X. (2024). Carbonized-nitridated Mo_2N-MoC heterostructures towards universal-medium hydrogen evolution reaction. *Chem. Eng. J. 500*, 156908. https://doi.org/10.1016/j.cej.2024.156908

7 Broersen, P. J. L., Paschalidou, V., & Garcia, A. C. (2025). Electrosynthesis of benzyl-*tert*-butylamine *via* nickel-catalyzed oxidation of benzyl alcohol. *Green Chem. 27*, 1475–1487. https://doi.org/10.1039/d4gc05171h

8 Cui, X., Tang, T., Zhang, F., Sun, L., & Zhang, B. (2025). New benchmark for pure nickel-based oxygen-evolution electrocatalyst: Tailored large $NiMoO_4{\cdot}xH_2O$ monocrystals for complete reconstruction. *Appl. Catal. B: Environ. 366*, 125024. https://doi.org/10.1016/j.apcatb.2025.125024

9 Tello, A., Cataño, F. A., Arunachalam, A., Oyarzún, D., Henriquez, R., Valdivia, P., Viswanathan, M. R., & Gómez, H. (2025). Green hydrogen production by photovoltaic-assisted alkaline water electrolysis: A review on the conceptualization and advancements. *Int. J. Hydrog. Energy 107*, 378–395. https://doi.org/10.1016/j.ijhydene.2024.04.333

10 Guo, Z., Ga, L., Liu, M., & Ai, J. (2024). Progress in metal-organic frameworks for small molecule oxidative coupled hydrogen production. *Chem. Eng. J. 480*, 148365. https://doi.org/10.1016/j.cej.2023.148365

11 Zhuang, W., Huang, Z., Pei, D., Hu, W., Yu, Q., Wang, H., Li, X., Wu, Y., Du, Z., & Sun, Q. (2024). Achieving exceptional strength by multi-dimensional in-situ Al_3Ti, TiN, AlN reinforcements in Al matrix composites. *J. Alloys Compd. 997*, 174880. https://doi.org/10.1016/j.jallcom.2024.174880

12 Li, K., Guo, C., Ma, X., Wu, J., Yu, J., Qi, Y., Tao, Z., Zheng, J., Xue, J., & He, Y. (2025). N-doped biomass-derived molybdenum carbide hybrids: Mo/C stoichiometry-engineered electrocatalysts for synergistic HER-OER water splitting. *Int. J. Hydrog. Energy 170*, 151242. https://doi.org/10.1016/j.ijhydene.2025.151242

13 Zhao, J., Gong, J., Yuan, J., Liu, Z., Zhang, C., & Hu, J. (2025). Iron-doped molybdenum carbide for improved reactants adsorption in oxygen evolution reaction. *Appl. Catal. A: Gen. 704*, 120419. https://doi.org/10.1016/j.apcata.2025.120419

14 Zhang, H., Li, Y., Liu, X., Wang, J., Wang, X., Xue, H., & He, Y. (2025). Exploration of multiscale design strategies of electrocatalysts for efficient electrochemical hydrogen and oxygen evolution reactions. *Chem. Eng. J. 522*, 167757. https://doi.org/10.1016/j.cej.2025.167757

15 Morales, E. C., Kazakova, M. A., Selyutin, A. G., Golubtsov, G. V., Morales, D. M., & Robledo, A. M. (2024). Competition between the oxygen evolution reaction and the electrooxidation of alcohols on heteroatom-functionalized multi-walled carbon nanotubes-supported Ni oxide catalysts. *Surf. Interfaces 46*, 104026. https://doi.org/10.1016/j.surfin.2024.104026

3

Single-atom Electrocatalysts for Alcohol-assisted Water Electrolysis

P. Keerthana[1,], Soorya S. Raj[2], and M. L. Aruna Kumari[2,*]*

[1]Department of Science, Alliance University, Bengaluru, India
[2]Department of Chemistry, The Oxford College of Science, Bengaluru, India

*Corresponding authors: keerthana.p@alliance.edu.in; drarunakumariml@gmail.com

3.1 Introduction

In the midst of the rapid growth of the global economy, the energy crisis and environmental degradation caused by the extensive use of nonrenewable energy sources, such as coal, oil, and other fossil fuels, have emerged as the principal barriers to sustainable human progress [1]. Developing environment-friendly energy sources is crucial to addressing global energy and environmental issues [2–4]. Hydrogen has several advantages, including its abundant supply, high calorific value ($143\,MJ{\cdot}kg^{-1}$), and zero emissions [5]. Currently, hydrocarbon reforming is the most common method for producing sustainable hydrogen. These hydrogen-generating systems continue to face the constraints of limited fossil fuel resources and the environmental effects of carbon dioxide (CO_2) as an inevitable by-product. Therefore, synthesizing sustainable hydrogen is of great importance to addressing the challenges posed by fossil fuels.

Water electrolysis is an electrochemical process that breaks down water into hydrogen and oxygen when a certain amount of voltage is applied. The hydrogen evolution reaction (HER) occurs at the cathode, while the oxygen evolution reaction (OER) takes place at the anode. The significant overpotential (>1.23 V) associated with the OER, characterized by its slow kinetics, plays a major role in driving up the costs of hydrogen production during water electrolysis [6, 7]. Different reactive oxygen intermediates (*OH, *O, and OOH*) produced during the OER process tend to interact with protons in the solution, which inhibits hydrogen evolution and consequently decreases faradaic efficiency (FE) [8]. Utilizing active oxygen intermediates with strong oxidation capabilities as an oxygen source in electrooxidation processes holds the promise of enabling the green synthesis of valuable bulk chemicals [9–13]. This approach is anticipated to lower anodic oxidation potential and overall energy consumption, thereby promoting hydrogen production on an industrial scale [14–16]. Also, the OER and the HER are closely connected, which can cause gas crossover and make an explosive combination (O_2/H_2). When oxygen gas crosses, it makes an explosive and also makes the reaction less efficient since it may easily turn into hydroxide anion or water on the cathode side. Moreover, O_2 presence may facilitate the generation of reactive oxygen species (ROS) that might harm the membrane [17, 18].

To overcome these constraints, hybrid electrolyzers must be developed that integrate water splitting with thermodynamically advantageous oxidation processes to support the industrial expansion of global chemicals. Alcohol-assisted water electrolysis (AAWE) is a new method that has become popular to overcome the OER bottleneck. AAWE uses an alcohol oxidation reaction instead of the high-potential OER since it is better thermodynamically and kinetically. By utilizing easily accessible or discarded biomass-derived alcohols (such as methanol, ethanol, or glycerol) as anodic feedstocks, AAWE accomplishes two objectives: reduced voltage and also alcohols can be oxidized by active oxygen species, resulting in the formation of aldehydes and acids [19, 20]. Although AAWE provides a distinct thermodynamic benefit by substituting the slow OER, the alcohol oxidation reaction continues to be a complicated process involving multiple electron transfers. The inherent complexity leads to notable kinetic bottlenecks, indicating that the reaction requires elevated overpotentials to advance at a rate suitable for commercial viability. Consequently, the search is focused on creating highly efficient electrocatalysts that can drastically reduce the activation energy barrier (ΔE_{act}) of the rate-determining phase.

Single-atom catalysts (SACs) have become the best way to get around this kinetic problem by giving isolated, electrically tailored active sites for precise regulation of intermediates [21]. The structure consists of discrete metal atoms (M) firmly attached to suitable supports, allowing for the optimal utilization of the functional material, as every atom is reachable and participates in the reaction [22]. This approach clearly leads to a significant reduction in the quantity of metal used, effectively addressing key economic issues linked to the elevated expenses of conventional nanocatalysts and bulk catalysts [23]. Furthermore, the implementation of SACs has significantly enhanced our comprehension of catalytic mechanisms while also connecting heterogeneous and homogeneous catalyses through the provision of uniformly distributed, atomically dispersed active sites [24]. The interactions between metal and support in SACs are crucial in influencing their activity, selectivity, and stability [25]. SACs' unsaturated coordination environments stabilize isolated atoms, adjust electronic structures, and regulate reactant activation and adsorption energies, favoring one reaction pathway over another [26, 27].

Consequently, atomic-scale confinement of active centers in SACs presents opportunities to enhance turnover rates, which refer to the essential catalytic efficiency of each metal site, thereby enabling the design of electrocatalysts with exceptional activity and selectivity for a specific reaction. Nevertheless, SACs encounter a variety of challenges; for example, the high surface free energy of SAs makes them prone to aggregation into nanoparticles (NPs), leading to a loss of many advantages associated with individual atoms [28, 29]. Therefore, it is crucial to formulate robust strategies for securing SAs on substrates to confirm stability during the reactions. Furthermore, the single-site arrangement of SACs may limit their effectiveness in intricate reactions, where the synergistic effects resulting from the interaction of diverse active sites and NPs and nanocomposites could be crucial [30, 31].

Despite these challenges, the incorporation of SACs into AAWE systems offers a promising pathway for advancing hydrogen production in future technologies. Recent developments in synthetic methodologies, including defect engineering, heteroatom doping, and the establishment of robust metal–support interactions, have greatly enhanced the stability and catalytic efficiency of SACs in electrochemical environments. Furthermore, operating spectroscopic methods and theoretical modeling have provided enhanced understanding of reaction mechanisms, facilitating the rational design of active sites with optimized adsorption energies for crucial intermediates. Consequently, SAECs are gaining recognition as strong candidates for accelerating alcohol oxidation and simultaneously reducing cell voltage, which in turn boosts hydrogen productivity. Ongoing investigation in this area is anticipated to close current knowledge gaps and realize the complete potential of SACs for effective, low-energy, and sustainable AAWE-driven hydrogen production.

3.2 Fundamentals of AAWE

OER at the anode is the conventional basis for water electrolysis (H_2 generation). However, OER is slow to react and needs large overpotentials, which means the cell voltage has to be higher, and the total energy efficiency is reduced. An alternate method replaces OER at the anode with electrochemical oxidation of alcohols. Alcohol oxidation reaction has been investigated as a half-anodic reaction within direct alcohol fuel cells, considering its relevance to the conversion of energy and storage applications. These oxidations of alcohol are particularly significant for converting a wide range of feedstocks, including those from petroleum and biomass, into compounds of higher value. For instance, ethanol, phenyl methanol, 5-hydroxymethylfurfural (HMF), and glycyl alcohol can oxidize and produce useful compounds, such as the building blocks employed in the pharmaceutical and polymer industries. It shows the clear benefit of alcohol-assisted H_2 evolution over conventional OER and chemical-assisted hydrogen evolution processes [32–34]. Recently, the process of hydrogen synthesis using an alcohol-assisted method at a lower voltage than that needed for traditional water electrolysis has been discovered. This process is known as “alcohol-assisted water splitting” or “electrochemical alcohol reforming,” as illustrated in Figure 3.1.

In AAWE, hydrogen is produced through the HER at the cathode and the alcohol oxidation reaction at the anode. This technique significantly reduces the total cell voltage and energy consumption by replacing the slow, energy-demanding conventional OER with the more thermodynamically advantageous alcohol oxidation reaction. The cathode facilitates the HER through the reduction of protons in acidic

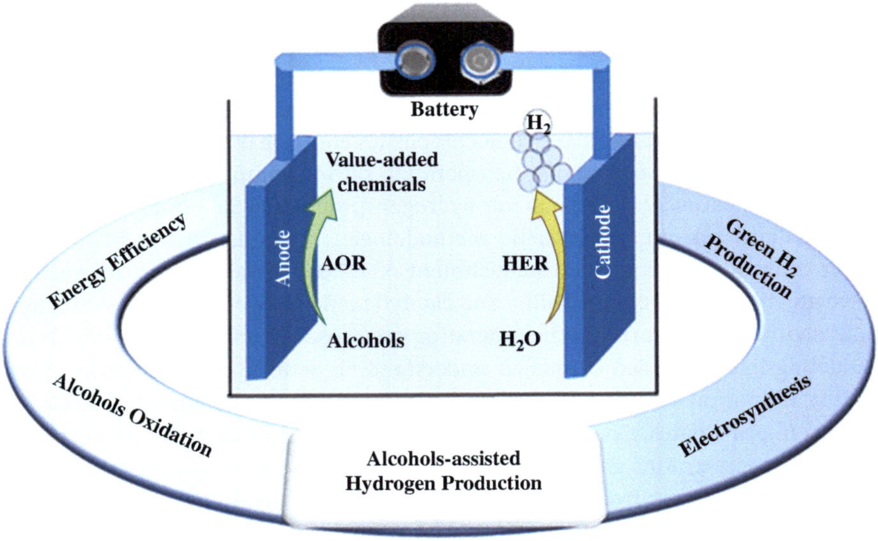

Figure 3.1 Schematic illustration of AAWE. *Source*: Adapted from [35].

media or water in alkaline/neutral media. This process occurs in sequential steps, ultimately important to the development of molecular hydrogen. The effectiveness of HER is determined by the catalyst's capacity to effectively bind and release hydrogen intermediates (H*), thereby ensuring minimal loss of energy.

At the same time, the alcohol oxidation reaction encompasses the oxidation of small alcohols like methanol, ethanol, or glycerol, which serve as electron donors. In contrast to the four-electron conventional OER process, the alcohol oxidation reaction involves fewer electron transfers and takes place at reduced overpotentials. This process facilitates the coproduction of value-added chemicals such as formate or acetate, which vary based on the substrate and catalyst used. Nevertheless, the alcohol oxidation reaction mechanism involves intricate proton-coupled electron-transfer (PCET) processes and the breaking of C—C, C—H, and O—H bonds, requiring catalysts that exhibit exceptional activity, durability, and selectivity toward the reaction. The integrated HER- alcohol oxidation reaction system enhances energy efficiency while providing a sustainable pathway for hydrogen production from renewable alcohols, in harmony with the objectives of a low-carbon, circular energy economy. Ongoing progress in catalyst engineering and system integration is essential to unlock the complete potential of this hybrid electrolysis approach [36–40].

3.2.1 Methanol-assisted Hydrogen Production

Methanol is an alcohol that is a widely available and cost-effective resource. Oxidation of methanol is a good alternative to conventional OER since it has favorable thermodynamics. However, electrochemical methanol oxidation has largely been done using noble-metal-based catalysts. The methanol electrooxidation in acidic conditions results in the formation of CO_2 and H_2O, which do not constitute value-added products. Formic acid, which costs $0.6–0.7 per kilogram, is one of the most significant chemicals that can be made by partially oxidizing methanol. It is utilized in the leather, textiles, pharmaceutical, rubber, and insecticides industries. So, electrooxidation produces a very desirable product, but it is still extremely hard to make electrocatalysts that are very active, selective, and long-lasting to help change methanol into partial-oxidation products. The oxidation of methanol typically involves a sequence of dehydrogenation steps, resulting in intermediates formation such as CH_2OH^*, $HCHO^*$, and CO^*, which exhibit strong adsorption characteristics. When CO^* is present on the surface, it has the potential to inhibit noble metal catalysts. For the facilitation of CO oxidation, the presence of surface OH^* or oxophilic components is essential. The mechanism involves PCETs, which can result in either partial-oxidation products, such as formaldehyde or formic acid, or complete oxidation to CO_2, contingent on the catalyst and the applied voltage [41–44]:

$$CH_3OH + * \rightarrow CH_3O^* + H^*$$
$$CH_3O^* \rightarrow CH_2O^* + H^*$$
$$CH_2O^* \rightarrow CHO^* + H^*$$
$$CHO^* \rightarrow CO^* + H^{**}$$

Electron-transfer process:

$$CO^* + 2OH^- \rightarrow CO_2 + H_2O + 2e^- - 4H^* + 4OH^- \rightarrow 4H_2O + 4e^-$$

3.2.2 Ethanol-assisted Hydrogen Production

The main advantage of generating H_2 by the electrooxidation of biomass-derived alcohols demonstrates that hydrogen production using an ethanol electrolyzer in aqueous medium requires up to 92% less energy than alkaline water approaches. A revolutionary anode design made electrochemical oxidation possible at current densities similar to those used in present proton electrolyte membrane (PEM) water electrolysis systems [45]. The process of ethanol oxidation initiates with the creation of an ethoxy intermediate, which is subsequently followed by dehydrogenation to yield acetaldehyde (CH_3CHO) or complete oxidation to acetate (CH_3COO^-) or CO_2. Acetaldehyde can further oxidize to acetic acid or experience C—C bond scission under more oxidative conditions. On precious metal catalysts like Pt (platinum) or Pd, ethanol generally experiences dissociative adsorption, resulting in the formation of surface-bound ethoxy species ($CH_3CH_2O^*$) and hydrogen atoms (H^*). The step that governs the rate typically involves the breaking of the C—H bond, resulting in the creation of acetaldehyde, which can then be further oxidized into more stable products as shown in Figure 3.2. The degree of selectivity is significantly influenced by pH levels, the characteristics of the catalyst surface, and the reaction potential, enabling a range of adjustments between partial and complete oxidation. Its mechanism is characterized by a dual-pathway approach governed by whether the C—C bond is cleaved [15, 46].

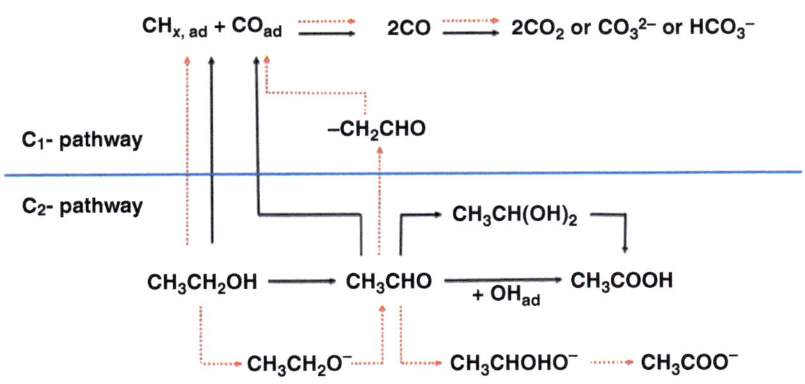

Figure 3.2 Reaction pathways for the oxidation of ethanol. *Source*: [47]. MDPI. CC BY 4.0.

3.2.3 Propanol-assisted Hydrogen Production

The oxidation of propanol follows the essential steps of PCET observed in lower alcohols; however, it involves more complex pathways due to the longer carbon chain and the possibility of multiple oxidation sites. The first step in the dehydrogenation of 1-propanol occurs as the hydroxyl group termination, resulting in the production of propanal (CH_3CH_2CHO) as the primary intermediate. This aldehyde can undergo further oxidation, resulting in the formation of propionic acid or, in some cases, shorter-chain carboxylates through C—C bond cleavage, depending on the ability of catalysts to enable deep oxidation. On the other hand, 2-propanol oxidation involves dehydrogenation at the secondary carbon, leading to the production of acetone (CH_3-CO-CH_3). This compound exhibits enhanced stability and is expected to endure additional oxidation unless the catalyst shows a strong oxidative capability. The kinetics of propanol oxidation are notably influenced by the catalyst composition, the surface's oxophilicity, and the applied potential. When compared to methanol and ethanol, propanol exhibits slower oxidation rates because it requires higher energy for the activation of the β-C—H bond, coupled with the enhanced stability of the resulting intermediates against C—C cleavage. The oxidation of propanol offers considerable promise in alcohol-assisted electrolysis, especially as it can yield valuable intermediates such as propanal, acetone, and propionic acid [48, 49]:

$$C_3H_7OH + H_2O \rightarrow C_3H_6O + 2H^+ + 2e^-$$

3.2.4 Glycerol-assisted Hydrogen Production

The oxidation of glycerol exhibits significant complexity, due to its three hydroxyl groups that create multiple reactive sites and branching pathways. The oxidation process can initiate at either the primary or the secondary alcohol groups, yielding significant intermediates such as glyceraldehyde and dihydroxyacetone through PCET-driven dehydrogenation. The intermediates undergo a sequence of oxidation, isomerization, or cleavage of C—C bond processes that are shaped by the properties of the catalyst, leading to a variety of products including glyceric acid, tartronic acid, glycolic acid, oxalic acid, and possibly formic acid or CO_2 under extreme conditions. The arrangement of catalysts, encompassing materials like Pt, Pd, Au, and Ni, as well as facet orientation and pH, plays a crucial role in determining selectivity. For example, catalysts utilizing Au tend to favor partial oxidation, resulting in C_3 acids, whereas those employing Pt are more prone to produce C—C cleavage products. The oxidation of glycerol, a biomass-derived and cost-effective polyol, effectively reduces the anodic overpotential in water electrolysis while simultaneously generating high-value chemicals applicable in pharmaceuticals, polymer precursors, and fine chemicals manufacturing. The unique properties of glycerol make it an exceptionally promising fuel for AAWE, improving hydrogen generation efficiency while also yielding valuable coproducts [50–52]:

$$C_3H_8O_3 + H_2O \rightarrow C_3H_6O_3 + 2H^+ + 2e^-$$

3.3 SAECs: Concept and Design Principles

SACs represent a novel advancement in the field of catalysis science. The material has metal atoms that are spread out on the surface of the solid support, and these atoms are held by strong interactions between the metal and the support. The attached metal atoms do not connect, which gives them electronic states that are different from those of metal clusters or particles, which have a lot of metal–metal orbital overlap, as shown in Figure 3.3. A conventional heterogeneous catalyst is usually a combination of metal particles of different sizes. Active species are only particles of certain sizes; the rest are inactive or may cause adverse effects. The NPs' active sites are usually surrounded by atoms that possess a small fraction of the

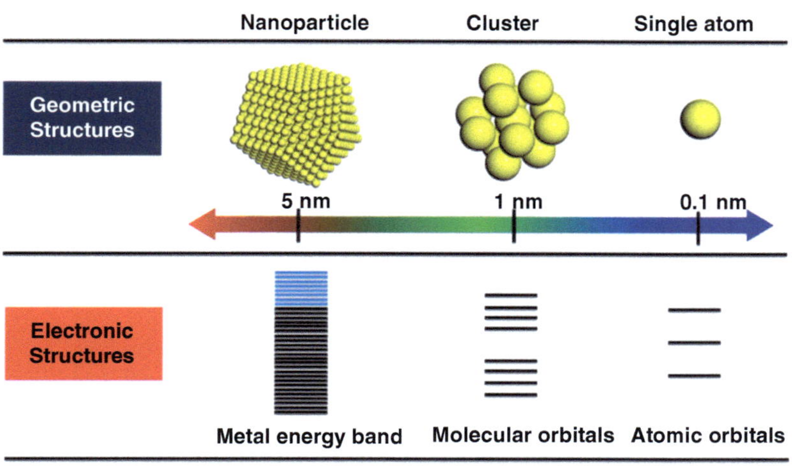

Figure 3.3 Geometric and electronic structures of SACs, NPs, and clusters. *Source*: [54]. Springer Nature. CC BY 4.0.

surface area. Consequently, conventional heterogeneous catalysts typically exhibit low atom utilization efficiency and selectivity, resulting in inadequate mass activity and diminished FE. Homogeneous catalysts possess well-defined active sites and adjustable coordination environments, resulting in high activity and exclusive selectivity for particular reactions. Reducing particle size increases surface-area-to-volume ratio and active-site utilization, exposing more active sites. Nanoengineering decreases particle size to the nanoscale, improving properties owing to size effects. Following this technique, the final downscaling will reach the atomic size and SAC. SACs bridge heterogeneous and homogeneous catalyses, combining their benefits.

The exploration of SACs for electrolysis is a pioneering domain within the realm of sustainable conversion of energy and storage applications. SACs function as catalysts where individual metal atoms are distributed on a support material such as carbon or metal oxides, presenting compelling benefits regarding catalytic activity, selectivity, and stability. The distribution characteristics of SACs are enhanced when metal atoms interact on the surface, resulting in a steady platform. The significant dispersion of metal is heavily dependent on the synthesis techniques and the electronic structure characteristics. Grasping the structure, synthesis, and geometry (specifically the size of the metal atom) is essential for evaluating catalytic performance. The synthesis encompasses various critical elements, including the careful choice of materials, the pathways for synthesis, the precise control at the atomic level for the distribution of metal atoms, and the formulation of precursor solutions. SACs exhibit remarkable catalytic performance, showcasing high active sites and selectivity for hydrogenation and oxidation reactions. SACs present numerous benefits over conventional bulk metal catalysis, including increased catalytic activity, remarkable stability, and superior selectivity [53].

Due to the increasing demand for conversion of energy, SACs have emerged as important catalysts. They are frequently used in electrochemical reduction due to their promising synthesis paths and excellent physical properties. However, electrochemical reduction processes depend on catalyst site stability and activity. Compositional stability and well-established supports allow SACs to thrive in an ecosystem. A range of innovative and traditional approaches have been utilized to develop SACs. The high surface energies associated with SAs present a significant challenge in the synthesis of SACs aimed at reducing carbon dioxide emissions. So far, the main approach to synthesis has concentrated on atomically thin materials (ATMs) limited to SACs, such as graphene with satisfactory rates, C_3N_4-based SACs, transition-metal sulfides like MoS_2 and WS_2, as well as materials containing carbides like MXene.

3.3.1 NM-SACs

Heterogeneous catalysts based on precious metals are regarded as the ultimate in numerous significant viable thermochemical and electrochemical processes, including oxidation of CO, the water–gas shift reaction, OER, NRR, HER, and ORR (oxygen reduction reaction). Nonetheless, the widespread commercialization of traditional nanomaterial-based catalysts faces significant challenges due to their limited availability and elevated costs. Enhancing the precious metals' atom efficiency could serve as an effective strategy to maintain strong catalyst performance while achieving a favorable balance of cost competitiveness through the reduction of precious metal usage. To minimize the amount used in heterogeneous catalysts, metals can be made strategically discrete on support materials' surface at the atomic level to form metal SACs.

SA metals can be applied to different metallic surfaces to form SA alloys. SA alloys arise when the dimensions and guest atoms' bond-formation energy closely align with those of the host metal atoms, accompanied by a relatively minor addition of guest metal. The modification of metallic surfaces' chemical properties was achieved by adjusting the electronic structure and geometric configuration via integration of SA precious metals into the alloys. Nanomaterial-based SA alloys exhibit considerable promise in facilitating selective hydrogenation, dehydrogenation, Ullmann reactions, and other processes. Although they show great potential in thermal catalytic and photocatalytic reactions, the vast majority of reported nanomaterial-based SACs (NM-SACs) supported on metal oxides and metallic surfaces remain. The chemical properties of metallic surfaces were achieved by adjusting the electronic structure and geometric configuration through the integration of SA precious metals into unexamined as viable electrocatalysts for clean energy conversion [55–63].

Recently, precious metal SACs supported on electronic conductive materials have emerged to facilitate electrochemical reactions. Platinum and ruthenium (Ru) metals have been atomically dispersed on various carbon-based materials, such as graphene, porous carbon, and graphitic carbon nitride (CN) [64, 65]. The defects, such as the edge and doped N, S, and O, function as binding sites for the immobilization of SA precious metals. Consequently, it is possible to attain a greater metal loading on carbon-based supports in comparison to metal oxide supports. The carbon-based SACs doped with precious metals demonstrated outstanding electrocatalytic performance, rivaling or even surpassing the Pt/C catalyst for both HER and ORR. Additionally, individual nanomaterials were effectively anchored onto various conductive substrates, including CoP, MoS_2, TiC, and TiN, utilized for HER and ORR applications [66–69].

3.3.2 SACs@ATMs

In water electrolysis, SACs play a crucial role in minimizing both the HER and the OER overpotentials by increasing atomic utilization efficacy and providing adjustable electronic structures. ATMs are an excellent supporting material for SACs due to their atomic thickness, enhanced specific surface area, and high electronic properties. The methods presented here efficiently address the shortcomings of traditional ports, such as black carbon and oxides of metal, which frequently encounter issues including suppressed active sites, low electrical conductivity, and insufficient stability. ATM substrates such as graphene, TMDs like MoS_2, metal nitrides, and MXenes offer high-binding-energy surface sites for individual atoms via defect engineering (including vacancies and heteroatom doping) and interfacial electronic modulation. The support materials act as a crucial platform for surpassing the performance limits of water electrolysis by facilitating the synergistic catalytic effects. The primary benefit of SACs at ATMs lies in the engineering of metal–support interactions, which allows for atomic-level refinement of reaction paths, representing a significant progression from conventional supports that function solely as passive frameworks. Individual atom–ATM interactions present a variety of unique advantages: (i) atomic dispersion stability; (ii) providing favorable local response environments that optimize atomic utilization; (iii) electronic structure adjustment to facilitate transfer of charge and increase essential activity; (iv) support for catalytic activity; and (v) tunability for functional optimization [70–73].

Single metal atoms demonstrate significantly large surface energy and a high tendency to aggregate and migrate into NPs, leading to the loss of their unique SA features. The surfaces of ATMs display defects, dopants, and precise functional groups that form strong chemical bonds, such as covalent, ionic, or coordination bonds with individual metal atoms, effectively anchoring them to the supports (Figure 3.4a). The strong interactions significantly elevate the migration energy barriers of the metal atoms, thereby effectively hindering their aggregation. In contrast to traditional supports, ATMs (e.g., graphene, MXenes, and TMDs) create numerous defect sites for the SA anchoring, including vacancies and edge sites. For example, excessive Ti vacancies on the surface of $Ti_3C_2T_x$ MXene to anchor Ni SACs create a robust interaction between Ni atoms and adjacent C atoms, notably reducing the energy of activation for the reaction, improving adsorption energies, and consequently speeding up the reaction kinetics. Moreover, the distinctive layered configurations of ATMs can trap SAs within the intermediate layer regions (Figure 3.4b). The unique characteristics of these catalysts guarantee that they preserve SA dispersion even in demanding electrocatalytic conditions, including elevated potentials and acidic or alkaline environments. This optimization leads to enhanced atomic utilization and sustained large activity

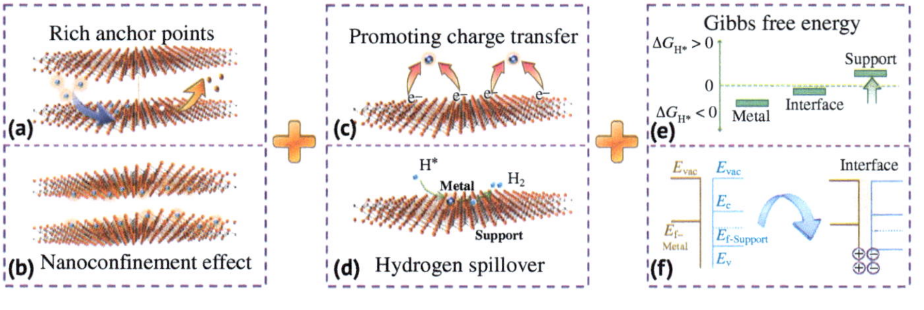

Figure 3.4 Illustrative representation of the interaction mechanism between metals and supports utilizing SACs@ATMs. (a) ATMs serve as plentiful anchor sites for individual metal atoms. (b) The distinct ATM layer exhibits a nanoconfinement effect, which stabilizes metal SA active sites. (c) Electron transfer between the metal atom and the supporting layer. (d) Metal individual atoms engage with the substrate, facilitating the transfer of hydrogen. (e, f) Electronic arrangements at the interface between metal SAs and support. *Source*: [76] with permission of Elsevier.

during the entire reaction. For instance, Pt atomically thin nanolayers were attached within the multilayered Mo_2TiC_2 MXene through Pt—Mo strong bonds at the interface, resulting in the enhanced platinum-atom electronic structure. This demonstrates that the ATM layer supports the creation of a distinct spatial confinement effect. In contrast to conventional bulk metal oxide, the robust interactions between metals and supports in SACs@ATMs increase the acquaintance and metal active-site stability, and also reach theoretically 100% atomic utilization. This indicates that very minimal quantities of noble metal atoms (such as platinum, iridium, and ruthenium) can achieve significantly high catalytic activity, leading to a substantial decrease in catalyst expenses. Remarkably high mass activities can also be achieved with non-noble metal SA catalysts, such as Fe, Co, Ni, and Cu [74, 75].

The electronic structure of SACs significantly enhances the catalytic activity, especially the positioning of the center of the d-band and the charge density. The electronic state of the metal sites can be greatly impacted by the interaction of electron transport between the metal atoms and the support. Supporting materials have the capacity to either donate or remove electrons from the individual metal atoms, thus changing their oxidation states and local charge densities (Figure 3.4c). Based on ATMs, the supporting atoms directly interact with the metal atoms, leading to a distinct spatial arrangement, and also have a direct impact on the electronic structure and metal centers' orbital hybridization. This specific electronic modulation enhances the adsorption energy of reaction intermediates on the active sites, lowers the energy barrier of the rate-determining step, and improves the essential catalytic activity, frequently surpassing the performance of equivalent metal atoms supported on bulk materials.

Recent work by Sun et al. reported platinum SAs attached to two-dimensional (2D) NiFe layered double hydroxide (LDH). In this system, Pt was negatively charged, and the strong metal–support with synergetic electrical connection improved electrocatalytic HER performance. Atomically scattered Pt sites connected to the support material by nonmetal bonds have increased oxidation states, which prevent water-splitting intermediates from adsorbing. Reducing Pt's oxidation state increases H* adsorption and HER activity. This Pt-derived system transferred electrons from Ni to Pt via Pt—Ni chemical bonds rather than Pt—O bonds. The oxygen vacancies in NiFe LDH increased Ni–Pt electron transport. The catalyst's hydrophilicity increased due to Ni's raised valence state, decreasing the energy barrier for H_2O dissociation [77, 78].

Functional groups or defect sites of the ATMs can also engage in the adsorption or activation of reactants, collaborating effectively with the metal atoms. Robust chemical bonding enables an effective route for the transfer of electrons between the metal atoms and the support. For example, a mechanism involving hydrogen spillover facilitated by surface hydroxyl (—OH) groups was utilized to greatly enhance the active hydrogen coverage on the catalyst surface, resulting in elevated electrocatalytic activity. Since electrocatalysis fundamentally relies on interfacial electron-transfer processes, conductive ATMs play a crucial role in facilitating efficient charge transport, thereby accelerating reaction kinetics. ATMs' hydrophilic or hydrophobic properties can also alter reactant and product mass transfer at active areas. Interactions between metal and support can enhance the adsorption and activation of reactants, speed up the conversion of intermediates, or aid in the desorption of products, ultimately improving the overall efficiency of the reaction (Figure 3.4d). In SACs@ATMs systems, it is crucial to emphasize that the support functions extend beyond a simple passive framework. Certain ATMs exhibit intrinsic catalytic activity for the HERs or OERs, thereby collaborating with SAs to enhance the efficiency of overall water splitting. The interactions between the individual metal and support create a synergistic effect, resulting in an enhancement that surpasses the mere sum of its components. To improve catalytic performance, it is crucial to improve the properties of the catalyst appropriately. The SACs' characteristics at ATMs, along with the resulting electronic structure effects, can be precisely tailored through specific support engineering strategies, such as support modification, defect introduction, and surface functionalization. Accurate alignment of the Fermi levels between the metal and the support can notably decrease interfacial charge accumulation (Figure 3.4e,f), thereby substantially improving interfacial hydrogen spillover [79].

The SACs@ATMs design possesses coordination and electronic structures that can be tailored to fulfill the exact requirements for HER and OER performances. The main benefit is the stabilization of individual atoms through robust interactions, which maximize atomic utilization by accurately modifying the electronic structures and coordination environments of the SACs, thereby greatly increasing both the inherent activity and stability of the catalyst. Furthermore, the metal–support interactions enhance the charge-transfer efficiency and can utilize the inherent properties of the support to achieve cooperative catalysis. The highly tunable interactions offer unprecedented opportunities for the development of efficient, stable, and low-cost electrocatalysts for water splitting. A comprehensive grasp and accurate control of metal–support interactions are essential for the thoughtful design and enhancement of SACs@ATMs' performance.

3.4 Synthesis Strategies of SAECs

The advancement of highly efficient catalysts and the investigation of their extensive applications contribute to the precise design and synthesis of well-defined SACs. Nonetheless, the propensity for metal atom fusion and aggregation during the synthetic process significantly obstructs the development of highly dispersed SAs, presenting a substantial challenge that requires attention. A variety of synthetic methods, such as top-down and bottom-up methods, have been put out thus far to create SACs, which stimulate materials science and expand its applications.

3.4.1 Bottom-up Methods

3.4.1.1 Atomic Layer Deposition Method

A thin-film deposition method called atomic layer deposition (ALD) enables precise control over the composition and thickness of the film at the atomic level. To create consistent coatings on substrates, gaseous precursors are used one after the other. A technique called ALD makes it possible to deposit several thin materials from the vapor phase layer by layer with atomic-level accuracy. The deposition of homogeneous metal atoms on substrates is facilitated by the ALD process's self-limiting nature. By enabling the deep dispersion of metal atoms into the absorbent substrate, this vapor-phase deposition approach maximizes the loading of SA active sites. (i) Being exposed to the first precursor vapor, (ii) elimination of extra precursor molecules and by-products, (iii) being exposed to a second precursor vapor, and (iv) elimination of extra precursor molecules and by-products are the four phases of this deposition process within a single cycle [80].

Metals may be gradually accumulated from SAs into NPs by adjusting the number of ALD cycles. The anchor sites must react with metal precursors during the ALD process. According to Sun and coworkers, Pt active sites may emerge on the carbon-based graphene substrate as a result of the interaction between the graphene nanosheets and (methylcyclopentadienyl)trimethyl platinum ($MeCpPtMe_3$), which is followed by exposure to O_2 via the ALD process [81]. Stambula et al. demonstrated that the nitrogen-doped functional groups serve as securing and nucleation sites for the deposition of SA Pt, utilizing nitrogen-doped graphene as a support material [82]. Metal oxides or metallic NPs may also interact with metal precursor vapors. Wang et al. reported a Pt_1/CeO_2 SAC, in which Pt atoms are particularly positioned on the cerium rows of the (110) and (100) surfaces of CeO_2 [83]. At 393 K, a temperature much lower than that used in the traditional ALD method using the identical precursors, Cao et al. revealed the Fe precursor, ferrocene ($FeCp_2$), and O_2 to deposit SA/$Fe_1(OH)_x$ selectively and exclusively on the Pt NPs of Pt/SiO_2 (Figure 3.5a) [84]. The ALD technique offers precise control in the production of SACs; however, its high cost restricts large-scale and commercial applications. Additionally, the process necessitates particular metal precursors, and achieving a high loading of SA metals remains difficult due to aggregation issues.

3.4.1.2 Electrodeposition Method

Electrodeposition refers to the process of depositing a material onto an electrode through electrochemical reactions. This technique facilitates the rapid, straightforward, and regulated SACs synthesis under slight operational conditions, eliminating the need for supplementary reducing or capping agents. By changing potential, density of the current, composition of the electrolyte, and deposition time, one may precisely regulate the thermodynamics and kinetics of the deposition of metal during the electrodeposition process, which supplies electrons to the metal at the surface of the electrode. When SAs are electrodeposited, low quantities of metal precursors are frequently used to prevent metal growth [85]. Zhang et al. used electrodeposition to develop an atomically distributed Pt catalyst on the surface of NiFe LDH. The catalyst maintained the single Pt sites on the support by containing an axial chlorine ligand that could be reversibly switched with a hydroxide ligand [86]. Lei et al. introduced a two-step electrodeposition method for loading SA Ir onto $Ni_{3-x}Fe_xS_x$ with a high surface density (Ir_1/NFS). To maximize iridium utilization, the researchers decoupled the substrate construction from the iridium

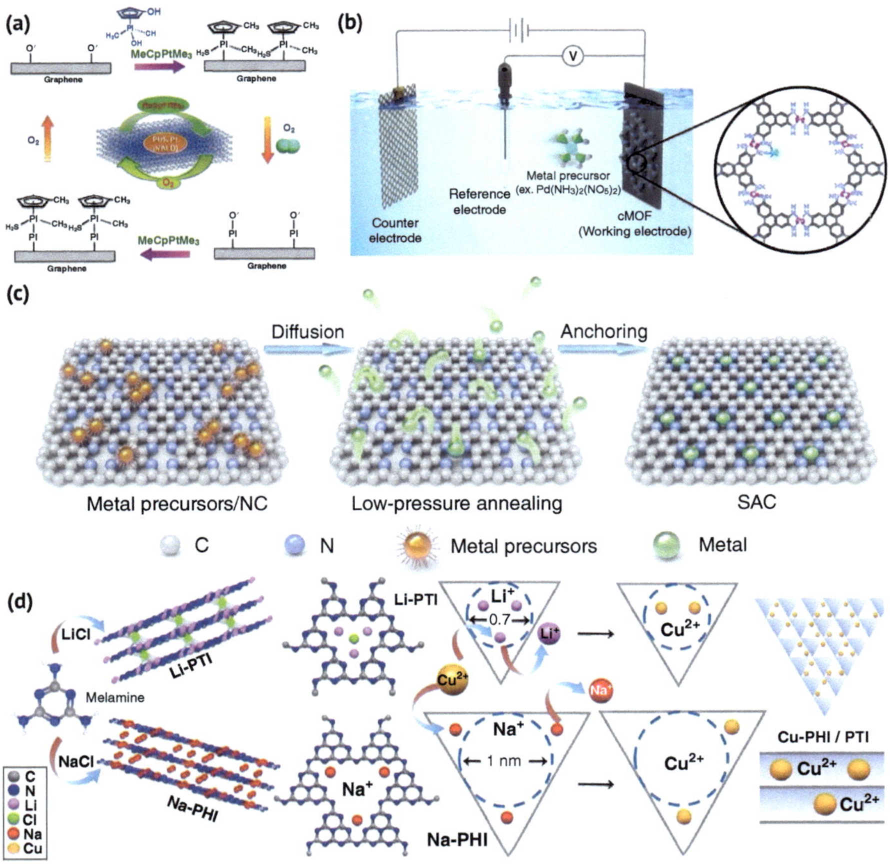

Figure 3.5 SACs synthesized via bottom-up approaches: (a) ALD method, (b) electrodeposition method, (c) pyrolysis method, and (d) wet-chemical synthesis method.

deposition, which resulted in the creation of Ir SA on the substrate surface [87]. Park et al. recently established the metal–organic framework (MOF)-derived SACs using this deposition method. Pd SAs introduced into the MOF, consisting of $Cu_3(HITP)_2$ (HITP = 2,3,6,7,10,11-hexaimino triphenylene), were achieved through deposition in the cathode, utilizing an extensive range from 0.513 to 0.013 V vs. Ag/AgCl (Figure 3.5b) [88].

3.4.1.3 Pyrolysis-assisted Method

Carbon-based SACs are mostly produced from metal–organic composites using the pyrolysis process. Metal-anchoring ligands, such as MOFs and organic polymers, primarily serve as precursors [89]. These precursors retain the metal coordination areas, and they are reduced into bulk carbon materials. Research has been conducted between active-site structure and catalytic activity through the use of organic precursors or inorganic additives that contain several metal-anchoring ligands. Although organic precursors act as a barrier to prevent metal sintering or agglomeration throughout the reaction, the SACs synthesis with large metal loadings is still challenging because bulk metal particles unintentionally develop at high temperatures, requiring further acid treatment [90].

SACs with ultrahigh metal loadings have recently been created using a variety of techniques, including slow-release synthesis (SRS), massive organic ligands, and additions containing metal-anchoring sites. Xia et al. synthesized Ir^{-} (41.6%), Pt^{-} (32.3%), and Ni—N—C (15%) with metal concentrations, respectively, using graphene-based quantum dots with surface functionalization as basic precursors, which have several metal-coordinating areas and retain structural integrity after pyrolysis [91]. Gu et al. approached to synthesize a Cu SAC, achieving a high metal utilization of 97% and a precisely controlled coordination structure. The researchers employed natural montmorillonite (MMT) as a precursor for Cu^{2+}, combined it with melamine ($C_3H_6N_6$), and subjected the precursors to a pyrolysis-assisted method. Melamine produced C_3N_4, which emitted NH_3, leading to the gradual release of Cu^{2+} from the Cu^{2+}-exchanged MMT. Cu–N_4 was synthesized by controlling the coordination structure between the released Cu^{2+} and C_3N_4. Post-synthetic techniques for creating SACs by annealing a combination of supporting materials and metal precursors are also available, in addition to the previously stated techniques for the simultaneous synthesis of SA sites and carbonaceous supports [92].

Hai et al. demonstrated that the SACs preparation with ultrahigh density and high metal loadings reached 23 wt.% through a two-step pyrolysis method. Precursors such as metals were initially introduced to the supporting materials through wet impregnation. Following the drying of the solvents, the mixture underwent annealing at a temperature below the decomposition point of the metal precursors. This process aimed to optimize metal exposure and facilitate the removal of additional species through the washing process. The lower-temperature annealing process effectively prevented metal sintering in the following high-temperature annealing phase, which was designed to remove residual ligands and stabilize metal atoms. The annealing method performed under reduced pressure has been recognized as a proficient technique for the preparation of SACs featuring elevated metal loadings [93]. Al-Hilfi et al. demonstrated that pressure reduction can enhance the SA metal loading by a factor of 3.2 relative to ambient pressure during the pyrolysis method for synthesizing SACs from several metal precursors, including Fe, Pd, Ni, Ru, Cu, Mo, Co, and Mn, as well as porous nitrogen-doped carbon (Figure 3.5c). Theoretical studies indicated that the metal atom diffusion was markedly increased at reduced pressure, attributed to the extended path of the atoms, which reduced agglomeration [94]. Wang et al. recorded a similar trend around the same period, which explains the suggested strategy's validity. This process has benefits, including being easy to use and using cheap precursors. It is also suitable for making SACs with large SA loadings and a variety of ligand environments. Making NPs or nanocomposites at high temperatures is still a big problem [95].

3.4.1.4 Wet-chemical Synthesis

Wet-chemical methods are extensively employed for the fabrication of SACs on support materials, owing to their operational simplicity and feasibility for large-scale production. In the wet-chemical process, metal ions adsorb onto the surfaces of support materials through coordination bonds with ligands or electrostatic interactions, subsequently undergoing drying, reduction, or stabilization [96]. Roy et al. synthesized Cu–N_2 SACs on crystalline 2D-CNs using this method (Figure 3.5d). Two 2D-CNs of poly-heptazine and poly-triazine imides were prepared, incorporating metal cations like Na^+ and Li^+, respectively. All the cations from the polymers were replaced by Cu^{2+} via cation exchange in aqueous media, with Cu^{2+} ions coordinating to the nitrogen atoms of the 2D-CNs [97].

Giulimondi et al. established a standardized approach utilizing emerging wetness impregnation on carbon substrates. Coprecipitation is a straightforward and rapid method for preparing SACs by the combination of all precursors, like metal and support materials, during this synthesis process. This process is employed to synthesize SACs supported on metal oxides, MOFs, metals, and high-entropy SA materials [98]. Zhang et al. made a Pt_1/Pd alloy for Li–O_2 batteries by reducing Pt(acetylacetonate)$_2$ and Pd(acetylacetonate)$_2$ in oleylamine solvent. They used ascorbic acid as the reducing agent and (acetylacetonato)dicarbonylrhodium(I) as the structure-directing agent. This process is a cheap and easy way to make SACs. However, making SACs with high metal loadings considered as hard and many functional groups or flaws on the support materials that can hold the metal precursors in place [99].

3.4.2 Top-down Method

Top-down synthetic strategies for SACs focus on the precise deconstruction of bulk metal precursors, including metallic foils or NPs, to produce isolated metal atoms that are then anchored onto an appropriate support material. This method presents an alternative to bottom-up techniques, frequently enabling synthesis at room temperature and utilizing readily available, cost-effective metal sources. The main methods in this category are thermal atomization, anodic stripping, and mechanical ball milling.

The thermal atomization method utilizes elevated temperatures to disrupt the robust cohesive bonds between metal atoms, facilitating the conversion of metal NPs into individual atoms through diffusion or evaporation processes. Jones et al. demonstrated a significant NP-to-SA-Pt strategy by vaporizing Pt below oxidizing conditions, resulting in the formation of volatile PtO_2, which was subsequently trapped and stabilized on CeO_2 supports at approximately 800 °C. This method utilizes the support's capacity to reduce surface energy and create robust trapping sites, thereby alleviating the reaggregation of dissociated atoms [100]. Comparable techniques have been effectively utilized for other noble metals, such as Pd, Pt, and Au, on N-doped carbon obtained from ZIF-8, as well as for base metals like Ni through thermal diffusion in defect-containing N-doped carbon [101]. This strategy encounters notable challenges, such as the elevated operational temperature necessary for breaking metallic bonds, a lack of universality, and the ongoing thermodynamic issue of attaining a high degree of atomization, constrained by Ostwald ripening effects [102].

An efficient room-temperature alternative is anodic stripping, an electrochemical technique that employs an electrolytic cell in which a sacrificial bulk metal counter-electrode consistently provides metal ions to the electrolyte. The anodic dissolution of this electrode facilitates the release of target metal ions, which are then transported to the working electrode. At this site, the ions are adsorbed and reduced, resulting in the formation of isolated metal atoms anchored at designated locations. The concentration of dissolved metal ions, which directly influences the final density of the SAs, is precisely regulated by adjusting parameters such as anode voltage, pH of the electrolyte, and the purging gas employed. Wang et al. employed a filtration method using a graphene-based oxide membrane (GOM) to facilitate the transport of Fe^{2+} ions from a Fe bulk foil, thereby enhancing the prevention of NP aggregation. GOM significantly condensed the Fe^{2+} diffusion rate, thus inhibiting nucleation and growth on the working electrode [103]. Qi et al. demonstrated high selectivity by initially assembling Co nanodisks onto MoS_2 nanosheets through ultrasonication to promote Co—S bond formation. This was followed by electrochemical cyclic voltammetry treatment, which leached the Co nanodisks, resulting in the formation of SA Co arrays [104].

Top-down ball milling offers a scalable mechanical method for the synthesis of SACs through the high-energy co-milling of a bulk metallic precursor with a support material. The mechanical energy dissociates surface atoms from the bulk metal, and the resulting adatoms are captured by strong anchoring sites on the support, typically nitrogen-doped carbon. The formation of the SA is facilitated by the notable disparity in bonding energy, exemplified by the strength of the Co—N bond compared to the Co—Co bond in bulk metal atoms. Wang et al. initially demonstrated this by co-milling bulk Co powder with sulfur and nitrogen co-doped carbon (SNC) [105]. Han et al. subsequently developed a high-efficiency method utilizing bulk Fe balls and graphite in an N_2 atmosphere. In this process, Fe balls were added into SAs, while the N_2 gas experienced dissociation catalytic effect on the galvanized Fe surfaces, resulting in the formation of essential N-anchoring sites on the graphite. The strategy demonstrates versatility across multiple metals (Co, Ni, and Cu) and supports (MgO, CeO_2), establishing top-down ball milling as a viable technique for the economical, large-scale synthesis of diverse SAC architectures [106].

3.5 Integration of SAECs into AAWE

SACs offer distinct benefits for alcohol oxidation in AAWE. The isolated metal centers, often characterized by M—N—C motifs for transition metals like Fe, Co, Ni, and Mn, and SA noble metals such as Pt_1 or Pd_1 anchored on conductive carbon supports, create distinct electronic structures and localized coordination environments. These factors significantly influence the adsorption energies of crucial intermediates, including alkoxide, dehydrogenated species, carbonyl intermediates, and carboxylates. The interactions specific to these sites lead to enhanced selectivity and activity for partial-oxidation pathways that conclude with aldehydes, carboxylates, or various other oxygenates, influenced by the identity of the catalyst and the pH level. SAECs enhance the number of exposed catalytically active centers per unit mass of precious metal, making them particularly appealing when costly metals are necessary to achieve high intrinsic turnover frequencies for challenging oxidation processes. The SA design also facilitates alloying, heteroatom coordination tuning, and support engineering to mitigate side reactions like C—C cleavage or deep oxidation to CO_2. The process of alcohol electrooxidation on SAECs generally involves steps of deprotonation and hydride-coupled/proton-coupled electron transfer. The pathway for primary alcohols typically includes the formation of an alkoxide species, which is followed by β-hydride elimination (or concerted PCET) to produce an aldehyde. This aldehyde can then undergo further oxidation to yield a carboxylic acid or its salt (such as formate, acetate, or lactate), contingent upon the specific conditions and design of the catalyst.

Zhu et al. reported that substituting the alcohol oxidation reaction with a thermodynamically advantageous EOR is considered a capable strategy to achieve both energy-efficient H_2 evolution and the production of high-value chemicals. This study presented the design of SA indium-doped platinum nanowires (SA In–Pt NWs) as high-performance electrocatalysts for the HER and EOR across a range of pH conditions. SA In–Pt NWs/C can be utilized to combine HER with EOR, effectively mitigating the significant overpotential associated with sluggish OER. This integration necessitates a reduced voltage of 0.62 V to achieve 10 $mA \cdot cm^{-2}$, in contrast to the 2.07 V required for water splitting. The reaction demonstrates a remarkable FE exceeding 93% in the conversion of ethanol to valuable acetate within the anodic cell, as shown in Figure 3.6. Mechanistic investigations reveal that the integration of the ultrathin 1D structure and SA In decoration optimally increases the number of active sites and efficiently activates Pt atoms for catalytic processes. The calculations conducted using density functional theory provide additional evidence that doped indium can significantly enhance the HER, while simultaneously facilitating the conversion of ethanol to acetate. Furthermore, utilizing SA In–Pt NWs/C as electrocatalysts allows for the incorporation of various other alcohols as anodic feedstock to facilitate coupled electrolysis [107].

Zhu et al. reported a thermolysis ligand-confining approach to synthesize a ruthenium-based SAC (Ru-SA/NSC) on N and S co-doped carbon. The effect of alcohol on the Ru-SA/NSC electrocatalytic oxidation was explored by modulating substituent groups in the adsorption mode (Figure 3.7). The horizontal adsorption via the O atom on Ru-SA/NSC greatly enhances the oxidation of phenyl methanol, achieving an ultralow potential of 0.97 V vs. reversible hydrogen electrode (RHE) at 10 $mA \cdot cm^{-2}$, along with high yield (~96%), selectivity (~99%), and FE (~100%) for aldehyde production. The vertical adsorption via the N atom in pyridine methanol over Ru-SA/NSC shows no response to the reaction. Moreover, in the coupling device for alcohol oxidation and HER, H_2 production is realized at a low potential of 1.21 V at 10 $mA \cdot cm^{-2}$, exceeding the performance of the benchmark Pt/C||IrO_2 (1.56 V), and offers valuable insights into the design of nanocatalysts aimed at the systematic conversion of organic molecules into value-added chemicals while simultaneously generating clean energy carriers [19].

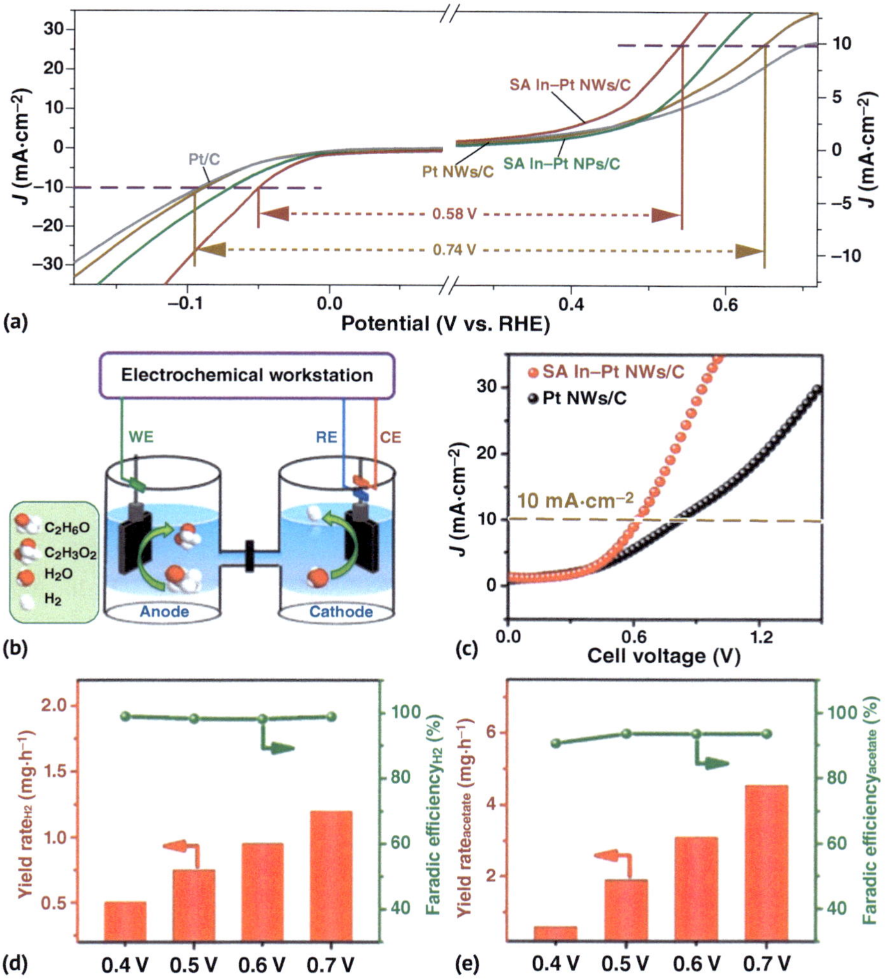

Figure 3.6 Performance analysis of coupled HER/EOR reactions using SA In–Pt NWs. (a) Values of ΔE derived from the HER and EOR polarization curves of SA In–Pt NWs/C, SA In–Pt NPs/C, Pt NWs/C, and Pt/C. The scanning rate was measured at 5 $mV{\cdot}s^{-1}$. (b) Diagram illustrating the two-electrode cell electrolyzer utilizing SA In–Pt NWs/C for both the anode and cathode components. (c) Polarization curves for the coupled HER/EOR reaction in the two-electrode cell electrolyzer utilizing SA In–Pt NWs/C and Pt NWs/C. Yield rates and FEs associated with the production of (d) H_2 in the cathodic cell and (e) acetate in the anodic cell utilizing SA In–Pt NWs/C as electrocatalysts. *Source*: [107]/with permission of John Wiley & Sons.

Ge et al. investigated the electrooxidation of alcohols produced from biomass in a unique study on biomass valorization using SAC. This represents a main strategy for the sustainable generation of chemicals that hold significant industrial value, including aldehydes. This process is typically carried out with non-noble metal catalysts functioning in base electrolytes. However, in these studies, side reactions

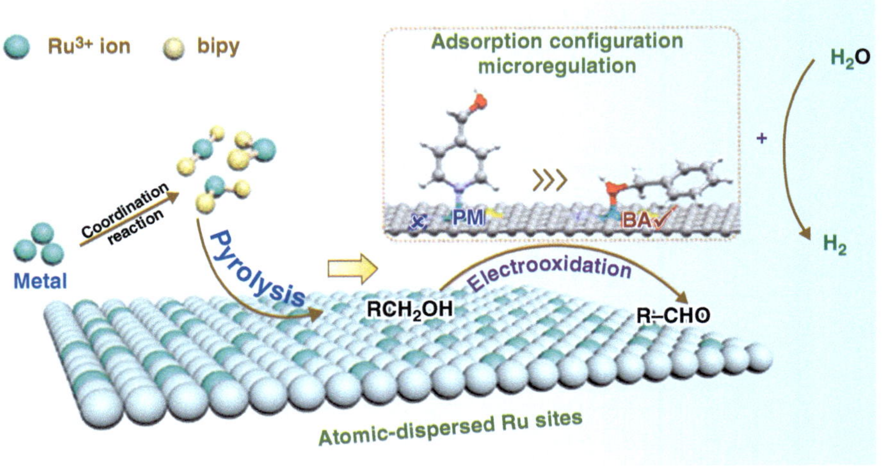

Figure 3.7 Schematic representation of the synthesis process for ruthenium SAECs and their utilization in the coupled alcohol electrooxidation reaction and HER.

such as dimerization or aldehyde oxidation to carboxylates take place, significantly decreasing the yield of aldehyde formation. The oxidation of alcohol response in a neutral medium is very beneficial in light of these factors. To obtain this outcome, Ru SAs were applied to nickel oxide-based nanosheets through wet impregnation followed by calcination. The SAC facilitated the electrooxidation of 5-HMF to 2,5-diformylfuran (DFF), which serves as a significant intermediate in the pharmacological and chemical sectors, achieving a current density of 10 $mA\cdot cm^{-2}$ at 1.283 V potential vs. RHE in a neutral electrolyte (1 M phosphate buffer). Additionally, it demonstrated a 90% remarkable selectivity and a 70% FE for DFF production at 1.5 V vs. RHE. A thorough examination of the reaction mechanism uncovered the significant role of the SA Ru sites in facilitating water dissociation, leading to a higher generation of *OH species in comparison to bare NiO nanosheets in a neutral medium, which subsequently interact with 5-HMF to yield DFF. Conversely, in an alkaline environment, the oxidation of the aldehyde is favored kinetically, hindering the controlled oxidation of 5-HMF to DFF. The effectiveness of the catalyst was further illustrated by the selective oxidation of different biomass-derived alcohols, achieving yields of up to 66.6% for benzaldehyde in a neutral electrolyte [108].

The introduction of SAECs in AAWE represents a significant advancement in the field of green electrochemistry. SAECs provide a significant kinetic benefit compared to traditional OER catalysts by finely adjusting the electronic structure (d-band shift), which prevents catalyst poisoning and speeds up the rate-limiting steps of the alcohol oxidation reaction. This capability allows for the concurrent, energy-efficient production of H_2 at low cell voltages alongside the precise and high-yield synthesis of valuable chemicals such as formate and acetate. The SAEC–AAWE system stands as a significant, multifaceted technology essential for progressing renewable energy infrastructure and creating a sustainable, electrochemical circular economy.

3.6 Challenges and Perspectives

SAECs, defined by their optimal use of precious metal atoms fixed onto conductive supports, signify a groundbreaking shift in the field of catalysis. By guaranteeing that each metal atom serves as an active site, SAECs attain remarkable atom efficiency and exhibit exceptional intrinsic activity and selectivity in crucial electrochemical reactions, including the ORR, HER, and intricate oxidation processes like the alcohol oxidation reaction. This outstanding performance profile, frequently exceeding that of conventional NP catalysts, has established SAECs as essential elements for advanced fuel cells, water electrolyzers, and sustainable synthesis technologies. Nonetheless, the transition of this technology from regulated laboratory environments to large-scale industrial use is significantly hindered by ongoing issues related to structural integrity, long-term performance under operational pressures, and the intrinsic complexities associated with mass production.

The primary challenge in the practical application of SAECs lies in ensuring their structural integrity within harsh electrochemical conditions. The seclusion of the active site, which contributes to its elevated activity, also renders it particularly susceptible to deactivation. The primary failure mechanism observed is metal leaching, or dissolution, in which the isolated metal atom undergoes oxidation at elevated anodic potentials (positive voltages), resulting in a weakening and eventual rupture of the M-support bond. This process is fundamentally represented by the reaction $M_{site} \rightarrow M_{aq}^{n+} + ne^-$, leading to the irreversible loss of active material into the electrolyte. Moreover, functioning under non-steady-state conditions, especially with rapid potential cycling or increased temperatures, enhances atomic mobility.

This results in aggregation and sintering, during which individual atoms move across the support surface and merge into inactive clusters (M^n) or NPs, compromising the high surface area and atom efficiency essential for SAEC performance. The operational reality of the electrochemical cell frequently complicates these deactivation pathways.

Alongside the direct loss of metal, the carbon-based supports frequently employed, like nitrogen-doped carbon (M—N—C), experience anodic corrosion at elevated potentials, particularly during the potential spikes that are typical during system startup and shutdown. The oxidation of the carbon framework ($C + 2H_2O \rightarrow CO_2 + 4H^+ + 4e^-$) results in the degradation of the support structure and the elimination of anchoring sites, such as nitrogen atoms. This process subsequently hastens the release of the metal atom, thereby initiating secondary leaching and sintering. Furthermore, in processes such as alcohol oxidation reaction, the formation of strongly adsorbed carbonaceous intermediates, including CO_{ads}, presents a considerable challenge. Although SAECs typically demonstrate superior CO tolerance relative to bulk metals, the presence of a single active site is significantly compromised by the obstruction caused by one poison molecule. This leads to a marked decrease in the catalyst's turnover frequency, highlighting the need for effective strategies to improve durability. To tackle these challenges, it is essential to adopt a comprehensive engineering strategy focused on reinforcing the metal–support interaction and improving the inherent stability of the host material. A highly effective strategy involves utilizing the strong metal–support interaction effect, where the metal atom is anchored onto intrinsically stable host materials such as transition-metal oxides (TiO_2, CeO_2) or carbides (TiC). This results in a more covalent and electronically stable bond, which increases the energy barrier for both atomic migration and dissolution. Moreover, advanced techniques in defect and edge engineering of the support material are utilized to establish controlled, high-density structural defects. The presence of these defects creates deep potential wells or atomic traps, which notably enhance the binding energy (E_b) of the individual atom, thereby safeguarding it from electrochemical or thermal stress.

In addition to supporting modification, strategies involving physical confinement provide a robust defense against sintering. Encapsulating the SA within the protective channels or cages of porous frameworks, such as MOFs or zeolites, effectively prevents atom migration and coalescence by imposing a physical barrier. This physical confinement safeguards the delicate SAEC structure from thermal degradation and mechanical shock during operation, all while ensuring access for reactants. Furthermore, managing the immediate coordination environment is essential; transitioning to high-coordinate architectures (e.g., $M-N_4$) offers numerous robust binding points, thereby improving stability compared to low-coordinate sites ($M-N_2$). The transition to non-carbonaceous supports, including hexagonal boron nitride (h-BN) and specific metal phosphides and borides, effectively addresses the main long-term failure mode by removing the susceptible carbon backbone completely, thus ensuring structural stability even in the presence of extreme anodic potentials. While there have been significant advancements in stabilizing individual SAEC sites, transitioning to industrial production introduces challenges related to scale-up that are both synthetic and economic.

The existing synthesis methods typically depend on high-temperature pyrolysis of precursors, representing sensitive batch processes. Minor fluctuations in temperature, gas flow, or heating rate can result in variations in active-site density and performance consistency. Achieving complete SA dispersion without cluster formation becomes significantly more challenging when scaling from grams to kilograms of catalyst material, resulting in expensive reproducibility issues between batches. A notable economic challenge is the generally low metal loading (often < 1 wt.%) necessary for achieving optimal SA activity. Although this approach is atom-efficient, it results in the processing of significant amounts of precursor and support materials to produce a relatively small quantity of the final product, which increases manufacturing costs and leads to considerable waste.

Future development must therefore concentrate on fundamentally transforming synthesis from batch to continuous manufacturing protocols. Techniques like continuous spray pyrolysis and continuous flow chemistry are under investigation to efficiently and consistently produce SAEC materials, ensuring high throughput and dependable quality control. Moreover, the incorporation of these catalysts into functional electrode systems presents further complexities. The intricate SAEC structure may be at risk during the electrode fabrication process, as the mechanical and chemical stresses involved in mixing the catalyst into an ink and applying it onto a gas diffusion layer can lead to premature sintering. Ultimately, in high-power electrolyzer systems, the performance is frequently constrained not by the inherent activity of the SAEC, but by the mass transport of reactants and products within the porous catalyst layer. Creating a highly conductive, thin SAEC layer that reduces mass-transport resistance while ensuring mechanical strength under continuous flow conditions presents a significant engineering challenge. Effectively addressing these interconnected challenges—from atomic-level stabilization to industrial-scale manufacturing—is crucial for unlocking the transformative potential of SAECs in the worldwide pursuit of sustainable energy conversion.

3.7 Conclusion

The integration of AAWE with SAECs marks a significant breakthrough in the field of electrochemical hydrogen production. By replacing the kinetically slow OER with energetically favorable alcohol oxidation reactions, AAWE effectively reduces the overall cell voltage and improves hydrogen production efficiency. SAECs are pivotal in realizing this potential because of their distinctive atomic-level active sites, which provide remarkable catalytic precision and high selectivity for value-added chemical products. The capacity to adjust intermediate binding, inhibit deep oxidation, and stabilize crucial reactive species facilitates customized pathways for generating oxygenates like formate, acetate, and glycolate. Compared to traditional OER-driven systems, AAWE offers significant energetic and economic benefits by integrating hydrogen generation with the simultaneous production of industrially relevant chemicals. Nonetheless, obstacles persist, such as ensuring the long-term stability of SAECs, refining alcohol feedstocks, and developing reactor configurations that can effectively manage multiphase transport and product separation. Ongoing progress in characterization, catalyst engineering, membrane design, and system integration is anticipated to significantly improve the practicality of this technology. In summary, the integration of SAEC with AAWE presents considerable potential for sustainable hydrogen production and chemical valorization, positioning it as a leader in the evolution of electrolysis technologies.

References

1 Siegelman, R. L., Milner, P. J., Kim, E. J., Weston, S. C., & Long, J. R. (2019). Challenges and opportunities for adsorption-based CO_2 capture from natural gas combined cycle emissions. *Energy Environ. Sci. 12*(7), 2161–2173. https://doi.org/10.1039/C9EE00505F

2 Chen, M., Kitiphatpiboon, N., Feng, C., Abudula, A., Ma, Y., & Guan, G. (2023). Recent progress in transition-metal-oxide-based electrocatalysts for the oxygen evolution reaction in natural seawater splitting: A critical review. *eScience 3*(2), 100111. https://doi.org/10.1016/J.ESCI.2023.100111

3 Wang, L., Chen, W., Zhang, D., Du, Y., Amal, R., Qiao, S., Wu, J., & Yin, Z. (2019). Surface strategies for catalytic CO_2 reduction: From two-dimensional materials to nanoclusters to single atoms. *Chem. Soc. Rev. 48*, 5310–5349. https://doi.org/10.1039/c9cs00163h

4 Wang, X. S., Li, L., Li, D., & Ye, J. (2020). Recent progress on exploring stable metal–organic frameworks for photocatalytic solar fuel production. *Sol. RRL 4*(8), 1900547. https://doi.org/10.1002/SOLR.201900547

5 Mazloomi, K. & Gomes, C. (2012). Hydrogen as an energy carrier: Prospects and challenges. *Renew. Sustain. Energy Rev. 16*(5), 3024–3033. https://doi.org/10.1016/J.RSER.2012.02.028

6 Wang, X., Xiao, H., Li, A., Li, Z., Liu, S., Zhang, Q., Gong, Y., Zheng, L., Zhu, Y., Chen, C., Wang, D., Peng, Q., Gu, L., Han, X., Li, J., & Li, Y. (2018). Constructing NiCo/Fe_3O_4 heteroparticles within MOF-74 for efficient oxygen evolution reactions. *J. Am. Chem. Soc. 140*(45), 15336–15341. https://doi.org/10.1021/JACS.8B08744

7 She, Z. W., Kibsgaard, J., Dickens, C. F., Chorkendorff, I., Nørskov, J. K., & Jaramillo, T. F. (2017). Combining theory and experiment in electrocatalysis: Insights into materials design. *Science 355*(6321), eaad4998. https://doi.org/10.1126/SCIENCE.AAD4998

8 Mu, X., Yuan, Y., Yu, M., Hu, Y., Zeng, W., Peng, W., Zhang, Y., Liu, X., Liu, S., & Mu, S. (2024). Robust water/seawater-electrolysis hydrogen production at industrial-scale current densities by modulating built-in-outer electric field of catalytic substance. *Nano Energy 131*, 110216. https://doi.org/10.1016/J.NANOEN.2024.110216

9 Zhang, H., Li, C., Lu, Q., Cheng, M. J., & Goddard, W. A. (2021). Selective activation of propane using intermediates generated during water oxidation. *J. Am. Chem. Soc. 143*(10), 3967–3974. https://doi.org/10.1021/JACS.1C00377

10 Tang, H. T., Zhou, H. Y., Pan, Y. M., Zhang, J. L., Cui, F. H., Li, W. H., & Wang, D. (2024). Single-atom manganese-catalyzed oxygen evolution drives the electrochemical oxidation of silane to silanol. *Angew. Chem. Int. Ed. 63*(3), e202315032. https://doi.org/10.1002/ANIE.202315032

11 Liu, F., Gao, X., Shi, R., Guo, Z., Tse, E. C. M., & Chen, Y. (2023). Concerted and selective electrooxidation of polyethylene-terephthalate-derived alcohol to glycolic acid at an industry-level current density over a Pd–$Ni(OH)_2$ catalyst. *Angew. Chem. Int. Ed. 62*(11), e202300094. https://doi.org/10.1002/ANIE.202300094

12 Zhang, Y., Mu, X., Liu, Z., Zhao, H., Zhuang, Z., Zhang, Y., Mu, S., Liu, S., Wang, D., & Dai, Z. (2024). Twin-distortion modulated ultra-low coordination PtRuNi-O_x catalyst for enhanced hydrogen production from chemical wastewater. *Nat. Commun. 15*(1), 10149. https://doi.org/10.1038/s41467-024-54513-6

13 Mu, X., Zhang, X., Chen, Z., Gao, Y., Yu, M., Chen, D., Pan, H., Liu, S., Wang, D., & Mu, S. (2024). Constructing symmetry-mismatched $Ru_xFe_{3-x}O_4$ heterointerface-supported Ru clusters for efficient hydrogen evolution and oxidation reactions. *Nano Lett. 24*(3), 1015–1023. https://doi.org/10.1021/ACS.NANOLETT.3C04690

14 Zhao, Y., Deng, C., Tang, D., Ding, L., Zhang, Y., Sheng, H., Ji, H., Song, W., Ma, W., Chen, C., & Zhao, J. (2021). α-Fe_2O_3 as a versatile and efficient oxygen atom transfer catalyst in combination with H_2O as the oxygen source. *Nat. Catal.* *4*(8), 684–691. https://doi.org/10.1038/s41929-021-00659-1

15 Chen, Y. X., Lavacchi, A., Miller, H. A., Bevilacqua, M., Filippi, J., Innocenti, M., Marchionni, A., Oberhauser, W., Wang, L., & Vizza, F. (2014). Nanotechnology makes biomass electrolysis more energy efficient than water electrolysis. *Nat. Commun.* *5*(1), 4036. https://doi.org/10.1038/ncomms5036

16 Zhou, H., Li, Z., Xu, S.-M., Lu, L., Xu, M., Ji, K., Ge, R., Yan, Y., Ma, L., Kong, X., Zheng, L., & Duan, H. (2021). Selectively upgrading lignin derivatives to carboxylates through electrochemical oxidative C(OH)–C bond cleavage by a Mn-doped cobalt oxyhydroxide catalyst. *Angew. Chem.* *133*(16), 9058–9064. https://doi.org/10.1002/ANGE.202015431

17 Carmo, M., Fritz, D. L., Mergel, J., & Stolten, D. (2013). A comprehensive review on PEM water electrolysis. *Int. J. Hydrog. Energy* *38*(12), 4901–4934. https://doi.org/10.1016/J.IJHYDENE.2013.01.151

18 Du, L., Sun, Y., & You, B. (2021). Hybrid water electrolysis: Replacing oxygen evolution reaction for energy-efficient hydrogen production and beyond. *Mater. Rep. Energy* *1*(1), 100004. https://doi.org/10.1016/J.MATRE.2020.12.001

19 Zhu, P., Shen, Y., Zhang, Z. M., Wang, D., Xi, B., An, X., Xiong, S., & An, C. (2024). Modulating alcohol adsorption modes for boosting electrooxidation-assisted hydrogen production. *ACS Catal.* *14*(10), 7674–7683. https://doi.org/10.1021/ACSCATAL.4C01078

20 Coutanceau, C. & Baranton, S. (2016). Electrochemical conversion of alcohols for hydrogen production: A short overview. *WIREs Energy Environ.* *5*(4), 388–400. https://doi.org/10.1002/WENE.193

21 Urso, M., Ju, X., Nittoor-Veedu, R., Lee, H., Zaoralová, D., Otyepka, M., & Pumera, M. (2025). Single atom engineering for electrocatalysis: Fundamentals and applications. *ACS Catal.* *15*(13), 11617–11663. https://doi.org/10.1021/ACSCATAL.4C08027

22 Yang, X. F., Wang, A., Qiao, B., Li, J., Liu, J., & Zhang, T. (2013). Single-atom catalysts: A new frontier in heterogeneous catalysis. *Acc. Chem. Res.* *46*(8), 1740–1748. https://doi.org/10.1021/AR300361M

23 Zhu, C., Fu, S., Shi, Q., Du, D., & Lin, Y. (2017). Single-atom electrocatalysts. *Angew. Chem. Int. Ed.* *56*(45), 13944–13960. https://doi.org/10.1002/ANIE.201703864;ISSUE:ISSUE:DOI

24 Zhou, Y., Jiang, Y., Ji, Y., Lang, R., Fang, Y., & Wu, C. D. (2023). The opportunities and challenges in single-atom catalysis. *ChemCatChem* *15*(5), e202201176. https://doi.org/10.1002/CCTC.202201176

25 Ren, Y., Wang, J., Zhang, M., Wang, Y., Cao, Y., Kim, D. H., Liu, Y., & Lin, Z. (2024). Strategies toward high selectivity, activity, and stability of single-atom catalysts. *Small* *20*(22), 2308213. https://doi.org/10.1002/SMLL.202308213;PAGE:STRING:ARTICLE/CHAPTER

26 Lu, B., Liu, Q., & Chen, S. (2020). Electrocatalysis of single-atom sites: Impacts of atomic coordination. *ACS Catal.* *10*(14), 7584–7618. https://doi.org/10.1021/ACSCATAL.0C01950

27 Wang, Y., Su, H., He, Y., Li, L., Zhu, S., Shen, H., Xie, P., Fu, X., Zhou, G., Feng, C., Zhao, D., Xiao, F., Zhu, X., Zeng, Y., Shao, M., Chen, S., Wu, G., Zeng, J., & Wang, C. (2020). Advanced electrocatalysts with single-metal-atom active sites. *Chem. Rev. 120*(21), 12217–12314. https://doi.org/10.1021/ACS.CHEMREV.0C00594

28 Zhang, Q. & Guan, J. (2020). Single-atom catalysts for electrocatalytic applications. *Adv. Funct. Mater. 30*(31), 2000768. https://doi.org/10.1002/ADFM.202000768

29 Humayun, M., Israr, M., Khan, A., & Bououdina, M. (2023). State-of-the-art single-atom catalysts in electrocatalysis: From fundamentals to applications. *Nano Energy 113*, 108570. https://doi.org/10.1016/J.NANOEN.2023.108570

30 Li, W., Guo, Z., Yang, J., Li, Y., Sun, X., He, H., Li, S., & Zhang, J. (2022). Advanced strategies for stabilizing single-atom catalysts for energy storage and conversion. *Electrochem. Energy Rev. 5*(3), 9. https://doi.org/10.1007/S41918-022-00169-Z

31 Luo, G., Song, M., Zhang, Q., An, L., Shen, T., Wang, S., Hu, H., Huang, X., & Wang, D. (2024). Advances of synergistic electrocatalysis between single atoms and nanoparticles/clusters. *Nano-Micro Lett. 16*(1), 241. https://doi.org/10.1007/S40820-024-01463-9

32 Badalyan, A. & Stahl, S. S. (2016). Cooperative electrocatalytic alcohol oxidation with electron-proton-transfer mediators. *Nature 535*(7612), 406–410. https://doi.org/10.1038/nature18008

33 Zhang, Z. & Deng, K. (2015). Recent advances in the catalytic synthesis of 2,5-furandicarboxylic acid and its derivatives. *ACS Catal. 5*(11), 6529–6544. https://doi.org/10.1021/ACSCATAL.5B01491

34 Lv, G., Wang, H., Yang, Y., Deng, T., Chen, C., Zhu, Y., & Hou, X. (2015). Graphene oxide: A convenient metal-free carbocatalyst for facilitating aerobic oxidation of 5-hydroxymethylfurfural into 2,5-diformylfuran. *ACS Catal. 5*(9), 5636–5646. https://doi.org/10.1021/ACSCATAL.5B01446

35 Arshad, F., Haq, T. u., Hussain, I., & Sher, F. (2021). Recent advances in electrocatalysts toward alcohol-assisted, energy-saving hydrogen production. *ACS Appl. Energy Mater. 4*(9), 8685–8701. https://doi.org/10.1021/ACSAEM.1C01932

36 Wagner, L. P., Reinpold, L. M., Kilthau, M., & Fay, A. (2023). A systematic review of modeling approaches for flexible energy resources. *Renew. Sustain. Energy Rev. 184*, 113541. https://doi.org/10.1016/J.RSER.2023.113541

37 Bender, M. T., Yuan, X., & Choi, K. S. (2020). Alcohol oxidation as alternative anode reactions paired with (photo)electrochemical fuel production reactions. *Nat. Commun. 11*(1), 4594. https://doi.org/10.1038/s41467-020-18461-1

38 Kuang, M., Wang, Y., Fang, W., Tan, H., Chen, M., Yao, J., Liu, C., Xu, J., Zhou, K., & Yan, Q. (2020). Efficient nitrate synthesis via ambient nitrogen oxidation with Ru-doped TiO_2/RuO_2 electrocatalysts. *Adv. Mater. 32*(26), 2002189. https://doi.org/10.1002/ADMA.202002189

39 Chadderdon, X. H., Chadderdon, D. J., Pfennig, T., Shanks, B. H., & Li, W. (2019). Paired electrocatalytic hydrogenation and oxidation of 5-(hydroxymethyl)furfural for efficient production of biomass-derived monomers. *Green Chem. 21*, 6210–6219. https://doi.org/10.1039/c9gc02264c

40 Chen, B., Li, X., Rui, P., Ye, Y., Ye, T., Zhou, R., Li, D., Carter, J. H., & Hutchings, G. J. (2022). The reaction pathways of 5-hydroxymethylfurfural conversion in a continuous flow reactor using copper catalysts. *Catal. Sci. Technol. 12*, 3016–3027. https://doi.org/10.1039/d1cy02197d

41 Bellotti, D., Cassettari, L., Mosca, M., & Magistri, L. (2019). RSM approach for stochastic sensitivity analysis of the economic sustainability of a methanol production plant using renewable energy sources. *J. Clean. Prod. 240*, 117947. https://doi.org/10.1016/J.JCLEPRO.2019.117947

42 Tomboc, G. M., Abebe, M. W., Baye, A. F., & Kim, H. (2019). Utilization of the superior properties of highly mesoporous PVP modified $NiCo_2O_4$ with accessible 3D nanostructure and flower-like morphology towards electrochemical methanol oxidation reaction. *J. Energy Chem. 29*, 136–146. https://doi.org/10.1016/J.JECHEM.2018.08.009

43 Klein, J., Argast, F., Engstfeld, A. K., Brimaud, S., & Behm, R. J. (2019). Electro-oxidation of methanol on Ru-core Pt-shell type model electrodes. *Electrochim. Acta 311*, 244–254. https://doi.org/10.1016/J.ELECTACTA.2019.04.081

44 Spurgeon, J. M. & Kumar, B. (2018). A comparative technoeconomic analysis of pathways for commercial electrochemical CO_2 reduction to liquid products. *Energy Environ. Sci. 11*(6), 1536–1551. https://doi.org/10.1039/C8EE00097B

45 Yang, Y. Y., Ren, J., Li, Q. X., Zhou, Z. Y., Sun, S. G., & Cai, W.-B. (2014). Electrocatalysis of ethanol on a Pd electrode in alkaline media: An in situ attenuated total reflection surface-enhanced infrared absorption spectroscopy study. *ACS Catal. 4*(3), 798–803. https://doi.org/10.1021/CS401198T

46 Bambagioni, V., Bevilacqua, M., Bianchini, C., Filippi, J., Lavacchi, A., Marchionni, A., Vizza, F., & Shen, P. K. (2010). Self-sustainable production of hydrogen, chemicals, and energy from renewable alcohols by electrocatalysis. *ChemSusChem 3*(7), 851–855. https://doi.org/10.1002/CSSC.201000103

47 Wang, Y., Zou, S., & Cai, W.-B. (2015). Recent advances on electro-oxidation of ethanol on Pt- and Pd-based catalysts: From reaction mechanisms to catalytic materials. *Catalysts 5*(3), 1507–1534. https://doi.org/10.3390/CATAL5031507

48 Zhao, Y., Xing, S., Meng, X., Zeng, J., Yin, S., Li, X., & Chen, Y. (2019). Ultrathin Rh nanosheets as a highly efficient bifunctional electrocatalyst for isopropanol-assisted overall water splitting. *Nanoscale 11*(19), 9319–9326. https://doi.org/10.1039/C9NR02153A

49 Wu, T., Zhu, X., Wang, G., Zhang, Y., Zhang, H., & Zhao, H. (2017). Vapor-phase hydrothermal growth of single crystalline NiS_2 nanostructure film on carbon fiber cloth for electrocatalytic oxidation of alcohols to ketones and simultaneous H_2 evolution. *Nano Res. 11*(2), 1004–1017. https://doi.org/10.1007/S12274-017-1714-0

50 Nomanbhay, S., Ong, M. Y., Chew, K. W., Show, P. L., Lam, M. K., & Chen, W. H. (2020). Organic carbonate production utilizing crude glycerol derived as by-product of biodiesel production: A review. *Energies 13*(6), 1483. https://doi.org/10.3390/EN13061483

51 Simões, M., Baranton, S., & Coutanceau, C. (2012). Electrochemical valorisation of glycerol. *ChemSusChem 5*(11), 2106–2124. https://doi.org/10.1002/CSSC.201200335

52 Garcia, A. C., Birdja, Y. Y., Tremiliosi-Filho, G., & Koper, M. T. M. (2017). Glycerol electro-oxidation on bismuth-modified platinum single crystals. *J. Catal. 346*, 117–124. https://doi.org/10.1016/J.JCAT.2016.12.013

53 Wang, X., Sun, L., Zhou, W., Yang, L., Ren, G., Wu, H., & Deng, W. Q. (2022). Iron single-atom catalysts confined in covalent organic frameworks for efficient oxygen evolution reaction. *Cell Rep. Phys. Sci. 3*(3), 100804. https://doi.org/10.1016/j.xcrp.2022.100804

54 Sun, M., Wu, T., & Huang, B. (2020). Designing the future atomic electrocatalyst for efficient energy systems. *Eng. Rep. 2*(12), e12327. https://doi.org/10.1002/ENG2.12327;WGROUP:STRING:PUBLICATION

55 Wei, H., Liu, X., Wang, A., Zhang, L., Qiao, B., Yang, X., Huang, Y., Miao, S., Liu, J., & Zhang, T. (2014). FeO_x-supported platinum single-atom and pseudo-single-atom catalysts for chemoselective hydrogenation of functionalized nitroarenes. *Nat. Commun. 5*(1), 5634. https://doi.org/10.1038/ncomms6634

56 Chen, Y., Lin, J., Li, L., Qiao, B., Liu, J., Su, Y., & Wang, X. (2018). Identifying size effects of Pt as single atoms and nanoparticles supported on FeO_x for the water-gas shift reaction. *ACS Catal. 8*(2), 859–868. https://doi.org/10.1021/ACSCATAL.7B02751

57 Liu, W., Zhang, L., Liu, X., Liu, X., Yang, X., Miao, S., Wang, W., Wang, A., & Zhang, T. (2017). Discriminating catalytically active FeN_x species of atomically dispersed Fe–N–C catalyst for selective oxidation of the C–H bond. *J. Am. Chem. Soc. 139*(31), 10790–10798. https://doi.org/10.1021/JACS.7B05130

58 Nguyen, L., Zhang, S., Wang, L., Li, Y., Yoshida, H., Patlolla, A., Takeda, S., Frenkel, A. I., & Tao, F. (2016). Reduction of nitric oxide with hydrogen on catalysts of singly dispersed bimetallic sites Pt_1Co_m and Pd_1Co_n. ACS Catal. *6*(2), 840–850. https://doi.org/10.1021/ACSCATAL.5B00842

59 Wang, Z. T., Hoyt, R. A., El-Soda, M., Madix, R. J., Kaxiras, E., & Sykes, E. C. H. (2017). Dry dehydrogenation of ethanol on Pt–Cu single atom alloys. *Top. Catal. 61*(5), 328–335. https://doi.org/10.1007/S11244-017-0856-3

60 Xiong, H., Lin, S., Goetze, J., Pletcher, P., Guo, H., Kovarik, L., Artyushkova, K., Weckhuysen, B. M., & Datye, A. K. (2017). Thermally stable and regenerable platinum–tin clusters for propane dehydrogenation prepared by atom trapping on ceria. *Angew. Chem. 129*(31), 9114–9119. https://doi.org/10.1002/ANGE.201701115

61 Aich, P., Wei, H., Basan, B., Kropf, A. J., Schweitzer, N. M., Marshall, C. L., Miller, J. T., & Meyer, R. (2015). Single-atom alloy Pd–Ag catalyst for selective hydrogenation of acrolein. *J. Phys. Chem. C 119*(32), 18140–18148. https://doi.org/10.1021/ACS.JPCC.5B01357

62 Zhang, L., Wang, A., Miller, J. T., Liu, X., Yang, X., Wang, W., Li, L., Huang, Y., Mou, C. Y., & Zhang, T. (2014). Efficient and durable Au alloyed Pd single-atom catalyst for the Ullmann reaction of aryl chlorides in water. *ACS Catal. 4*(5), 1546–1553. https://doi.org/10.1021/CS500071C

63 Pei, G. X., Liu, X. Y., Yang, X., Zhang, L., Wang, A., Li, L., Wang, H., Wang, X., & Zhang, T. (2017). Performance of Cu-alloyed Pd single-atom catalyst for semihydrogenation of acetylene under simulated front-end conditions. *ACS Catal. 7*(2), 1491–1500. https://doi.org/10.1021/ACSCATAL.6B03293

64 Feng, Q., Zhao, S., Wang, Y., Dong, J., Chen, W., He, D., Wang, D., Yang, J., Zhu, Y., Zhu, H., Gu, L., Li, Z., Liu, Y., Yu, R., Li, J., & Li, Y. (2017). Isolated single-atom Pd sites in intermetallic nanostructures: High catalytic selectivity for semihydrogenation of alkynes. *J. Am. Chem. Soc. 139*(21), 7294–7301. https://doi.org/10.1021/JACS.7B01471

65 Kyriakou, G., Boucher, M. B., Jewell, A. D., Lewis, E. A., Lawton, T. J., Baber, A. E., Tierney, H. L., Flytzani-Stephanopoulos, M., & Sykes, E. C. H. (2012). Isolated metal atom geometries as a strategy for selective heterogeneous hydrogenations. *Science 335*(6073), 1209–1212. https://doi.org/10.1126/SCIENCE.1215864

66 Zhang, R.-Q., Lee, T.-H., Yu, B.-D., Stampfl, C., & Soon, A. (2012). The role of titanium nitride supports for single-atom platinum-based catalysts in fuel cell technology. *Phys. Chem. Chem. Phys. 14*, 16552–16557. https://doi.org/10.1039/c2cp41392b

67 He, F., Li, K., Yin, C., Wang, Y., Tang, H., & Wu, Z. (2017). Single Pd atoms supported by graphitic carbon nitride, a potential oxygen reduction reaction catalyst from theoretical perspective. *Carbon 114*, 619–627. https://doi.org/10.1016/J.CARBON.2016.12.061

68 Zhang, C., Sha, J., Fei, H., Liu, M., Yazdi, S., Zhang, J., Zhong, Q., Zou, X., Zhao, N., Yu, H., Jiang, Z., Ringe, E., Yakobson, B. I., Dong, J., Chen, D., & Tour, J. M. (2017). Single-atomic ruthenium catalytic sites on nitrogen-doped graphene for oxygen reduction reaction in acidic medium. *ACS Nano 11*(7), 6930–6941. https://doi.org/10.1021/ACSNANO.7B02148

69 Cheng, N., Stambula, S., Wang, D., Banis, M. N., Liu, J., Riese, A., Xiao, B., Li, R., Sham, T. K., Liu, L. M., Botton, G. A., & Sun, X. (2016). Platinum single-atom and cluster catalysis of the hydrogen evolution reaction. *Nat. Commun. 7*(1), 13638. https://doi.org/10.1038/ncomms13638

70 Fu, H., Liu, H., Wang, X., Zhang, W., Zhang, H., Luo, Y., Deng, X., King, G., Chen, N., Wang, L., & Wu, Y. A. (2024). Reverse hydrogen spillover on metal oxides for water-promoted catalytic oxidation reactions. *Adv. Mater. 36*(36), 2407534. https://doi.org/10.1002/ADMA.202407534

71 Wu, R., Zhang, H., Ma, H., Zhao, B., Li, W., Chen, Y., Liu, J., Liang, J., Qin, Q., Qi, W., Chen, L., Li, J., Li, B., & Duan, X. (2024). Synthesis, modulation, and application of two-dimensional TMD heterostructures. *Chem. Rev. 124*(17), 10112–10191. https://doi.org/10.1021/ACS.CHEMREV.4C00174

72 Qi, Y., Sun, L., & Liu, Z. (2024). Super graphene-skinned materials: An innovative strategy toward graphene applications. *ACS Nano 18*(6), 4617–4623. https://doi.org/10.1021/ACSNANO.3C11971

73 Wang, H., Li, X., Deng, Y., Jiang, J., Ma, H., & Zou, J. (2025). Advances in MXene-based single-atom catalysts for electrocatalytic applications. *Coord. Chem. Rev. 529*, 216462. https://doi.org/10.1016/J.CCR.2025.216462

74 Rong, C., Su, T., Li, Z., Chu, T., Zhu, M., Yan, Y., Zhang, B., & Xuan, F. Z. (2024). Elastic properties and tensile strength of 2D $Ti_3C_2T_x$ MXene monolayers. *Nat. Commun. 15*(1), 1566. https://doi.org/10.1038/s41467-024-45657-6

75 Binder, J., Dabrowska, A. K., Tokarczyk, M., Rousseau, A., Valvin, P., Bozek, R., Nogajewski, K., Kowalski, G., Pacuski, W., Gil, B., Cassabois, G., Stepniewski, R., & Wysmolek, A. (2024). Homoepitaxy of boron nitride on exfoliated hexagonal boron nitride flakes. *Nano Lett. 24*(23), 6990–6996. https://doi.org/10.1021/ACS.NANOLETT.4C01310

76 Chen, X., Yuan, D., Rong, C., Wu, C., Liu, P., Liu, H. K., Wang, D., Dou, S. X., & Dou, Y. (2025). Single-atom catalysts supported on atomically thin materials for water splitting. *Adv. Powder Mater. 4*(5), 100330. https://doi.org/10.1016/J.APMATE.2025.100330

77 Sun, H., Chen, H. C., Humayun, M., Qiu, Y., Ju, J., Zhang, Y., Bououdina, M., Xue, X., Liu, Q., Pang, Y., & Wang, C. (2024). Unlocking the catalytic potential of platinum single atoms for industry-level current density chlorine tolerance hydrogen generation. *Adv. Funct. Mater. 34*(49), 2408872. https://doi.org/10.1002/ADFM.202408872

78 Zhang, Y., Yang, J., Ge, R., Zhang, J., Cairney, J. M., Li, Y., Zhu, M., Li, S., & Li, W. (2022). The effect of coordination environment on the activity and selectivity of single-atom catalysts. *Coord. Chem. Rev. 461*, 214493. https://doi.org/10.1016/J.CCR.2022.214493

79 Zhang, Y., Guo, L., Tao, L., Lu, Y., & Wang, S. (2019). Defect-based single-atom electrocatalysts. *Small Methods 3*(9), 1800406. https://doi.org/10.1002/SMTD.201800406

80 Pakkala, A. & Putkonen, M. (2010). Atomic layer deposition. In *Handbook of Deposition Technologies for Films and Coatings: Science, Applications and Technology*, 3e (ed. P. M. Martin), 364–391. Boston: William Andrew Publishing. https://doi.org/10.1016/B978-0-8155-2031-3.00008-9

81 Zhang, L., Banis, M. N., & Sun, X. (2018). Single-atom catalysts by the atomic layer deposition technique. *Natl. Sci. Rev. 5*(5), 628–630. https://doi.org/10.1093/NSR/NWY054

82 Stambula, S., Gauquelin, N., Bugnet, M., Gorantla, S., Turner, S., Sun, S., Liu, J., Zhang, G., Sun, X., & Botton, G. A. (2014). Chemical structure of nitrogen-doped graphene with single platinum atoms and atomic clusters as a platform for the PEMFC electrode. *J. Phys. Chem. C 118*(8), 3890–3900. https://doi.org/10.1021/JP408979H

83 Wang, C., Gu, X. K., Yan, H., Lin, Y., Li, J., Liu, D., Li, W. X., & Lu, J. (2016). Water-mediated Mars–Van Krevelen mechanism for CO oxidation on ceria-supported single-atom Pt_1 catalyst. *ACS Catal. 7*(1), 887–891. https://doi.org/10.1021/ACSCATAL.6B02685

84 Cao, L., Liu, W., Luo, Q., Yin, R., Wang, B., Weissenrieder, J., Soldemo, M., Yan, H., Lin, Y., Sun, Z., Ma, C., Zhang, W., Chen, S., Wang, H., Guan, Q., Yao, T., Wei, S., Yang, J., & Lu, J. (2019). Atomically dispersed iron hydroxide anchored on Pt for preferential oxidation of CO in H_2. *Nature 565*(7741), 631–635. https://doi.org/10.1038/s41586-018-0869-5

85 Shi, Y., Lee, C., Tan, X., Yang, L., Zhu, Q., Loh, X., Xu, J., & Yan, Q. (2022). Atomic-level metal electrodeposition: Synthetic strategies, applications, and catalytic mechanism in electrochemical energy conversion. *Small Struct.* *3*(3), 2100185. https://doi.org/10.1002/SSTR.202100185

86 Zhang, T., Jin, J., Chen, J., Fang, Y., Han, X., Chen, J., Li, Y., Wang, Y., Liu, J., & Wang, L. (2022). Pinpointing the axial ligand effect on platinum single-atom-catalyst towards efficient alkaline hydrogen evolution reaction. *Nat. Commun.* *13*(1), 6875. https://doi.org/10.1038/s41467-022-34619-5

87 Lei, Z., Cai, W., Rao, Y., Wang, K., Jiang, Y., Liu, Y., Jin, X., Li, J., Lv, Z., Jiao, S., Zhang, W., Yan, P., Zhang, S., & Cao, R. (2022). Coordination modulation of iridium single-atom catalyst maximizing water oxidation activity. *Nat. Commun.* *13*(1), 24. https://doi.org/10.1038/s41467-021-27664-z

88 Park, C., Shin, H., Jeon, M., Cho, S. H., Kim, J., & Kim, I. D. (2024). Single-atom catalysts in conductive metal–organic frameworks: enabling reversible gas sensing at room temperature. *ACS Nano.* *18*(38), 26066–26075. https://doi.org/10.1021/ACSNANO.4C05815

89 Cui, M., Xu, B., Shi, X., Zhai, Q., Dou, Y., Li, G., Bai, Z., Ding, Y., Sun, W., Liu, H., & Dou, S. (2024). Metal–organic framework-derived single-atom catalysts for electrocatalytic energy conversion applications. *J. Mater. Chem. A* *12*(30), 18921–18947. https://doi.org/10.1039/D4TA03518F

90 Gao, F., Huang, J., Ruan, Y., Li, H., Gong, P., Wang, F., Tang, Q., & Jiang, Y. (2023). Unraveling the structure transition and peroxidase mimic activity of copper sites over atomically dispersed copper-doped carbonized polymer dots. *Angew. Chem. Int. Ed.* *62*(7), e202214042. https://doi.org/10.1002/ANIE.202214042

91 Xia, C., Qiu, Y., Xia, Y., Zhu, P., King, G., Zhang, X., Wu, Z., Kim, J. Y. (Timothy), Cullen, D. A., Zheng, D., Li, P., Shakouri, M., Heredia, E., Cui, P., Alshareef, H. N., Hu, Y., & Wang, H. (2021). General synthesis of single-atom catalysts with high metal loading using graphene quantum dots. *Nat. Chem.* *13*(9), 887–894. https://doi.org/10.1038/s41557-021-00734-x

92 Gu, C. H., Wang, S., Zhang, A. Y., Liu, C., Jiang, J., & Yu, H. Q. (2023). Slow-release synthesis of Cu single-atom catalysts with the optimized geometric structure and density of state distribution for Fenton-like catalysis. *Proc. Natl. Acad. Sci. USA* *120*(43), e2311585120. https://doi.org/10.1073/PNAS.2311585120;ISSUE:ISSUE:DOI

93 Hai, X., Xi, S., Mitchell, S., Harrath, K., Xu, H., Akl, D. F., Kong, D., Li, J., Li, Z., Sun, T., Yang, H., Cui, Y., Su, C., Zhao, X., Li, J., Pérez-Ramírez, J., & Lu, J. (2021). Scalable two-step annealing method for preparing ultra-high-density single-atom catalyst libraries. *Nat. Nanotechnol.* *17*(2), 174–181. https://doi.org/10.1038/s41565-021-01022-y

94 Al-Hilfi, S. H., Jiang, X., Heuer, J., Akula, S., Tammeveski, K., Hu, G., Yang, J., Wang, H. I., Bonn, M., Landfester, K., Müllen, K., & Zhou, Y. (2024). Single-atom catalysts through pressure-controlled metal diffusion. *J. Am. Chem. Soc.* *146*(29), 19886–19895. https://doi.org/10.1021/JACS.4C03066/SUPPL_FILE/JA4C03066_SI_004.MP4

95 Wang, Y., Li, C., Han, X., Bai, J., Wang, X., Zheng, L., Hong, C., Li, Z., Bai, J., Leng, K., Lin, Y., & Qu, Y. (2024). General negative pressure annealing approach for creating ultra-high-loading single atom catalyst libraries. *Nat. Commun.* *15*(1), 5675. https://doi.org/10.1038/s41467-024-50061-1

96 Chen, Y., Ji, S., Chen, C., Peng, Q., Wang, D., & Li, Y. (2018). Single-atom catalysts: Synthetic strategies and electrochemical applications. *Joule 2*(7), 1242–1264. https://doi.org/10.1016/J.JOULE.2018.06.019

97 Roy, S., Li, Z., Chen, Z., Mata, A. C., Kumar, P., Sarma, S. C., Teixeira, I. F., Silva, I. F., Gao, G., Tarakina, N. V., Kibria, M. G., Singh, C. V., Wu, J., & Ajayan, P. M. (2024). Cooperative copper single-atom catalyst in 2D carbon nitride for enhanced CO_2 electrolysis to methane. *Adv. Mater. 36*(13), 2300713. https://doi.org/10.1002/ADMA.202300713

98 Giulimondi, V., Vanni, M., Damir, S., Zou, T., Mitchell, S., Krumeich, F., Ruiz-Ferrando, A., López, N., Gata-Cuesta, J. J., Guillén-Gosálbez, G., Smit, J. J., Johnston, P., & Pérez-Ramírez, J. (2024). Convergent active site evolution in platinum single atom catalysts for acetylene hydrochlorination and implications for toxicity minimization. *ACS Catal. 14*(18), 13652–13664. https://doi.org/10.1021/ACSCATAL.4C03533

99 Zhang, E., Dong, A., Yin, K., Ye, C., Zhou, Y., Tan, C., Li, M., Zheng, X., Wang, Y., Gao, X., Li, H., Wang, D., & Guo, S. (2024). Electron localization in rationally designed Pt_1Pd single-atom alloy catalyst enables high-performance Li–O_2 batteries. *J. Am. Chem. Soc. 146*(4), 2339–2344. https://doi.org/10.1021/JACS.3C12734

100 Jones, J., Xiong, H., DeLaRiva, A. T., Peterson, E. J., Pham, H., Challa, S. R., Qi, G., Oh, S., Wiebenga, M. H., Hernández, X. I. P., Wang, Y., & Datye, A. K. (2016). Thermally stable single-atom platinum-on-ceria catalysts via atom trapping. *Science 353*(6295), 150–154. https://doi.org/10.1126/SCIENCE.AAF8800

101 Wei, S., Li, A., Liu, J. C., Li, Z., Chen, W., Gong, Y., Zhang, Q., Cheong, W. C., Wang, Y., Zheng, L., Xiao, H., Chen, C., Wang, D., Peng, Q., Gu, L., Han, X., Li, J., & Li, Y. (2018). Direct observation of noble metal nanoparticles transforming to thermally stable single atoms. *Nat. Nanotechnol. 13*(9), 856–861. https://doi.org/10.1038/s41565-018-0197-9

102 Yang, J., Qiu, Z., Zhao, C., Wei, W., Chen, W., Li, Z., Qu, Y., Dong, J., Luo, J., Li, Z., & Wu, Y. (2018). In situ thermal atomization to convert supported nickel nanoparticles into surface-bound nickel single-atom catalysts. *Angew. Chem. Int. Ed. 57*(43), 14095–14100. https://doi.org/10.1002/ANIE.201808049

103 Wang, Z., Yang, J., Cao, J., Chen, W., Wang, G., Liao, F., Zhou, X., Zhou, F., Li, R., Yu, Z. Q., Zhang, G., Duan, X., & Wu, Y. (2020). Room-temperature synthesis of single iron site by electrofiltration for photoreduction of CO_2 into tunable syngas. *ACS Nano 14*(5), 6164–6172. https://doi.org/10.1021/ACSNANO.0C02162

104 Qi, K., Cui, X., Gu, L., Yu, S., Fan, X., Luo, M., Xu, S., Li, N., Zheng, L., Zhang, Q., Ma, J., Gong, Y., Lv, F., Wang, K., Huang, H., Zhang, W., Guo, S., Zheng, W., & Liu, P. (2019). Single-atom cobalt array bound to distorted 1T MoS_2 with ensemble effect for hydrogen evolution catalysis. *Nat. Commun. 10*(1), 5231. https://doi.org/10.1038/s41467-019-12997-7

105 Wang, H., Wang, X., Pan, J., Zhang, L., Zhao, M., Xu, J., Liu, B., Shi, W., Song, S., & Zhang, H. (2021). Ball-milling induced debonding of surface atoms from metal bulk for construing high-performance dual-site single-atom catalysts. *Angew. Chem. 133*(43), 23338–23342. https://doi.org/10.1002/ANGE.202109356

106 Han, G. F., Li, F., Rykov, A. I., Im, Y. K., Yu, S. Y., Jeon, J. P., Kim, S. J., Zhou, W., Ge, R., Ao, Z., Shin, T. J., Wang, J., Jeong, H. Y., & Baek, J. B. (2022). Abrading bulk metal into single atoms. *Nat. Nanotechnol.* *17*(4), 403–407. https://doi.org/10.1038/s41565-022-01075-7

107 Zhu, Y., Zhu, X., Bu, L., Shao, Q., Li, Y., Hu, Z., Chen, C.-T., Pao, C. W., Yang, S., & Huang, X. (2020). Single-atom in-doped subnanometer Pt nanowires for simultaneous hydrogen generation and biomass upgrading. *Adv. Funct. Mater.* *30*(49), 2004310. https://doi.org/10.1002/ADFM.202004310

108 Ge, R., Wang, Y., Li, Z., Xu, M., Xu, S.-M., Zhou, H., Ji, K., Chen, F., Zhou, J., & Duan, H. (2022). Selective electrooxidation of biomass-derived alcohols to aldehydes in a neutral medium: Promoted water dissociation over a nickel-oxide-supported ruthenium single-atom catalyst. *Angew. Chem.* *134*(19), e202200211. https://doi.org/10.1002/ANGE.202200211

4

Metal Alloys Electrocatalysts for Alcohol-assisted Water Electrolysis

Rafael M. P. Araujo[1], Sidney C. Z. Costa[1,2], M. Janete Giz[1,], and Giuseppe A. Camara[1,*]*

[1]*Laboratory of Electrochemistry and Advanced Materials (LEMA), Institute of Chemistry, Federal University of Mato Grosso do Sul, Campo Grande, Brazil*
[2]*Federal Institute of Mato Grosso do Sul, Corumbá, Brazil*

*Corresponding authors: janete.giz@ufms.br; giuseppe.silva@ufms.br

4.1 Fundamentals of Water Electrolysis

A general classification of electrochemical cells is based on their categorization as either galvanic devices or electrolysis cells. By supplying electrical energy from an external voltage source, i.e., by applying a potential, electrons are delivered to the device, allowing the direction of electrode reactions to be altered and enabling the conversion of electrical energy into chemical energy. Electrolyzers are devices that drive electrolytic processes, enabling the decomposition of compounds through the passage of electric current. There are several types of electrolyzers, which can be classified based on the type of cell and the type of electrolyte used, for example, alkaline water electrolyzer (AWE), proton exchange membrane (PEM), solid oxide electrolyzer cell (SOEC), and anion exchange membrane (AEM) [1, 2].

Electrolyzers have a wide range of applications, with the most relevant being the production of green hydrogen. Hydrogen gas (H_2) is considered one of the most promising energy carriers for the future due to its high energy density and absence of pollutant emissions [3]. Water electrolysis is an established commercial technology and a widely used electrochemical method for obtaining pure hydrogen under moderate conditions. However, it still presents many limitations. Notably, it has a high electrical energy consumption (the practical potential to drive water electrolysis is much higher than the theoretical one, of 1.23 V), mainly caused by the sluggishness of the oxygen evolution reaction (OER) [4]. This condition is particularly critical when one considers that the valueless oxygen is mainly responsible for the electrical losses during conventional water electrolysis. Furthermore, there is an inherent risk of forming explosive mixtures of oxygen and hydrogen due to gas crossover, since both substances are produced in the same device [4–6]. Regarding oxygen, there is the additional drawback that it can be converted back into water at the cathode of the electrolyzer, which implies additional energy consumption [4].

Various strategies have already been proposed to overcome these obstacles, such as reducing ohmic losses (*iR*), among others. Nevertheless, the low stability and high cost of membranes remain significant barriers to their large-scale application. Another strategy has been the development of catalysts with high activity and stability. Although the literature reports the existence of such materials, water electrolyzers still require an operating voltage between 1.6 and 2 V, mainly because the kinetically slow oxygen evolution must be driven by enhanced overpotentials [7].

4.2 Hybrid Water Electrolysis

In order to circumvent these limitations, recent advances have led to the construction of hybrid water electrolyzers, which combine water splitting with thermodynamically more favorable anodic reactions. In this context, the chemically assisted hydrogen evolution reaction (CAHER) has recently emerged as a practical and innovative approach for hydrogen production, aiming to make the process more efficient through the addition of oxidizable substances to the electrolytes, thereby reducing its energy consumption and increasing its practical viability [4, 7].

The CAHER approach is similar to the conventional electrochemical water-splitting process, with the only difference being the replacement of the aqueous solution by reductive chemical solutions in the electrolytes. Hydroxyl or aldehyde groups of small organic molecules are oxidized at the anode instead of oxygen, resulting in the formation of value-added chemicals or inert gases [4, 7]. The best candidates should be cost-effective, water-soluble at room temperature, oxidize at lower potentials than water, produce nontoxic substances, and, if possible, generate valuable products [4, 7]. Alcohols, amines, urea, hydrazine, and some carbohydrates meet these criteria and are good candidates for alternative anodic oxidation half-reactions in the electrolysis processes [4]. At the cathode, water molecules are reduced, and hydrogen is produced by receiving electrons from the external circuit. Therefore, replacing the anodic half-reaction (OER) by the oxidation of small organic molecules becomes thermodynamically more favorable and also prevents the formation of reactive oxygen species (ROS). CAHER cells have shown a promising pathway to overcome the limitations of traditional water electrolysis, as they are more energy-efficient, can produce value-added products, and offer enhanced safety by avoiding the generation of gaseous oxygen. Additionally, they present reduced costs due to the elimination of the need for separation membranes, since there is no generation of oxygen [7, 8].

4.3 Alcohol-assisted Water Electrolysis

Within this broader context, alcohol-assisted water electrolysis (AAWE) has recently gained relevance, since the typical theoretical oxidation potentials of alcohols are a small fraction of the OER (Figure 4.1). Furthermore, depending on their molecular

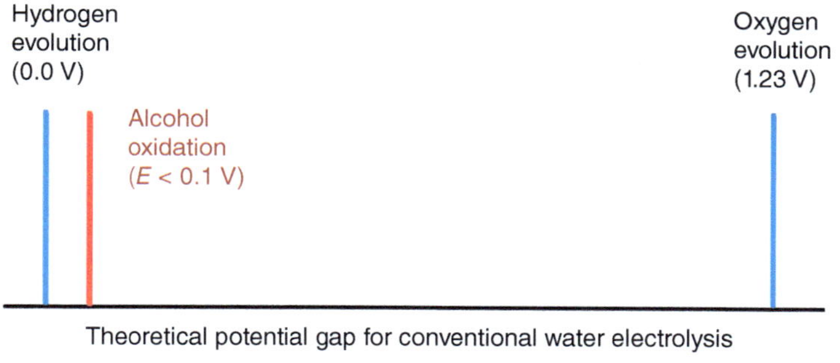

Figure 4.1 Thermodynamic equilibrium potentials for water electrolysis compared to alcohol oxidation. The potentials of the oxidation reactions of alcohols represent only an approximation of the specific potentials.

structure, they can be oxidized to harmless substances (many of which are liquids at room temperature) [8] and even generate value-added oxidation products, whose low production costs (compared to traditional synthetic routes) can compensate for possible low reaction yields.

4.4 Metal Alloys for Alcohol Electrooxidation

Let us now focus on the alcohol-assisted hydrogen evolution reactions (AAHERs), specifically those catalyzed by metal alloys, which are the subject of this chapter. Two main advantages of such an approach are that alcohols oxidize in potentials much lower than water, reducing the energy consumption, and depending on the starting molecule, they can generate value-added substances [8].

Historically, alcohols and associated molecules are considered as potential candidates for driving a revolution in the ways we produce and consume energy. The electrooxidation of alcohols has been investigated as the anodic branch of direct alcohol fuel cells in the recent decades and these fuel cells are emerging as candidates for the implementation of electrochemical energy conversion and storage systems with low environmental impact [9–11].

When conventional water electrolysis and AAWE are compared, an important drawback must be considered. In the first case, the reduction of oxygen is the major reaction, although hydrogen peroxide can also be observed [7, 8], while the electrooxidation of alcohols involves multi-electronic and branched reactions.

In this context, platinum (Pt) emerges as the quintessential monometallic catalyst for the electrooxidation of small-chain alcohols—such as methanol and ethanol—in acidic environments [9–11]. However, its selectivity for CO_2 (the pathway which generates most electrons per oxidized molecule) is still low, and depends strongly on the structure of the electrode surface [9, 12, 13]. Also, platinum faces another major problem, which is the surface poisoning by strongly adsorbed species [9, 12], with emphasis on carbon monoxide (CO), which is a ubiquitous intermediate during the electrooxidation of organic molecules [9, 14].

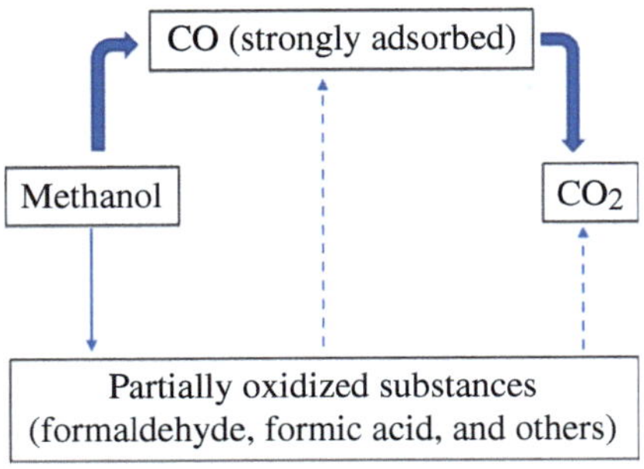

Figure 4.2 Schematic representation of the "dual-pathway" mechanism for the electrooxidation of methanol to carbon dioxide in an acidic environment.

The presence of CO can be understood in terms of the dual pathway established during the electrooxidation of methanol, which we summarize in the scheme shown in Figure 4.2. Namely, there is an indirect pathway that produces a strong adsorbed intermediate, consensually addressed as carbon monoxide. The other path (the direct one) involves the presence of species loosely adsorbed and even dissolved in the surroundings of the electrode. The main identified species are formaldehyde and formic acid, but there may be others [9].

Besides being a common intermediate in many (electro)oxidation reactions of carbonaceous molecules, CO is also a by-product of hydrocarbon reforming (in which hydrogen and carbon dioxide are generated). Due to its electronic structure, CO is a highly reactive species. In the presence of noble metals such as platinum, CO blocks active sites, can occupy small surface domains, and even displace previously adsorbed molecules, as recently demonstrated for the electrooxidation of glycerol [9, 14].

The slow kinetics of this indirect pathway has much to do with the difficulty of oxidizing this strongly adsorbed CO. Discussions about the mechanism of electrochemical oxidation of carbon monoxide are based on the proposal suggested by Gilman [15]. In it, CO is adsorbed on a free site on the metal surface:

$$CO + \; * \rightleftarrows CO_{ads} \tag{4.1}$$

where * denotes a free site on the platinum surface. The oxygenated species that reacts with CO is generally accepted to be the surface-bound OH generated by the oxidation of water:

$$H_2O + \; * \rightleftarrows OH_{ads} + H^+ + e^- \tag{4.2}$$

A Langmuir–Hinshelwood reaction between CO_{ads} and OH_{ads} leads to the formation of CO_2:

$$CO_{ads} + OH_{ads} \Rightarrow CO_2 + H^+ + e^- + 2* \tag{4.3}$$

As the reaction proceeds, new free sites are slowly generated by the reaction of OH_{ads} with CO_{ads}. Thus, the structural sensitivity of CO electrooxidation is attributed to the preferential formation of these OH_{ads} at the sites of the steps and defects generated by the oxidation of water. The oxidation of CO manifests through co-adsorptions sensitive to the structure of certain spectator species, such as an anion, where due to the preferential interaction of catalytic intermediates with step or defect sites, the reaction at the steps or defects completely dominates the reaction on the terrace [12]. However, its specific behavior in single-crystal electrodes varies depending on the orientation of the steps and terraces and the pH of the electrolyte (acidic or alkaline) [12]. With an increase in the OH adsorption rate, consequently, there is an increase in the overall CO oxidation rate and in the currents associated with it.

The difficulties related to the complete oxidation of alcohols become more challenging as their main chain grows, since in this case carbon–carbon bonds are also involved. For example, when one tries to electrooxidize ethanol, the formation of CO_2 passes through two adsorbed intermediates C_{1ads} and C_{2ads}, representing fragments with one and two carbon atoms, respectively. The cleavage of the C—C bond in this molecule represents another major problem in ethanol electrocatalysis. Again, platinum is poisoned by strongly adsorbed ethanol oxidation intermediate species, as in the case of methanol [16], but in this case, they are not limited to CO:

$$CH_3CH_2OH \Rightarrow CH_3CH_2OH_{ads} \Rightarrow C_{1ads}, C_{2ads} \Rightarrow CO_2 \text{ (total oxidation)} \quad (4.4)$$

$$CH_3CH_2OH \Rightarrow CH_3CH_2OH_{ads} \Rightarrow CH_3CHO \Rightarrow CH_3COOH \text{ (partial oxidation)} \quad (4.5)$$

In this case, the main substances partially oxidized are acetaldehyde and acetic acid. Here, the main problem does not refer to a poisoning effect, since both species can desorb from the surface. However, the oxidation pathways generating acetaldehyde and acetic acid deliver two and four electrons, respectively, while the full oxidation for CO_2 responds for 12 electrons per molecule. Hence, the potential energetic content of the reaction is strongly affected by the presence of these parallel reactions [17].

Therefore, monometallic catalysts (such as platinum) do not meet the requirements for reactions with the complexity of those treated here. These are multielectronic processes, whose mechanisms depend on steps such as dissociative adsorption, breaking of C—C bonds, oxidation of functional groups, and fast desorption of eventual oxidation products [14].

One of the many approaches to overcome these barriers and mitigate platinum (and other noble metals) poisoning has focused on the addition of metals or metal oxides that act as cocatalysts. Concerning the electrooxidation of methanol and ethanol, ruthenium (Ru) and tin are of particular interest [10, 16, 18, 19]. According to this sequence of steps (classically known as bifunctional mechanism), the main catalyst (Pt) is responsible for the steps of adsorption and dehydrogenation, while the second metal (or metal oxide) acts as the source of oxygen required to oxidize those adsorbed species and release the catalytic sites to a new cycle of adsorption–oxidation of the alcohol [17, 19, 20].

A way to illustrate such behavior is to observe the role of Pt–Ru alloys during the electrooxidation of methanol [20]. In such a case, when Ru is combined with Pt a steep increase in the methanol oxidation rate is observed, which is attributed to the ability of Ru to adsorb OH groups at lower potentials than those observed for Pt. Namely, methanol is adsorbed and dissociated over Pt sites, generating CO and formyl-like species (—CHO). On the other hand, water molecules are broken over Ru domains, producing adsorbed OH. Finally, both species react, thus forming CO_2 [20]. The sequence of steps is illustrated below:

$$\begin{aligned} &\text{Pt} + \text{CH}_3\text{OH}_{\text{sol}} \Rightarrow \text{Pt—CH}_3\text{OH}_{\text{ads}} \Rightarrow \text{Pt—COH}_{\text{ads}} + 3\text{H}^+ + 3\text{e}^- \\ &\Rightarrow \text{Pt—CO}_{\text{ads}} + \text{H}^+ + \text{e}^- \end{aligned} \tag{4.6}$$

$$\text{Ru} + \text{H}_2\text{O} \Rightarrow \text{Ru—OH}_{\text{ads}} + \text{H}^+ + \text{e}^- \tag{4.7}$$

$$\text{Pt—COH}_{\text{ads}} + \text{Ru—OH}_{\text{ads}} \rightarrow \text{Pt} + \text{Ru} + \text{CO}_2 + 2\text{H}^+ + 2\text{e}^- \tag{4.8}$$

$$\text{Pt—CO}_{\text{ads}} + \text{Ru—OH}_{\text{ads}} \rightarrow \text{Pt} + \text{Ru} + \text{CO}_2 + \text{H}^+ + \text{e}^- \tag{4.9}$$

Besides the classical Pt–Ru catalyst, a plethora of bimetallic alloys, as Pt–Sn, Pt–Ni (nickel), Pt–Co, Pt–W, and Pt–Mo (molybdenum), has been investigated to assist Pt in alcohol oxidation reactions [17].

The main metal alloys used are typically based on combinations of noble metals (such as Pd [palladium], Pt, and Au [gold]) with other metals or metalloids, often in multicomponent or high-entropy forms. The most prominent alloys are high-entropy alloys (HEAs), Pd-based bimetallic and trimetallic alloys, and Pt-based alloys, often supported on carbon materials [17, 20–26].

The crucial role in improving the catalytic activity in alcohol oxidation reactions is mainly due to two pivotal effects: the bifunctional action and the electronic effect. The first one occurs whenever secondary metals, such as ruthenium [18–20], nickel [17], or gold [26], provide oxygenated species (mainly OH) at relative low potentials. These species help in the oxidation of intermediates that poison the catalyst surface, as is the case already discussed. In turn, the electronic effect results from the modification of the electronic structure of the main metal—such as platinum or palladium—due to the alloying with other metals, which weakens the adsorption of undesirable species and improves the overall efficiency of the reaction. Alloys such as PdCuBP (boron, phosphorus) and PdMB (M = Cu, Pt) further tune the electronic and bifunctional effects, improving the catalytic performance and the stability [24, 25].

The structural and chemical stability of a catalyst is also a key factor to consider when designing novel materials. Strategies such as Pt segregation to the surface—as observed in Pt–Ni alloys—help to protect non-noble metals from dissolution. Furthermore, ordered alloys, such as Pt–Ni intermetallic phases, show significantly higher resistance to degradation compared to disordered structures. These approaches contribute to greater durability of catalytic systems [17]. In addition to platinum-based alloys, recently less expensive alternatives have also been successfully explored. The Pd–Au alloy, for example, presents a catalytic activity approximately 2.3 times higher than that of carbon-supported Pd (Pd/C) in the oxidation of ethanol in alkaline media, being a promising solution to reduce costs [27]. Pd–Cu, Pd–Ag, and Pd–Sn alloys also showed improved performance compared to pure Pd, with Pd–Ag showing better stability and Pd–Cu showing moderate activity [26].

Alloys composed of five or more metals, such as the high-entropy compound FeCoNiAlMo alloy, demonstrated high electrocatalytic activity for benzyl alcohol (BA) oxidation. The high-entropy effect, lattice distortion, and synergistic interactions between multiple metals improved charge transfer and reaction kinetics [21]. PdPtCuAgAu nanowires exhibit high activity for ethanol oxidation and high stability and tolerance to CO_2, the high-entropy effect and structural advantages (e.g., anisotropic morphology) contribute to their performance. They outperformed simpler alloys due to synergistic effects and improved surface properties [28].

Table 4.1 presents the most common alloying elements investigated in the literature due to their distinct catalytic properties. The information presented is considered when used as an alloying element for Pt, in the context of its application in the electrooxidation of ethanol. This approach is justified by the diversity of compositions and combinations between the elements observed in the different studies.

Table 4.2 expands the studies with metal alloys to other molecules besides ethanol. In it, we present the conditions under which the data were obtained and establish the usage limit for each material.

Table 4.1 Some important metal alloys designed for ethanol electrooxidation and their corresponding properties.

Electrode	Electrolyte	Alloy features	Starting potential (EOR)	Ref.
Pt–Co	0.1 M NaOH + 1 M ethanol	Provides sites for OH^- adsorption at lower potentials, facilitating the oxidation of toxic intermediates (such as CO)	0.39 vs. RHE	[29]
Pt–Au	0.1 M $HClO_4$ + 1 M ethanol	Weakens the adsorption of toxic intermediates (such as CO) and increases the availability of active sites for oxidation	0.07 vs. RHE	[30]
Pt–Ru	0.5 M H_2SO_4 + 1 M ethanol	Forms oxygenated species (Ru—OH) at low potentials, which aid in the oxidation of toxic intermediates (such as CO)	~0.30 vs. RHE	[31]
Pt–Ni	0.1 M $HClO_4$ + 0.2 M ethanol	Forms adsorbed OH species (OH_{ads}), which are essential for the complete oxidation of intermediates such as CO to CO_2	~0.45 V vs. RHE	[32]
SnO_2@Pt	0.5 M H_2SO_4 + 1 M ethanol	Provides oxygenated species, facilitating the oxidation of toxic intermediates such as adsorbed CO	0.23 vs. RHE	[33]
Pt–WO_3	0.5 M H_2SO_4 + 1 M ethanol	Hydrogen spillover effect, acting as an intermediary in the anodic oxidation of hydrogen	0.3 vs. RHE	[34]

Table 4.2 Summary of metallic alloys currently used for alcohol electrooxidation and key parameters compared.

Catalyst	Electrolyte	Starting potential (vs. RHE)	Current density	Best for	Ref.
PdPtCuAgAu nanowires	1 M KOH	~0.444 V	7.7 A·mg_{Pd+Pt}^{-1}	Ethanol and methanol	[28]
FeCoNiAlMo/CNT	1 M KOH	~1.1 V	100 mA·cm^{-2}	BA	[21]
Pd–Cu nanosheets	1 M KOH	~0.566 V	2,518 mA·cm^{-2}	Ethanol	[26]
Pd–Ag nanosheets	1 M KOH	~0.566 V	2,226 mA·cm^{-2}	Ethanol	[26]
Pd–Sn nanosheets	1 M KOH	~0.566 V	1,598 mA·cm^{-2}	Ethanol	[26]
PdMoPt	1 M KOH	~–0.20 V	10 mA·cm^{-2}	Methanol	[22]
Pd–Ni/Ni_{RDE} (rotating disk electrode)	1 M NaOH	~0.02 V	793 mA·cm^{-2}	Glycerol	[23]
Pd_{10}–Au_{90}	0.5 M H_2SO_4	~0.1 V	1.45 mA·μg_{Pd+Au}^{-1}	Ethanol (alkaline)	[27]
Pt-Pd@TDI/rGO	0.1 M $HClO_4$	~0.58 V	~2,590 mA·mg_{Pt}^{-1}	Methanol	[35]
PdBP NWs@N–G	1 M KOH	~0.2 V	1.76 A·mg_{Pd}^{-1}	Ethanol	[24]

4.5 Metal Alloys for Hydrogen Evolution Reaction

Metals alloys also are a key focus in the search for efficient, stable, and cost-effective electrocatalysts for the hydrogen evolution reaction (HER), which is central to water electrolysis and green hydrogen production [36, 37]. Many noble metal catalysts have good HER performance in acidic environments. However, the pathways of HER are different in different pH environments. Under alkaline conditions, the first step of HER is to cleave the adsorbed H_2O to produce adsorbed H*, while this process does not happen in an acidic environment. Also, Pt has a poor ability to dissociate water. As a result, HER activity in alkaline media is about two to three orders of magnitude lower than in acidic media. Alloying Pt with transition metals (e.g., Pt–Fe, Pt–Co, and Pt–Ni) enhances water adsorption and dissociation, generating a proton-rich environment near the catalyst. This association overcomes the limitations of pure Pt in alkaline media, where water dissociation is rate-limiting, and improves the overall HER kinetics by facilitating both the Volmer (water dissociation) and Heyrovsky/Tafel (hydrogen formation) steps [36, 38–40].

While platinum remains the benchmark for HER activity, alloying strategies—especially those involving non-noble metals—are being developed to reduce costs and improve performance. In this context, Ni–Mo-based amorphous alloys and HEAs composed of transition metals (e.g., Ni, Fe, Mo, Co, and Cr [chromium]) offer high activity and stability, with some achieving performance comparable to platinum [41–48].

Transition-metal alloys, such as nickel–molybdenum [41, 42] and nickel–chromium [45], have demonstrated high catalytic activity for HER, especially under alkaline conditions. Their performance is attributed to favorable electronic structures and synergistic effects between constituent metals. For instance, Ni–Mo alloys exhibit enhanced HER activity due to the rougher electrode surfaces and porous structures, which optimizes hydrogen adsorption and desorption dynamics. Chen et al. [46] showed that the active sites for HER are primarily Mo atoms within the Co_3Mo alloy. The combination of Co and Mo provides abundant active sites and intrinsic catalytic activity, while the hierarchical structure (nanoparticles on nanosheets) increases the surface area and the accessibility for reactants [46].

Lin et al. [47] prepared ternary $Ir_{1-x}Rh_xSb$ intermetallic alloys and showed that the electron transfer from antimony (Sb) to iridium/rhodium (Ir/Rh) creates negatively charged, electron-rich Ir/Rh sites. These sites optimize hydrogen adsorption, facilitating the HER by making hydrogen adsorption and desorption more efficient.

Recently, Wu et al. reported that doping Cu with early transition metals (like Mn) modifies the electronic structure, particularly the d-orbitals, enhancing coupling with O-2p orbitals. This adjustment lowers the energy barrier for H_2O dissociation and optimizes the adsorption/desorption of H^+, improving HER activity [48].

HEAs, composed of multiple principal elements such as RhRuPtIrPd thin film on glass carbon catalyst, have emerged as promising candidates for HER electrocatalysis. Their unique atomic structures with subsurface amorphization (disordered atomic arrangement near the surface) provide a high density of active sites and enhanced HER activity by facilitating hydrogen adsorption and evolution in an acidic electrolyte, but the HER performance in alkaline solution was not as good due to sluggish reactions [49].

Despite the progress, challenges remain in optimizing the stability, activity, and scalability of metal alloy electrocatalysts for HER. Future research should focus on:

- developing synthesis methods for uniform and defect-free alloy nanostructures;
- understanding the mechanistic pathways of HER on alloy surfaces through advanced characterization techniques;
- exploring the synergistic effects in multicomponent alloys to tailor electronic properties for enhanced HER activity;
- integrating computational modeling to predict and design new alloy compositions with optimal HER performance.

4.6 Summary of the Current Landscape

The most prominent alloys are HEAs, Pd-based bimetallic and trimetallic alloys, and Pt-based alloys often supported on carbon materials. Bimetallic alloys present a trade-off between cost and performance. Palladium-based alloys stand out as having good activity, resistance to deactivation, and stability as a catalyst for alcohol oxidation [22–24, 27]. Pt–Ni alloys are stable and cost-effective alternatives to Pt–Ru (MOR) and Pt–Sn (EOR) alloys, with ternary Pt–M–Ni systems offering further improvements [17, 23]. However, there are still challenges in composition optimization; ternary alloys, such as PdMoPt, have a better synergistic effect and exhibit high resistance to CO poisoning and low cell voltage for hydrogen production [24]. HEAs have demonstrated high stability and activity; the high-entropy effect, lattice distortion, and synergistic interactions among the five metals contributed to the best catalytic performance, but at a high cost, despite the fact that alloys using non-noble metals provide a better cost–benefit ratio [21, 28]. Despite this, the study of high-entropy nanoalloy electrocatalysts is still in the early stages of catalytic applications; however, they are very promising and can be used in several reactions, such as, nitrogen reduction reaction (NRR), CO_2 reduction reaction (CO_2RR), OER, HER, oxygen reduction reaction (ORR), and alcohol oxidation reaction. Doping is another emerging area of study for HEAs and their derivatives which can increase the number of active sites in oxides, which when combined with theoretical calculations and experiments accelerates the design of high-performance electrocatalysts and provides valuable suggestions for the effective development of catalytic materials.

References

1 Scott, K. (2019). Introduction to electrolysis, electrolysers and hydrogen production. In *Electrochemical Methods for Hydrogen Production*, 1e (ed. K. Scott), 1–27. London: The Royal Society of Chemistry.

2 Brett, C. M. A. & Brett, A. M. O. (1993). *Electrochemistry: Principles, Methods, and Applications*. New York: Oxford University Press.

3 Saeedmanesh, A., Mac Kinnon, M. A., & Brouwer, J. (2018). Hydrogen is essential for sustainability. *Curr. Opin. Electrochem. 12*, 166–181. https://doi.org/10.1016/j.coelec.2018.11.009

4 Wang, H. Y., Sun, M. L., Ren, J. T., & Yuan, Z. Y. (2023). Circumventing challenges: Design of anodic electrocatalysts for hybrid water electrolysis systems. *Adv. Energy Mater. 13*(4), 2203568. https://doi.org/10.1002/aenm.202203568

5 Trinke, P., Bensmann, B., & Hanke-Rauschenbach, R. (2017). Current density effect on hydrogen permeation in PEM water electrolyzers. *Int. J. Hydrog. Energy 42*(21), 14355–14366. https://doi.org/10.1016/j.ijhydene.2017.03.231

6 Rausch, B., Symes, M. D., Chisholm, G., & Cronin, L. (2014). Decoupled catalytic hydrogen evolution from a molecular metal oxide redox mediator in water splitting. *Science 345*(6202), 1326–1330. https://doi.org/10.1126/science.1257443

7 Chen, L. & Shi, J. (2018). Chemical-assisted hydrogen electrocatalytic evolution reaction (CAHER). *J. Mater. Chem. A 6*(28), 13538–13548. https://doi.org/10.1039/c8ta03741h

8 Arshad, F., Haq, T. u., Hussain, I., & Sher, F. (2021). Recent advances in electrocatalysts toward alcohol-assisted, energy-saving hydrogen production. *ACS Appl. Energy Mater. 4*(9), 8685–8701. https://doi.org/10.1021/acsaem.1c01932

9 Koper, M. T. M., Lai, S. C. S., & Herrero, E. (2008). Mechanisms of the oxidation of carbon monoxide and small organic molecules at metal electrodes. In *Fuel Cell Catalysis: A Surface Science Approach* (eds. A. Wieckowski and M. T. M. Koper), 159–207. Hoboken, NJ: John Wiley & Sons.

10 Yaqoob, L., Noor, T., & Iqbal, N. (2021). A comprehensive and critical review of the recent progress in electrocatalysts for the ethanol oxidation reaction. *RSC Adv. 11*(27), 16768–16804. https://doi.org/10.1039/d1ra01841h

11 Zheng, Y., Wan, X., Cheng, X., Cheng, K., Liu, Z., & Dai, Z. (2020). Advanced catalytic materials for ethanol oxidation in direct ethanol fuel cells. *Catalysts 10*(2), 166. https://doi.org/10.3390/catal10020166

12 Koper, M. T. M. (2011). Structure sensitivity and nanoscale effects in electrocatalysis. *Nanoscale 3*(5), 2054–2073. https://doi.org/10.1039/c0nr00857e

13 Rizo, R., Ferre-Vilaplana, A., Herrero, E., & Feliu, J. M. (2023). Ethanol electro-oxidation reaction selectivity on platinum in aqueous media. *ACS Sustain. Chem. Eng. 11*(13), 4960–4968. https://doi.org/10.1021/acssuschemeng.2c02663

14 Carvalho, J. A., Gaiotti, A. C., Zanata, C. R., Fonseca, S., Giz, M. J., Pinto, L. M. C., & Camara, G. A. (2023). How does CO co-adsorption affect glycerol electro-oxidation on polycrystalline platinum? In situ FTIR and DFT approaches. *Electrochim. Acta 443*, 141978. https://doi.org/10.1016/j.electacta.2023.141978

15 Gilman, S. (1964). The mechanism of electrochemical oxidation of carbon monoxide and methanol on platinum. II. The "reactant-pair" mechanism for electrochemical oxidation of carbon monoxide and methanol. *J. Phys. Chem. 68*(1), 70–80. https://pubs.acs.org/sharingguidelines

16 Antolini, E. (2007). Catalysts for direct ethanol fuel cells. *J. Power Sources 170*(1), 1–12. https://doi.org/10.1016/j.jpowsour.2007.04.009

17 Antolini, E. (2017). Pt-Ni and Pt-M-Ni (M = Ru, Sn) anode catalysts for low-temperature acidic direct alcohol fuel cells: A review. *Energies 10*(1), 42. https://doi.org/10.3390/en10010042

18 Gasteiger, H. A., Marković, N., Ross Jr., P. N., & Cairns, E. J. (1994). Electro-oxidation of small organic molecules on well-characterized Pt-Ru alloys. *Electrochim. Acta 39*(11–12), 1825–1832. https://doi.org/10.1016/0013-4686(94)85171-9

19 Marković, N. M., Gasteiger, H. A., Ross Jr., P. N., Jiang, X., Villegas, I., & Weaver, M. J. (1995). Electro-oxidation mechanisms of methanol and formic acid on Pt-Ru alloy surfaces. *Electrochim. Acta 40*(1), 91–98. https://doi.org/10.1016/0013-4686(94)00241-R

20 Gojković, S. L., Vidaković, T. R., & Durović, D. R. (2003). Kinetic study of methanol oxidation on carbon-supported PtRu electrocatalyst. *Electrochim. Acta 48*(24), 3607–3614. https://doi.org/10.1016/S0013-4686(03)00481-X

21 Wang, Y., Ni, M., Yan, W., Zhu, C., Jiang, D., Yuan, Y., & Du, H. (2024). Supported high-entropy alloys for electrooxidation of benzyl alcohol assisted water electrolysis. *Adv. Funct. Mater. 34*(11), 2311611. https://doi.org/10.1002/adfm.202311611

22 Liu, J., Li, T., Wang, Q., Liu, H., Wu, J., Sui, Y., Li, H., Tang, P., & Wang, Y. (2024). Bifunctional PdMoPt trimetallene boosts alcohol-water electrolysis. *Chem. Sci. 15*(40), 16660–16668. https://doi.org/10.1039/d4sc04764h

23 White, J., Anil, A., Martín-Yerga, D., Salazar-Alvarez, G., Henriksson, G., & Cornell, A. (2022). Electrodeposited PdNi on a Ni rotating disk electrode highly active for glycerol electrooxidation in alkaline conditions. *Electrochim. Acta 403*, 139714. https://doi.org/10.1016/j.electacta.2021.139714

24 Lv, H., Sun, L., Xu, D., & Liu, B. (2020). Ternary metal-metalloid-nonmetal alloy nanowires: a novel electrocatalyst for highly efficient ethanol oxidation electrocatalysis. *Sci. Bull. 65*(21), 1823–1831. https://doi.org/10.1016/j.scib.2020.05.005

25 Wang, Y., Lv, H., Sun, L., Guo, X., Xu, D., & Liu, B. (2021). Ultrathin and wavy PdB alloy nanowires with controlled surface defects for enhanced ethanol oxidation electrocatalysis. *ACS Appl. Mater. Interfaces 13*(15), 17599–17607. https://doi.org/10.1021/acsami.1c02039

26 Yang, L., Li, Z., Chen, C., Wang, J., Yin, Q., Zhang, Y., & Guo, P. (2023). Assembly of alloyed PdM (Ag, Cu, and Sn) nanosheets and their electrocatalytic oxidation of ethanol and methanol. *Inorg. Chem. 62*(37), 15320–15328. https://doi.org/10.1021/acs.inorgchem.3c02558

27 Feng, Y. Y., Liu, Z. H., Xu, Y., Wang, P., Wang, W. H., & Kong, D. S. (2013). Highly active PdAu alloy catalysts for ethanol electro-oxidation. *J. Power Sources 232*, 99–105. https://doi.org/10.1016/j.jpowsour.2013.01.013

28 Fan, D., Guo, K., Zhang, Y., Hao, Q., Han, M., & Xu, D. (2022). Engineering high-entropy alloy nanowires network for alcohol electrooxidation. *J. Colloid Interface Sci. 625*, 1012–1021. https://doi.org/10.1016/j.jcis.2022.06.105

29 Santasalo-Aarnio, A., Sairanen, E., Arán-Ais, R. M., Figueiredo, M. C., Hua, J., Feliu, J. M., Lehtonen, J., Karinen, R., & Kallio, T. (2014). The activity of ALD-prepared PtCo catalysts for ethanol oxidation in alkaline media. *J. Catal. 309*, 38–48. https://doi.org/10.1016/j.jcat.2013.08.027

30 Zhou, W., Li, M., Zhang, L., & Chan, S. H. (2014). Supported PtAu catalysts with different nano-structures for ethanol electrooxidation. *Electrochim. Acta 123*, 233–239. https://doi.org/10.1016/j.electacta.2013.12.153

31 Spinacé, E. V., Neto, A. O., & Linardi, M. (2004). Electro-oxidation of methanol and ethanol using PtRu/C electrocatalysts prepared by spontaneous deposition of platinum on carbon-supported ruthenium nanoparticles. *J. Power Sources 129*(2), 121–126. https://doi.org/10.1016/j.jpowsour.2003.11.056

32 Sulaiman, J. E., Zhu, S., Xing, Z., Chang, Q., & Shao, M. (2017). Pt–Ni octahedra as electrocatalysts for the ethanol electro-oxidation reaction. *ACS Catal. 7*(8), 5134–5141. https://doi.org/10.1021/acscatal.7b01435

33 Silva, J. C. M., de Souza, R. F. B., Parreira, L. S., Neto, E. T., Calegaro, M. L., & Santos, M. C. (2010). Ethanol oxidation reactions using SnO_2@Pt/C as an electrocatalyst. *Appl. Catal. B: Environ. 99*(1–2), 265–271. https://doi.org/10.1016/j.apcatb.2010.06.031

34 Zhang, D. Y., Ma, Z. F., Wang, G., Konstantinov, K., Yuan, X., & Liu, H. K. (2006). Electro-oxidation of ethanol on Pt-WO_3/C electrocatalyst. *Electrochem. Solid-State Lett. 9*(9), A423. https://doi.org/10.1149/1.2211807

35 Tiwari, J. N., Lee, W. G., Sultan, S., Yousuf, M., Harzandi, A. M., Vij, V., & Kim, K. S. (2017). High-affinity-assisted nanoscale alloys as remarkable bifunctional catalyst for alcohol oxidation and oxygen reduction reactions. *ACS Nano 11*(8), 7729–7735. https://doi.org/10.1021/acsnano.7b01073

36 Cui, Z., Jiao, W., Huang, Z., Chen, G., Zhang, B., Han, Y., & Huang, W. (2023). Design and synthesis of noble metal-based alloy electrocatalysts and their application in hydrogen evolution reaction. *Small 19*, 2301465. https://doi.org/10.1002/smll.202301465

37 Chen, Z., Wei, W., Song, L., & Ni, B. J. (2022). Hybrid water electrolysis: A new sustainable avenue for energy-saving hydrogen production. *Sustain. Horiz. 1*, 100002. https://doi.org/10.1016/j.horiz.2021.100002

38 Yang, W., Cheng, P., Li, Z., Lin, Y., Li, M., Zi, J., Shi, H., Li, G., Lian, Z., & Li, H. (2022). Tuning the cobalt–platinum alloy regulating single-atom platinum for highly efficient hydrogen evolution reaction. *Adv. Funct. Mater. 32*, 2205920. https://doi.org/10.1002/adfm.202205920

39 Li, Q., Zhang, B., Sun, C., Sun, X., Li, Z., Du, Y., Liu, J. C., & Luo, F. (2024). Enhanced alkaline hydrogen evolution reaction via electronic structure regulation: Activating PtRh with rare earth Tm alloying. *Small 20*, 2400662. https://doi.org/10.1002/smll.202400662

40 Gao, G., Zhu, G., Chen, X., Sun, Z., & Cabot, A. (2023). Optimizing Pt-based alloy electrocatalysts for improved hydrogen evolution performance in alkaline electrolytes: A comprehensive review. *ACS Nano 17*, 20804–20824. https://doi.org/10.1021/acsnano.3c05810

41 Zhang, S., Huang, J., Ding, D., Tang, J., & Deng, X. (2024). Recent progress in NiMo-based amorphous alloys for electrocatalytic hydrogen evolution reaction. *Trans. Nonferr. Met. Soc. China 34*, 26–49. https://doi.org/10.1016/s1003-6326(23)66380-0

42 Raja, D., Cheng, C., Ting, Y., & Lu, S. (2023). NiMo-MOF-derived carbon-armored Ni_4Mo alloy of an interwoven nanosheet structure as an outstanding pH-universal catalyst for hydrogen evolution reaction at high current densities. *ACS Appl. Mater. Interfaces 15*, 20130–20140. https://doi.org/10.1021/acsami.3c01061

43 Xu, Y., Zhang, X., Liu, Y., Wang, R., Yang, Y., & Chen, J. (2023). A critical review of research progress for metal alloy materials in hydrogen evolution and oxygen evolution reaction. *Environ. Sci. Pollut. Res. 30*(5), 11302–11320. https://doi.org/10.1007/s11356-022-24728-5

44 Zhang, G., Ming, K., Kang, J., Huang, Q., Zhang, Z., Zheng, X., & Bi, X. (2018). High entropy alloy as a highly active and stable electrocatalyst for hydrogen evolution reaction. *Electrochim. Acta 279*, 19–23. https://doi.org/10.1016/j.electacta.2018.05.035

45 Nady, H., El-Rabiei, M. M., Samy, M., Deyab, M. A., & Abd El-Hafez, G. M. (2021). Novel Ni-Cr-based alloys as hydrogen fuel sources through alkaline water electrolytes. *Int. J. Hydrog. Energy 46*, 34749–34766. https://doi.org/10.1016/j.ijhydene.2021.08.056

46 Chen, J., Ge, Y., Feng, Q., Zhuang, P., Chu, H., Cao, Y., Smith, W. R., Dong, P., Ye, M., & Shen, J. (2019). Nesting Co_3Mo binary alloy nanoparticles onto molybdenum oxide nanosheet arrays for superior hydrogen evolution

reaction. *ACS Appl. Mater. Interfaces 11*, 9002–9010. https://doi.org/10.1021/acsami.8b19148

47 Lin, Z., Xiao, B., Huang, M., Yan, L., Wang, Z., Huang, Y., Shen, S., Zhang, Q., Gu, L., & Zhong, W. (2022). Realizing negatively charged metal atoms through controllable d-electron transfer in ternary $Ir_{1-x}Rh_xSb$ intermetallic alloy for hydrogen evolution reaction. *Adv. Energy Mater. 12*, 2200855. https://doi.org/10.1002/aenm.202200855

48 Wu, R., Liu, X., Li, Z., Li, S., Zhu, H., Zhong, H., Wang, H., Wu, Y., Jiang, S., Zhang, X., & Wang, X. (2025). Electronic mechanism for the enhanced hydrogen evolution reaction activity of Cu alloys. *Acta Mater. 284*, 120644. https://doi.org/10.1016/j.actamat.2024.120644

49 Zou, Y., Jing, L., Zhang, J., Luo, S., Wang, L., Li, Y., Goei, R., Tan, K. W., & Tok, A. I. Y. (2024). ALD-made noble metal high entropy alloy nanofilm with sub-surface amorphization for enhanced hydrogen evolution. *J. Mater. Chem. A 12*, 5668–5678. https://doi.org/10.1039/d3ta05908a

5

Metal Borides and Selenides for Alcohol-assisted Water Electrolysis

Yogapriya Selvaraj[1], *Senthilkumar Nangan*[2], *Kiruthika Paramasivam*[3], *Anbazhagan Venkattappan*[2], *Vijayakumar Elayappan*[4], *and Soorathep Kheawhom*[1,5,6,*]

[1]*Department of Chemical Engineering, Chulalongkorn University, Bangkok, Thailand*
[2]*Centre for Research and Development, Department of Chemistry, Vinayaka Mission's Kirupananda Variyar Arts & Science College, Vinayaka Mission's Research Foundation, Salem, India*
[3]*Materials Research and Product Laboratory, Department of Chemistry, Coimbatore Institute of Technology, Coimbatore, India*
[4]*Advanced Energy Materials R&D Division, Dongkwang Co. Ltd., Anyang-si, Republic of Korea*
[5]*Department of Chemical Technology, Chulalongkorn University, Bangkok, Thailand*
[6]*Center of Excellence on Advanced Materials for Energy Storage, Chulalongkorn University, Bangkok, Thailand*

*Corresponding authors: soorathep.k@chula.ac.th

5.1 Introduction

Addressing global energy challenges demands a decisive shift toward environmentally friendly technologies that can support a sustainable and low-carbon future. Extensive research efforts have been dedicated to harnessing renewable energy sources such as hydro, biobased, wind, solar, and geothermal power in order to reduce the world's reliance on fossil fuels. Despite their significant potential, the widespread integration of these renewables is often constrained by high capital costs and relatively expensive electricity generation. Moreover, the intermittent and weather-dependent characteristics of these energy sources highlight the pressing need for energy storage systems that are efficient, scalable, and economically viable to guarantee a stable and uninterrupted power supply [1, 2].

Hydrogen is gaining recognition as a promising clean energy carrier, thanks to its high energy density and its capacity to efficiently store and convert renewable energy. It possesses an energy content substantially higher than conventional fuels like gasoline and methane. Currently, approximately 65 million metric tons of hydrogen are produced globally each year, with an annual growth rate of around 6%. Nevertheless, over 90% of hydrogen is currently produced through fossil-fuel-based methods such as natural gas reforming and coal gasification. While these methods are well established and capable of large-scale production, they are also significant sources of carbon dioxide (CO_2) emissions. Presently, hydrogen is predominantly used in ammonia synthesis, petroleum refining, and, to a lesser extent, in the chemical, pharmaceutical, and aerospace industries. It also holds substantial promises for fuel-cell technologies in transportation, power generation, and portable applications, offering high energy efficiency and low emissions. Despite its potential, only a small fraction of hydrogen is currently produced through water electrolysis, a cleaner, albeit more energy-intensive and costlier, method, highlighting the need for more sustainable and economically viable hydrogen production pathways [3–8].

Enhancing the hydrogen energy cycle is essential to fully unlock hydrogen's potential as a clean and sustainable energy vector. A thorough enhancement of the entire hydrogen value chain from production and storage to transportation and utilization is essential to address current technical and economic challenges. Such advancements will play a pivotal role in enabling hydrogen's large-scale deployment across diverse sectors, including industry, transportation, and power. This transition is integral to achieving long-term carbon neutrality goals and building a resilient, low-carbon energy infrastructure.

Water electrolysis stands out as a clean and efficient technology for producing high-purity hydrogen by splitting water into hydrogen and oxygen. One of its most valuable features is its ability to directly integrate with renewable energy sources, offering a pathway to decarbonized hydrogen generation and flexible energy storage. Nonetheless, due to its relatively high energy demands, electrolysis currently accounts for only about 4% of global hydrogen production. As of 2015, the cost of producing hydrogen via distributed electrolysis facilities was estimated at $3.90 per kilogram, with cost-reduction targets aiming for $2.30 per kilogram by 2020. The associated electricity consumption was approximately 57–60 $kWh{\cdot}kg^{-1}$ of hydrogen, with ongoing technological improvements aiming to reduce this to 50 $kWh{\cdot}kg^{-1}$ by 2023 [9–12].

To lower the energy requirements of water electrolysis, researchers have explored the use of alternative feedstocks such as alcohols, particularly bio-derived alcohols, for assisted hydrogen production. Alcohol-assisted electrolysis offers potential energy savings and environmental benefits. Methanol has been widely studied in this context; however, its high toxicity poses serious health and environmental hazards. Consequently, ethanol has gained attention as a safer and more practical alternative. In addition to being nontoxic, ethanol possesses a higher lower heating value (LHV) of 29 $MJ{\cdot}kg^{-1}$ compared to methanol's 19.7 $MJ{\cdot}kg^{-1}$, making it more energy-dense and environmentally favorable for hydrogen generation in both transportation and stationary energy systems [13–16].

In recent years, significant research has focused on developing advanced catalysts to enhance the efficiency of alcohol-assisted water electrolysis. The primary objectives include reducing energy consumption, improving catalytic kinetics, and ensuring long-term system stability. Transition-metal phosphides such as Ni_2P, CoP, and FeP are known for their bifunctional catalytic activity and favorable intermediate binding energies. Similarly, metal carbides and nitrides like Mo_2C, WC, MoN, and TiN exhibit excellent conductivity, platinum (Pt)-like behavior, and electrochemical robustness. Additionally, transition-metal oxides and hydroxides, particularly mixed oxides (e.g., $NiCo_2O_4$, $FeCo_2O_4$) and layered double hydroxides (LDHs), have demonstrated high catalytic activity and large electrochemically active surface areas.

Although noble metals like Pt, Pd, and Ru remain performance benchmarks, their cost and scarcity necessitate the development of alternative materials. The rational design of nanostructured, doped, or composite catalysts supported on conductive carbon materials has become a key focus in making alcohol-assisted water electrolysis commercially viable [17–22].

Among the diverse materials studied, metal borides (e.g., NiB [nickel boron], CoB [cobalt boride], MoB, and FeB) and metal selenides (e.g., NiSe, CoSe, $MoSe_2$, and FeSe) have emerged as particularly promising catalyst candidates for alcohol-assisted water splitting. These materials offer a unique combination of physicochemical properties, including high electrical conductivity, tunable electronic structures, and strong chemical stability. Their ability to reduce the overpotentials required for both alcohol oxidation reactions at the anode and hydrogen evolution reaction (HER) at the cathode has been well documented. Furthermore, their long-term operational durability makes them attractive for practical deployment. This chapter aims to provide a comprehensive overview of recent advancements in the synthesis, structural design, and electrochemical performance of metal borides and selenides, with a focus on their role in enabling scalable, energy-efficient, and sustainable hydrogen production through alcohol-assisted electrolysis.

5.2 Electrochemical Reactions Involved in OER, HER, and Alcohol Oxidation Reaction at Electrode Material

5.2.1 Mechanism of OER

Oxygen evolution reaction (OER) plays a vital role as a half-cell process in water splitting and other electrochemical energy conversion systems. In contrast to the HER, OER follows a more intricate mechanism involving the transfer of four electrons, resulting in sluggish kinetics and higher energy demands. Although OER can take place in both acidic and alkaline environments, it is generally more efficient in alkaline media due to enhanced catalyst stability and reduced corrosion.

In alkaline environments, the OER mechanism typically initiates with the adsorption of hydroxide ions (OH^-) onto the electrochemically active sites of the catalyst surface, forming M—OH species. This is followed by a sequence of proton-coupled electron transfer (PCET) steps, resulting in the formation of surface-bound intermediates such as M=O and M—OOH. These species play a crucial role in the pathway toward molecular oxygen (O_2) generation. Two primary routes are generally proposed: (i) the direct coupling of two M=O species to release O_2, and (ii) the stepwise conversion of M—OH to M=O, then M—OOH, and finally O_2. The complete transformation restores the original active site, allowing for continued catalytic cycling. To drive OER, an overpotential above 1.23 V vs. RHE is needed. Differences in intermediate binding energies (e.g., M—OH, M=O, and M—OOH) between ideal and real catalysts affect the potential-limiting step and overall efficiency. Thus, understanding this energetics is key to developing efficient electrocatalysts. If kinetic limitations are negligible, an ideal catalyst facilitates OER without requiring any overpotential. However, for real catalysts, a higher potential (E_3) is often needed to drive all reaction steps energetically downhill. This is due to variations in the adsorption energies of intermediates—strong O_2 binding hampers M—OOH formation, while weak oxygen binding makes M—O formation difficult [23–29].

The OER half-reactions are:

Acidic medium: $2H_2O \rightarrow O_2 + 4H^+ + 4e^-$

Alkaline medium: $4OH^- \rightarrow O_2 + 2H_2O + 4e^-$

The stepwise OER mechanism in alkaline media includes:

$M + OH^- \rightarrow M{-}OH + e^-$

$M{-}OH + OH^- \rightarrow M{-}O + H_2O$

$2M{-}O \rightarrow 2M + O_2$

$M{-}O + OH^- \rightarrow M{-}OOH + e^-$

$M{-}OOH + OH^- \rightarrow M + O_2 + H_2O + e^-$

5.2.2 Mechanism of HER

The HER is an important electrochemical process that occurs at the cathode during water splitting, resulting in the production of hydrogen gas (H_2). HER typically involves a two-step mechanism, beginning with the Volmer step, in which water molecules are reduced at the electrode surface. In this step, a water molecule (H_2O) engages with the active metal site (M) on the catalyst, leading to a one-electron transfer that results in the adsorption of a hydrogen atom onto the surface, forming a metal–hydrogen intermediate (M—H*), while simultaneously releasing a hydroxide ion (OH^-) into the solution. Following this initial adsorption, hydrogen gas is generated through one of two pathways. The first is the Heyrovsky step, where the adsorbed hydrogen species (M—H*) reacts with another water molecule and an additional electron from the electrode. This reaction forms molecular hydrogen (H_2), regenerates the active metal site, and produces another hydroxide ion. Alternatively, hydrogen can be produced via the Tafel step, a recombination mechanism in which two adjacent adsorbed hydrogen atoms (M—H*) combine directly on the electrode surface to form hydrogen gas (H_2), releasing the catalyst sites for further reactions. These sequential steps, represented by the Volmer–Heyrovsky or Volmer–Tafel pathways, highlight the essential role of catalyst surface properties in determining the efficiency and rate of HER, particularly under alkaline conditions where water, rather than protons, serves as the hydrogen source [23, 29–32].

HER reaction steps in alkaline medium:

Volmer: $M + H_2O + e^- \rightarrow M{-}H^* + OH^-$

Heyrovsky: $M{-}H^* + H_2O + e^- \rightarrow M + H_2 + OH^-$

Tafel: $2M{-}H^* \rightarrow 2M + H_2$

5.2.3 Mechanism of Alcohol Oxidation Reaction

Scheme 5.1 illustrates the alcohol-assisted water electrolysis process. In this process, alcohol molecules (ROH) undergo electrooxidation at the anode, generating value-added carbonyl compounds and releasing protons/electrons with significantly lower anodic potential compared to the OER. In alcohol-assisted water electrolysis, hydrogen is generated via the HER at the cathode and the alcohol oxidation reaction at the anode. By substituting the sluggish, energy-demanding OER with the more thermodynamically favorable alcohol oxidation reaction, this method notably lowers the overall cell voltage and energy consumption. At the cathode, HER proceeds via proton (acidic media) or water (alkaline/neutral media) reduction through sequential steps—Volmer, followed by either Heyrovsky or Tafel—to form molecular hydrogen. The efficiency of HER is governed by the catalyst's ability to bind and release hydrogen intermediates (H*) effectively, ensuring minimal energy loss.

Simultaneously, the alcohol oxidation reaction involves the oxidation of small alcohols such as methanol, ethanol, or glycerol, which act as electron donors. Unlike the four-electron OER process, alcohol oxidation reaction proceeds via fewer electron transfers and occurs at lower overpotentials. It also enables the coproduction of value-added chemicals like formate or acetate, depending on the substrate and catalyst. However, the alcohol oxidation reaction mechanism includes complex PCET steps and the cleavage of C—C, C—H, and O—H bonds, necessitating catalysts with high activity, selectivity, and durability. The combined HER– alcohol oxidation reaction system not only improves energy efficiency but also offers a sustainable route for hydrogen generation using renewable alcohols, aligning with the goals of

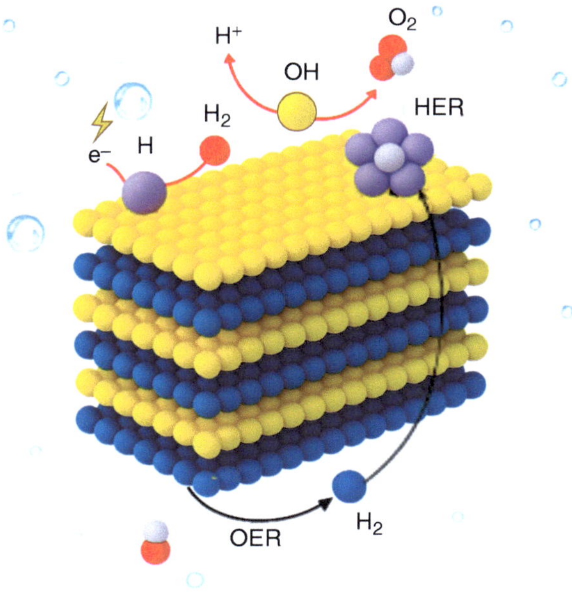

Scheme 5.1 Schematic diagram of alcohol-assisted electrolysis system.

a low-carbon, circular energy economy. Continued advancements in catalyst engineering and system integration are crucial to realize the full potential of this hybrid electrolysis strategy [23, 33–38].

The typical pathway for direct anodic oxidation involves the following mechanism:

$$MO_x + H_2O \rightarrow MO_x(^*OH) + H^+ + e^-$$
$$R + MO_x(^*OH) \rightarrow \text{Inorganic ions} + CO_2 + MO_x$$
$$MO_x(^*OH) \rightarrow MO_{(x+1)} + H^+ + e^-$$
$$MO_{(x+1)} + R \rightarrow MO_x + RO$$
$$MO_x(^*OH) \rightarrow MO_x + ½O_2 + H^{+}_{+} + e^-$$
$$MO_{(x+1)} \rightarrow ½O_2 + MO_x$$

5.3 Electrocatalytic Performance in Alcohol-assisted Electrolysis in Metal Borides and Selenides

5.3.1 Significance of HER

The cornerstone of the hydrogen economy is water electrolysis, which establishes itself as a green replacement for fossil fuels by providing a sustainable and clean method of producing hydrogen and oxygen through the electrochemical splitting of water [39]. Electrolysis contains two key half-reactions: HER at the cathode, and OER at the anode. In HER, hydrogen gas is generated by the reduction of protons in acidic media or water molecules in alkaline media. Meanwhile, at the anode, the OER oxidizes water molecules to produce oxygen gas [40]. HER plays a central role in converting electrical energy into storable chemical energy. As a core step in sustainable hydrogen production, its performance directly affects energy efficiency. The overall feasibility of green hydrogen relies heavily on the effectiveness of the HER process [41].

The development of advanced electrocatalysts is crucial to reducing energy barriers and quickening HER kinetics. Because of its high turnover frequency and ideal hydrogen adsorption energy, platinum continues to be the benchmark for HER; however, its high cost and limited supply prevent widespread use [42]. Despite Pt's exceptional catalytic performance in both acidic and alkaline environments, these drawbacks have led to a great deal of research into earth-abundant alternative [43]. Transition-metal-based compounds, including carbides, borides, selenides, sulfides, and phosphides, have become attractive options because they provide a balance between stability, catalytic efficiency, and economic feasibility. These substances seek to support the sustainability and scalability required for large-scale hydrogen production while achieving activity that is comparable to that of Pt.

Transition-metal borides and selenides have drawn a lot of interest as promising HER electrocatalysts among the different non-precious substitutes. Because of their special bonding structures that promote quick electron transfer and effective hydrogen adsorption, metal borides like NiB, CoB, and MoB have exceptional electrical conductivity, chemical stability, and catalytic activity [44]. Likewise, transition-metal selenides such as $MoSe_2$, $NiSe_2$, and CoSe have favorable HER kinetics, a large number of active sites, and tunable electronic characteristics. By using techniques like doping, interface engineering, and heterostructure formation, their catalytic performance can be further improved, allowing for effective operation in both acidic and alkaline environments [41, 45]. Recent developments have demonstrated that selenides can attain HER activities similar to those of noble metals when optimized structurally and compositionally, for example, by adding high-valent elements or creating hybrid interfaces. These developments are part of a growing movement to create scalable, cost-effective catalysts that can drive the world's switch to clean hydrogen energy. Researchers are creating next-generation electrolyzers that combine high performance and economic viability by utilizing the inherent benefits of borides and selenides [41].

5.3.2 HER Activity of Borides and Selenides

5.3.2.1 HER Activity of Borides

Transition-metal borides (TMBs) have attracted growing interest as effective and durable electrocatalysts for the HER, owing to their inherent metallic conductivity, excellent chemical stability, and advantageous hydrogen adsorption properties. The strong metal–boron (M—B) bonds in borides provide structural rigidity and promote efficient electron transfer, which are critical for fast HER kinetics across a wide pH range [45]. Nickel boride (Ni_2B) is widely recognized for its high electrical conductivity and excellent chemical stability, which make it a robust candidate for HER applications. Yuan et al. introduced nickel boride of three surface-oxidized electrocatalysts: Ni–B–O@Ni_3B, Ni–B–O@NiB_2, and Ni–B–O@Ni_2B. The naturally formed surface oxide layers significantly improved OER activity, although HER performance was slightly diminished. Boron plays a crucial role by promoting nickel oxidation and aiding the generation of active intermediates for OER. Among the three catalysts, Ni–B–O@Ni_3B demonstrated the highest overall performance. For overall water splitting, it achieved a current density of 10 $mA{\cdot}cm^{-2}$ at 1.58 V when employed as both the anode and cathode in an alkaline electrolyzer. It also exhibited excellent durability, comparable to that of noble-metal-based electrodes [46]. The synthesis of nickel boron nanoparticles supported on Vulcan carbon, as reported by Huang et al., was achieved through a simple and scalable two-step process involving chemical reduction and subsequent annealing. The optimized NiB-400 catalyst showed excellent HER performance in 1 M KOH, with low overpotentials (114 mV at 10 $mA{\cdot}cm^{-2}$ and 215 mV at 40 $mA{\cdot}cm^{-2}$ [Figure 5.1]) and stable operation at 7.6 $mA{\cdot}cm^{-2}$ for 20 hours. Its high activity is attributed to small particle size and an expanded Ni lattice, which enhance hydrogen adsorption and electrocatalytic efficiency [47].

Shi et al. reported that cobalt boride, typically in an amorphous form, offers a high surface area and numerous active sites due to its structural disorder. The strong Co–B interactions help stabilize HER intermediates and facilitate efficient electron transport. A notable example is the crystalline/amorphous CoP@CoB nanorod structure,

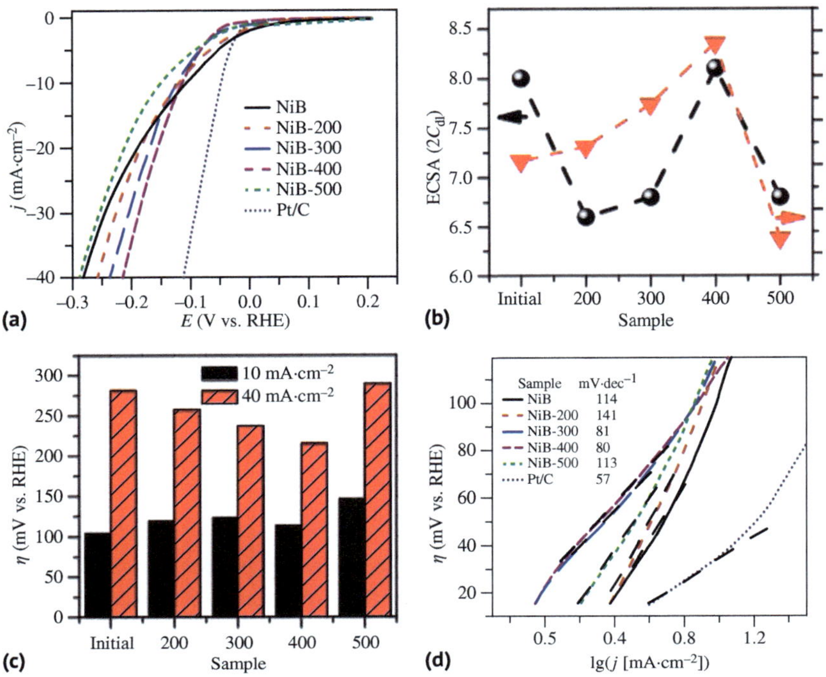

Figure 5.1 (a) Polarization curves of NiB series and Pt/C; (b) ESCA ($2C_{dl}$) and current density at 200 mV vs. Ni loading; (c) overpotentials at 10 and 40 mA·cm^{-2}; and (d) corresponding Tafel plots. *Source*: [47]/with permission of Elsevier.

which exhibits impressive HER performance with low overpotentials of 56.3 mV in alkaline and 81.2 mV in acidic media at a current density of 10 mA·cm^{-2}, along with excellent long-term [48]. Guang et al. developed a self-supporting nanoporous CoBP electrocatalyst via dealloying, highlighting the synergistic role of Co, B, and P in enhancing HER activity by improving hydrogen desorption. Its bi-continuous structure boosts active surface area and transport efficiency. The optimized CoBP (B/P = 1:3) showed a low overpotential of 42 mV at 10 mA·cm^{-2}, a Tafel slope of 39.8 mV·dec^{-1}, and stable performance over 20 hours in alkaline media (Figure 5.2) [49].

5.3.2.2 HER Activity of Selenides

Transition-metal selenides have emerged as promising HER electrocatalysts due to their tunable electronic structures, excellent chemical stability, and cost-effectiveness compared to noble metals like platinum. Transition-metal dichalcogenides (TMDs), such as $MoSe_2$ and WSe_2 (tungsten selenide), feature graphene-like layered structures, where metal atoms are sandwiched between selenium atoms in trigonal prismatic or octahedral coordination [50]. Other selenides, including NiSe and $CoSe_2$, display various crystal structures such as cubic or orthorhombic and exhibit metallic or semimetallic properties that enhance their catalytic behavior [51]. These materials demonstrate low overpotentials, high durability in both alkaline and

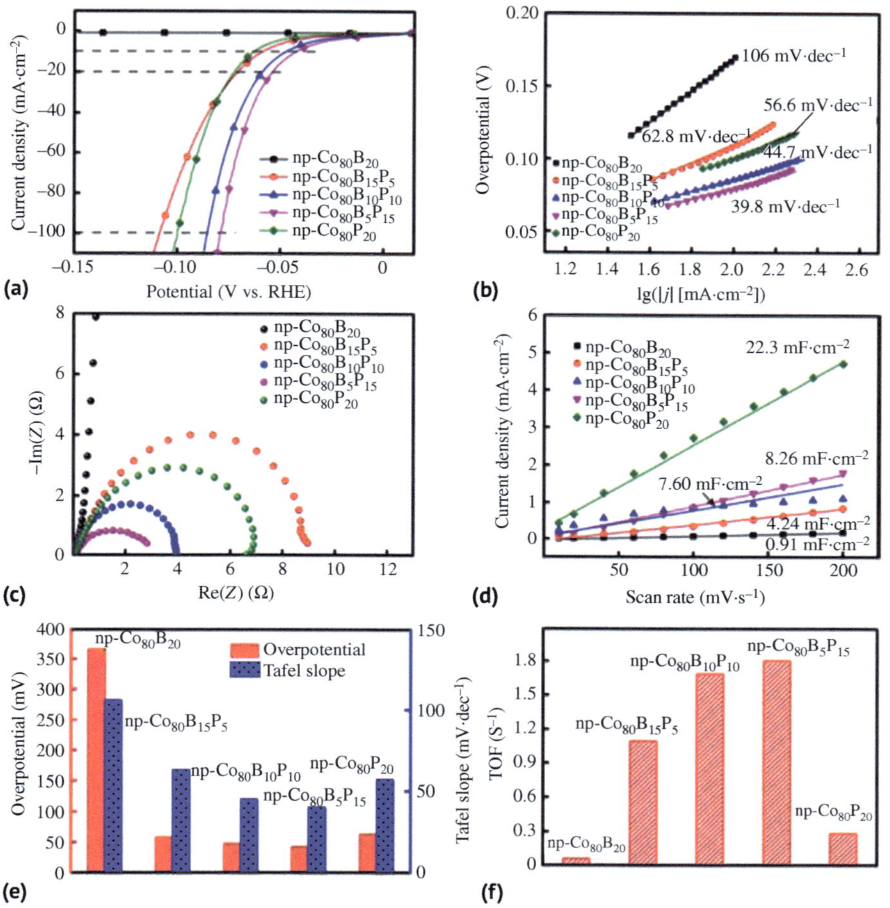

Figure 5.2 HER performance of np-CoBP catalysts in 1 M KOH: (a) LSV curves, (b) Tafel plots, (c) EIS spectra, (d) double-layer capacitance, (e) overpotentials at 10 mA·cm^{-2} and Tafel slopes, and (f) TOF values. *Source*: [49]/with permission of John Wiley & Sons.

acidic electrolytes, and offer the flexibility to tailor active sites for improved HER performance [52].

TMDs like $MoSe_2$ and WSe_2 have a layered structure with metal atoms between chalcogen layers, held by weak van der Waals forces, enabling easy exfoliation into high-surface-area 2D nanosheets [53]. For HER, the catalytic efficiency of TMDs mainly depends on edge-located active sites, as the basal planes are largely inactive [53]. The basal planes in the 2H-phase TMDs are typically catalytically inert due to their high hydrogen adsorption free energy ($\Delta G_H \approx$ 2 eV), while edge sites—particularly those of metal and chalcogen atoms—exhibit adsorption energies closer to thermoneutral values ($\Delta G_H \approx 0$ eV), making them catalytically favorable [53].

In contrast to MoS_2, where the molybdenum edges are primarily catalytically active, $MoSe_2$ exhibits improved performance because both molybdenum and selenium edges contribute to HER activity [54]. The HER performance of WSe_2 is limited in comparison to $MoSe_2$ because it primarily uses its selenium edge, which binds hydrogen less strongly. These materials' naturally layered structure also permits different structural alterations, like defect engineering and heterostructure formation, which can activate the otherwise inert basal planes and enhance overall catalytic performance [55–57].

MoS_2 is a promising Pt-free HER catalyst, but its activity is limited by few active sites. This study enhances performance by activating basal planes through a hydrothermal method and low-temperature H_2O_2 etching, introducing oxygen doping and sulfur vacancies (dual defects). These modified MoS_2 nanosheets showed significantly improved HER performance, requiring only 143 mV of overpotential to reach 10 $mA{\cdot}cm^{-2}$ in 0.5 M H_2SO_4 [58]. Tungsten selenide, a 2D semiconductor, shows great promise for HER. In this study, few-layered, graphene-like WSe_2 nanosheets were synthesized via a simple one-pot solvothermal method. The catalyst exhibited a low onset potential (~150 mV), a Tafel slope of 78 $mV{\cdot}dec^{-1}$, and excellent stability. Its high activity stems from a porous structure and abundant exposed edge sites [59]. Similarly, in the case of WSe_2, nickel doping has been found to significantly improve HER activity in acidic media. The introduction of Ni modifies the electronic structure, enhancing electrical conductivity by promoting effective hybridization between Ni-3d and Se-4p orbitals, which accelerates charge-transfer kinetics [60].

5.3.3 Synergistic Effects in Bimetallic or Hybrid Systems

Bimetallic and hybrid systems combining borides or selenides with other materials have attracted interest for HER due to synergistic effects that boost active sites, conductivity, and performance. Mu et al. developed an in situ-grown $NiFeSe_4/NiSe_2$ catalyst, achieving 10 $mA{\cdot}cm^{-2}$ at a low overpotential of 121 mV. It also demonstrated remarkable stability, operating continuously for 36 hours across varying current densities without any noticeable performance degradation, confirming its high durability for HER applications [61]. A 3D hollow $CoSe_2$@$MoSe_2$ core–shell structure showed excellent HER activity, with a low overpotential of 108 mV at 10 $mA{\cdot}cm^{-2}$ in 1 M KOH. Its performance stems from abundant interfaces and optimized Gibbs free energies for water splitting and hydrogen adsorption [62].

The NiCoS/CNTs catalyst showed strong HER performance, reaching 100 $mA{\cdot}cm^{-2}$ at 327 mV with a Tafel slope of 64 $mV{\cdot}dec^{-1}$. EIS revealed a low charge-transfer resistance (1.94 Ω), indicating efficient electron transport. The enhanced activity is due to the synergy between NiCoS and CNTs, boosting conductivity and active-site density [63]. Additionally, an aerogel with a Mo-doped $NiSe_2$–$CoSe_2$ heterostructure showed similar HER activity, with an overpotential of 57 mV at 10 $mA{\cdot}cm^{-2}$. The improved water adsorption qualities and increased electropositivity brought about by Mo incorporation are responsible for the improved performance [64].

The Co_9S_8/MoS_2 hybrid embedded in N, S, and O-doped carbon (NSOC) shows excellent HER performance, with overpotentials of 194 mV (1 M KOH) and 233 mV (0.5 M H_2SO_4) at 10 $mA{\cdot}cm^{-2}$. Its activity is boosted by the synergy between MoS_2 and Co_9S_8 and the doped porous carbon matrix, enhancing conductivity and active-site exposure [65]. similarly, a hydrothermally synthesized $NiCo_2Se_4/NiTe_2$ nanorod array achieves 163 mV at 50 $mA{\cdot}cm^{-2}$ due to efficient charge and mass transfer at heterointerfaces [66].

5.4 Tafel Slope, Onset Potential, and Overpotential

5.4.1 Tafel Slope

The HER on Pt surfaces remains one of the most extensively studied electrochemical processes, yet its pH-dependent kinetics and detailed mechanism are still not fully understood. The HER typically involves three fundamental steps:

1. Volmer step—electrochemical adsorption of hydrogen:

   Acidic medium: $H^+ + e^- + * \rightarrow H^*$
   Alkaline medium: $H_2O + e^- \rightarrow H^* + OH^-$

2. Tafel step—chemical combination of two adsorbed hydrogen atoms:

   $H^* + H^* \rightarrow H_2$

3. Heyrovsky step—electrochemical desorption of hydrogen:

   Acidic medium: $H^* + H^+ + e^- \rightarrow H_2$
   Alkaline medium: $H^* + H_2O + e^- \rightarrow H_2 + OH^-$

Tafel analysis is used to study HER kinetics, with the slope indicating the voltage needed for a 10-fold increase in current. It reveals the reaction mechanism and rate-determining step: ~120 $mV{\cdot}dec^{-1}$ for the Volmer step, ~40 $mV{\cdot}dec^{-1}$ for the Heyrovsky step, and ~30 $mV{\cdot}dec^{-1}$ for the Tafel step [67].

Wu et al. developed a strategy to boost HER performance by tuning surface electron density. Nitrogen-doped $NiCo_2S_4$ nanowires showed excellent activity with a 41 mV overpotential at 10 $mA{\cdot}cm^{-2}$ and a 37 $mV{\cdot}dec^{-1}$ Tafel slope in alkaline media, rivaling Pt/C. Nitrogen doping enhances the electronic structure and active-site density, supporting a Volmer–Heyrovsky mechanism [68]. In alkaline media, $Co_{1-x}Fe_xB$ catalysts show high Tafel slopes (105–177 $mV{\cdot}dec^{-1}$), indicating the Volmer step is rate-limiting [69]. In contrast, amorphous CoB and CoWB nanoparticles exhibit lower slopes (71 and 69 $mV{\cdot}dec^{-1}$), suggesting a Volmer–Heyrovsky mechanism driven by efficient hydrogen adsorption and charge transfer [70]. The general schematic mechanism of HER is shown in Figure 5.3. These examples demonstrate that bimetallic or hybrid systems often yield lower Tafel slopes than single-component catalysts, approaching the performance of platinum (~30 $mV{\cdot}dec^{-1}$).

5.4.2 Onset Potential

The onset potential in HER refers to the potential at which a noticeable increase in cathodic current is first observed, signaling the initiation of hydrogen gas generation. A lower (less negative) onset potential indicates a more efficient catalyst,

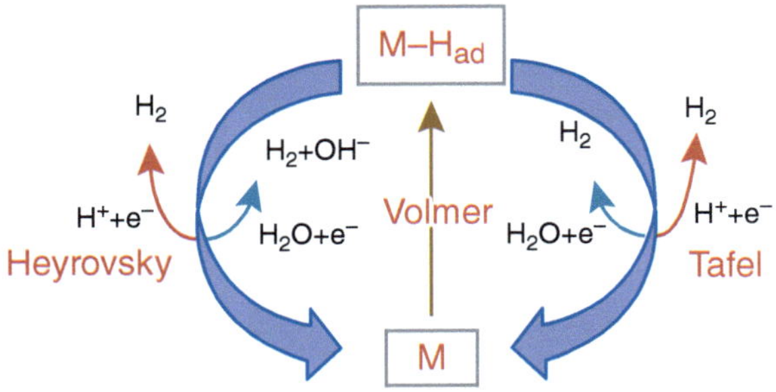

Figure 5.3 Schematic of HER. *Source*: [41]/with permission of Springer Nature.

as it means the reaction starts at a lower energy input highlighting the material's ability to drive HER more easily [71]. For example, a Mo-doped $NiSe_2$–$CoSe_2$ heterostructure aerogel exhibited an onset potential close to 0 V vs. RHE in acidic media, attributed to Mo doping enhancing electropositivity and facilitating water molecule adsorption [64]. A low-cost solution-based method was used to synthesize vertically aligned $MoSe_2$ thin films, enhancing charge transport and exposing active edge sites for efficient HER. The 15-minute-grown film showed the lowest onset overpotential of −88 mV at 10 mA·cm^{-2} [72]. A unique 1T/2H $MoSe_2$ structure was synthesized via a simple hydrothermal method and grown on carbon cloth as binder-free HER electrodes. Cu-doped $MoSe_2$ showed an onset potential of −113 mV and a Tafel slope of 87.2 mV·dec^{-1}, with enhanced activity due to copper doping, phase transformation, and improved charge transfer [73]. These low onset potentials in hybrid systems highlight the role of heterointerfaces and bimetallic synergism in lowering the energy barrier for HER initiation compared to single-component selenides or borides (e.g., $MoSe_2$ with onset potentials of ~100–150 mV).

5.4.3 Overpotential

In the HER, overpotential (η) refers to the extra voltage that must be applied beyond the theoretical (equilibrium) potential to initiate and sustain the reaction at a practical rate. It arises due to kinetic barriers such as slow reaction kinetics, charge-transfer resistance, and mass-transport limitations. A lower overpotential indicates a more efficient electrocatalyst, as it requires less energy to attain a given current density commonly evaluated at 10 mA·cm^{-2}, which approximates real-world operating conditions [74, 75]. For instance, a $NiCo_2Se_4$/$NiTe_2$ nanorod array attained an overpotential of 163 mV at 50 mA·cm^{-2} in acidic media, driven by enhanced electron-mass transfer at heterointerfaces and bimetallic synergism [66]. In another study, a MoS_2/Co_9S_8@NSOC hybrid exhibited an overpotential of 194 mV at 10 mA·cm^{-2} in 0.5 M H_2SO_4, benefiting from synergistic interactions between Co_9S_8 and MoS_2 within a conductive carbon matrix [76]. For boride systems, molybdenum diboride (MoB_2) nanospheres have demonstrated promising HER performance, with an onset overpotential of just 154 mV vs. RHE. This relatively low overpotential suggests favorable reaction kinetics and a lower energy barrier for hydrogen evolution. The exceptional activity is ascribed to the exposed surface terminations and abundant grain boundaries, which provide active sites and promote efficient charge transfer during the reaction [77].

The combined evaluation of Tafel slope, onset potential, and overpotential underscores the enhanced HER performance of bimetallic and hybrid systems based on selenides and borides, outperforming their single-component counterparts. The observed lower Tafel slopes and decreased overpotentials reflect the synergistic benefits of strategies such as heterostructuring, elemental doping, and the use of conductive supports. These modifications effectively tune the electronic structure, increase the density of active sites, and facilitate faster charge transfer, making such systems highly attractive for efficient and economically viable HER catalysis.

5.5 Alcohol Oxidation Reaction Mechanism of Ethanol, Methanol, and Glycerol Oxidation

The electrochemical alcohol oxidation reaction plays a vital role in sustainable energy systems, particularly in DAFCs and in electrolytic hydrogen production, where it offers a lower-energy alternative to the OER. Alcohol oxidation reaction enables the oxidation of alcohols such as methanol, ethanol, and glycerol into value-added carbonyl compounds while simultaneously generating electrons for energy applications [78]. In DAFCs, alcohol oxidation reaction occurs at the anode and is typically catalyzed by noble metals like Pt, Pd, and Au. However, non-precious metals such as nickel and cobalt have emerged as promising alternatives, especially in alkaline media, where they exhibit enhanced catalytic activity and improved reaction kinetics. Alkaline conditions also mitigate catalyst poisoning and allow more efficient utilization of low-cost metals. The Tafel slope, onset potential, and overpotential are critical metrics for evaluating alcohol oxidation reaction electrocatalysts [79, 80].

The electrochemical alcohol oxidation reaction refers to the anodic oxidation of alcohols such as methanol, ethanol, and glycerol in applications like DAFCs and electrolytic hydrogen production. This process not only generates value-added products such as carbonyl compounds and carboxylic acids but also promotes energy-efficient hydrogen evolution. The specific reaction mechanisms are influenced by factors including the type of alcohol, the catalyst material, and the electrolyte environment (typically acidic or alkaline). These mechanisms involve distinct pathways for C—H bond activation, C—C bond cleavage, and multi-electron transfer processes [80, 81].

5.5.1 Ethanol Oxidation

In alkaline media, ethanol oxidation generally follows a dual-pathway mechanism, resulting in either partial oxidation to acetaldehyde (CH_3CHO) or complete oxidation to acetate (CH_3COO^-) or carbon dioxide, depending on the catalyst and

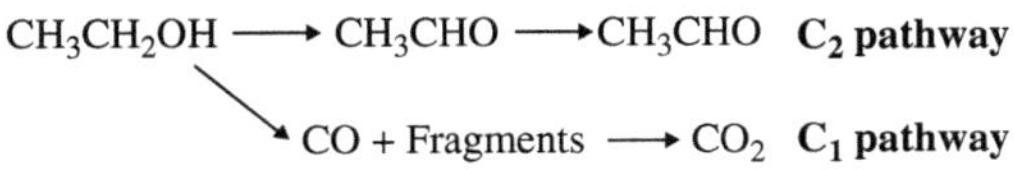

Scheme 5.2 Mechanism of ethanol oxidation on Pt electrode in acidic medium: C_1 and C_2 routes. *Source*: [82]. American Chemical Society.

operating conditions, as shown in Scheme 5.2. On precious metal catalysts such as Pt or Pd, ethanol typically undergoes dissociative adsorption, forming surface-bound ethoxy species ($CH_3CH_2O^*$) and hydrogen atoms (H*). The rate-determining step is usually the cleavage of the C—H bond, leading to the formation of acetaldehyde, which can subsequently be oxidized to more stable products [82].

For example, a Pt–Ni seed–core–frame nanostructure has been shown to significantly improve C—H bond cleavage in alkaline media, achieving a low overpotential of approximately 0.3 V at a current density of 100 $mA{\cdot}cm^{-2}$. This enhanced activity is attributed to Ni's ability to modulate the electronic structure of Pt, thereby reducing CO poisoning [83]. Platinum nanoparticles (~20 wt.%) supported on calcined Fe_2O_3–carbon composites demonstrated improved catalytic activity for ethanol electrooxidation (EOR) in acidic media. Calcination and metal oxide incorporation enhanced Pt dispersion, reduced particle size to 3.09 nm, and strengthened metal–support interactions. Among the catalysts, Pt/5% Fe_2O_3–C showed the best performance, featuring a low onset potential (0.2 V vs. Ag/AgCl), high current density (0.52 $mA{\cdot}cm^{-2}$), and good long-term stability. Dual calcination further increased the mass activity by 2.3 times compared to commercial Pt/C, owing to synergistic electronic effects and the ability of Fe_2O_3 to promote C—C bond cleavage during ethanol oxidation [84].

5.5.2 Methanol Oxidation

Methanol oxidation is a complex electrocatalytic process that is highly influenced by the electrolyte medium, acidic or alkaline, and the catalyst employed. In acidic media, platinum-based catalysts are widely used due to their strong ability to adsorb methanol. The reaction mechanism begins with the adsorption of methanol onto the Pt surface, where it undergoes dehydrogenation to form a surface-bound "methoxy intermediate" $(CH_3O)^*$:

$$CH_3OH + Pt \rightarrow Pt{-}CH_3O^* + H^+ + e^-$$

This methoxy species then undergoes further dehydrogenation to yield a "formyl intermediate" (CHO)*:

$$Pt{-}CH_3O^* \rightarrow Pt{-}CHO^* + 2H^+ + 2e^-$$

Subsequently, the CHO* intermediate is further oxidized to produce adsorbed carbon monoxide (CO_{ads}):

$$Pt{-}CHO^* \rightarrow Pt{-}CO_{ads} + H^+ + e^-$$

However, the strong binding of CO_{ads} to the Pt surface leads to catalyst poisoning, as it blocks active sites and inhibits further methanol oxidation. To address this issue, bimetallic catalysts such as Pt–Au or Pt–Ru are utilized. These materials enhance the formation of surface-bound hydroxyl species (OH*) at lower onset potentials (around 0.4 V vs. RHE), which facilitate the oxidative removal of CO_{ads}:

$$Pt\text{—}CO_{ads} + Au\text{—}OH^* \rightarrow CO_2 + Pt + Au + H^+ + e^-$$

This improves the reaction kinetics, as evidenced by a lower Tafel slope (~60 $mV{\cdot}dec^{-1}$), and reduces the overpotential required for methanol oxidation [85].

The overall reaction for complete oxidation of methanol in acidic media is as follows:

$$CH_3OH + H_2O \rightarrow CO_2 + 6H^+ + 6e^-$$

In alkaline media, nickel-based MOR catalysts naturally form a $Ni(OH)_2$ layer on the surface through reaction with the solution. This $Ni(OH)_2$ layer is catalytically significant and can transform into other active phases. Notably, Barakat et al. observed that $Ni(OH)_2$ can directly facilitate methanol oxidation at lower potentials without converting to nickel oxyhydroxide (NiOOH), suggesting its catalytic activity [86]. However, at higher anodic potentials, $Ni(OH)_2$ oxidizes to NiOOH, which is widely recognized as the primary active species in the conventional MOR pathway.

Methanol oxidation on Ni-based catalysts occurs via two interconnected cyclic pathways. At low potentials, $Ni(OH)_2$ directly participates, while at higher potentials, NiOOH dominates, enhancing methanol conversion efficiency. Most research indicates that NiOOH drives the majority of catalytic activity in alkaline MOR systems.

The phase transformations involved are:

$$Ni(OH)_2 \rightarrow NiOOH + H^+ + e^-$$
$$Ni(OH)_2 + OH^- \rightarrow Ni(OH)_3 + e^-$$
$$Ni(OH)_3 \rightarrow NiOOH + H_2O$$

Once formed, NiOOH oxidizes methanol through a stepwise process involving surface-bound intermediates:

$$NiOOH + CH_3OH_{ads} \rightarrow CH_3O_{ads} + Ni(OH)_2$$
$$NiOOH + CH_3O_{ads} \rightarrow CH_2O_{ads} + Ni(OH)_2$$
$$NiOOH + CH_2O_{ads} \rightarrow CHO_{ads} + Ni(OH)_2$$
$$NiOOH + CHO_{ads} \rightarrow CO_{ads} + Ni(OH)_2$$
$$NiOOH + CO_{ads} \rightarrow CO_2 + Ni(OH)_2$$

This sequential oxidation mechanism underscores the bifunctional nature of Ni-based catalysts. The continuous cycling between $Ni(OH)_2$ and NiOOH sustains methanol oxidation, with intermediates like CH_3O^-, CH_2O, CHO, and CO progressively oxidized to CO_2 (Figure 5.4) [87].

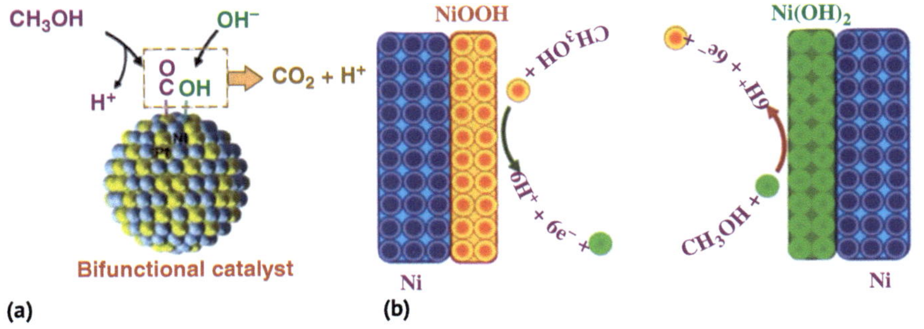

Figure 5.4 (a) Bifunctional mechanism of Ni-based catalysts for CO_{ads} oxidation. (b) Transformation pathway during methanol oxidation. *Source*: [87]/with permission of Elsevier.

5.5.3 Glycerol Oxidation

Glycerol oxidation is a multifaceted electrochemical process owing to the presence of three hydroxyl groups, which can undergo selective or complete oxidation, leading to a variety of products such as glyceraldehyde, dihydroxyacetone, glyceric acid, tartronic acid, and C—C bond cleavage products like formate and CO_2. In alkaline media, catalysts such as Au/CoOOH enable selective oxidation toward glyceric acid and tartronic acid, operating at a low overpotential (~0.34 V at 10 $mA{\cdot}cm^{-2}$). This efficiency is attributed to the generation of electrophilic OH* species on the CoOOH surface, which facilitate the cleavage of C—H and O—H bonds [88]. The alkaline oxidation mechanism begins with glycerol adsorption onto the catalyst surface through one or more hydroxyl groups, followed by stepwise deprotonation and oxidation (Figure 5.5). This results in intermediates such as glyceraldehyde or dihydroxyacetone, which can then be further oxidized to glyceric acid and, in some cases, tartronic acid through continued oxidation of the secondary alcohol group. For example, a $Ni_{0{\cdot}67}V_{0{\cdot}33}$ LDH combined with an aminoxyl radical achieved ~91% selectivity for tartronic acid at an onset potential of ~1.2 V vs. RHE, where synergistic interactions between Ni and V were found to stabilize the C—C bonds and suppress unwanted over-oxidation [89].

The stepwise mechanism in alkaline media can be outlined as follows:

Adsorption of glycerol onto the catalyst surface:

$$HOCH_2{-}CHOH{-}CH_2OH\ (\text{glycerol}) \rightarrow \text{Catalyst}{-}CHOH_{ads}$$

Oxidation to glyceraldehyde or dihydroxyacetone:

$$CHOH_{ads} + OH^- \rightarrow CHO\ (\text{glyceraldehyde}) + H_2O + e^-$$

Further oxidation to glyceric acid:

$$CHO + OH^- \rightarrow COOH\ (\text{glyceric acid}) + H_2O + e^-$$

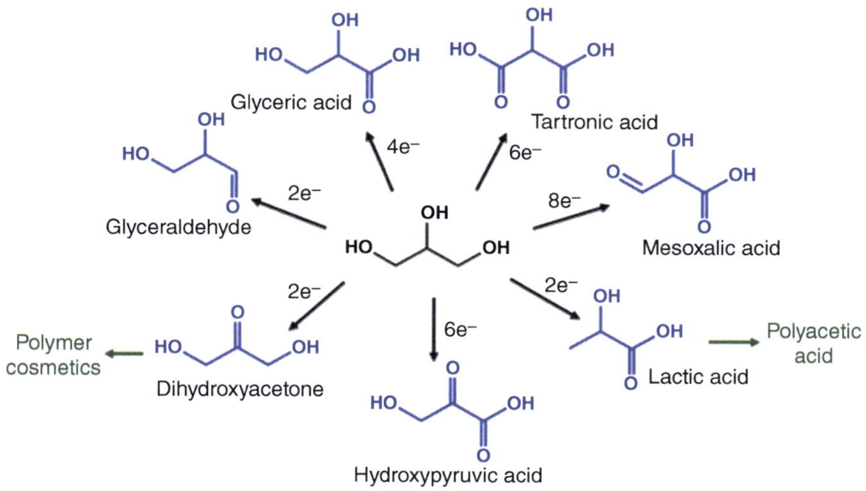

Figure 5.5 Overview of alcohol electrooxidation reactions, key products, and applications. *Source*: [91]/with permission of American Chemical Society.

Oxidation of glyceric acid to tartronic acid occurs depending on catalyst properties and applied potential. In contrast, acidic media typically favor Pt-based catalysts, which promote C—C bond cleavage, resulting in the formation of formate and CO_2. However, these reactions require significantly higher overpotentials (~0.8 V vs. RHE) due to stronger adsorption of intermediates and the risk of CO poisoning [90]. The acidic pathway involves:

Strong adsorption of glycerol on the Pt surface:

$$OHCH_2{-}CHOH{-}CH_2OH + Pt \rightarrow Pt{-}CH_2OH + H^+ + e^-$$

Dehydrogenation to produce surface-bound CO:

$$Pt{-}CH_2OH \rightarrow Pt{-}(CO) + H^+ + e^-$$

Oxidation of CO to CO_2, typically at higher potentials:

$$Pt{-}(CO) + H_2O \rightarrow CO_2 + 2H^+ + e^-$$

Thus, the reaction environment and catalyst composition play critical roles in determining the selectivity and efficiency of glycerol oxidation. While acidic conditions favor deeper oxidation and bond breaking, alkaline media, especially with non-noble, bimetallic, or hybrid catalysts, allow for milder, more controlled oxidation, favoring the production of valuable oxygenated compounds. Such advancements are crucial for the electrocatalytic valorization of biomass-derived polyols and the development of sustainable energy conversion technologies [89, 90].

5.5.4 Borides and Selenides as Anodic Catalysts

Recent developments have showcased the strong potential of boride- and selenide-based materials as efficient anodic electrocatalysts in alcohol oxidation reaction -coupled electrolysis systems (Scheme 5.3).

For example, Hong et al. synthesized a $NiCoB@NiCoWO_4$ heterostructure through a one-step chemical reduction process. This amorphous catalyst exhibited outstanding activity for ethanol oxidation in alkaline media. When paired with a Pt/C cathode for hydrogen evolution, the full cell attained a remarkably low operating voltage of 1.50 V, producing hydrogen at a rate 19 times greater than conventional water electrolysis, with a faradaic efficiency exceeding 97% for acetate production [92]. In a separate study, Braun et al. examined the performance of CoNiB catalysts subjected to different thermal treatments. The catalyst annealed at 500 °C demonstrated superior activity, reducing the overpotential for alcohol oxidation by approximately 224 mV compared to the OER at 10 mA·cm^{-2}. The system also showed excellent product selectivity, delivering 85–90% faradaic efficiency for acetate and 60–80% for formate, indicating its effectiveness for both ethanol and glycerol oxidation, as shown in Figure 5.6 [93]. Qi et al. reported that a Ni_3B/Ni heterostructure served as a highly efficient and stable catalyst for methanol oxidation in alkaline media. It achieved nearly 100% faradaic efficiency toward formate at a low overpotential of ~147 mV (at 10 mA·cm^{-2}), and maintained long-term durability. In-situ characterization confirmed that reactive oxygen species (O*/OH*) initiate the methanol oxidation process before the catalyst itself undergoes oxidation. Density functional theory (DFT) calculations revealed that CH_2O formation is

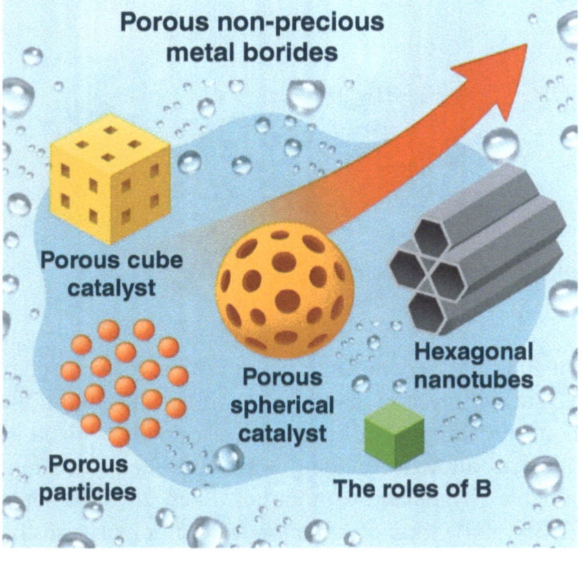

Scheme 5.3 Schematic illustration of porous metal borides integrated into an alcohol-assisted electrolysis system.

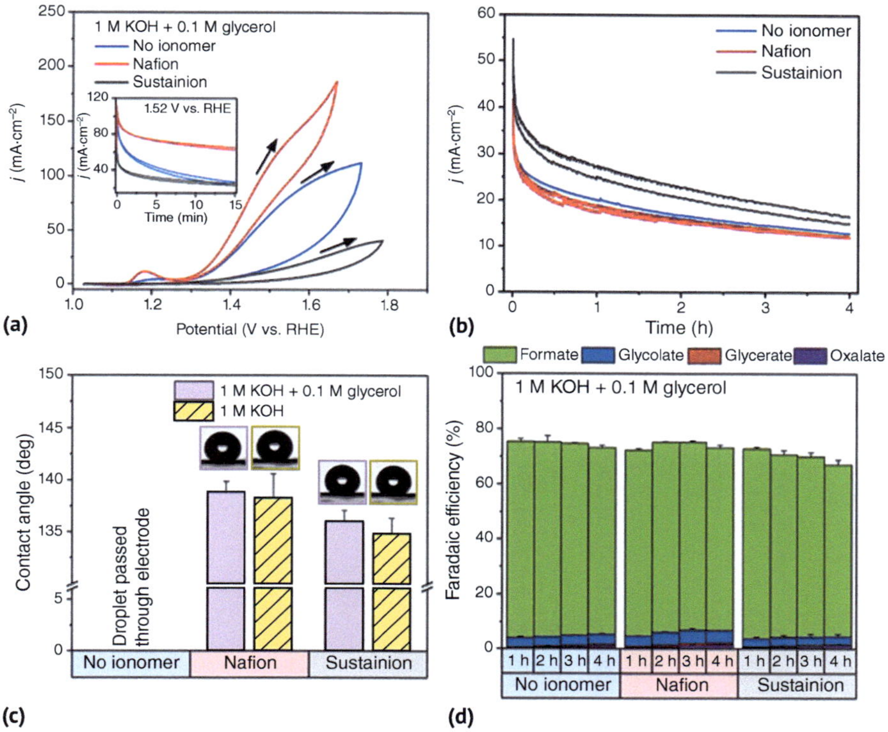

Figure 5.6 Effects of ionomer and cell setup on glycerol oxidation: (a) CVs of CoNiB:BO1:0500 °C in Ar-saturated 1 M KOH + 0.1 M glycerol (inset: chronoamperometry at 1.52 V vs. RHE). (b) The catalyst was tested in a flow cell at 1.52 V vs. RHE for 4 hours; duplicate runs confirmed stable performance and the FEs are shown. (c) Contact angle measurements. (d) FEs of corresponding system. *Source*: [93]. IOP Publishing. CC BY 4.0.

the rate-determining step, further supporting the catalyst's high selectivity and activity [94].

In the case of selenides, Yao et al. introduced high-entropy selenides (HESes) as efficient catalysts for glycerol electrooxidation, coupled with hydrogen evolution in an alkali–acid hybrid electrolytic cell. This system required only ~0.55 V to operate at 10 $mA{\cdot}cm^{-2}$ and delivered ~92% faradaic efficiency for formate and ~99% for hydrogen, while remaining stable for over 12 days of continuous operation (Figure 5.7). The multielement composition of HESes contributed to a rich distribution of active sites and high structural integrity, enabling selective and energy-efficient electrochemical conversions [95]. Furthermore, Ramakrishnan et al. developed a hybrid electrocatalyst composed of ultrafine Pt nanoparticles anchored on MoS_2/N-doped reduced graphene oxide (Pt@MoS_2/NrGO) [96]. This material demonstrated exceptional activity for alcohol oxidation, achieving a mass activity of 448 $mA{\cdot}mg_{pt}^{-1}$ for methanol—significantly outperforming commercial Pt/C. It also showed two to three times higher activity for ethylene glycol and glycerol oxidation, with over 80% activity retained after 500 cycles. For the oxygen reduction reaction (ORR),

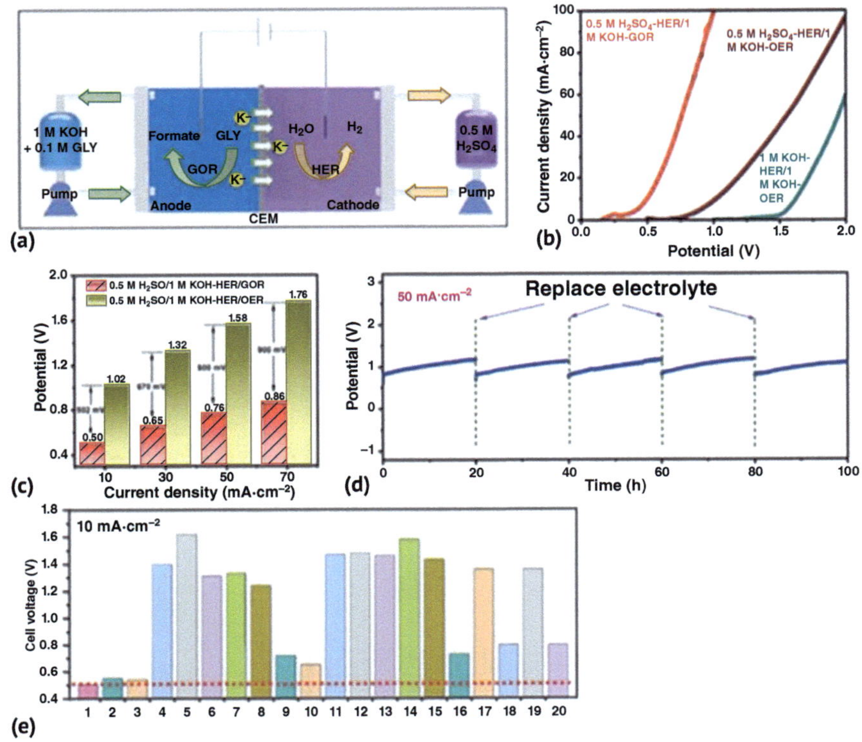

Figure 5.7 Electrochemical performance of (CoNiCuMnMo)Se/CF//20%-Pt/C. (a) Schematic of the asymmetric-electrolyte electrolyzer setup. (b) LSV curves comparing HER–GOR and HER–OER performances. (c) Cell voltages for HER–GOR and traditional HER–OER electrolyzers at various current densities using asymmetric electrolytes. (d) Long-term stability test of the hybrid alkaline–acidic electrolyzer at 50 mA·cm^{-2}. (e) Observed cell voltage at 10 mA cm^{-2}. *Source*: [95]/with permission of Springer Nature.

the catalyst achieved a half-wave potential of 0.895 V vs. RHE, and maintained excellent durability with negligible performance loss even after 30,000 cycles, making it a strong candidate for direct alcohol fuel-cell applications.

5.5.5 Selectivity and Product Distribution

The selectivity of alcohol electrooxidation reactions is a key factor that depends heavily on the type of catalyst, the alcohol being oxidized, and the electrolyte environment, particularly whether it is acidic or alkaline. Recent studies by Shi et al. and Zhao et al. showcased the strong performance of non-precious metal catalysts, including transition-metal oxides, borides, and LDHs, especially in alkaline conditions. Catalysts like NiOOH, CoOOH, and NiFe LDH can form high-valent metal centers (e.g., Ni^{3+} or Co^{3+}) that stabilize reactive oxygen species (OH* or O*), which selectively activate C—H and O—H bonds while preserving the C—C framework of alcohols [80, 97]. This mechanism promotes the formation of value-added chemicals, such as acetate from ethanol and glycerate or tartronate from glycerol, with faradaic efficiencies often surpassing 85–90%. For instance, $Ni_{0.67}V_{0.33}$ LDH exhibits around 91% selectivity for tartronic acid due to the synergistic electronic interaction between Ni and V, which enhances electron transfer and intermediate stability.

In contrast, noble metal catalysts like Pt and Pd, particularly in acidic media, often show reduced selectivity due to their tendency to break C—C bonds, leading to full oxidation products like CO_2 and formate. These catalysts also suffer from CO poisoning and generally operate at higher overpotentials, with lower selectivity for partial-oxidation products such as acetaldehyde or acetic acid. Acidic environments further hinder selective oxidation by limiting the availability of surface OH species required for effective bond activation.

Overall, selectivity in alcohol electrooxidation plays a crucial role in determining the efficiency, sustainability, and product value of electrochemical systems. High selectivity allows for the production of valuable chemicals while minimizing carbon loss and energy consumption. By carefully designing catalysts and optimizing reaction conditions, especially in alkaline media, researchers can steer reaction pathways toward desirable products, making alcohol oxidation reaction a more viable strategy for green hydrogen generation and biomass upgrading.

Selectivity in alcohol oxidation reaction is highly dependent on the catalyst composition, the structure of the alcohol, and the electrolyte environment, with alkaline media generally offering superior selectivity for value-added products. For example, ascorbic acid; a simple organic molecule has been explored as a non-precious metal catalyst for methanol oxidation in alkaline conditions, achieving nearly 95% selectivity toward formate at a low overpotential of around 0.3 V. This performance is attributed to its ability to promote C—H bond cleavage while avoiding CO poisoning [98]. In ethanol oxidation, NiO-based catalysts show a strong preference for acetate production (around 90% selectivity) over acetaldehyde, driven by a hydrogen atom transfer (HAT) mechanism involving Ni^{3+}—$(OH)_{ads}$ species [97]. For glycerol oxidation, Au/CoOOH catalysts deliver about 85% selectivity for glyceric acid at low potentials [99], while $Ni_{0.67}V_{0.33}$ LDH catalysts achieve approximately 91% selectivity for tartronic acid, effectively maintaining the three-carbon structure of glycerol [89].

In acidic environments, however, selectivity tends to decrease due to the enhanced likelihood of C—C bond cleavage. Pt-based catalysts under acidic conditions typically produce CO_2 and formate as dominant products, with lower selectivity (about 60–70%) for partial-oxidation products like acetaldehyde [100]. On the other hand, Au nanoparticles supported on $LaMnO_3$ perovskite have shown excellent selectivity for glycerol oxidation in alkaline media. After 24 hours of reaction, the catalyst achieved around 88% selectivity toward tartronic acid with complete (100%) glycerol conversion. In the early reaction phase, glyceric acid appeared as a key intermediate, with a selectivity of approximately 66%, which gradually converted to tartronic acid. Compared to TiO_2-supported systems that favor dehydration and produce lower selectivity, the $LaMnO_3$ support enables efficient stepwise oxidation while minimizing C—C bond cleavage and side reactions [101]. Similarly, selenide-based systems such as $NiCo_2Se_4/NiTe_2$ nanorods show ~100% selectivity for formate in methanol oxidation, benefiting from strong Ni–Co synergistic effects [102]. Overall, alkaline electrolytes enhance selectivity by stabilizing hydroxide intermediates, enabling controlled partial oxidation, while acidic media favor complete oxidation, thereby limiting the formation of desirable intermediates.

5.5.6 Overall Electrolysis Performance

The overall electrolysis performance of systems employing borides and selenides as anodic catalysts for the alcohol oxidation reaction coupled with the HER at the cathode is a critical aspect of developing efficient and sustainable electrochemical systems for hydrogen production and value-added chemical synthesis. Full-cell configurations, cell voltage, energy consumption, and durability are key metrics for assessing the practical viability of these electrocatalysts in DAFCs or hybrid electrolysis setups.

Recent studies on bifunctional electrocatalysts for methanol-assisted water electrolysis revealed significant improvements in full-cell performance. The system utilizing in-situ crystallized NiOOH/CoOOH heterostructures on velutipes-like CoSe/NiSe nanorods attained a low cell voltage of 1.47 V at 10 $mA\cdot cm^{-2}$, demonstrating strong catalytic synergy and effective hydroxide ion adsorption [103]. A novel selenide-based composite, serving as a dual-function catalyst for both OER and methanol oxidation, further lowered the cell voltage to 1.42 V at the same current density, emphasizing its potential for energy-efficient hydrogen production [104].

Additionally, a cobalt-sulfo-boride catalyst exhibited excellent high-current performance, delivering 1.58 V at 500 $mA\cdot cm^{-2}$ while retaining structural integrity and catalytic activity over prolonged use. Nickel selenides enabled a coupled methanol oxidation and HER with an exceptionally low full-cell voltage of just 0.84 V, indicating significantly reduced energy requirements [105]. Collectively, these systems exhibit notably lower energy consumption compared to conventional water electrolysis and maintain high durability, preserving over 95% of their original performance after 50–100 hours of continuous operation, underscoring their promise for real-world applications.

5.6 Challenges and Future Perspective

Metal borides and selenides have emerged as promising bifunctional electrocatalysts for alcohol-assisted water electrolysis, owing to their excellent electrical conductivity, tunable surface chemistry, and notable catalytic activity. Despite these advantages, several key challenges must be addressed to enable their successful transition from laboratory research to real-world applications. One of the primary concerns is their stability during long-term operation. These materials often undergo surface oxidation, leaching of active components, or phase transformations under harsh electrochemical conditions, which significantly affect their durability and performance. Enhancing their structural robustness and corrosion resistance is therefore critical for sustained activity over prolonged use. Another major challenge lies in scaling up and integrating these materials into practical electrolyzer systems. Most existing studies are confined to small-scale, half-cell experiments, and translating this performance to full-cell or commercial configurations remains difficult. Issues such as achieving uniform large-scale synthesis, incorporating catalysts into stable and conductive electrode systems, and maintaining efficiency at high current densities must be overcome. Additionally, while metal borides and selenides are generally more cost-effective than noble metals, a detailed economic assessment is necessary to determine their viability in industrial settings. Exploring options for catalyst recyclability and regeneration will further enhance their sustainability. Furthermore, significant research gaps exist in understanding their mechanistic behavior. The precise nature of active sites, the evolution of surface species under reaction conditions, and the control of product selectivity in alcohol oxidation (e.g., favoring formate or acetate over CO_2) require deeper investigation. In situ and operando characterization techniques, supported by computational modeling, are essential tools for unraveling these complexities and guiding rational catalyst design. Addressing these challenges will pave the way for efficient, scalable, and economically viable hydrogen production through alcohol-assisted water electrolysis using metal boride and selenide catalysts [106–110].

5.7 Conclusion

Metal borides and selenides have demonstrated significant potential as next-generation electrocatalysts for alcohol-assisted water electrolysis. Their intrinsic properties—such as high electrical conductivity, thermal stability, tunable surface chemistry, and abundant active sites—make them suitable candidates for driving both the HER and alcohol oxidation reaction. These materials exhibit remarkable catalytic activity, structural robustness, and the ability to operate efficiently under diverse electrochemical conditions. Alcohol-assisted water electrolysis, as an alternative to conventional water splitting, offers key advantages including reduced energy input due to the lower overpotentials associated with alcohol oxidation reaction, and the generation of value-added by-products such as formate or acetate. This dual benefit of enhanced energy efficiency and economic return makes the integration of metal borides and selenides in such systems particularly appealing. Furthermore, coupling this approach with renewable or bio-derived alcohol feedstocks aligns well with sustainability goals and circular economy principles. Moving forward, advancing the practical application of metal borides and selenides in alcohol-assisted electrolysis will require addressing challenges such as catalyst stability under long-term operation, surface reconstruction dynamics, mechanistic understanding of intermediate formation, and scale-up for industrial deployment. A strategic roadmap should focus on (i) atomic-level engineering of catalyst surfaces to improve selectivity and durability, (ii) development of cost-effective synthesis routes, (iii) integration into scalable electrolyzer architectures, and (iv) life cycle assessment and recycling strategies. With continued interdisciplinary research and innovation, metal borides and selenides hold the promise to drive efficient, low-carbon hydrogen production and become central to future sustainable energy technologies.

Acknowledgments

Our team expresses gratitude to the Thailand Science Research and Innovation Fund Chulalongkorn University (IND_FF_68_169_2100_028), the Program Management Unit for Human Resources and Institutional Development, Research, and Innovation (B42G680030), and the Chulalongkorn Academic Advancement into the Second Century Project Post-Doctoral Fellowship (C2F).

References

1 Holladay, J. D., Hu, J., King, D. L., & Wang, Y. (2009). An overview of hydrogen production technologies. *Catal. Today 139*, 244–260. https://doi.org/10.1016/j.cattod.2008.08.039

2 Levin, D. B. & Chahine, R. (2010). Challenges for renewable hydrogen production from biomass. *Int. J. Hydrog. Energy 35*, 4962–4969. https://doi.org/10.1016/j.ijhydene.2009.08.067

3 Ju, H. K., Badwal, S., & Giddey, S. A. (2018). Comprehensive review of carbon and hydrocarbon assisted water electrolysis for hydrogen production. *Appl. Energy 231*, 502–533.

4 IEA (2007). Hydrogen production & distribution. IEA Energy Technology Essentials No. ETE 05. https://webstore.iea.org/iea-energy-technology-essentials-hydrogenproduction-distribution

5 Krishna, R. H. (2023). Review of research on production methods of hydrogen: Future fuel. *Eur. J. Biotechnol. Biosci. 1*, 84–93.

6 Acar, C. & Dincer, I. (2014). Comparative assessment of hydrogen production methods from renewable and non-renewable sources. *Int. J. Hydrog. Energy 39*, 1–12. https://doi.org/10.1016/j.ijhydene.2013.10.060

7 Mueller-Langer, F., Tzimas, E., Kaltschmitt, M., & Peteves, S. (2007). Techno-economic assessment of hydrogen production processes for the hydrogen economy for the short and medium term. *Int. J. Hydrog. Energy 32*, 3797–3810. https://doi.org/10.1016/j.ijhydene.2007.05.027

8 Stiegel, G. J. & Ramezan, M. (2006). Hydrogen from coal gasification: An economical pathway to a sustainable energy future. *Int. J. Coal Geol. 65*, 173–190. https://doi.org/10.1016/j.coal.2005.05.002

9 Carmo, M., Fritz, D. L., Mergel, J., & Stolten, D. (2013). A comprehensive review on PEM water electrolysis. *Int. J. Hydrog. Energy 38*, 4901–4934.

10 Çelik, D. & Meltem, Y. (2017). Investigation of hydrogen production methods in accordance with green chemistry principles. *Int. J. Hydrog. Energy 42*, 23395–23401.

11 IEA Hydrogen (2017). Global trends and outlook for hydrogen. Special Reports. http://ieahydrogen.org/PUBLICATIONS,-REPORTS-PRESENTATIONS/Special-Reports.aspx

12 Buttler, A. & Spliethoff, H. (2018). Current status of water electrolysis for energy storage, grid balancing and sector coupling via power-to-gas and power-to-liquids: A review. *Renew. Sustain. Energy Rev. 82*, 2440–2454.

13 Turner, J. A. (2004). Sustainable hydrogen production. *Int. J. Hydrog. Energy 29*(7), 783–793. https://doi.org/10.1016/j.ijhydene.2003.09.015

14 Marshall, A. T. & Haverkamp, R. G. (2008). Production of hydrogen by the electrochemical reforming of glycerol–water solutions in a PEM electrolysis cell. *Int. J. Hydrog. Energy 33*, 4649–4654.

15 Sasikumar, G., Muthumeenal, A., Pethaiah, S. S., Nachiappan, N., & Balaji, R. (2008). Aqueous methanol electrolysis using proton conducting membrane for hydrogen production. *Int. J. Hydrog. Energy 33*, 5905–5910.

16 Silveira, J. L., Braga, L. B., de Souza, A. C. C., Antunes, J. S., & Zanzi, R. (2009). The benefits of ethanol use for hydrogen production in urban transportation. *Renew. Sustain. Energy Rev. 13*, 2525–2534.

17 Mahdavi, M. A., L'Esperance, G., & Savadogo, O. (1996). Hydrogen evolution reaction on nickel-boron electrodes in aqueous and methanolic/ethanolic alkaline media. *Can. J. Chem. 74*(4), 678–684. https://doi.org/10.1139/v96-043

18 Tang, Y., Zhang, L., & Li, X. (2024). Amorphous Ni–Co boride nanostructures coupled with $NiCoWO_4$ for efficient ethanol electrooxidation. *Catal. Sci. Technol. 14*, 2777–2787. https://doi.org/10.1039/D4CY00277F

19 Chen, F.-Y., Wu, Z.-Y., Zhu, Z., Yu, S.-H., & Li, Y. (2021). Recent advances in transition-metal boride electrocatalysts for water splitting. *ACS Mater. Au 1*(3), 237–259. https://doi.org/10.1021/acsmaterialsau.1c00079

20 Zhu, S., Xia, J., & Guo, S. (2023). Recent advances in transition metal selenide electrocatalysts for water splitting. *ACS Omega 8*, 36274–36285. https://doi.org/10.1021/acsomega.3c07911

21 Li, Y., Wang, H., Xie, L., Liang, Y., Hong, G., & Dai, H. (2020). MoS_2 nanoparticles grown on graphene: An advanced catalyst for the hydrogen evolution reaction. *Nano Res. 13*, 2618–2635. https://doi.org/10.1007/s12274-020-2618-y

22 Zhao, X., Fan, K., Li, X., & Sun, L. (2024). Transition metal borides, carbides, phosphides, pnictides, and chalcogenides for oxygen evolution electrocatalysis: From fundamentals to applications. *Chem. Rev. 124*(5), 2768–2856.

23 Veeramani, K., Janani, G., Kim, J., Surendran, S., Lim, J., Jesudass, S. C., Mahadik, S., Lee, H., Kim, T.-H., Kim, J. K., & Sim, U. (2023). Hydrogen and value-added products yield from hybrid water electrolysis: A critical review on recent developments. *Renew. Sustain. Energy Rev. 177*, 113541. https://doi.org/10.1016/j.rser.2023.113541

24 Yan, Y., Xia, B. Y., Zhao, B., & Wang, X. (2016). A review on noble-metal-free bifunctional heterogeneous catalysts for overall electrochemical water splitting. *J. Mater. Chem. 4*(45), 17587–17603.

25 Hu, C., Zhang, L., & Gong, J., (2019). Recent progress made in the mechanism comprehension and design of electrocatalysts for alkaline water splitting. *Energy Environ. Sci. 12*(9), 2620–2645.

26 Luo, J., Vermaas, D. A., Bi, D., Hagfeldt, A., Smith, W. A., & Gratzel, M. (2016). Bipolar membrane assisted solar water splitting in optimal pH. *Adv. Energy Mater.* *6*(13), 1600100.

27 Vesborg, P. C. K. & Jaramillo, T. F. (2012). addressing the terawatt challenge: Scalability in the supply of chemical elements for renewable energy. *RSC Adv.* *2*(21), 7933–7947.

28 Chatenet, M., Pollet, B. G., Dekel, D. R., Dionigi, F., Deseure, J., Millet, P., Braatz, R. D., Bazant, M. Z., Eikerling, M., Staffell, I., Balcombe, P., Shao-Horn, Y., & Schäfer, H. (2022). Water electrolysis: From textbook knowledge to the latest scientific strategies and industrial developments. *Chem. Soc. Rev.* *51*(11), 4583–4762.

29 Arshad, M. A., Irfan, R. M., Hwang, Y., & Park, J. Y. (2021). Recent advances in electrocatalysts toward alcohol-assisted energy-saving hydrogen production. *Nano Today 41*, 101295. https://doi.org/10.1016/j.nantod.2021.101295

30 Zhang, H., Maijenburg, A. W., Li, X., Schweizer, S. L., & Wehrspohn, R. B. (2020). Bifunctional heterostructured transition metal phosphides for efficient electrochemical water splitting. *Adv. Funct. Mater.* *30*(34), 2003261.

31 Artero, V., Chavarot-Kerlidou, M., & Fontecave, M. (2011). Splitting water with cobalt. *Angew. Chem. Int. Ed.* *50*(32), 7238–7266.

32 Surendran, S., Shanmugapriya, S., Sivanantham, A., Shanmugam, S., & Kalai Selvan, R. (2018). Electrospun carbon nanofibers encapsulated with NiCoP: A multifunctional electrode for supercapattery and oxygen reduction, oxygen evolution, and hydrogen evolution reactions. *Adv. Energy Mater.* *8*(20), 1800555.

33 Li, Y., Wei, X., Han, S., Chen, L., & Shi, J. (2021). MnO_2 electrocatalysts coordinating alcohol oxidation for ultra-durable hydrogen and chemical productions in acidic solutions. *Angew. Chem. Int. Ed.* *60*(39), 21464–21472.

34 Bender, M. T., Yuan, X., & Choi, K. S. (2020). Alcohol oxidation as alternative anode reactions paired with (photo)electrochemical fuel production reactions. *Nat. Commun.* *11*(1), 4594.

35 Kuang, M., Wang, Y., Fang, W., Tan, H., Chen, M., Yao, J., Liu, C., Xu, J., Zhou, K., & Yan, Q. (2020). Efficient nitrate synthesis via ambient nitrogen oxidation with Ru-doped TiO_2/RuO_2 electrocatalysts. *Adv. Mater.* *32*(26), 2002189.

36 Chadderdon, X. H., Chadderdon, D. J., Pfennig, T., Shanks, B. H., & Li, W. (2019). Paired electrocatalytic hydrogenation and oxidation of 5-(hydroxymethyl) furfural for efficient production of biomass-derived monomers. *Green Chem.* *21*(22), 6210–6219.

37 Chen, B., Li, X., Rui, P., Ye, Y., Ye, T., Zhou, R., Li, D., Carter, J. H., & Hutchings, G. J. (2022). The reaction pathways of 5-hydroxymethylfurfural conversion in a continuous flow reactor using copper catalysts. *Catal. Sci. Technol.* *12*(9), 3016–3027.

38 Li, K. & Sun, Y. (2018). Electrocatalytic upgrading of biomass-derived intermediate compounds to value-added products. *Chem. Eur. J.* *24*(69), 18258–18270.

39 Wang, S., Lu, A., & Zhong, C. J. (2021). Hydrogen production from water electrolysis: Role of catalysts. *Nano Converg.* *8*, 4. https://doi.org/10.1186/s40580-021-00254-x

40 Zeng, C. T., Dai, L., Jin, Y. H., & Wang, H. (2021). Design strategies toward transition metal selenide-based catalysts for electrochemical water splitting. *Sustain. Energy Fuels 5*, 1347. https://doi.org/10.1039/D0SE01722A

41 Jiang, Y., Gao, S., & Liu, X. (2024). Recent achievements in selenium-based transition metal electrocatalysts for pH-universal water splitting. *Nano Res. 17*, 5763–5785. https://doi.org/10.1007/s12274-024-6485-9.

42 Hansen, J. N., Prats, H., Toudahl, K. K., Secher, N. M., Chan, K., Kibsgaard, J., & Chorkendorff, I. (2021). Is there anything better than Pt for HER? *ACS Energy Lett. 6*, 1175–1180. https://doi.org/10.1021/acsenergylett.1c00246

43 Terlouw, T., Bauer, C., McKenna, R., & Mazzotti, M. (2022). Large-scale hydrogen production via water electrolysis: A techno-economic and environmental assessment. *Energy Environ. Sci. 15*, 3583. https://doi.org/10.1039/d2ee01023b

44 Jiang, Y. & Lu, Y. (2020). Designing transition-metal-boride-based electrocatalysts for applications in electrochemical water splitting. *Nanoscale 12*, 9327–9351. https://doi.org/10.1039/D0NR01279C

45 Peng, X., Yan, Y., Jin, X., Huang, C., Jin, W., Gao, B., & Chu, P. K. (2020). Recent advance and prospectives of electrocatalysts based on transition metal selenides for efficient water splitting. *Nano Energy 78*, 105234. https://doi.org/10.1016/j.nanoen.2020.105234

46 Yuan, W., Zhao, X., Hao, W., Li, J., Wang, L., Ma, X., & Guo, Y. (2019). The performance of surface oxidized Ni_3B, Ni_2B and NiB_2 electrocatalysts for overall water splitting. *ChemElectroChem 6*(3), 764–779. https://doi.org/10.1002/celc.201801354

47 Huang, T., Shen, T., Gong, M., Deng, S., Lai, C., Liu, X., Zhao, T., Teng, L., & Wang, D. (2019). Ultrafine Ni-B nanoparticles for efficient hydrogen evolution reaction. *Chin. J. Catal. 40*, 1867–1873. https://doi.org/10.1016/S1872-2067(19)63331-0

48 Shi, P., Zhang, Y., Zhang, G., Zhu, X., Wang, S., & Wang, A. (2021). A crystalline/amorphous CoP@CoB hierarchical core–shell nanorod array for enhanced hydrogen evolution. *J. Mater. Chem. A 9*, 19719–19724. https://doi.org/10.1039/D1TA02783B

49 Guang, H. L., Zhu, S. L., & Liang, Y. Q. (2021). Highly efficient nanoporous CoBP electrocatalyst for hydrogen evolution reaction. *Rare Met. 40*, 1031–1039. https://doi.org/10.1007/s12598-020-01697-7

50 Chhowalla, M., Shin, H., & Eda, G. (2013). The chemistry of two-dimensional layered transition metal dichalcogenide nanosheets. *Nature Chem. 5*, 263–275. https://doi.org/10.1038/nchem.1589

51 Feng, W., Pang, W., Xu, Y., Guo, A., Gao, X., Qiu, X., & Chen, W. (2020). Transition metal selenides for electrocatalytic hydrogen evolution reaction. *ChemElectroChem 7*(1), 31–54. https://doi.org/10.1002/celc.201901623

52 Ji, Q., Zhang, Y., Shi, J., Sun, J., Zhang, Y., & Liu, Z. (2016). Morphological engineering of CVD-grown transition metal dichalcogenides for efficient electrochemical hydrogen evolution. *Adv. Mater. 28*(29), 6207–6212. https://doi.org/10.1002/adma.201504762

53 Tsai, C., Chan, K., Pedersen, F., & Norskov, J. K. (2014). Active edge sites in $MoSe_2$ and WSe_2 catalysts for the hydrogen evolution reaction: a density functional study. *Phys. Chem. Chem. Phys. 16*, 13156–13164. https://doi.org/10.1039/C4CP01237B

54 Rai, R. K., Sarkar, B., Ram, R., Nanda, K. K., & Ravishankar, N. (2022). Designed synthesis of a hierarchical $MoSe_2$@WSe_2 hybrid nanostructure as a bifunctional electrocatalyst for total water-splitting. *Sustain. Energy Fuels 6*, 1708–1718. https://doi.org/10.1039/D1SE01843D

55 Shu, H., Zhou, D., Li, F., Cao, D., & Chen, X. (2017). Defect engineering in $MoSe_2$ for hydrogen evolution reaction: From point defects to edges. *ACS Appl. Mater. Interfaces 9*(49), 42688–42698. https://doi.org/10.1021/acsami.7b12478

56 Patel, A. B., Sumesh, C. K., Vaghasiya, J. V., Chauhan, P., Patel, V., Soni, S. S., Patel, K. D., Garg, P., Solankia, G. K., & Pathak, V. M. (2022). Synergistic 2D $MoSe_2$@WSe_2 nanohybrid hetero structure toward superior hydrogen evolution and flexible supercapacitor. *Nanoscale 14*, 6636–6647. https://doi.org/10.1039/D2NR00632D

57 Wazir, M. B., Daud, M., Safeer, S., Almarzooqi, F., & Qurashi, A. (2022). Review on 2D molybdenum diselenide ($MoSe_2$) and its hybrids for green hydrogen (H_2) generation applications. *ACS Omega 7*, 16856–16865. https://doi.org/10.1021/acsomega.2c00330

58 Zhao, H., Zhang, H., & Huang, R. (2024). Fabrication of MoS_2 with dual defects of O-doping and S-vacancies for high-efficiency hydrogen production. *Electrocatalysis 15*, 20–28. https://doi.org/10.1007/s12678-023-00850-x

59 Wang, X., Chen, Y., Zheng, B., Qi, F., He, J., Li, Q., Li, P., & Zhang, W. (2017). Graphene-like WSe_2 nanosheets for efficient and stable hydrogen evolution. *J. Alloys Compd. 691*, 698–704. https://doi.org/10.1016/j.jallcom.2016.08.305

60 Kadam, S. R., Enyashin, A. N., Houben, L., Bar-Ziv, R., & Bar-Sadan, M. (2020). Ni–WSe_2 nanostructures as efficient catalysts for electrochemical hydrogen evolution reaction (HER) in acidic and alkaline media. *J. Mater. Chem. A 8*, 1403–1416. https://doi.org/10.1039/C9TA10990K

61 Mu, L., Qiu, S., Zhao, G., Zhang, B., Liao, W., Zhao, N., & Xu, X. (2024). A high-efficiency $NiFeSe_4$/$NiSe_2$ bifunctional electrocatalyst with outstanding oxygen evolution reaction and overall water splitting performance. *J. Mater. Chem. A 12*, 1714–1724. https://doi.org/10.1039/D3TA06701G

62 Zhang, L., Lei, Y., & Zhou, D. (2022). Interfacial engineering of 3D hollow $CoSe_2$@ultrathin $MoSe_2$ core@shell heterostructure for efficient pH-universal hydrogen evolution reaction. *Nano Res. 15*, 2895–2904. https://doi.org/10.1007/s12274-021-3887-9

63 Yousaf, S., Abdou, S. N., & Rasheed, T. (2024). Designing NiCoS/CNTs composites for highly efficient bifunctional electrocatalyst in water splitting. *Ionics 30*, 7281–7292. https://doi.org/10.1007/s11581-024-05818-5

64 Liu, H., Yang, F., Chen, F., Che, S., Chen, N., Sun, S., Ta, N., Sun, Y., Wu, N., Sun, Y., & Li, Y. (2023). Interface and electronic structure regulation of Mo-doped $NiSe_2$-$CoSe_2$ heterostructure aerogel for efficient overall water splitting. *J. Colloid Interface Sci. 640*, 1040–1051. https://doi.org/10.1016/j.jcis.2023.02.154

65 Chen, T. T., Wang, R., Li, L., Li, Z., & Zang, S. (2020). MOF-derived Co_9S_8/MoS_2 embedded in tri-doped carbon hybrids for efficient electrocatalytic hydrogen evolution. *J. Energy Chem. 44*, 90–96. https://doi.org/10.1016/j.jechem.2019.09.018

66 Bhat, M. A., Islam, S. U., & Majid, K. (2024). Interfacial electronic structure engineering of $NiCo_2Se_4$ and $NiTe_2$ nanorods for enhanced hydrogen evolution reaction. *ChemPhysChem 25*(23), e202400454. https://doi.org/10.1002/cphc.202400454

67 Wan, C., Ling, Y., Wang, S., Pu, H., Huang, Y., & Duan, X. (2024). Unraveling and resolving the inconsistencies in Tafel analysis for hydrogen evolution reactions. *ACS Cent. Sci. 10*, 658–665. https://doi.org/10.1021/acscentsci.3c01439

68 Wu, Y., Liu, X., Han, D., Song, X., Shi, L., Song, Y., Niu, S., Xie, Y., Cai, J., Wu, S., Kang, J., Zhou, J., Chen, Z., Zheng, X., Xiao, X., & Wang, G. (2018). Electron density modulation of $NiCo_2S_4$ nanowires by nitrogen incorporation for highly efficient hydrogen evolution catalysis. *Nat. Commun. 9*, 1425. https://doi.org/10.1038/s41467-018-03858-w

69 Abeysinghe, J. P. & Gillan, E. G. (2024). Rapid solid-state metathesis reactions for the formation of cobalt–iron monoboride solid-solutions and investigation of their water splitting electrocatalytic activity. *Mater. Adv. 5*, 705. https://doi.org/10.1039/d3ma00728f

70 Chunduri, A., Bhide, A., Gupta, S., Mali, K. H., Bhagat, B. R., Dashora, A., Spreitzer, M., Fernandes, R., Patel, R., & Patel, N. (2023). Exploring the role of multi-catalytic sites in an amorphous Co–W–B electrocatalyst for hydrogen and oxygen evolution reactions. *ACS Appl. Energy Mater. 6*, 4630–4641. https://doi.org/10.1021/acsaem.2c04164

71 McAuley, C. B. (2022). Defining the onset potential. *Curr. Opin. Electrochem. 37*, 101176. https://doi.org/10.1016/j.coelec.2022.101176

72 Vikraman, D., Hussain, S., Akbar, K., Adaikalam, K., Lee, S. H., Chun, S. H., Jung, J., Kim, H. S., & Park, H. J. (2018). Facile synthesis of molybdenum diselenide layers for high performance hydrogen evolution electrocatalysts. *ACS Omega 3*(5), 5799–5807. https://doi.org/10.1021/acsomega.8b00459

73 Gautham Kumar, G., Balaji Bhargav, P., Balaji, C., & Shobana Priyanka, D. (2025). Triggering the phase transition of molybdenum di-selenide ($MoSe_2$) 1T@2H by introducing copper (Cu^+): Experimental insights and DFT analysis for the hydrogen evolution reaction. *Sustain. Energy Fuels 9*, 1338–1352. https://doi.org/10.1039/D4SE01623H

74 Schalenbach, M., Tesch, R., Kowalski, P. M., & Eichelad, R. (2024). The electrocatalytic activity for the hydrogen evolution reaction on alloys is determined by element-specific adsorption sites rather than d-band properties. *Phys. Chem. Chem. Phys. 26*, 14171–14185. https://doi.org/10.1039/d4cp01084a

75 Qadeer, M. A., Zhang, X., Farid, M. A., Tanveer, M., Yan, Y., Du, S., Huang, Z. F., Tahir, M., & Zou, J. (2024). A review on fundamentals for designing hydrogen evolution electrocatalyst. *J. Power Sources 613*, 234856. https://doi.org/10.1016/j.jpowsour.2024.234856

76 Pang, C., Ma, X., Wu, Y., Li, S., Xu, Z., Wang, M., & Zhu, X. (2022). Microflower-like $Co_9S_8@MoS_2$ heterostructure as an efficient bifunctional catalyst for overall water splitting. *RSC Adv. 12*, 22931–22938. https://doi.org/10.1039/D2RA04086G

77 Jothi, P. R., Zhang, Y., Scheifers, J. P., Park, H., & Fokwa, B. P. T. (2017). Molybdenum diboride nanoparticles as highly efficient electrocatalyst for the hydrogen evolution reaction. *Sustain. Energy Fuels 1*, 1928–1934. https://doi.org/10.1039/C7SE00397H

78 Monyoncho, E. A., Woo, T. K., & Baranova, E. A. (2019). Ethanol electrooxidation reaction in alkaline media for direct ethanol fuel cells. *Electrochemistry 15*, 1–57. https://doi.org/10.1039/9781788013895-00001

79 Cermenek, B., Genorio, B., Winter, T., Wolf, S., Connell, J. G., Roschger, M., Letofsky-Papst, I., Kienzl, R., Bitschnau, B., & Hacker, V. (2020). Alkaline ethanol oxidation reaction on carbon supported ternary PdNiBi nanocatalyst using modified instant reduction synthesis method. *Electrocatalysis 11*, 203–214. https://doi.org/10.1007/s12678-019-00577-8

80 Shi, J., Ma, J., Ma, E., Li, J., Hu, Y., Fan, L., & Cai, W. (2024). Electrochemical alcohol oxidation reaction on precious metal-free catalysts: Mechanism, activity, and selectivity. *Carbon Neutral. 3*(2), 285–312. https://doi.org/10.1002/cnl2.116

81 Ozoemena, K. I. (2016). Nanostructured platinum-free electrocatalysts in alkaline direct alcohol fuel cells: Catalyst design, principles and applications. *RSC Adv. 6*, 89523–89550. https://doi.org/10.1039/c6ra15057h

82 Yang, Y. Y., Ren, J., Li, Q. X., Zhou, Z. Y., Sun, S. G., & Cai, W. B. (2014). Electrocatalysis of ethanol on a Pd electrode in alkaline media: An in situ attenuated total reflection surface-enhanced infrared absorption spectroscopy study. *ACS Catal. 4*(3), 798–803. https://doi.org/10.1021/cs401198t

83 Chen, S., Wu, H., Tao, J., Xin, H., Zhu, Y., & Chen, J. (2019). Pt–Ni seed-core-frame hierarchical nanostructures and their conversion to nanoframes for enhanced methanol electro-oxidation. *Catalysts 9*(1), 39. https://doi.org/10.3390/catal9010039

84 Altarawneh, R. M. (2025). Enhancing ethanol electrooxidation in acidic media using Pt nanoparticles supported on metal oxide-modified Vulcan XC72 nanocomposites. *Langmuir 41*(17), 11101–11112. https://doi.org/10.1021/acs.langmuir.5c00838

85 Chung, D. Y., Lee, K. J., & Sung, Y. E. (2016). Methanol electrooxidation on Pt surface: revisiting the cyclic voltammetry interpretation. *J. Phys. Chem. C 120*(17), 9028–9035. https://doi.org/10.1021/acs.jpcc.5b12303

86 Barakat, N. A. M., Yassin, M. A., Al-Mubaddel, F. S., & Amen, M. T. (2010). New electrooxidation characteristic for Ni-based electrodes for wide application in methanol fuel cells. *Appl. Catal. A: Gen. 555*(5), 148–154. https://doi.org/10.1016/j.apcata.2018.02.016

87 Liu, C., Yang, F., Schechter, A., & Feng, L. (2023). Recent progress of Ni-based catalysts for methanol electrooxidation reaction in alkaline media. *Adv. Sens. Energy Mater. 2*, 100055. https://doi.org/10.1016/j.asems.2023.100055

88 Antolini, E. (2019). Glycerol electro-oxidation in alkaline media and alkaline direct glycerol fuel cells. *Catalysts 9*(12), 980. https://doi.org/10.3390/catal9120980

89 Dong, L., Chang, G. R., Feng, Y., Yao, X. Z., & Yu, X. Y. (2022). Regulating Ni site in NiV LDH for efficient electrocatalytic production of formate and hydrogen by glycerol electrolysis. *Rare Met. 41*, 1583–1594. https://doi.org/10.1007/s12598-021-01881-3

90 Gomes, J. F. & Filho, G. T. (2011). Spectroscopic studies of the glycerol electro-oxidation on polycrystalline Au and Pt surfaces in acidic and alkaline media. *Electrocatalyst 2*, 96–105. https://doi.org/10.1007/s12678-011-0039-0

91 Arshad, F., Haq, T., Hussain, I., & Sher, F. (2021). Recent advances in electrocatalysts toward alcohol-assisted, energy-saving hydrogen production. *ACS Appl. Energy Mater. 4*, 8685–8701. https://doi.org/10.1021/acsaem.1c01932

92 Hong, A., Du, X., Qian, K., Fang, Z., Duan, Y., Sui, J., Wei, W., & Li, R. (2024). Amorphous cobalt–nickel borides boost electrocatalytic ethanol oxidation coupled with energy-saving hydrogen production. *Catal. Sci. Technol. 14*, 4007–4018. https://doi.org/10.1039/D4CY00277F

93 Braun, M., Chatwani, M., Kumar, P., Hao, Y., Sanjuan, I., Apostoleri, A., Brix, A. C., Morales, D. C., Hagemann, U., & Heidelmann, M. (2023). Cobalt nickel boride as electrocatalyst for the oxidation of alcohols in alkaline media. *J. Phys. Energy 5*, 024005. https://doi.org/10.1088/2515-7655/acbb2a

94 Qi, Y., Zhang, Y., Yang, L., Zhao, Y., Zhu, Y., Jiang, H., & Li, C. (2022). Insights into the activity of nickel boride/nickel heterostructures for efficient methanol electrooxidation. *Nat. Commun. 13*, 4602. https://doi.org/10.1038/s41467-022-32443-5

95 Yao, H., Wang, Y., Zheng, Y., Yu, X., Ge, J., Zhu, Y., & Guo, X. (2023). High-entropy selenides: A new platform for highly selective oxidation of glycerol to formate and energy-saving hydrogen evolution in alkali-acid hybrid electrolytic cell. *Nano Res. 16*, 10832–10839. https://doi.org/10.1007/s12274-023-5842-4

96 Ramakrishnan, S., Karuppannan, M., Vinothkannan, M., Ramachandran, K., Kwon, O. J., & Yoo, D. J. (2019). Ultrafine Pt nanoparticles stabilized by MoS_2/N-doped reduced graphene oxide as durable electrocatalyst for alcohol oxidation and oxygen reduction reactions. *ACS Appl. Mater. Interfaces 11*(13), 12504–12515. https://doi.org/10.1021/acsami.9b00192

97 Zhao, M., Yang, X., Fu, Z., Wang, W., Wen, W., Xiao, H., Zhang, L., Zhang, J., Lv, B., & Jia, J. (2024). Facile electrolysis-solvothermal synthesis of NiO_x/graphene for enhanced ethanol oxidation to acetate. *Dalton Trans. 53*, 4237–4242. https://doi.org/10.1039/d3dt03963c

98 Hasan, M. M. (2023). Environment-friendly ascorbic acid fuel cell. *Electrochem 4*(1), 31–41. https://doi.org/10.3390/electrochem4010003

99 Carrettin, S., McMorn, P., Johnston, P., Griffin, K., & Hutchings, G. J. (2002). Selective oxidation of glycerol to glyceric acid using a gold catalyst in aqueous sodium hydroxide. *Chem. Commun. 7*, 696–697. https://doi.org/10.1039/B201112N

100 Wang, Y., Zou, S., & Cai, W. B. (2015). Recent advances on electro-oxidation of ethanol on Pt- and Pd-based catalysts: From reaction mechanisms to catalytic materials. *Catalysts 5*(3), 1507–1534. https://doi.org/10.3390/catal5031507

101 Evans, C. D., Bartley, J. K., Taylor, S. H., Hutchings, G. J., & Kondrat, S. A. (2023). Perovskite supported catalysts for the selective oxidation of glycerol to tartronic acid. *Catal. Lett. 153*, 2026–2035. https://doi.org/10.1007/s10562-022-04111-2

102 Li, J., Yu, H., Chi, J., Luo, X., Li, T., & Shao, Z. (2024). An efficient $NiCoSe_4$/NiCo-LDH/CF catalyst for the co-production of value-added formate and hydrogen via selective methanol electro-oxidation. *J. Mater. Chem. A 12*, 25791–25800. https://doi.org/10.1039/d4ta04581e

103 Du, J., You, S., Li, X., Tang, B., Jiang, B., Yu, Y., Cai, Z., Ren, N., & Zou, J. (2020). In situ crystallization of active NiOOH/CoOOH heterostructures with hydroxide ion adsorption sites on velutipes-like CoSe/NiSe nanorods as catalysts for oxygen evolution and cocatalysts for methanol oxidation. *ACS Appl. Mater. Interfaces 12*(1), 686–697. https://doi.org/10.1021/acsami.9b16626

104 Jalali, F., Sheikhi, S., & Hassani, N. (2025). Novel selenide composite as an effective bifunctional electrocatalyst for energy-saving hydrogen production through methanol-assisted water electrolysis. *Energy Fuels 39*(1), 992–1004. https://doi.org/10.1021/acs.energyfuels.4c04146

105 Suryawanshi, A., John, R. A. B., Bhide, A., Gupta, S., Spreitzer, M., Patel, R., Fernandes, R., & Patel, N. (2024). Designing bifunctional electrocatalysts based on complex cobalt sulfo-boride compound for high-current-density alkaline water electrolysis. *Energy Fuels 38*, 18965–18975. https://doi.org/10.1021/acs.energyfuels.4c03171

106 Gupta, S., Patel, M. K., Miotello, A., & Patel, N. (2020). Metal boride-based catalysts for electrochemical water-splitting: A review. *Adv. Funct. Mater. 30*(1), 1906481.

107 Morales, D. M., Jambrec, D., Kazakova, M. A., Braun, M., Sikdar, N., Koul, A., Brix, A. C., Seisel, S., Andronescu, C., & Schuhmann, W. (2022). Electrocatalytic conversion of glycerol to oxalate on Ni oxide nanoparticles-modified oxidized multiwalled carbon nanotubes. *ACS Catal. 12*(2), 982–992.

108 Woo, J., Moon, B. C., Lee, U., Oh, H.-S., Chae, K. H., Jun, Y., Min, B. K., & Lee, D. K. (2022). Collaborative electrochemical oxidation of the alcohol and aldehyde groups of 5-hydroxymethylfurfural by NiOOH and $Cu(OH)_2$ for superior 2,5-furandicarboxylic acid production. *ACS Catal. 12*(7), 4078–4091.

109 Hu, L., Zhong, P., Zhu, J., Wang, J., Zheng, Y., Lin, X., Zhang, Y., & Yang, H. (2024). Bimetallic metal–organic framework-derived cobalt selenide-based composites as bifunctional electrocatalysts for both hydrogen evolution and mono-alcohol oxidation. *CrystEngComm 26*, 4804–4811.

110 Saxena, A., Kapila, S., Medvedeva, J. E., & Nath, M. (2023). Copper cobalt selenide as a bifunctional electrocatalyst for the selective reduction of CO_2 to carbon-rich products and alcohol oxidation. *ACS Appl. Mater. Interfaces 15*(11), 14433–14446.

6

Non-oxide Electrocatalysts for Glycerol Oxidation-coupled Water Electrolysis

Phonnapha Tangthuam[1,2], *Vipada Aupama*[1,2], *and Soorathep Kheawhom*[1,3,4,*]

[1]*Department of Chemical Engineering, Chulalongkorn University, Bangkok, Thailand*
[2]*Department of Chemical Technology, Chulalongkorn University, Bangkok, Thailand*
[3]*Bio-Circular-Green Economy Technology and Engineering Center, Department of Chemical Technology, Chulalongkorn University, Bangkok, Thailand*
[4]*Center of Excellence on Advanced Materials for Energy Storage, Chulalongkorn University, Bangkok, Thailand*

*Corresponding author: soorathep.k@chula.ac.th

6.1 Introduction and Scope

Water electrolysis provides a direct route to convert renewable electricity into molecular hydrogen. In conventional devices, the cathode catalyzes the hydrogen evolution reaction (HER), while anode catalyzes the oxygen evolution reaction (OER). Although HER can proceed on many catalysts at relatively low overpotentials, OER is intrinsically sluggish. It involves multi-electron, proton-coupled electron-transfer steps and O—O bond formation. As a result, practical electrolyzers operate with significant anodic overpotentials and cell voltages well above the thermodynamic minimum. At the anode, a large fraction of electrical energy is dissipated, and the anodic product, oxygen, often has limited value and imposes additional safety and gas-handling requirements.

Hybrid water electrolysis aims to overcome the energy penalty of conventional electrolysis by replacing the anodic OER with an alternative oxidation reaction that is thermodynamically more favorable and often kinetically faster [1]. Rather than evolving oxygen, the anode oxidizes a suitable feedstock, such as small organic molecules, biomass-derived polyols, or components of waste streams, to yield value-added oxygenated products [2]. If this anodic reaction requires a lower anode potential than OER at the same current density, the full-cell voltage decreases and the electrical energy consumption per kilogram of hydrogen is reduced. At the same time, the anode becomes a coproduction platform for chemicals, so the process delivers hydrogen and upgraded organics in parallel. In Figure 6.1a,b, this concept is illustrated by comparing a conventional HER–OER cell with a glycerol-assisted

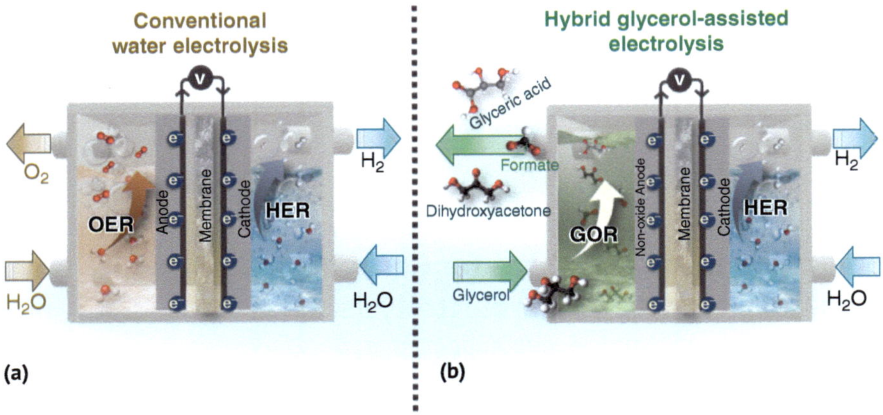

Figure 6.1 (a) Conventional water electrolysis (HER coupled to OER) and (b) hybrid glycerol-assisted electrolysis (HER coupled to glycerol oxidation), highlighting the reduced anodic requirement and the coproduction of hydrogen and glycerol-derived oxygenates.

HER–GOR cell. The latter operates at a lower voltage while producing liquid oxidation products at the anode.

Glycerol is a particularly strategic substrate in this context. It is produced in large volumes as a by-product of the biodiesel and oleochemical industries, often in forms and quantities that exceed existing market demand. Glycerol contains three hydroxyl groups on a flexible C_3 backbone, which enables multiple oxidation pathways. In alkaline media, these routes can generate families of partial-oxidation products, including glyceraldehyde (GLAD), dihydroxyacetone (DHA), glyceric acid, and tartronic acid, as well as more deeply oxidized C_2 and C_1 species, such as glycolate, oxalate, and formate [3, 4]. Many of the corresponding half-reactions have equilibrium potentials lower than that of OER. In practice, glycerol oxidation on active catalysts often begins at potentials where OER remains kinetically suppressed.

The complexity of glycerol oxidation is both an opportunity and a challenge [5]. The reaction network is multistep and multi-electron. It involves numerous competing pathways that differ in which hydroxyl group is activated, the sequence of C—H and O—H bond activation, and the extent of C—C bond cleavage. Catalysts must steer this network toward desirable products while maintaining low overpotentials and stable performance at relevant current densities [6]. The anode should preferentially promote glycerol oxidation over OER, avoid uncontrolled over-oxidation of intermediates, and remain resilient under strongly alkaline and organic-rich conditions.

Non-oxide electrocatalysts are well suited to this task. Metal chalcogenides, phosphides, carbides, nitrides, and their hybrids generally exhibit high electronic conductivity and flexible electronic structures [7]. These properties facilitate efficient charge transport and allow tuning of adsorption energies for key intermediates. Under anodic polarization in alkaline glycerol-containing electrolytes, many non-oxide phases do not remain in their as-synthesized form. Instead, they undergo in situ reconstruction, forming thin layers of metal oxyhydroxide-like species at the surface while retaining a conductive non-oxide core [8]. The resulting core–shell or gradient architectures combine active oxyhydroxide motifs at the interface with highly conductive backbones. This combination is advantageous for demanding multi-electron processes such as glycerol oxidation.

This dynamic behavior provides a flexible design space. By tuning the composition, defect chemistry, and nanostructure of non-oxide catalysts, the extent and nature of surface reconstruction can be systematically controlled [9]. This control governs the accessible oxidation states and the distribution of active sites that favor C_3 partial oxidation over deeper C_2 or C_1 pathways [10]. Non-oxide catalysts can therefore be engineered for both high current density and tailored selectivity.

Within this chapter, earlier sections have examined non-oxide catalysts from the perspectives of their intrinsic properties and their roles in water splitting and alcohol oxidation. Here, non-oxide catalysts are treated as a collective toolbox for glycerol oxidation-coupled electrolysis, with emphasis on principles that span different material families. The central questions address reconstruction under glycerol oxidation conditions, the tuning of electronic and structural features to favor glycerol oxidation over OER and control selectivity, and the integration of these materials into hybrid electrolyzers using glycerol as the anodic substrate.

The chapter progresses from foundational concepts to design principles and, finally, to system-level implementation. It first outlines the thermodynamic and mechanistic basis of glycerol oxidation in alkaline media and defines performance descriptors that are specific to glycerol-assisted hybrid electrolysis, supported by a reaction network and a qualitative potential–pH operating map. It then surveys key non-oxide electrocatalyst families and the structural motifs used to engineer them for glycerol oxidation, framed by a conceptual materials landscape and representative catalyst systems. Building on this framework, the discussion develops mechanistic insights that connect surface evolution, adsorption energetics, and oxidation-state accessibility to activity and selectivity, and distills these links into qualitative design rules. The chapter next addresses durability and poisoning, and explains how these phenomena shape performance in alkaline–alkaline and hybrid alkali–acid electrolyzer configurations. It closes by highlighting operando characterization strategies and future opportunities in data-driven discovery, operation with realistic glycerol streams, and integration into practical devices.

6.2 Fundamentals of Glycerol Oxidation in Hybrid Electrolysis

Design of effective non-oxide catalysts and hybrid electrolyzers requires a clear understanding of how glycerol oxidation differs thermodynamically from water oxidation. It also requires insight into the organization of the glycerol reaction network and the performance metrics most informative for glycerol-assisted systems [11]. In Figure 6.2, the key elements are summarized. It shows a reaction network and a qualitative map of how potential and pH influence the dominant pathways.

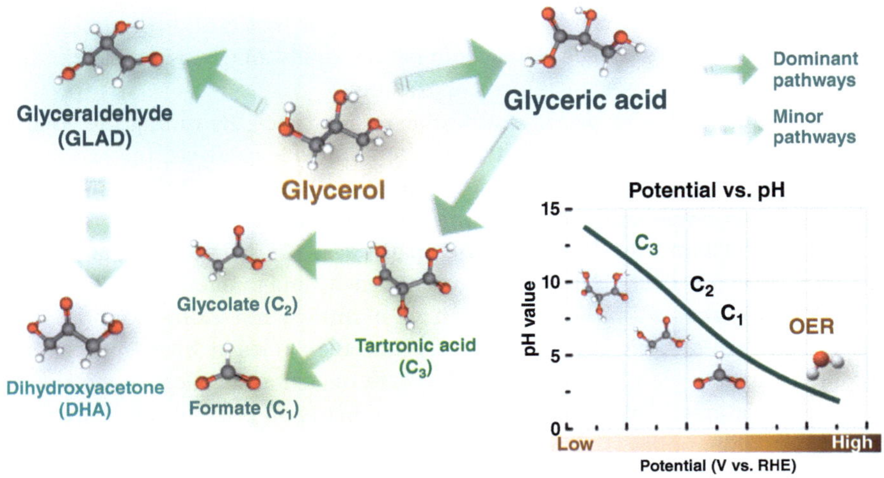

Figure 6.2 Schematic illustration of the glycerol oxidation reaction network, showing primary pathway GLAD, with DHA as a minor product formed subsequently from GLAD. Further oxidation leads to C_3 acids (glyceric and tartronic acids), and, at higher driving forces, to C_2 and C_1 products (glycolate and formate). Solid arrows indicate dominant pathways, while dashed arrows represent minor pathways.

6.2.1 Thermodynamic Comparison of OER and Glycerol Oxidation

In alkaline water electrolysis, the anode performs OER. The equilibrium potential is well defined, but the reaction sequence is complex and requires the formation and cleavage of O—O bonds. Practical operation at tens to hundreds of $mA{\cdot}cm^{-2}$ requires substantial overpotentials. Cell voltages in real devices typically range from 1.6 to 1.8 V or higher, depending on the catalyst, electrolyte, and temperature [12].

Glycerol oxidation, by contrast, involves an organic substrate with preexisting C—C bonds and multiple hydroxyl groups. There is no single half-reaction, since different products correspond to different stoichiometries [13]. Many relevant oxidation pathways from glycerol to GLAD, DHA, C_3 acids, or C_2 and C_1 products have equilibrium potentials substantially lower than that of the OER in alkaline media [14]. For catalytically active materials, glycerol oxidation can initiate at potentials where water oxidation remains kinetically slow, allowing both partial- and deeper-oxidation products to form while oxygen evolution is suppressed.

If the cathode operates near the HER equilibrium potential, replacing OER with glycerol oxidation at the anode directly affects the full-cell voltage. At a given current density, the required anode potential is reduced by the difference between the overpotentials for glycerol oxidation and OER. The overall cell voltage therefore decreases, lowering the electrical energy consumption per kilogram of hydrogen. The potential window in which glycerol oxidation dominates typically lies below that in which OER becomes significant. The magnitude of this advantage depends on the specific glycerol oxidation pathway and the catalyst. In particular, it depends on whether the system is biased toward partial C_3 oxidation or toward more deeply oxidized C_2 and C_1 products.

6.2.2 Glycerol Oxidation Reaction Network

Glycerol contains two primary and one secondary hydroxyl groups. In alkaline solution, deprotonation yields glycerolate species that adsorb on the catalyst surface. The subsequent reaction network is rich and branched. Two main C_3 oxidation routes are commonly distinguished [10]. In one route, oxidation begins at a primary hydroxyl group to yield GLAD, which can be further oxidized to glyceric acid and, under stronger driving forces, to higher-valent C_3 acids such as tartronic acid [4]. In the second route, the secondary hydroxyl group is oxidized to a carbonyl group, forming DHA, which can subsequently be oxidized or rearranged to C_3 acids. These pathways preserve the C_3 backbone and are therefore particularly attractive when targeting higher-value C_3 products [11]. In parallel, C—C bond cleavage becomes increasingly prominent at higher potentials or on highly oxidizing catalyst surfaces. This process leads to the formation of C_2 species such as glycolate and oxalate and C_1 species such as formate. These products are more thermodynamically oxidized and often predominate when high-valent metal sites or higher potentials are present.

The prevailing branch depends on catalyst composition and structure, surface oxidation state, potential and pH, glycerol concentration, and mass transport. These parameters determine how glycerolate binds and which hydroxyl group is preferentially activated. They also govern how readily C—C bonds are cleaved and when the OER becomes competitive. Together, they define regions in potential–pH space where C_3 partial oxidation is favored and regions where deeper C_2 and C_1 oxidation and OER dominate.

6.2.3 Performance Metrics for Glycerol-assisted Electrolysis

Hybrid glycerol-assisted electrolysis employs standard descriptors of electrocatalytic performance, complemented by metrics that capture product distribution and carbon utilization. At the cell level, the primary energetic descriptor is the cell voltage as a function of current density. For a given hydrogen production rate, the reduction in cell voltage relative to an OER-based reference directly indicates the energetic gain. This advantage can be expressed as an absolute voltage difference or as electrical energy consumption per unit of hydrogen produced.

At the anode, performance must be characterized by both rate and selectivity. For each anodic product, a partial current density can be defined from its formation rate and electron stoichiometry. FE is then the fraction of the total current used to form that product. Reporting FE for hydrogen at the cathode and for the dominant glycerol oxidation products at the anode indicates the efficiency with which current is used. Because glycerol contains carbon, carbon efficiency is also essential. It may be defined as the fraction of input carbon that appears in chosen target products, normalized to the total carbon detected in all products. High carbon efficiency toward a specific product implies minimal over-oxidation or diversion into undesired species. Product yield rates, such as moles of product per hour per electrode area or per reactor volume in flow cells, combine current density and selectivity into a productivity measure for both hydrogen and organics.

Hybrid systems must also contend with competition between glycerol oxidation and OER. At a given potential, the total anodic current is shared between these reactions. A useful concept is the potential window during which glycerol oxidation dominates, and OER remains suppressed. Within this window, the current–potential curve for glycerol oxidation is steep while that for OER is relatively flat. Outside it, OER increasingly consumes current, eroding voltage savings and influencing product distributions. Designing catalysts and operating conditions that broaden and stabilize this window is a central objective in the subsequent sections. Taken together, cell voltage, partial current densities, FE, carbon efficiency, product yield rates, and the extent of the glycerol-dominated potential window form a framework for comparing catalysts and hybrid cell designs.

6.3 Non-oxide Electrocatalyst Families and Design Strategies for Glycerol Oxidation

With fundamentals in place, attention turns to the materials toolbox. Here, the focus is on non-oxide electrocatalysts based on metal chalcogenides, phosphides, carbides, nitrides, and their hybrids. These materials are treated as members of a shared design space characterized by high electronic conductivity, flexible electronic structures, and dynamic reconstruction under anodic conditions. In Figure 6.3, this shared design space is outlined. In Table 6.1, representative catalyst systems, operating conditions, and key performance metrics are summarized.

6.3.1 Comparative Overview Under Glycerol Oxidation Conditions

Metal chalcogenides, particularly sulfides and selenides of Ni, Co, Fe, and Mo, have been widely investigated for alkaline electrocatalysis [26]. They often exhibit high electronic conductivity and flexible valence states, making them attractive precursors for glycerol oxidation. Under anodic polarization in alkaline glycerol solutions, these materials undergo partial anion leaching and oxidation of metal centers, forming thin, oxyhydroxide-like surface layers while retaining a chalcogenide-rich, conductive core. The resulting structures can be viewed as core–shell or graded architectures. In these architectures, the surface provides redox-active metal–oxygen sites for dehydrogenation and C—O or C—C bond activation, while the core ensures efficient electron transport and mechanical stability.

Metal phosphides exhibit related yet distinct behavior. Phosphorus incorporation modifies the metal's electronic structure and can increase electron density at active sites, thereby altering adsorption energies for hydroxide and organic intermediates [27]. During glycerol oxidation, many phosphides transform to mixed-metal oxyhydroxide–phosphate surfaces. Phosphate species at or near the interface influence local proton and hydroxide activities and the hydrogen-bonding environment, which in turn affects adsorption and stability of glycerolate and carboxylate intermediates. Compared with chalcogenides, phosphides offer additional tuning parameters via phosphorus content and distribution, enabling adjustments to activity, selectivity, and corrosion resistance.

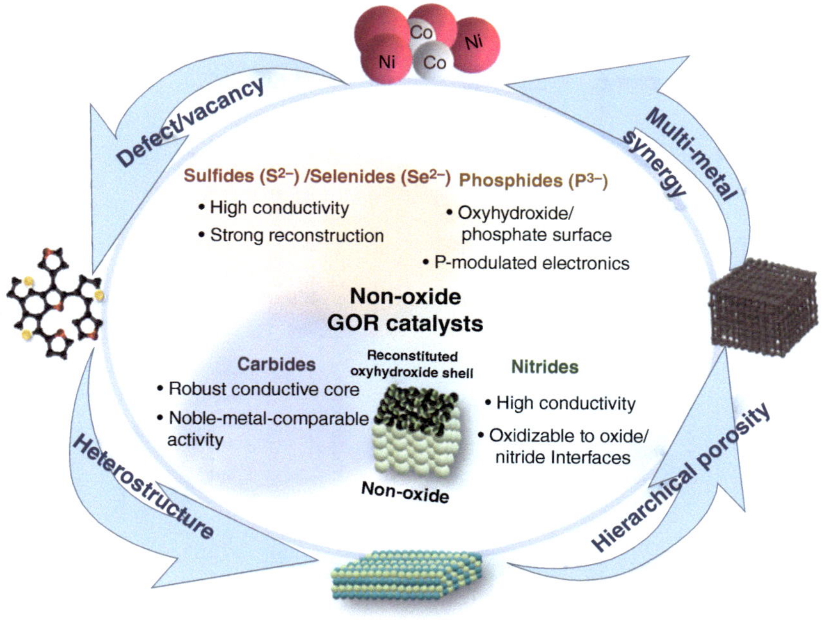

Figure 6.3 Materials landscape and reconstructed active states of non-oxide GOR catalysts. Sulfides/selenides, phosphides, carbides, and nitrides are positioned on a conceptual map, with hybrids located at the overlaps. Each family is depicted as a conductive non-oxide core with a thin reconstructed oxyhydroxide shell that governs GOR under anodic bias. Key attributes are highlighted: strong reconstruction and frequent GOR/OER activity (sulfides/selenides), P-modulated oxyhydroxide/phosphate surfaces (phosphides), robust conductive cores (carbides), and oxidizable oxide/nitride interfaces (nitrides).

Metal carbides and nitrides form a third major class. They often exhibit metallic conductivity and robust metal–nonmetal bonding. Under strongly oxidizing conditions, they can evolve into oxide-rich surfaces. Within the glycerol oxidation potential window, they commonly act as conductive cores supporting thin oxyhydroxide or mixed oxide–non-oxide layers. In many reported systems, their primary role is to provide low-resistance electron pathways and mechanical support for nanostructured architectures [28]. At the same time, the immediate reaction environment is governed by oxidized metal species that mediate glycerol oxidation.

In practice, hybrid architectures that combine multiple non-oxide families and carbon supports are common. Examples include sulfide–phosphide heterostructures, carbide-supported chalcogenides, and nitride–phosphide composites. In such systems, one component may provide a high density of redox-active sites, another may stabilize desirable oxidation states or defect configurations, and a third may enhance conductivity or control wettability. These hybrids occupy the intersections of the materials landscape and often achieve performance that exceeds that of the individual phases.

Across all families, a common pattern emerges. Under glycerol oxidation conditions, non-oxide catalysts are best regarded as dynamically evolving core–shell or

Table 6.1 Representative non-oxide electrocatalysts for glycerol oxidation-coupled water electrolysis. For each system, the table lists catalyst composition and structure, non-oxide family, e.g., sulfide, phosphide, carbide, nitride, and hybrid, synthesis route, electrolyte conditions, dominant anodic products, typical potentials or cell voltages at specified current densities, FE for hydrogen and key products, and stability metrics. The entries illustrate the diversity of material choices and performance profiles across the non-oxide design space.

Catalyst system	Electrolyte	Performance descriptor	Major products and selectivity	Durability	Reference
Ni–Mo–S/NF (sulfide; coral-like, bifunctional anode and cathode)	1 M KOH + 0.1 M glycerol (alkaline)	$E_{anode} \approx 1.31$ V vs. RHE at 10 mA·cm^{-2}; $V_{cell} \approx$ 1.37 V at 10 mA·cm^{-2}	Formate: FE $\approx$ 95%; H_2: FE $\approx$ 99.6%	25 h @ 10 mA·cm^{-2} (anodic GOR); 10-cycle FE stability	[15]
NiOOH/Ni_3S_2/NF (sulfide-derived oxyhydroxide skin on 3D scaffold)	1 M KOH + 0.1 M glycerol	$E_{anode} = 1.227$ V vs. RHE @ 10 mA·cm^{-2}	Formate: FE $\approx$ 97.7%	24 h @ 10 mA·cm^{-2}	[16]
Ni_3N/Co_3N nanowires (nitride heterostructure, conductive core)	1 M KOH + 0.1 M glycerol	$E_{anode} = 1.26$ V vs. RHE @ 50 mA·cm^{-2}	Formate: FE $\approx$ 94.6%	12 h (reported at 1.35 V vs. RHE)	[17]
$CoSe_2$/CC (selenide; reconstructed surface under GOR)	1 M KOH + 0.5 M glycerol	$E_{anode} = 1.23$ V vs. RHE @ 10 mA·cm^{-2}	Formate: FE $\approx$ 88%	90 h (reported at 1.48 V vs. RHE)	[18]
Mn–$CoSe_2$/CFC (selenide; Mn-doped $CoSe_2$ on carbon-fiber cloth)	1 M KOH + 0.1 M glycerol	$E_{anode} = 1.27$ V vs. RHE @ 10 mA·cm^{-2}	Formate: FE > 90%	12 h @ 10 mA·cm^{-2}	[19]
Mn–CoN@C/NF (nitride; carbon-shelled nanoarrays on Ni foam)	1 M KOH + 0.3 M glycerol	$E_{anode} = 1.19$ V vs. RHE @ 10 mA·cm^{-2}	Formate: FE $\approx$ 97.7%	70 h (reported at 1.4 V vs. RHE)	[20]
Co–Mn–S/NF (sulfide or oxysulfide-derived; reconstructed surface)	1 M KOH + 0.1 M glycerol	$E_{anode} = 1.347$ V vs. RHE at 100 mA·cm^{-2}	Formate: FE $\approx$ 90%	60 h @ 10 mA·cm^{-2}	[21]
$CoCuN_{0.6}$/CP (Cu-modified Co nitride on conductive paper)	1 M KOH + 0.1 M glycerol	$E_{anode} = 1.07$ V vs. RHE @ 10 mA·cm^{-2}	Formate: FE $\approx$ 90%	60 h (reported at 1.43 V vs. RHE)	[22]
HEA–CoNiCuMnMo/CC (high-entropy alloy; compositional tuning)	1 M KOH + 0.1 M glycerol	$E_{anode} = 1.25$ V vs. RHE @ 100 mA·cm^{-2}	Formate: FE > 90%	12 h @ 10 mA·cm^{-2}	[23]
$FeCoNiCrMnS_2$/CC (high-entropy sulfide on carbon cloth)	1 M KOH + 0.1 M glycerol	$E_{anode} = 1.10$ V vs. RHE @ 10 mA·cm^{-2}	Formate: FE > 90%	12 h @ 100 mA·cm^{-2}	[24]
(CoNiCuMnMo)Se/CF (high-entropy selenide on carbon felt)	1 M KOH + 0.1 M glycerol	$E_{anode} = 1.20$ V vs. RHE @ 10 mA·cm^{-2}	Formate: FE > 90%	25 h @ 10 mA·cm^{-2}	[25]

gradient materials. The non-oxide core provides conductivity and structural support, whereas the reconstructed oxyhydroxide-rich surface defines the immediate reaction environment. Anion species such as S, P, N, and C influence this environment indirectly through electronic and structural effects rather than acting as static active sites.

6.3.2 Design Motifs Across Non-oxide Families

Several structural and compositional motifs recur across non-oxide catalyst families when the objective is to optimize glycerol oxidation. One prominent motif is multimetal synergy. Incorporating two or more transition metals (Ni–Co, Ni–Fe, and Co–Fe) into non-oxide frameworks yields ensembles of active sites with distinct oxidation states and ligand environments. Under operating conditions, these multimetal non-oxides typically reconstruct into mixed-metal oxyhydroxide surfaces, whose redox potentials and adsorption properties can be systematically tuned through compositional control. This tunability enables deliberate adjustment of the balance between C_3 partial oxidation and deeper C_2 or C_1 oxidation, as well as control over the potential window in which glycerol oxidation dominates over the OER.

Nanostructuring and hierarchical porosity form a second motif. High surface area alone is not sufficient in viscous glycerol-containing electrolytes. Pore size distribution and architecture strongly influence the mass transport of glycerol, intermediates, and hydroxide. Mesopores facilitate molecular diffusion. Macropores enhance electrolyte circulation and bubble removal, particularly in flow or gas-evolving devices. Nanosheets, vertically aligned nanowires, hollow structures, and three-dimensional networks reduce diffusion distances and minimize ohmic losses. These architectures must also accommodate increased viscosity and product accumulation without rapid pore blockage.

A third motif is defect and vacancy engineering. Anion vacancies in sulfides and selenides, and analogous defects in phosphides, carbides, and nitrides, create undercoordinated metal sites with modified electronic environments. Such sites can show altered adsorption energies for glycerolate, hydroxide, and carbonyl species, often lowering barriers for key bond-breaking and bond-forming steps. Excessive defect densities, however, undermine structural integrity, accelerate dissolution, and promote over-oxidation to inactive phases. Accordingly, optimal defect levels represent a compromise between enhanced reactivity and sufficient stability.

Heterostructures represent a fourth motif. Interfaces between non-oxide and oxide phases, or between non-oxide and carbon, often exhibit enhanced reactivity because charge redistribution at the interface creates regions with tailored electron density and binding properties. Non-oxide–oxide interfaces combine the high conductivity of non-oxides with the favorable surface chemistry of oxides or oxyhydroxides. Non-oxide–carbon heterostructures improve electron transport, wettability, and local reactant concentrations. These motifs define the practical design space across non-oxide families. The summary table can be read not only as a list of materials and performance data but also as a catalogue of how these motifs have been implemented in practice.

6.4 Mechanistic Insights: Activity and Selectivity Tuning on Non-oxide Surfaces

To progress from qualitative design motifs to predictive catalyst design, it is necessary to understand how non-oxide catalysts operate under glycerol oxidation conditions. Key questions concern the identity of the active phase, the descriptors that correlate with catalytic activity, and the levers that govern selectivity both between glycerol oxidation and the OER and within the glycerol oxidation network itself. In Table 6.2, the resulting design rules are organized by linking structural factors to mechanistic effects and practical implications.

6.4.1 Active Phase and In Situ Reconstruction

Under glycerol oxidation conditions, the as-synthesized non-oxide phase is rarely identical to the active state. In alkaline glycerol solutions, metal sulfides and selenides undergo partial anion leaching and metal oxidation to form oxyhydroxide-like surfaces. Phosphides generate surface phosphate and oxyhydroxide species. Carbides and nitrides develop mixed oxide–non-oxide interfaces. The outer few nanometers are dominated by metal–oxygen coordination environments, whereas the interior retains non-oxide character.

This reconstructed surface hosts the redox-active metal sites that mediate glycerolate adsorption, O—H and C—H dehydrogenation, C—C cleavage, and product desorption. The underlying non-oxide core ensures efficient electron transport to and from these sites, influencing their electronic structure through charge transfer and strain. Reconstruction depth and continuity are crucial. Thin, continuous layers on conductive cores enable rapid charge transfer and high turnover frequencies. In contrast, thick, resistive oxides increase charge-transfer resistance and shift operation into regimes where OER is more favorable. From a design standpoint, controlling reconstruction emerges as a central lever. Composition, dopants, and potential limits should stabilize a reconstructed state that is sufficiently developed to provide the metal–oxygen motifs required for glycerol oxidation. This state must remain thin and conductive to preserve low resistance and maintain a wide glycerol-dominated potential window.

6.4.2 Activity Descriptors

Activity in glycerol oxidation depends on how surface sites interact with glycerolate and related species. A small set of qualitative descriptors, derived from electronic structure and adsorption behavior, is useful for organizing trends. One descriptor is the effective position of surface d-states relative to the Fermi level. Higher-lying d-bands tend to yield stronger binding of oxygenated intermediates. This descriptor reflects the effective position of surface d-states relative to the Fermi level. Lower d-bands correspond to weaker binding and potentially higher activation barriers. In multimetal systems, composition controls this d-band position and therefore provides a way to tune adsorption strength.

Table 6.2 Mechanistic design rules for non-oxide glycerol oxidation electrocatalysts.

Design factor	Mechanistic effect	Implication for activity	Implication for selectivity	Implication for durability
Multimetal synergy (Ni–Co, Ni–Fe, and Co–Fe)	Tunes redox potentials and adsorption in the reconstructed oxyhydroxide shell; provides complementary sites for dehydrogenation and oxygenate stabilization	Lowers GOR onset and raises current via faster redox cycling and improved charge transfer	Broadens the GOR-dominant window and delays OER; composition can shift products from C_3 partial oxidation to deeper C_1/C_2 oxidation	Distributes redox load and slows leaching and over-oxidation, stabilizing the active shell
Defect and vacancy density (S/Se/P/N/C vacancies)	Creates undercoordinated sites and modifies local charge; influencing glycerolate activation and reconstruction kinetics	Enhances higher site density and lower barriers when defects are moderate	Moderate defect densities can favor C_3-selective pathways; excessive defects promote strong oxidation, C—C cleavage, and earlier OER	Excessive defect densities accelerate dissolution and uncontrolled oxide thickening; optimized defect levels improve stability
Reconstruction depth and continuity	Forms an oxyhydroxide-rich surface on conductive core; shell thickness controls resistance and access to high-valent sites	Thin, continuous shells yield low charge-transfer resistance; thick shells penalize kinetics	Controlled reconstruction favors GOR over OER; thick oxides narrow the window and promote OER and deep oxidation	Self-limiting shells resist drift; progressive thickening increases voltage and accelerates decay
Hydroxide vs. glycerolate affinity	Competitive adsorption of OH-derived species and glycerolate determines surface coverage and dominant channel	Balanced binding avoids site blocking and supports sustained reaction rate	Preferential glycerolate adsorption suppresses OER; overly strong binding increases poisoning and selectivity drift	Tailored affinity reduces fouling and preserves wettability; strong carboxylate binding accelerates deactivation
Oxidation-state accessibility and redox mediator behavior	Ni/Co/Fe redox couples mediate dehydrogenation and oxygen insertion; very-high-valent states promote C—C cleavage and OER	Rapid reversible redox cycling increases turnover at practical current densities	Intermediate oxidation states favor C_3 oxygenates and acids; higher oxidizing states increase C_2/C_1 products and OER competition	Operation in a stable redox range reduces corrosion and irreversible phase change

(Continued)

Table 6.2 (Continued)

Design factor	Mechanistic effect	Implication for activity	Implication for selectivity	Implication for durability
Nanostructuring and hierarchical porosity	Improves transport in viscous glycerol; enhances ionic access and bubble release	Higher utilization at high current with lower transport losses	Stabilizing local pH can reduce over-oxidation; poorly designed pores can increase deep oxidation via long residence time	Robust 3D scaffolds reduce delamination; suitable pores limit blockage by products and oligomers
Heterostructures and conductive supports (non-oxide/oxide, non-oxide/carbon)	Interfacial charge redistribution and strain tune adsorption; improve conductivity and shell anchoring	Lower ohmic losses and higher apparent activity via improved electron pathways and optimized interfacial sites	Interfaces can bias adsorption geometry and suppress early OER; mixed motifs can steer primary vs. secondary hydroxyl activation	Strong anchoring limits sintering and detachment; carbon supports require potential control to limit corrosion
Electrolyte and operating protocol (KOH, glycerol, potential cycling, and flow)	Controls glycerolate speciation, OH availability, and reconstruction rate; flow governs product removal and local gradients	Optimized alkalinity and glycerol improve kinetics; flow reduces transport limitation	Potential cycling and pulsing help maintain the GOR window and suppress OER; flow reduces secondary reactions and improves carbon efficiency	Pretreatment and flushing reduce poisoning; realistic feeds can accelerate fouling and membrane issues, requiring system mitigation

A second descriptor is the effective oxidation state or charge density on surface metal centers. Sites that are too reduced may lack sufficient oxidizing power to drive deeper dehydrogenation. Highly oxidized sites, by contrast, can encourage over-oxidation and C—C cleavage and may also promote the OER. Many high-performing catalysts operate with a distribution of intermediate oxidation states, which can be viewed as a compromise between these extremes.

A third descriptor concerns the relative affinity of the surface for glycerolate vs. hydroxide. Sites that bind hydroxide much more strongly than glycerolate tend to favor OER, because water-derived intermediates occupy the surface more easily. Sites that bind glycerolate preferentially within a suitable potential window support glycerol oxidation at lower overpotentials. If glycerolate binding is too strong, coverage can become excessive and impede turnover. Design factors such as hydroxide affinity and glycerolate binding strength summarize these relationships and are consolidated in the design rules table. These descriptors are usually inferred from calculations and experimental trends rather than measured directly. Nevertheless, they provide a useful framework for interpreting how changes in composition, defect density, or heterostructure influence activity.

6.4.3 Selectivity: Glycerol Oxidation vs. OER and Within the Glycerol Network

Selectivity must be considered at two levels. The first is the competition between glycerol oxidation and OER. For the given catalyst and electrolyte, both reactions may be accessible, and the applied potential determines their relative rates. If the reconstructed surface is highly active toward OER and binds OER intermediates strongly, oxygen evolution can initiate at potentials only slightly higher than those required for glycerol oxidation. This behavior narrows the potential window in which glycerol oxidation dominates. In that case, a large fraction of the current at higher potentials is consumed by OER, reducing the energy advantage of hybrid operation.

Favoring glycerol oxidation over OER requires a region where adsorbed glycerolate and intermediates can be activated with lower barriers than water-derived species. This selectivity window typically corresponds to moderate oxidation states and surface chemistries that stabilize glycerol-derived intermediates without excessively stabilizing OER intermediates. Composition, defect density, and reconstruction depth influence this balance by shaping local electronic structure and coordination environments.

The second level concerns selectivity within the glycerol oxidation network. The surface effectively determines which hydroxyl group is attacked first, whether C—C bonds are cleaved, and how far carbon atoms are oxidized. Adsorption geometry is central. Surfaces that favor binding through a primary hydroxyl group support the GLAD pathway. Surfaces that stabilize binding through the secondary hydroxyl group promote formation of DHA. Local orientation and steric constraints govern which C—H bonds are accessible for abstraction and which transition states are favored.

Relative barriers for C—H vs. C—C activation are equally important. Strong metal–oxygen bonds and highly oxidized sites can lower barriers for C—C cleavage and drive the system toward C_2 and C_1 products. Surfaces that moderate this tendency favor C_3 partial oxidation. Alloying, doping, and heterostructure formation adjust these barriers by modulating local charge density and orbital overlap. Co-adsorption of hydroxide, glycerolate, intermediates, and products further shapes the reaction landscape. High hydroxide coverage can compete with glycerolate for adsorption and steer the system toward OER, while accumulation of intermediates can block active sites or promote secondary reactions. Qualitative rules emerging from these considerations include the following. Multimetal compositions and intermediate oxidation states can widen the potential window in which glycerol oxidation dominates. Surfaces that favor primary or secondary hydroxyl activation can be used to steer selectivity toward different C_3 products. Structural features that suppress C—C cleavage promote C_3-selective operation. Features that stabilize highly oxidized intermediates promote deeper C_2 and C_1 oxidation.

6.4.4 Consolidated Design Rules

Reported catalyst systems illustrate these principles. Some Ni–Co sulfide or phosphide precursors, after reconstruction, form mixed Ni–Co oxyhydroxide surfaces that display a broad potential window in which glycerol oxidation currents are high while OER remains comparatively suppressed. These systems often yield C_3 products such as GLAD, DHA, and glyceric acid with high FE at moderate potentials [29]. Mechanistically, the combination of Ni and Co tunes d-band position and oxidation state distributions so that glycerolate activation is facile, C—C cleavage barriers remain relatively high, and hydroxide affinity is balanced. Other systems, for example, certain Ni–Fe or Co–Fe non-oxides with strongly oxidizing reconstructed surfaces, promote deeper oxidation to C_2 and C_1 products and show earlier OER onset. Here, higher average oxidation states and different metal–oxygen bonding patterns lower C—C cleavage barriers and stabilize more oxidized intermediates, at the expense of a narrower glycerol-dominated potential window.

Between these extremes are catalysts designed to sustain strong glycerol oxidation while delaying OER onset. These systems often employ multimetallic compositions, controlled defect densities, and carefully constructed heterostructures that moderate hydroxide affinity and maintain a thin, conductive reconstructed layer. These insights are consolidated in the design rules table by mapping design factors, such as multimetal synergy, reconstruction depth, defect density, hydroxide and glycerolate affinity, nanostructure, and heterostructure, onto mechanistic effects and practical implications for activity, selectivity, and stability.

6.5 Durability, Poisoning, and Hybrid Electrolyzer Architectures

This section examines degradation and poisoning phenomena specific to non-oxide glycerol oxidation catalysts and connects them to alkaline–alkaline and hybrid alkali–acid electrolyzer architectures. Catalysts designed according to the principles

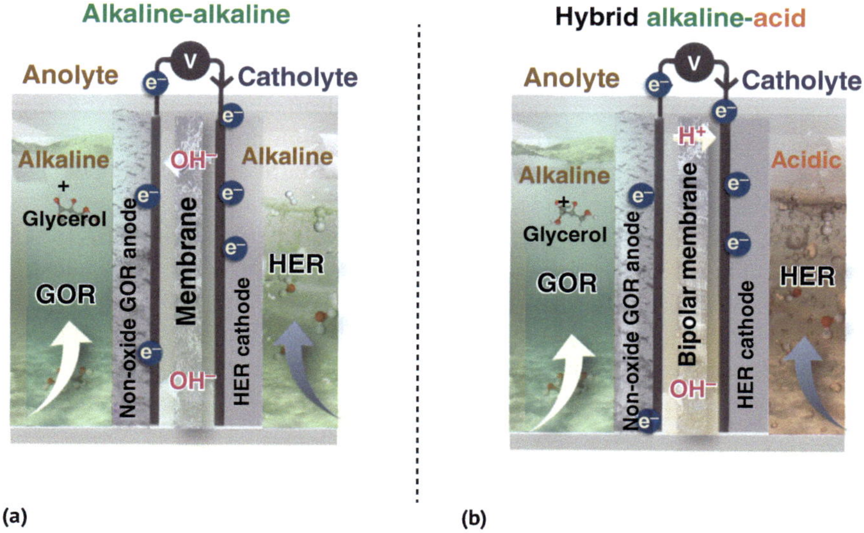

Figure 6.4 Hybrid electrolyzer architectures for glycerol-assisted hydrogen production. (a) Alkaline–alkaline cell with a 3D porous non-oxide GOR anode and an alkaline HER cathode; an AEM or separator conducts OH^-/anions and separates hydrogen from liquid products. (b) Hybrid alkali–acid cell with an alkaline glycerol anolyte and acidic catholyte separated by a bipolar or ion-exchange membrane, maintaining a pH gradient and further reducing cell voltage via electro-neutralization.

outlined previously must ultimately operate in complete cells. In practice, this often requires stable performance over extended periods and at high current densities. In Figure 6.4a,b, two archetypal configurations are depicted.

6.5.1 Degradation Modes

Several degradation processes are particularly relevant for non-oxide glycerol oxidation anodes. Anion and metal leaching are common in chalcogenides and phosphides. Oxidative conditions can drive dissolution of S, Se, or P species and, to some extent, of less noble metal cations. Over time, selective leaching modifies local composition and defect structure in the reconstructed layer and, in severe cases, in the core. The catalyst may drift away from the optimized region of composition space, leading to lower activity and changes in selectivity. Uncontrolled thickening of oxide or oxyhydroxide layers is another degradation pathway. Although a thin, conductive oxyhydroxide shell is beneficial, continued oxidation can produce thicker, less conductive films that increase charge-transfer resistance and reduce accessible surface area. The operating potential may then be forced into a region where OER is more favorable, diminishing the hybrid advantage.

Morphological instability is also important. Nanostructured architectures such as nanosheets, nanowires, and hollow structures can restructure, sinter, or fracture under redox cycling, local Joule heating, and gas bubble evolution.

Structural collapse reduces effective surface area and disrupts electron and ion percolation pathways. When catalysts are supported on carbon, local oxygen generation and high potentials can also promote carbon corrosion and delamination. Organic fouling is specific to organic oxidation. Strongly adsorbed carboxylates and oligomeric or polymeric species derived from aldehyde intermediates can accumulate on catalyst surfaces and within pores. In addition, impurities in technical or crude glycerol, such as soaps, salts, and residual organics, can contribute to fouling. These species can block active sites, alter wettability, and hinder mass transport. In flow systems, fouling can reduce porosity and increase pressure drop. These degradation modes can interact. Oxide thickening can alter susceptibility to fouling. Organic films may either temporarily shield the underlying catalyst or create local conditions that accelerate corrosion. Understanding these couplings is vital for achieving durable operation.

6.5.2 Strategies for Durability and Poisoning Resistance

Mitigation strategies operate at three levels: material design, electrode architecture, and operating protocol. At the material design level, alloying and judicious doping can stabilize the active state against excessive leaching and over-oxidation. Incorporating more noble elements or specific stabilizers into non-oxide frameworks can reduce the driving force for dissolution and slow conversion to thick, inactive oxides. Multimetal compositions can share redox load across different cations, reducing the likelihood that any single species reaches very high oxidation states that favor OER and corrosive pathways. These strategies aim to maintain a controlled reconstruction depth and favorable oxidation state distribution.

At the electrode architecture level, robust conductive supports and three-dimensional current collectors improve mechanical and electronic stability. High-surface-area metal foams, meshes, or corrosion-resistant carbons can accommodate mechanical stresses and preserve low-resistance pathways even as the catalyst layer undergoes moderate restructuring. When non-oxide catalysts are grown directly on such scaffolds, integrated electrodes that are less prone to delamination are obtained. Architectures with appropriate porosity and flow paths can also limit local accumulation of intermediates and products, reducing fouling and diffusion limitations.

Operating protocols provide another layer of control. Limiting the upper potential in galvanostatic operation or applying potential ceilings in potentiostatic operation can avoid regimes dominated by OER and aggressive oxidation. Pulsed or intermittent operation can allow adsorbed species to desorb or be oxidized without driving deep structural changes. Periodic flushing with fresh electrolyte can reduce organic fouling, although cleaning procedures must be chosen to avoid damage to the catalyst or membrane. For cells treating crude glycerol, upstream purification, for example, removal of soaps and particulates, can significantly improve stability. These strategies are most effective when developed together. Material choices and electrode architectures should match the anticipated operating window, and operating protocols should stabilize the mechanistically favorable states identified in previous sections.

6.5.3 Alkaline–Alkaline and Hybrid Alkali–Acid Architectures

Non-oxide glycerol oxidation anodes can be integrated into different electrolyzer configurations. Two illustrative cases are the alkaline–alkaline and hybrid alkali–acid architectures. In the alkaline–alkaline configuration, both anolyte and catholyte are alkaline. Glycerol is dissolved in the anolyte and oxidized at the non-oxide anode. Hydrogen is evolved at a cathode operating in the same pH domain. An anion-exchange membrane or porous separator separates the compartments and allows hydroxide or anion transport. This arrangement is chemically simple. Non-oxide catalysts that reconstruct to oxyhydroxides are generally stable in alkaline environments, and the same broad electrolyte type can be used on both sides. The reduction in cell voltage depends on how effectively the catalyst sustains glycerol oxidation at potentials where OER remains suppressed, and electrode and flow design must accommodate the higher viscosity of glycerol-containing electrolytes.

In the hybrid alkali–acid configuration, the anolyte remains alkaline and contains glycerol, whereas the catholyte is acidic. A membrane or bipolar interface separates the compartments and maintains a pH gradient while allowing ion transport. This setup combines low anodic potentials for glycerol oxidation in alkaline media with favorable HER kinetics in acid, potentially lowering the overall cell voltage further. This configuration can be interpreted as an electro-neutralization scheme in which protons and hydroxide ions are consumed across the membrane. For non-oxide anodes, this architecture retains the alkaline environment at the anode but modifies ion-transport and potential distributions; the membrane must tolerate contact with both alkaline glycerol and acidic solutions and resist crossover of organics and metal ions.

Both architectures can be implemented in batch or flow mode. In batch cells, anolyte composition evolves as glycerol is consumed and products accumulate, which influences kinetics, selectivity, and fouling. Flow cells continuously supply fresh glycerol-containing electrolyte and remove products, enabling more stable operation and better control of concentration profiles. Non-oxide catalysts grown on three-dimensional, flow-through electrodes are particularly suitable for such devices because they provide a large interfacial area and strong convective transport.

6.5.4 Device-level Benchmarking

Device-level benchmarking is required to compare hybrid cells and evaluate their practicality. The primary energetic metric is the cell voltage as a function of current density, compared with an OER-based reference. The voltage difference at relevant current densities can be converted into electrical energy savings per kilogram of hydrogen. It is essential to report these values under conditions where both systems are optimized. FE for hydrogen and major glycerol oxidation products should be measured under the same operating conditions. Combined with current density, they determine product yield rates and carbon efficiency. Reporting the distribution of C_3, C_2, and C_1 products and highlighting specific targets provides a clearer view of process relevance than lumped oxidation data.

Stability at high current densities is also crucial. Long-term tests at currents relevant for practical applications should monitor cell voltage drift, FE, and product distributions. Such measurements reveal the cumulative effects of degradation modes outlined above. Stability metrics, such as hours of operation at fixed current density with limited performance loss, are valuable for comparison across catalysts and cell designs.

Benchmarking inevitably involves trade-offs. Systems optimized for deep oxidation to C_1 products with significant voltage savings may be appropriate when glycerol is used as a sacrificial feed. Systems designed for high C_3 selectivity at moderate current density and more modest voltage reduction may be better suited for integrated chemical coproduction. The mechanistic framework in Section 6.4 helps interpret these trade-offs and identify routes for further optimization.

6.6 Operando Characterization and Future Directions

Non-oxide glycerol oxidation catalysts are inherently dynamic. Their active state is shaped by reconstruction, leaching, fouling, and changes in the local environment during operation. Ex-situ characterization captures only before-and-after snapshots and may miss the configurations that control activity and selectivity. This section outlines how operando and in situ tools can reveal the true active state and how such insights can be combined with data-driven methods and realistic feedstock studies to guide future developments. In Figure 6.5, a schematic of an operando platform and feedback loop is presented.

6.6.1 Role of Operando and In Situ Methods

Operando and in situ methods probe catalysts during glycerol oxidation under realistic conditions, including relevant potentials or current densities, the presence of glycerol and supporting electrolyte, and device-relevant architectures. These techniques track changes in coordination environment, surface composition, adsorbates, and product distributions as functions of time and operating parameters. For non-oxide catalysts that reconstruct into core–shell architectures, such approaches are crucial for tracking the evolution of the active state and for revealing how defects, interfacial structures, and fouling layers develop and interact with the underlying catalyst. Correlating these structural and chemical changes with electrochemical and analytical data enables mechanistic hypotheses to be tested and refined.

6.6.2 Key Techniques

Several classes of operando and in situ techniques are particularly informative. X-ray absorption spectroscopy, including XANES and EXAFS, provides information on average oxidation state and local coordination of metal centers. Time-resolved operando XAS can quantify changes in oxidation state and coordination environment. It can also monitor the evolution from non-oxide precursors to oxyhydroxide-like

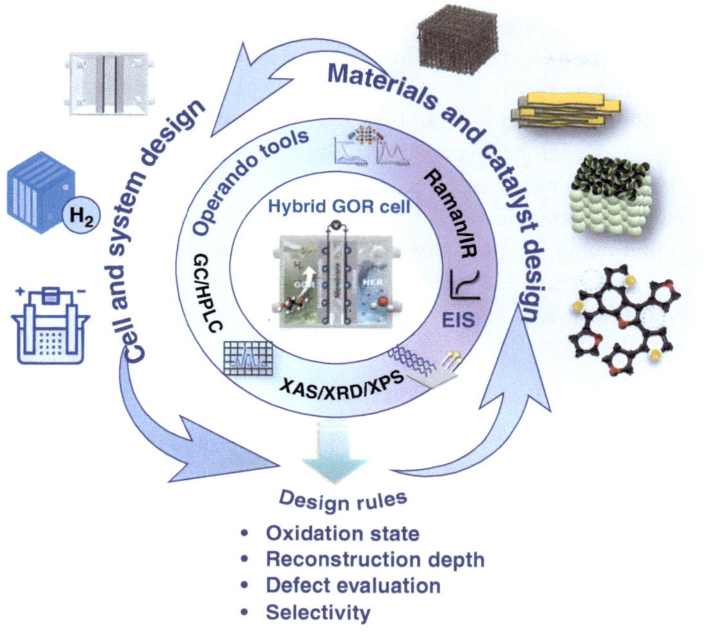

Figure 6.5 Operando characterization platform and design feedback loop for non-oxide GOR catalysts. A simplified hybrid glycerol oxidation cell is interrogated using complementary operando probes, including XAS/XRD/XPS, Raman/IR, EIS (electrochemical impedance spectroscopy), and GC/HPLC, to track oxidation state, structure, kinetics, and product distributions. Integrated data streams are used to refine catalyst composition, reconstruction control, electrode architecture, and operating windows.

surfaces and assess reconstruction depth and the stability of the active state. X-ray diffraction detects the formation or disappearance of crystalline phases and complements the local structural information from XAS. X-ray photoelectron spectroscopy, in near-ambient or quasi-operando configurations, can probe surface composition and the chemical states of nonmetal species such as S, P, N, and C.

Vibrational spectroscopy, particularly Raman and infrared spectroscopy, detects adsorbed intermediates and products near the catalyst surface. Raman bands associated with C—O and C=O groups and carboxylates, together with infrared signatures of glycerolate, aldehydes, and acids, can be tracked as functions of potential and time. By correlating Raman and infrared signals with electrochemical responses and product analysis, it is possible to identify which branches of the glycerol oxidation network are active. This approach also reveals how these pathways respond to changes in catalyst composition, defect density, and operating conditions.

EIS performed under operating conditions separates charge-transfer and mass-transport contributions. Changes in charge-transfer resistance can be linked to modifications in the reconstructed layer. Diffusion-related features may indicate pore blockage, viscosity changes, or bubble accumulation. Interpreting EIS in conjunction with structural probes connects morphological or structural changes to kinetic consequences. Online or at-line product analysis, for example, gas chromatography and high-performance liquid chromatography, completes the picture. By sampling effluents from anode and cathode compartments, one can quantify hydrogen evolution, glycerol conversion, and the distribution of C_3, C_2, and C_1 products. These data, together with electrochemical measurements, yield time-resolved faradaic and carbon efficiencies.

In an integrated operando platform, data streams from structural probes, spectroscopies, EIS, and product analysis feed into a common analysis framework, which then guides choices of composition, structure, and operating conditions.

6.6.3 From Operando Data to Design Rules

The primary value of operando studies is in linking observable changes to design variables. A few examples illustrate this connection. Operando XAS and XRD can determine whether high-performing catalysts share a common range of surface oxidation states and reconstruction depths. If catalysts that exhibit optimal activity and C_3 selectivity consistently show thin oxyhydroxide layers with specific average oxidation states, this supports explicit constraints on reconstruction depth and oxidation state. These constraints can then be incorporated into catalyst design rules.

Operando Raman and infrared spectra identify which intermediates dominate under given conditions and how their populations change with potential and time. Surfaces that favor primary hydroxyl activation may show strong signatures of GLAD and glyceric acid. In contrast, surfaces that promote deeper oxidation may exhibit stronger formate and oxalate features. Tracking these signatures as catalysts age or as operating conditions change reveals how defect evolution and fouling influence selectivity.

EIS measurements, combined with structural and compositional data, can show how charge-transfer resistance and diffusion behavior respond to structural changes and fouling. Significant increases in charge-transfer resistance may indicate oxide thickening or loss of conductive pathways. Evolving diffusion signatures can signal pore blockage or viscosity changes. These insights guide decisions on nanostructure, porosity, and cleaning protocols. By feeding such observations into an iterative design–measure–refine cycle, qualitative rules can be gradually converted into more quantitative guidelines. Composition ranges, defect densities, reconstruction depths, and operating potentials can be narrowed based on observed correlations between structure, intermediates, and performance.

6.6.4 Future Directions

Several research directions appear particularly promising in light of the mechanistic and operando insights discussed above. Single-atom and dual-site catalysts supported on non-oxide substrates are one frontier. Isolated metal atoms or well-defined site pairs on engineered supports could enable precise control over coordination environment, oxidation state, and adsorption geometry. This precision may allow selective activation of primary vs. secondary hydroxyl groups and fine control over the balance between C—H and C—C activation. Demonstrating stability of such sites under realistic glycerol oxidation conditions will require advanced operando characterization.

Data-driven discovery offers another route. The design space of non-oxide compositions, defect patterns, and heterostructures is large. High-throughput calculations can generate descriptors such as d-band centers, charge densities, and adsorption energies for glycerolate and hydroxide. Machine learning models can then be trained to identify promising combinations of metals, anions, and structural motifs that are likely to deliver favorable activity and selectivity. Experimental work, guided by operando diagnostics, can create a feedback loop that gradually improves model fidelity. Operation with real glycerol streams from biodiesel or biorefinery processes is a further priority. Such streams contain inorganic salts, soaps, and diverse organic impurities that strongly influence corrosion, fouling, and membrane performance. Studying non-oxide catalysts under these conditions, rather than only in pure glycerol solutions, is essential for realistic evaluation. These studies are likely to reveal additional degradation modes and may motivate adaptations in materials, reactor design, and feed pretreatment.

Finally, scale-up and system integration will determine whether glycerol-assisted hybrid electrolysis becomes a practical technology. Hybrid electrolyzers that pair non-oxide glycerol oxidation anodes with HER cathodes could be deployed at sites where glycerol is produced, enabling on-site coproduction of hydrogen and chemicals. Designing such systems requires attention to reactor engineering, heat and mass management, product separation, and long-term control of operating conditions. Lessons from pilot-scale demonstrations, monitored with operando and field-relevant diagnostics, can inform both materials development and process design.

In summary, non-oxide electrocatalysts for glycerol oxidation provide a rich and dynamic platform for hybrid water electrolysis. Combining thermodynamic understanding, mechanistic insights, structured design rules, durability strategies, and operando characterization enables systematic engineering of catalysts and systems that exploit the full potential of glycerol-assisted hydrogen production. Viewed in this way, non-oxide catalysts constitute a tunable platform for coupling renewable hydrogen generation with the valorization of waste glycerol streams.

References

1 Chen, Z., Wei, W., Song, L., & Ni, B.-J. (2022). Hybrid water electrolysis: A new sustainable avenue for energy-saving hydrogen production. *Sustain. Horiz.* *1*, 100002. https://doi.org/10.1016/j.horiz.2021.100002

2 Veeramani, K., Janani, G., Kim, J., Surendran, S., Lim, J., Jesudass, S. C., Mahadik, S., Lee, H., Kim, T.-H., Kim, J. K., & Sim, U. (2023). Hydrogen and value-added products yield from hybrid water electrolysis: A critical review on recent developments. *Renew. Sustain. Energy Rev. 177*, 113227. https://doi.org/10.1016/j.rser.2023.113227

3 Song, M., Bai, Y., Li, J., & Qi, X. (2025). Efficient strategies for the preparation of non-noble metal catalysts for electrocatalytic glycerol oxidation towards high-value-added chemicals. *RSC Adv. 15*, 20513–20529. https://doi.org/10.1039/d5ra02222c

4 Vitulano, F., Uggeri, F., Lattuada, L., Minguzzi, A., & Vertova, A. (2025). Tackling electrocatalytic oxidation of glycerol to dihydroxyacetone: A comprehensive review. *Curr. Opin. Electrochem. 51*, 101665. https://doi.org/10.1016/j.coelec.2025.101665

5 Wang, G., Chen, J., Qiao, F., Wang, J., & Wen, Z. (2025). Hybrid dual-electrolyte electrochemical cells for glycerol oxidation upgradation. *Chem. Sci. 16*, 12651–12678. https://doi.org/10.1039/d5sc02411k

6 Ye, H., Favero, S., Tyrrell, H., Plub-in, K., Hankin, A., Rao, R. R., Stephens, I. E. L., Titirici, M., & Luo, H. (2025). Progress and challenges in electrochemical glycerol oxidation: The importance of benchmark methods and protocols. *ChemCatChem 17*(13), e00152. https://doi.org/10.1002/cctc.202500152

7 Zhang, Y., Liao, L., Zhou, H., Qi, Y., Sun, J., Zhang, Y., Zhou, Q., Wang, Y., Tang, D., & Yu, F. (2025). Visualizing dynamic competitive reconstruction of trimetallic hybrid catalysts for stable hybrid water electrolysis at large current density. *Energy Environ. Sci. 18*, 7695–7707. https://doi.org/10.1039/d5ee00172b

8 Deng, B., Shen, J., Lu, J., Huang, C., Chen, Z., Peng, F., & Liu, Y. (2025). Ru doping triggering reconstruction of cobalt phosphide for coupling glycerol electrooxidation with seawater electrolysis. *J. Energy Chem. 100*, 317–326. https://doi.org/10.1016/j.jechem.2024.08.056

9 Wang, S., Yan, Y., Du, Y., Zhao, Y., Li, T., Wang, D., Schaaf, P., & Wang, X. (2024). Inhibiting the deep reconstruction of Ni-based interface by coordination of chalcogen anions for efficient and stable glycerol electrooxidation. *Adv. Funct. Mater. 34*(39), 2404290. https://doi.org/10.1002/adfm.202404290

10 Terekhina, I. & Johnsson, M. (2024). Tuneable C3 product selectivity of glycerol electrooxidation on cubic and dendritic Pt nanocatalysts. *Nanoscale 16*, 13000–13010. https://doi.org/10.1039/d4nr01127a

11 Wan, H., Dai, C., Jin, L., Luo, S., Meng, F., Chen, G., Duan, Y., Liu, C., Xu, Q., Lu, J., & Xu, Z. J. (2022). Electro-oxidation of glycerol to high-value-added C1–C3 products by iron-substituted spinel zinc cobalt oxides. *ACS Appl. Mater. Interfaces 14*, 14293–14301. https://doi.org/10.1021/acsami.2c02215

12 Zuo, Y., Bellani, S., Ferri, M., Saleh, G., Shinde, D. V., Zappia, M. I., Brescia, R., Prato, M., De Trizio, L., Infante, I., Bonaccorso, F., & Manna, L. (2023). High-performance alkaline water electrolyzers based on Ru-perturbed Cu nanoplatelets cathode. *Nat. Commun. 14*, 4680. https://doi.org/10.1038/s41467-023-40319-5

13 Ebeling, K. M., Bongartz, D., Mürtz, S., Palkovits, R., & Mitsos, A. (2024). Thermodynamic and economic potential of glycerol oxidation to replace oxygen evolution in water electrolysis. *Ind. Eng. Chem. Res. 63*, 8250–8260. https://doi.org/10.1021/acs.iecr.3c03615

14 Ma, J., Tian, Y., Li, J., Zhang, T., Sun, X., & Zhang, Q. (2024). Tuning electronic structure of NiFe layered double hydroxides with vanadium doping toward formate production from glycerol oxidation. *Int. J. Hydrog. Energy 82*, 810–816. https://doi.org/10.1016/j.ijhydene.2024.07.322

15 Liu, Y., Chen, B., Nie, H., Gui, S., & Deng, J. (2025). 3D coral-like Ni–Mo–S/NF bifunctional electrocatalyst for concurrent glycerol valorization to formate and energy-efficient H_2 evolution. *RSC Adv. 15*, 48487–48497. https://doi.org/10.1039/d5ra08754f

16 Xu, L., Yang, Y., Li, C., Ning, R., Ma, J., Yao, M., Geng, S., & Liu, F. (2024). Unveiling the mechanism of electrocatalytic oxidation of glycerol by in-situ electrochemical spectroscopy. *Chem. Eng. J. 481*, 148304. https://doi.org/10.1016/j.cej.2023.148304

17 Zhu, Y., Qian, Q., Chen, Y., He, X., Shi, X., Wang, W., Li, Z., Feng, Y., Zhang, G., & Cheng, F. (2023). Biphasic transition metal nitride electrode promotes nucleophile oxidation reaction for practicable hybrid water electrocatalysis. *Adv. Funct. Mater. 33*(25), 2300547. https://doi.org/10.1002/adfm.202300547

18 Wang, G., Chen, J., Li, K., Huang, J., Huang, Y., Liu, Y., Hu, X., Zhao, B., Yi, L., Jones, T. W., & Wen, Z. (2022). Cost-effective and durable electrocatalysts for Co-electrolysis of CO_2 conversion and glycerol upgrading. *Nano Energy 92*, 106751. https://doi.org/10.1016/j.nanoen.2021.106751

19 Fan, L., Ji, Y., Wang, G., Zhang, Z., Yi, L., Chen, K., Liu, X., & Wen, Z. (2022). Bifunctional Mn-doped $CoSe_2$ nanonetworks electrode for hybrid alkali/acid electrolytic H_2 generation and glycerol upgrading. *J. Energy Chem. 72*, 424–431. https://doi.org/10.1016/j.jechem.2022.04.027

20 Li, J., Meng, X., Song, X., Qi, J., Liu, F., Xiao, X., Du, Y., Xu, G., Jiang, Z., Ye, S., Huang, S., & Qiu, J. (2024). Valence engineering via manganese-doping on cobalt nitride nanoarrays for efficient electrochemically paired glycerol valorization and H_2 production. *Adv. Funct. Mater. 34*(33), 2316718. https://doi.org/10.1002/adfm.202316718

21 Fang, Y., Dai, C., Liu, X., Wang, Y., Ju, C., He, S., Shi, R., Liu, Y., Zhang, J., Zhu, Y., & Wang, J. (2024). Sulfur-doped manganese-cobalt hydroxide with promoted surface reconstruction for glycerol electrooxidation assisted hydrogen production. *Nano Energy 127*, 109754. https://doi.org/10.1016/j.nanoen.2024.109754

22 Shi, K., Si, D., Teng, X., Chen, L., & Shi, J. (2023). Enhanced electrocatalytic glycerol oxidation on $CuCoN_{0.6}$/CP at significantly reduced potentials. *Chin. J. Catal. 53*, 143–152. https://doi.org/10.1016/s1872-2067(23)64515-2

23 Fan, L., Ji, Y., Wang, G., Chen, J., Chen, K., Liu, X., & Wen, Z. (2022). High entropy alloy electrocatalytic electrode toward alkaline glycerol valorization coupling with acidic hydrogen production. *J. Am. Chem. Soc. 144*, 7224–7235. https://doi.org/10.1021/jacs.1c13740

24 Wang, P., Wang, G., Chen, K., Pan, W., Yi, L., Wang, J., Chen, Q., Chen, J., & Wen, Z. (2023). High-power hybrid alkali-acid fuel cell for synchronous glycerol valorization implemented by high-entropy sulfide electrocatalyst. *Nano Energy 118*, 108992. https://doi.org/10.1016/j.nanoen.2023.108992

25 Verma, S., Lu, S., & Kenis, P. J. A. (2019). Co-electrolysis of CO_2 and glycerol as a pathway to carbon chemicals with improved technoeconomics due to low electricity consumption. *Nat. Energy 4*, 466–474. https://doi.org/10.1038/s41560-019-0374-6

26 Ou, W., Ye, X., & Zhou, Y. (2023). Recent advances in Ni (oxy) hydroxides and Ni sulfides catalysts for oxygen evolution reactions. *Coord. Chem. Rev. 493*, 215274. https://doi.org/10.1016/j.ccr.2023.215274

27 Ma, L., Miao, Y., Yang, J., Fu, Y., Yan, Y., Zhang, Z., Li, Z., & Shao, M. (2024). Promoting electrocatalytic glycerol C—C bond cleavage to formate coupled with H_2 production over a $Cu_xNi_{2-x}P$ catalyst. *Adv. Energy Mater. 14*(34), 2401061. https://doi.org/10.1002/aenm.202401061

28 Tian, D., Denny, S. R., Li, K., Wang, H., Kattel, S., & Chen, J. G. (2021). Density functional theory studies of transition metal carbides and nitrides as electrocatalysts. *Chem. Soc. Rev. 50*, 12338–12376. https://doi.org/10.1039/d1cs00590a

29 Fernández-Caso, K., Molera, M., Andreu, T., Solla-Gullón, J., Montiel, V., Díaz-Sainz, G., Álvarez-Guerra, M., & Irabien, A. (2024). Coupling glycerol oxidation reaction using Ni-Co foam anodes to CO_2 electroreduction in gas-phase for continuous co-valorization. *Chem. Eng. J. 480*, 147908. https://doi.org/10.1016/j.cej.2023.147908

Part III

Hydrazine-assisted Water Electrolysis

7

Metal Sulfides as Electrocatalysts for Hydrazine-assisted Water Electrolysis

Aravindhan Selvaraj[1], *Hemalatha Parangusan*[2], *Dhanasekaran Vikraman*[3], *Hyun-Seok Kim*[3], *Akram Alfantazi*[4], *and K. Karuppasamy*[4,*]

[1]*Department of Chemistry, Dr. B. S. Abdur Rahman Crescent Institute of Science and Technology, Chennai, India*
[2]*Department of Physics, Mannar Thirumalai Naicker College, Madurai, India*
[3]*Division of Electronics and Electrical Engineering, Dongguk University, Seoul, Republic of Korea*
[4]*Department of Chemical and Petroleum Engineering, Khalifa University of Science and Technology, Abu Dhabi, United Arab Emirates*

*Corresponding author: karuppasamyiitb@gmail.com

7.1 Introduction

The current ecological situation is worsening because of the rapid diminution of fossil fuels and increasing severity [1, 2]. Civilization has been subjected to severe pressure due to the increasing demands for energy, ecological pollution, and the sudden depletion of fossil fuels [3, 4]. In energy-oriented research, it is essential to identify innovative energy systems that are green, inexpensive, and renewable. Hydrogen (H_2) is seen as one of the valuable potential contenders for alternative energy systems in the next generation because of its facile production and enormously large mass-based energy density [5, 6]. H_2 is primarily obtained through the steam reforming of natural gas, but the production cost is high, and its purity is low [6]. The pilot manufacturing process of extremely pure H_2 can be achieved through water electrolysis, which splits water into H_2 and O_2, releasing no CO_2 emissions during the process, and with excellent productivity [7, 8]. Oxygen evolution reaction (OER) and hydrogen evolution reaction (HER) are both significant half-reactions in which a four-proton-coupled electron transfer in OER achieves the formation of an O—O bond [9–12]. Overcoming the high thermodynamic potential on the OER electrode is crucial for commercializing H_2 formation through electrocatalytic water splitting, as it incorporates the HER with other oxidation reactions that have the lowest potential and swift reaction kinetics [10, 12]. The adequate replacement of OER is achieved through the oxidation of H_2, N_2H_4, or ethanol. The goal is to decrease the potential at the anode, leading to the production of H_2 at an affordable price [13].

The simultaneous synthesis of hydrogen and hydrazine oxidation is one such method that is both economically feasible and pollution-free [14]. Because of its capacity for full electrocatalytic oxidation to N_2 and H_2O, N_2H_4 has attracted great interest. In addition to reducing energy consumption, the hydrazine oxidation reaction (HzOR) eases the breakdown of N_2H_4 as a contaminant in industrial and agricultural wastewater, allowing it to be upgraded into products of higher value [15].

The high power density, low thermodynamic potential, and fast electrooxidation kinetics of the HzOR have made it a popular option [16, 17]. Although noble metals (Pt, Ir, and Ru) are excellent electrocatalysts for water splitting, their high cost, scarcity, and decreased stability make them less appropriate for large-scale applications [18]. The surface morphology, atomic structure, and composition are primary factors that determine the efficiency of the electrocatalyst. The use of transition-metal sulfides (TMSs) has been effective in electrocatalyzing HER, OER, and overall water splitting in recent years [19, 20]. In terms of structure, these materials often follow the X–M–X formula. This formula uses "X" to indicate a sulfur family element, like sulfur (S), on the central plane and "M" to indicate a transition metal, such as cobalt (Co) or molybdenum (Mo), on the same plane [21]. In two-dimensional transition-metal dichalcogenides (2D-TMDs), the adjacent layers are joined by comparatively lesser van der Waals forces, whereas strong ionic–covalent interactions distinguish the basal plane. These substances are perfect candidate materials for highly effective HER catalysts because of their structural characteristics, which provide them with a large specific surface area, excellent charge-transport efficiency, fine-tuned atomic-level control, and changeable physical and chemical properties [22].

This chapter summarizes the metal sulfide electrocatalysts for hydrazine-assisted water electrolysis.

7.2 A Basic Understanding of N_2H_4-assisted HER

To reduce the voltage essential for water electrolysis and conserve energy in H_2 formation, it is essential to replace the OER and construct an innovative electrochemical source for water splitting as a half-reaction (HER/HzOR) [2], which is also termed the N_2H_4-assisted HER system. This new system comprises HER and HzOR, which function as cathodic and anodic reactions, correspondingly. The N_2H_4-assisted HER was primarily performed in a KOH electrolyte, and overall reactions are provided in Eqs. (7.1)–(7.3):

$$\text{At cathode:}\quad 4H_2O + 4e^- \rightarrow 2H_2 + 4OH^- \tag{7.1}$$

$$\text{At anode:}\quad N_2H_4 + 4OH^- \rightarrow N_2 + 4e^- + 4H_2O \tag{7.2}$$

$$\text{Overall reaction:}\quad N_2H_4 \rightarrow N_2 + 2H_2 \tag{7.3}$$

7.2.1 Parameters Required to Estimate the Catalytic Performance

It is well known that onset potential refers to the potential at which electrochemical reaction begins, leading to an enhancement in current density. At a particular thermodynamic potential, all electrochemical reactions occurred without any obstacle [23, 24]. In an ideal system, there is an additional potential necessary to complete electrochemical reaction, which is termed overpotential (η). The η for the HER and OER is determined from Eqs. (7.4) and (7.5) [19]:

$$\eta^{\text{OER}} = \text{Potential obtained from experiment (V)} - 1.23\ \text{(V)} \tag{7.4}$$

$$\eta^{\text{HER}} = \text{Potential obtained from experiment (V)} - 0\ \text{(V)} \tag{7.5}$$

7.2.2 Exchange Current Density (j_0) and Tafel Slope

Tafel analysis is frequently used to correlate catalytic activity and to comprehend the mechanism of electrocatalytic events. Rate-determining step (RDS) and Tafel slope provide information about how sensitive the j response is to the applied voltage. To obtain the Tafel slope, one needs to plot the log of j vs. η obtained from iR-amended linear sweep voltammetry (LSV). When HER is carried out in an alkaline environment, the Volmer–Heyrovsky or Volmer–Tafel pathways (Eqs. 7.6–7.11) are thought to be the reaction mechanism [13, 25, 26]:

$$e^- + H_2O \rightarrow OH^- + H^* \ \text{(Volmer reaction)} \tag{7.6}$$

$$T_{SV} = \frac{\mathrm{d}\log(j)}{\mathrm{d}\eta} = \frac{2.303RT}{\alpha\eta F} \tag{7.7}$$

$$e^- + H_2O \rightarrow OH^- + H^* + H_2 \ \text{(Heyrovsky reaction)} \tag{7.8}$$

$$T_{SH} = \frac{2.303RT}{\alpha\eta F} \tag{7.9}$$

$$H^* + H^* \rightarrow H_2 \ \text{(Tafel reaction)} \tag{7.10}$$

$$T_{ST} = \frac{2.303RT}{\alpha\eta F} \tag{7.11}$$

where T_S and R are the Tafel slope and universal gas constant; T, F, and α are the reaction temperature, Faraday constant, and symmetry coefficient.

7.2.3 Turnover Frequency

One significant factor used to determine the intrinsic activity of electrocatalysts is turnover frequency (TOF), which measures a catalytic center's specific activity, determined by the number of molecules of the reactant that are converted per active site in a given amount of time [23]. As for heterogeneous electrocatalysts, their standards can be determined through the equation: TOF $= jA/(2Fn)$, where j indicates current density at a given η, A suggests the working electrode's surface area, and F and n are the Faraday constant and number of moles of active sites over the working electrode [27, 28]. A greater TOF value indicates an increased intrinsic activity of an electrocatalyst.

The relationship between total number of electrons spent from external circuit and moles of products generated over the electrode is revealed by faradaic efficiency (FE) [28]. FE indicates the ratio of H_2 or O_2 amount that is observed experimentally to the amount that is calculated theoretically. By integrating the chronoamperometric or chronopotentiometric analysis, theoretical values can be computed. The experimental results are determined by means of analyzing gas production using water–gas displacement process or gas chromatography process. A high FE indicates an actual potential for the electrode response to be industrialized since it is connected with acceptable energy usage. Usually, heat or by-products formed over the electrode result in decreased FE. Galvanostatic electrolysis by gas chromatography yields experimental and theoretical data for the H_2 production in hydrazine-assisted HER, respectively.

7.2.4 Stability

For water-splitting process, the catalytic activity is highly essential, however, prolonged durability is also crucial for practical applications of energy conversion. Measuring stability is mainly carried out using two methods: chronoamperometry or chronopotentiometry, along with different voltametric techniques, including LSV and CV [29, 30]. In CV, it is performed repetitively in a particular potential range, such as η_{onset}. After several thousand CV cycles, the polarization plot has been plotted, followed by η of same j after and before CV cycling being correlated. Excellent stability of electrocatalyst is demonstrated if the difference is negligible. Furthermore, concerning time-dependent processes, including galvanostatic (or potentiostatic) approaches, the potential fluctuation j at a fixed j (or potential) over time is examined. Less fluctuation means greater stability at 10 mA cm^{-2}. In contrast, HzOR stability tests usually use 100 mA cm^{-2}, because this value is widely recognized as the primary benchmark in the field of electrocatalysis.

7.3 Hydrazine-assisted HER Applications

7.3.1 Nickel-based Catalysts

HER assisted by hydrazine is most likely to be achieved using bifunctional electrocatalysts made up of Ni-based materials. In 2017, Sun and coworkers reported NiS_2 nanoarray by the sulfidation conversion method [31, 32], which exhibited prolonged durability in KOH solution with high catalytic activity toward HzOR. The better bifunctional catalyst for energy-saving electrolytic H_2 formation by substituting OER with HzOR is achieved by its high HER activity. To provide 50 mA cm^{-2}, two-electrode alkaline electrolyzer only needs 0.59 V, which is 1.34 V lower than the N_2H_4-free counterpart, and has almost 100% FE for HER (Figure 7.1) [32].

Zhao and coworkers reported the synthesis of a 3D nickel foam (NF)-anchored ultrathin Ni_3S_2 nanosheet electrocatalyst with a thickness of 9–14 nm using the vapor-phase hydrothermal (VPH) method [33]. The multifunctional electrocatalytic activities of Ni_3S_2/NF-2 have the potential to reduce energy consumption and increase H_2 evolution in alkaline media by substituting OER with HzOR (Figure 7.2). With a voltage of 0.867 V, the Ni_3S_2/NF-2 two-electrode system has excellent durability and can provide 100 mA cm^{-2} at a concentration of 0.2 M hydrazine [33].

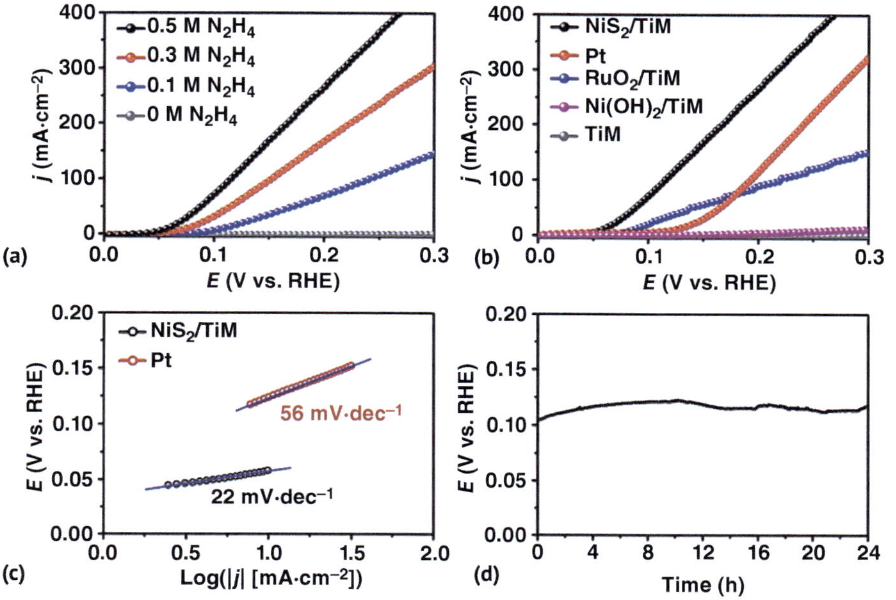

Figure 7.1 (a) LSV plot of NiS_2/TiM against various concentrations of N_2H_4, (b) LSV plot of pristine TiM, RuO_2/TiM, $Ni(OH)_2$/TiM, Pt, and NiS_2/TiM toward 0.5 M N_2H_4 at a sweep rate of 5 mV s^{-1} for HzOR, (c) Tafel graphs of NiS_2/TiM and Pt, and (d) chronopotentiometric plot at a fixed 100 mA cm^{-2} for NiS_2/TiM catalyst in 1 M KOH. *Source*: [32]/with permission of Elsevier.

To validate the electrolyzer cell for N_2H_4-assisted HER, Sun et al. constructed an N_2H_4 fuel cell. They demonstrated H_2 generation at a yield of 9.2 mol h^{-1} g_{cat}^{-1} on NiCo/MXene-based electrodes with 2.75 kW per m^3 H_2 power consumption at 500 mA cm^{-2}, reporting 48% less power than that of commercial water electrolyzers [34]. Furthermore, crystalline Ni_3S_2-supported NF with an Ni–Fe–OH amorphous layer improved corrosion resistance and catalytic OER performance in KOH medium, providing electrode stability for 50 hours at an industrial-scale current density of 1,000 mA cm^{-2}. In addition, electronic charge transfer can be significantly enhanced by augmenting electronic structure through the formation of heterostructures, which improves electrochemical performance. For improving the active sites and boosting the intrinsic activity of metal chalcogenides, dual cation doping is adopted to alter the electronic structure of catalysts and introduce active sites, which inserts two transition metals into metal chalcogenides, e.g., Cu/Fe co-doped Ni_3S_4, Fe/Mn co-doped Ni_3S_2, Fe/V co-doped Ni_3S_2, and Fe/Co co-doped Ni_3Se_4 are representatives of efficient catalysts for water electrolysis [35]. The electronic interaction between the host and transition-metal dopants in these catalysts helps to optimize the catalyst's electronic configuration, which lowers the reaction energy barrier. Furthermore, it is suggested that transition-metal dopants be added to increase electrical conductivity, which is advantageous for electrochemical reactions. Recently, Krishnamurthy et al. found that Fe and V dopants in Fe/V co-doped Ni_2Se_2 led to lattice distortion, and the enrichment of sulfur active edge sites contributes

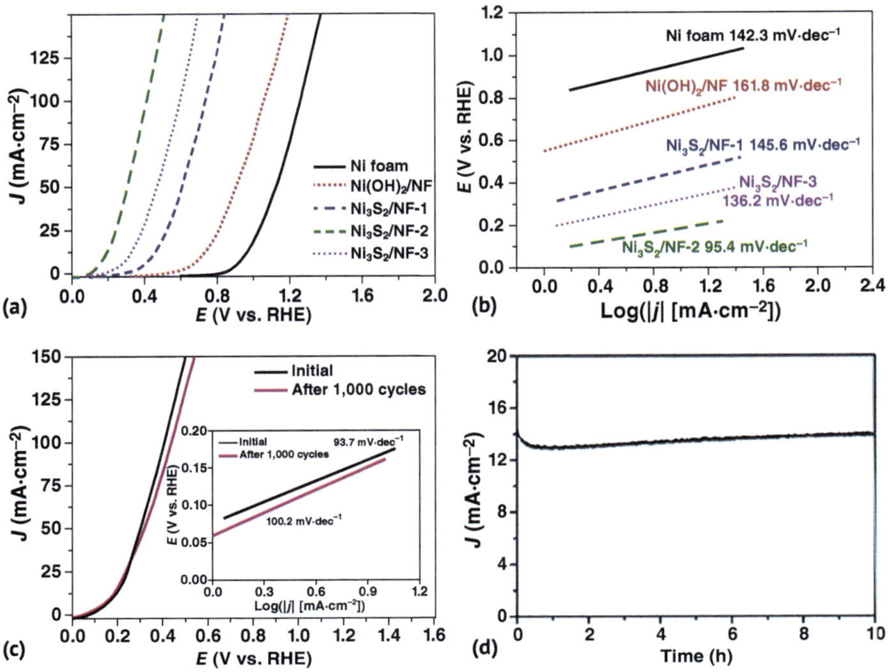

Figure 7.2 (a) LSV plot with *iR* drop-free compensation of NF, Ni_3S_2/NF-2, $Ni(OH)_2$/NF, Ni_3S_2/NF-1, and Ni_3S_2/NF-3 in 1 M KOH with 0.2 M N_2H_4, (b) its respective Tafel slopes, and (c, d) HzOR and durability results of the Ni_3S_2/NF-2 electrocatalyst with 0.2 M N_2H_4 in 1 M KOH. *Source*: [33]/with permission of The Royal Society of Chemistry.

to boosted catalytic performance for overall water splitting [36]. In addition to transition metals, dopants of the leading group (such as Al) have been employed to enhance host Ni sites. In contrast, Fe/Al dopants have been used to improve the adsorption strength of process intermediates. Furthermore, the catalyst's noticeable etching of Al in the alkaline solution results in rich defective sites and more electroactive sites.

7.3.2 Cobalt-based Catalysts

The superior conductivity, durability, and activity of Co-based compounds, particularly cobalt phosphides and cobalt sulfides, have gained increasing attention from researchers. The Co-based compounds are viewed as the potential electrocatalysts for industrial HER [37, 38]. Ma et al. reported a CoS_2 nanoarray-based catalyst for HzOR that is very effective and stable in basic conditions [39]. By replacing OER with HzOR, it could be employed as the bifunctional catalyst for energy-saving electrolytic H_2 production due to its high HER activity. To drive 100 mA cm^{-2}, the two-electrode system requires a cell potential of 0.81 V, and its electrochemical durability is strong and approximately 100% FE for H_2 evolution is achieved (Figure 7.3). This CoS_2 nanoarray has superior catalytic activity and stability, as well as scalability and

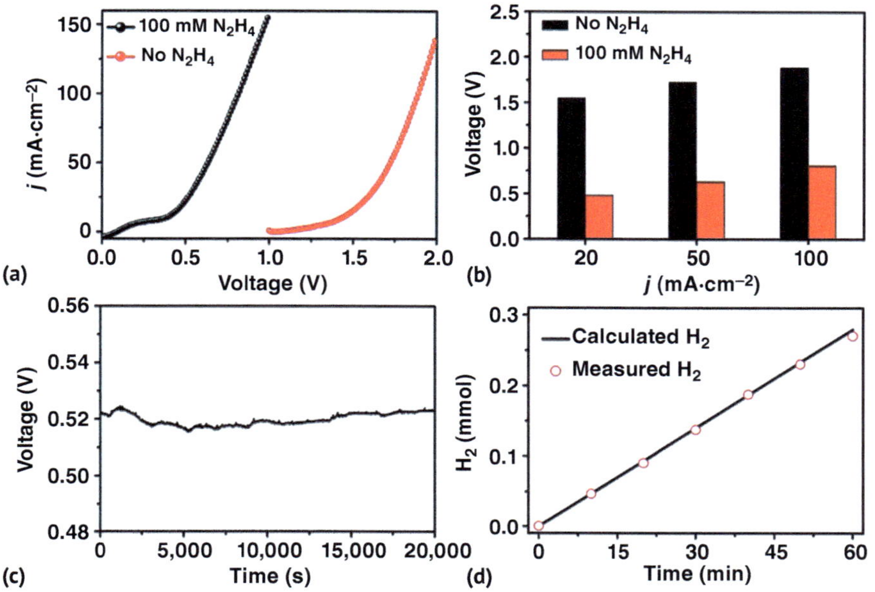

Figure 7.3 (a) LSV plots for N_2H_4 electrolysis and water electrolysis for CoS_2/TiM in two-electrode arrangement at 5 mV s^{-1}, (b) voltage correlation to attain various current densities ranging from 20 to 100 mA cm^{-2} in electrolyte in the presence and absence of 100 mM N_2H_4, (c) chronopotentiometric plot of symmetrical CoS_2/TiM||CoS_2/TiM electrode with fixed current density (25 mA cm^{-2}), and (d) the theoretical and experimentally calculated H_2 quantities measured vs. time for CoS_2/TiM||CoS_2/TiM electrolyzer in 1 M KOH. *Source*: [39]/with permission of The Royal Society of Chemistry.

a low-cost fabrication process. It is expected to be used for future applications as an attractive earth-abundant 3D catalyst electrode that can generate cost-effective and pilot-scale N_2H_4-assisted H_2 [21, 40].

Ding and coworkers developed thin Fe-doped CoS_2 nanosheet bifunctional catalyst for the HER and HzOR. The HER metrics produced were outstanding, with lower η values, smaller Tafel slopes, and high TOF values obtained [41]. The obtained nanosheets can deliver 100 mA cm^{-2} and have long-term stability, higher HzOR activities, and smaller potential of 129 mV (Figure 7.4). The system was able to function with a H_2 evolution rate of 9.95 mmol·h^{-1}, achieving 98% FE. This catalyst can withstand a 20-hour period at ambient temperature and pressure [41].

Very recently, Ostrikov and coworkers, using the electrochemical extraction method, followed by sonication, prepared an ~3-nm 2D ultrathin few-layer nanosheet from the bulk NbS_2 [42]. The NbS_2 nanosheets/CP showed superior electrocatalytic properties to HER with an overpotential of 90 mV and a low Tafel slope of 83 mV dec^{-1}. The characteristics include 0 mA cm^{-2} current density and a low Tafel slope of 83 mV dec^{-1}. Considering that few-layer NbS_2 nanosheets/CP have small overpotential of 90 mV, it is an exceptional performance compared to the earlier literature on NbS_2.

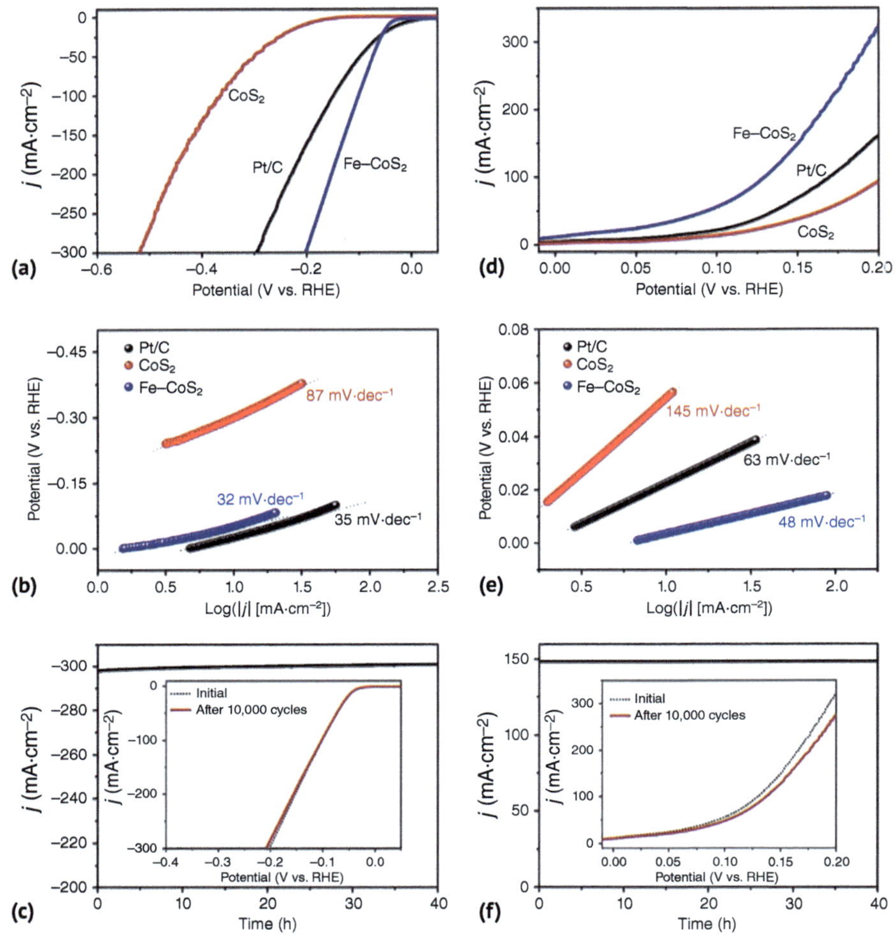

Figure 7.4 Fe–CoS_2 nanosheets' electrochemical performances for HER/HzOR in three-electrode cells with KOH at ambient temperature. (a) The Fe–CoS_2 and pristine CoS_2 nanosheets' HER polarization curves at 20 wt.% Pt/C. (b) Tafel plots (c) chronoamperometric curve recorded on the Fe–CoS_2 nanosheets for 40 hours at a constant working potential of −0.2 V for HER, with an onset current density value of 298 mA cm^{-2}, which is remarkably similar to the one (296 mA cm^{-2}) at −0.2 V in panel (a). Although panels (d)–(f) are for HzOR, their semantics are identical to those of panels (a)–(c). In panel (f), 0.15 V is the working potential that is utilized. The Fe–CoS_2 nanosheets' HER and HzOR polarization curves are shown in the insets in panels (c) and (f) before and following 10,000 LSV cycles, respectively. 1 M KOH and 1 M KOH/0.1 M hydrazine were used for HER and HzOR as the electrolytes. *Source*: [41]/Springer Nature. CC BY 4.0.

7.4 Design and Mechanism

The difficult task of identifying and pinpointing the active sites of electrocatalysts may be behind sluggish development in engineering effective and advanced electrocatalysts. To successfully address this subject, it is essential to use both experimental and theoretical results to shed more focus on feasible mechanism that governs the electrocatalytic process [33]. When designing bifunctional electrocatalysts for hydrazine-assisted hydrogen systems, it is vital to consider the catalysts' support, crystal growth in specific direction, multilevel structures, heteroatoms introduction, and structural defects formation [43]. As discussed in nickel-based catalysts, Zheng and coworkers discovered that modifying the reaction time of VPH can control the growth of the (110) plane of Ni_3S_2. Prolonged treatment leads to more crystal development that exposes (110) planes. These results suggest that controlling crystal growth in a specific direction during the preparation process is an effective method to enhance catalytic behavior [44].

7.5 Conclusion

To encourage H_2 production commercially through electrochemical processes, there is a need for concentrated research on high-performing and energy-saving electrocatalysts. A cheaper way to achieve large-scale H_2 production is by combining HER and HzOR, which have the smallest thermodynamic oxidation potential (−0.33 V), as a substitute for OER. Several reports have been published recently about TM-based electrocatalysts, which demonstrate their larger potential for effectively catalyzing the N_2H_4-assisted HER. Their unique morphological features and optimized physicochemical characteristics enhance the electrochemical performance. The catalytic performance of electrocatalysts is highly effective for HER and HzOR, which results in a significant reduction in cell voltage necessary for water splitting. To deliver 10 mA cm^{-2} or even up to 500 mA cm^{-2}, a voltage of less than 1 V is required, which indicates a high economic value. To summarize, industrial wastewater that contains hydrazine can be recycled and transformed into a green economy for industry, which allows waste to be turned into energy.

References

1 Yuan, C. Z., Hui, K. S., Yin, H., Zhu, S., Zhang, J., Wu, X. L., Hong, X., Zhou, W., Fan, X., Bin, F., Chen, F., & Hui, K. N. (2021). Regulating intrinsic electronic structures of transition-metal-based catalysts and the potential applications for electrocatalytic water splitting. *ACS Mater. Lett. 3*(6), 752–780. https://doi.org/10.1021/acsmaterialslett.0c00549

2 Majhi, K. C. & Yadav, M. (2023). Transition metal-based chalcogenides as electrocatalysts for overall water splitting. *ACS Eng. Au 3*(5), 278–284. https://doi.org/10.1021/acsengineeringau.3c00014

3 Jiao, Y., Zheng, Y., Jaroniec, M., & Qiao, S. Z. (2015). Design of electrocatalysts for oxygen-and hydrogen-involving energy conversion reactions. *Chem. Soc. Rev. 44*(8), 2060–2086. https://doi.org/10.1039/C4CS00470A

4 Ran, J., Zhang, J., Yu, J., Jaroniec, M., & Qiao, S. Z. (2014). Earth-abundant cocatalysts for semiconductor-based photocatalytic water splitting. *Chem. Soc. Rev. 43*(22), 7787–7812. https://doi.org/10.1039/C3CS60425J

5 Ali, M. S., Khan, M. S. H., Tuhin, R. A., Kabir, M. A., Azad, A. K., & Farrok, O. (2024). Hydrogen energy storage and transportation challenges: A review of recent advances. In *Hydrogen Energy Conversion and Management* (eds. M. M. K. Khan, A. K. Azad, and A. M. T. Oo), 255–287. Amsterdam: Elsevier.

6 Angelico, R., Giametta, F., Bianchi, B., & Catalano, P. (2025). Green hydrogen for energy transition: A critical perspective. *Energies 18*(2), 404.

7 Akyüz, E. S., Telli, E., & Farsak, M. (2024). Hydrogen generation electrolyzers: Paving the way for sustainable energy. *Int. J. Hydrog. Energy 81*, 1338–1362. https://doi.org/10.1016/j.ijhydene.2024.07.175

8 Wang, S., Lu, A., & Zhong, C.-J. (2021). Hydrogen production from water electrolysis: Role of catalysts. *Nano Converg. 8*(1), 4. https://doi.org/10.1186/s40580-021-00254-x

9 Vikraman, D., Hussain, S., Karuppasamy, K., Feroze, A., Kathalingam, A., Sanmugam, A., Chun, S.-H., Jung, J., & Kim, H.-S. (2020). Engineering the novel $MoSe_2$-Mo_2C hybrid nanoarray electrodes for energy storage and water splitting applications. *Appl. Catal. B: Environ. 264*, 118531. https://doi.org/10.1016/j.apcatb.2019.118531

10 Jothi, V. R., Karuppasamy, K., Maiyalagan, T., Rajan, H., Jung, C.-Y., & Yi, S. C. (2020). Corrosion and alloy engineering in rational design of high current density electrodes for efficient water splitting. *Adv. Energy Mater. 10*(24), 1904020. https://doi.org/10.1002/aenm.201904020

11 Vikraman, D., Hussain, S., Abbas, Z., Karuppasamy, K., Santhoshkumar, P., Jung, J., & Kim, H.-S. (2023). Density functional theory approximations and experimental investigations on $Co_{1-x}Mo_xTe_2$ alloy electrocatalysts tuning the overall water splitting reactions. *ACS Appl. Mater. Interfaces 15*(22), 26893–26909. https://doi.org/10.1021/acsami.3c05055

12 Karuppasamy, K., Bose, R., Vikraman, D., Ramesh, S., Kim, H. S., Alhseinat, E., Alfantazi, A., & Kim, H.-S. (2023). Revealing the effect of various organic ligands on the OER activity of MOF-derived 3D hierarchical cobalt oxide @ carbon nanostructures. *J. Alloys Compd. 934*,167909. https://doi.org/10.1016/j.jallcom.2022.167909

13 Karfa, P., Majhi, K. C., & Madhuri, R. (2018). Shape-dependent electrocatalytic activity of iridium oxide decorated erbium pyrosilicate toward the hydrogen evolution reaction over the entire pH range. *ACS Catal. 8*(9), 8830–8843. https://doi.org/10.1021/acscatal.8b01363

14 Li, S., Hou, Y., Jiang, L., Feng, G., Ge, Y., & Huang, Z. (2025). Progress in hydrazine oxidation-assisted hydrogen production. *Energy Rev. 4*(1), 100105. https://doi.org/10.1016/j.enrev.2024.100105

15 Sharma, P. J., Joshi, K. K., Sahatiya, P., Sumesh, C. K., & Pataniya, P. M. (2025). Functional nickel iron sulphide/hydroxide catalysts for hydrazine oxidation and energy saving hydrogen production. *J. Mater. Chem. A 13*(14), 9865–9877. https://doi.org/10.1039/D4TA08796H

16 Miao, R. & Compton, R. G. (2021). The electro-oxidation of hydrazine: A self-inhibiting reaction. *J. Phys. Chem. Lett. 12*(6), 1601–1605. https://doi.org/10.1021/acs.jpclett.1c00070

17 Vorms, E. A., Papaefthymiou, V., Faverge, T., Bonnefont, A., Chatenet, M., Savinova, E. R., & Oshchepkov, A. G. (2024). Mechanism of the hydrazine hydrate electrooxidation reaction on metallic Ni electrodes in alkaline media as revealed by electrochemical methods, *online* DEMs and *ex situ* XPS. *Electrochim. Acta 507*, 145056. https://doi.org/10.1016/j.electacta.2024.145056

18 Wang, H.-Y. & Yuan, Z.-Y. (2025). Hydrazine-assisted water electrolysis system: Performance enhancement and application expansion. *Mater. Horiz. 12*(14), 5123–5148. https://doi.org/10.1039/D5MH00118H

19 Liu, H., Liu, Y., Li, M., Liu, X., & Luo, J. (2020). Transition-metal-based electrocatalysts for hydrazine-assisted hydrogen production. *Mater. Today Adv. 7*, 100083. https://doi.org/10.1016/j.mtadv.2020.100083

20 Lee, J., Lee, S. A., Lee, T. H., & Jang, H. W. (2025). Unlocking the potential of chemical-assisted water electrolysis for green hydrogen production. *Ind. Chem. Mater. 3*(3), 277–310. https://doi.org/10.1039/D4IM00163J

21 Sethulakshmi, N., Nellaiappan, S., Kechanda Prasanna, P., Das, T., Irusta, S., Chakraborty, S., & Sharma, S. (2022). Nanocoral architecture for enhanced hydrazine assisted water oxidation: Insight from experiment and theory. *J. Electroanal. Chem. 922*, 116776. https://doi.org/10.1016/j.jelechem.2022.116776

22 Wang, H.-Y., Wang, L., Ren, J.-T., Tian, W., Sun, M., Feng, Y., & Yuan, Z.-Y. (2023). Taking advantage of potential coincidence region: Advanced self-activated/propelled hydrazine-assisted alkaline seawater electrolysis and Zn–hydrazine battery. *ACS Nano 17*(11), 10965–10975. https://doi.org/10.1021/acsnano.3c03095

23 Zhao, Y., Jia, N., Wu, X.-R., Li, F.-M., Chen, P., Jin, P.-J., Yin, S., & Chen, Y. (2020). Rhodium phosphide ultrathin nanosheets for hydrazine oxidation boosted electrochemical water splitting. *Appl. Catal. B: Environ. 270*, 118880. https://doi.org/10.1016/j.apcatb.2020.118880

24 Huang, X., Wang, Y., Zhu, Q., Zhou, K., Zhi, H., & Yang, J. (2021). High quality synthesis of Rh nanocubes and their application in hydrazine hydrate oxidation assisted water splitting. *Inorg. Chem. Commun. 134*, 109023. https://doi.org/10.1016/j.inoche.2021.109023

25 Shen, W., Yin, J., Jin, J., Hu, Y., Hou, Y., Xiao, J., Zhao, Y.-Q., & Xi, P. (2022). Progress in in situ research on dynamic surface reconstruction of electrocatalysts for oxygen evolution reaction. *Adv. Energy Sustain. Res. 3*(8), 2200036. https://doi.org/10.1002/aesr.202200036

26 Shinagawa, T., Garcia-Esparza, A. T., & Takanabe, K. (2015). Insight on Tafel slopes from a microkinetic analysis of aqueous electrocatalysis for energy conversion. *Sci. Rep. 5*(1), 13801. https://doi.org/10.1038/srep13801

27 Luo, F., Pan, S., Xie, Y., Li, C., Yu, Y., Bao, H., & Yang, Z. (2023). Hydrazine-assisted acidic water splitting driven by iridium single atoms. *Adv. Sci. 10*(32), 2305058. https://doi.org/10.1002/advs.202305058

28 Guo, R., Zhang, Y., Zhang, X., Yang, H., & Hu, T. (2024). Synthesis of $Co_{0.52}Cu_{0.48}$/Cu@S–C arrays for the enhanced performance of hydrazine-assisted hydrogen production. *Int. J. Hydrog. Energy 68*, 472–480. https://doi.org/10.1016/j.ijhydene.2024.04.180

29 Yin, J., Jin, J., Zhang, H., Lu, M., Peng, Y., Huang, B., Xi, P., & Yan, C.-H. (2019). Atomic arrangement in metal-doped NiS_2 boosts the hydrogen evolution reaction in alkaline media. *Angew. Chem. Int. Ed. 58*(51), 18676–18682. https://doi.org/10.1002/anie.201911470

30 Yin, J., Li, Y., Lv, F., Lu, M., Sun, K., Wang, W., Wang, L., Cheng, F., Li, Y., Xi, P., & Guo S. (2017). Oxygen vacancies dominated NiS_2/CoS_2 interface porous nanowires for portable Zn–air batteries driven water splitting devices. *Adv. Mater. 29*(47), 1704681. https://doi.org/10.1002/adma.201704681

31 Tang, C., Zhang, R., Lu, W., Wang, Z., Liu, D., Hao, S., Du, G., Asiri, A. M., & Sun, X. (2017). Energy-saving electrolytic hydrogen generation: Ni_2P nanoarray as a high-performance non-noble-metal electrocatalyst. *Angew. Chem. Int. Ed. 56*(3), 842–846. https://doi.org/10.1002/anie.201608899

32 Wang, J., Ma, X., Liu, T., Liu, D., Hao, S., Du, G., Kong, R., Asiri, A. M., & Sun, X. (2017). NiS_2 nanosheet array: A high-active bifunctional electrocatalyst for hydrazine oxidation and water reduction toward energy-efficient hydrogen production. *Mater. Today Energy 3*, 9–14. https://doi.org/10.1016/j.mtener.2017.02.002

33 Liu, G., Sun, Z., Zhang, X., Wang, H., Wang, G., Wu, X., Zhang, H., & Zhao, H. (2018). Vapor-phase hydrothermal transformation of a nanosheet array structure $Ni(OH)_2$ into ultrathin Ni_3S_2 nanosheets on nickel foam for high-efficiency overall water splitting. *J. Mater. Chem. A 6*(39), 19201–19209. https://doi.org/10.1039/C8TA07162D

34 Sun, F., Qin, J., Wang, Z., Yu, M., Wu, X., Sun, X., & Qiu, J. (2021). Energy-saving hydrogen production by chlorine-free hybrid seawater splitting coupling hydrazine degradation. *Nat. Commun. 12*(1), 4182. https://doi.org/10.1038/s41467-021-24529-3

35 Chen, Z., Han, N., Wei, W., Chu, D., & Ni, B.-J. (2024). Dual doping: An emerging strategy to construct efficient metal catalysts for water electrolysis. *EcoEnergy 2*(1), 114–140. https://doi.org/10.1002/ece2.29

36 Krishnamurthy, P., Kumar, A., Alqarni, S. A., Silambarasan, S., & Maiyalagan, T. (2024). Iron and vanadium co-doped Ni_3S_2 flower like nanosheets as an efficient electrocatalysts for water splitting. *Surf. Interfaces 44*, 103694. https://doi.org/10.1016/j.surfin.2023.103694

37 Liu, X., Xi, W., Li, C., Li, X., Shi, J., Shen, Y., He, J., Zhang, L., Xie, L., Sun, X., Wang, P., Luo, J., Liu, L.-M., & Ding, Y. (2018). Nanoporous Zn-doped Co_3O_4 sheets with single-unit-cell-wide lateral surfaces for efficient oxygen evolution and water splitting. *Nano Energy 44*, 371–377. https://doi.org/10.1016/j.nanoen.2017.12.016

38 Liu, X., Chang, Z., Luo, L., Xu, T., Lei, X., Liu, J., & Sun, X. (2014). Hierarchical $Zn_xCo_{3-x}O_4$ nanoarrays with high activity for electrocatalytic oxygen evolution. *Chem. Mater. 26*(5), 1889–1895. https://doi.org/10.1021/cm4040903

39 Ma, X., Wang, J., Liu, D., Kong, R., Hao, S., Du, G., Asiri, A. M., & Sun, X. (2017). Hydrazine-assisted electrolytic hydrogen production: CoS_2 nanoarray as a superior bifunctional electrocatalyst. *New J. Chem. 41*(12), 4754–4757. https://doi.org/10.1039/C7NJ00326A

40 Ren, X. & Tong, Y. (2024). The self-reconstruction of Co-modified bimetallic hydroxysulfide nanosheet arrays for efficient hydrazine assisted water splitting. *Int. J. Hydrog. Energy 49*, 489–497. https://doi.org/10.1016/j.ijhydene.2023.08.109

41 Liu, X., He, J., Zhao, S., Liu, Y., Zhao, Z., Luo, J., Hu, G., Sun, X., & Ding, Y. (2018). Self-powered H_2 production with bifunctional hydrazine as sole consumable. *Nat. Commun. 9*(1), 4365. https://doi.org/10.1038/s41467-018-06815-9

42 Si, J., Zheng, Q., Chen, H., Lei, C., Suo, Y., Yang, B., Zhang, Z., Li, Z., Lei, L., Hou, Y., & Ostrikov, K. (2019). Scalable production of few-layer niobium disulfide nanosheets via electrochemical exfoliation for energy-efficient hydrogen evolution reaction. *ACS Appl. Mater. Interfaces 11*(14), 13205–13213. https://doi.org/10.1021/acsami.8b22052

43 Guo, R., Gao, L., Zhang, Y., Zhang, X., Ma, M., & Hu, T. (2023). Enhancing hydrazine-assisted hydrogen production by constructing CoP-Co_2P bifunctional catalysts. *Appl. Surf. Sci. 617*, 156602. https://doi.org/10.1016/j.apsusc.2023.156602

44 Zhao, Y., Sun, Y., Li, H., Zeng, S., Li, R., Yao, Q., Chen, H., Zheng, Y., & Qu, K. (2023). Highly enhanced hydrazine oxidation on bifunctional Ni tailored by alloying for energy-efficient hydrogen production. *J. Colloid Interface Sci. 652*, 1848–1856. https://doi.org/10.1016/j.jcis.2023.09.003

8

Metal Phosphides Electrocatalysts for Hydrazine-assisted Water Electrolysis

Rahul D. Kerkar

Government College of Arts, Science and Commerce, Khandola-Marcel, India

Corresponding author: rahulkerkar27@gmail.com

8.1 Introduction

The growing global demand for clean, sustainable, and renewable energy has made hydrogen a central figure in the pursuit of a carbon-neutral energy economy. Sustainable means of energy production is highly demanding because of the energy crisis due to the depletion of fossil fuels [1–4]. Hydrogen is a versatile energy carrier with a high energy-to-weight ratio that can be converted efficiently into electricity through fuel cells without emitting greenhouse gases. In current industrial practice, hydrogen is predominantly obtained from fossil-fuel-derived feedstocks through processes such as steam methane reforming, which release large quantities of carbon dioxide and thus compromise the sustainability of hydrogen as a clean energy source [5]. Therefore, it is essential to develop hydrogen production technologies that are scalable, energy-efficient, and environmentally sustainable. Among the various approaches, water electrolysis has emerged as one of the most promising methods, which involves the electrochemical splitting of water into hydrogen and oxygen using electrical energy from renewable sources such as solar or wind power. Thus, with such clean energy inputs, water electrolysis represents a truly green and sustainable route for hydrogen generation [6, 7]. However, a major challenge lies in the oxygen evolution reaction (OER) at the anode, which is considered to be both a thermodynamically unfavorable and kinetically sluggish reaction. The OER requires a large overpotential and involves a complex four-electron-transfer pathway, making it the rate-determining step in conventional water electrolysis [8–10]. This significantly increases the overall cell voltage and the cost of hydrogen production.

Thus, to overcome these critical issues, researchers have explored thermodynamically favorable alternatives that can replace anodic OER reaction. One promising alternative is replacing the anodic side with hydrazine oxidation reaction (HzOR) during conventional water electrolysis due to its lower potential requirement [11–13]. The HzOR proceeds at a much lower potential of −0.33 V vs. the reversible hydrogen electrode (RHE) compared to the OER's potential of +1.23 V vs. RHE [14]. This significant difference in potential helps in decreasing the overall cell voltage required for hydrogen generation. Furthermore, hydrazine oxidation produces nitrogen as the sole by-product, thereby avoiding the carbon footprint responsible for the greenhouse effect. Collectively, water splitting with hydrazine makes the process highly energy-efficient and environmentally friendly for high-yield hydrogen production. Although the concept of hydrazine-assisted electrolysis is attractive, its practical implementation requires development of efficient, stable, and cost-effective electrocatalysts that can facilitate both the hydrogen evolution reaction (HER) at the cathode and the HzOR at the anode. The key material, catalytic, and system-level factors governing hydrazine-assisted electrolysis are summarized in Figure 8.1.

Noble metals such as platinum, iridium, and ruthenium have demonstrated excellent catalytic activity for both HER and HzOR, but their high cost and limited availability hinder their large-scale application. Therefore, extensive research is going on to develop non-noble-metal-based catalysts that can achieve comparable activity and durability by lowering the cost of the material. In this context, metal phosphides are observed to be a highly promising class of electrocatalysts for electrocatalytic application. Metal phosphides, particularly those from transition-metal series such as nickel, cobalt, iron, molybdenum, etc., have been extensively investigated in recent years due to their unique physicochemical properties [15–17]. Metal phosphides exhibit unique

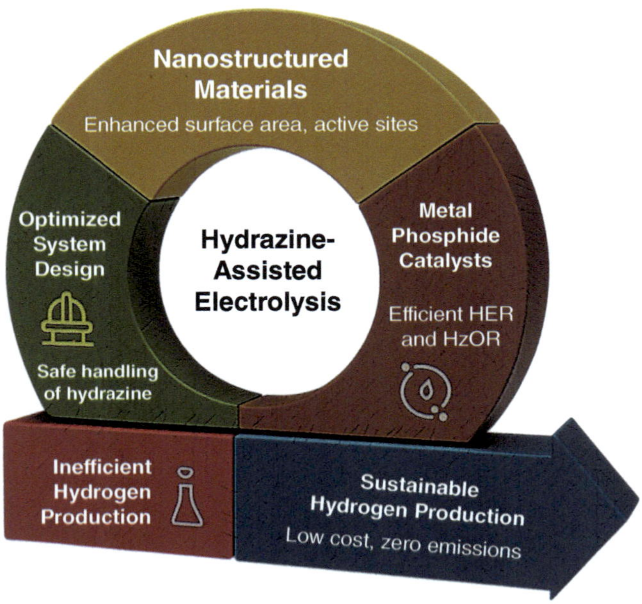

Figure 8.1 Schematic of hydrazine-assisted electrolysis highlighting key materials, catalyst design, and system factors enabling sustainable hydrogen production.

metal-like character, which results in high electrical conductivity, favorable and tunable electronic structures that are required to promote electrolysis. When phosphorus (P) is incorporated into a metal, it induces electron redistribution by modifying the metal's d-band center, which in turn helps in improved catalytic properties [18, 19]. Moreover, with good structural stability and resistance to corrosion under alkaline conditions, metal phosphides are now a better platform for long-term electrochemical applications. For instance, nickel phosphide (Ni_2P) has been reported to exhibit high activity for both HER and HzOR, outperforming many other non-noble metal catalysts [13]. Similarly, cobalt phosphide (CoP) and iron phosphide (FeP) have shown promising activity and durability under electrochemical testing [17, 20–23]. These studies suggested that metal phosphides possess inherent dual functionality, catalyzing both cathodic and anodic reactions during hydrogen production. Thus, introducing metal phosphides and hydrazine at the anode in practical water electrolysis applications makes it a highly promising and cost-effective strategy.

Despite the promising activity and durability of metal phosphides in hydrazine-assisted electrolysis, several challenges must be systematically addressed for their widespread deployment. The most common demerits include catalyst poisoning, incomplete hydrazine oxidation, and stability during prolonged operation [24, 25]. Additionally, hydrazine itself is a toxic and hazardous chemical, which raises safety concerns for large-scale industrial applications. Therefore, ongoing research aims not only to improve the catalytic performance of metal phosphides but also to optimize system design, reaction conditions, and safety protocols to ensure practical viability. Recent advances in materials science have further enabled the development of nanostructured metal phosphides with enhanced surface area, abundant exposed active sites, and tailored morphologies. Strategies such as heteroatom doping, surface modification, and hybridization with conductive carbon supports have further boosted their catalytic efficiency and stability. For example, N–P–Ni synthesized through a multifaceted heteroatom-doping strategy exhibits a remarkably low overpotential of 25.8 mV at 10 $mA{\cdot}cm^{-2}$, closely approaching the performance of commercial Pt/C catalysts [26]. Notably, C and N co-doped $NiPS_3$ achieves an ultralow overpotential of 53.2 mV at 10 $mA{\cdot}cm^{-2}$, while carbon-supported Ni_2P shows enhanced HER activity with an onset overpotential of 88 mV and a Tafel slope of 53 $mV{\cdot}dec^{-1}$; additionally, Co–Ni_5P_4/C demonstrates excellent bifunctional activity for both HER and OER [27–29], thus demonstrating the versatile nature of metal phosphides for designing next-generation electrocatalysts.

The scope of this chapter is to provide a comprehensive overview of the recent progress in the development of metal phosphide electrocatalysts for hydrazine-assisted water electrolysis. The introduction lays out the background and significance of hydrogen production, the limitations of conventional OER-based water splitting, and the advantages of adopting hydrazine oxidation as an alternative anodic reaction. Subsequent sections will focus on the design principles, synthesis methods, catalytic mechanisms, and performance evaluations of various metal phosphides. The chapter will also discuss the challenges that remain, such as long-term stability, catalyst poisoning, and the safe handling of hydrazine, as well as potential future directions for research and application.

8.2 Fundamentals of Hydrazine-assisted Water Electrolysis

Hydrazine-assisted water electrolysis for hydrogen generation has given a promising pathway, where the conventional OER at the anode is replaced by the HzOR. Studies have shown that this substitution lowers the energy demand compared to traditional water electrolysis, thus making the overall process energy-efficient. Furthermore, the overall hydrogen generation process in hydrazine-assisted water electrolysis can be understood through the following half-cell reactions under alkaline conditions.

In alkaline conditions, the half-cell reactions proceed as follows:

Cathodic half-reaction (HER):

$$4H_2O + 4e^- \rightarrow 2H_2 + 4OH^- \tag{8.1}$$

Here, water molecules are reduced at the cathode to produce hydrogen gas and hydroxide ions;

Anodic half-reaction (HzOR):

$$N_2H_4 + 4OH^- \rightarrow N_2 + 4H_2O + 4e^- \tag{8.2}$$

At the anode, hydrazine undergoes oxidation to form nitrogen gas and water while releasing electrons;

Overall cell reaction:

$$N_2H_4 + H_2O \rightarrow N_2 + 3H_2 \tag{8.3}$$

The combination of these half-reactions represents the net chemical transformation occurring during hydrazine-assisted water electrolysis.

Thus, hydrazine acts as a sacrificial agent, lowering the energy barrier while simultaneously producing molecular nitrogen alongside hydrogen, as illustrated by Eqs. (8.1)–(8.3) [2]. Further, a schematic illustration of the hydrazine-assisted electrolysis cell is shown in Figure 8.2.

The mechanistic pathway of HzOR involves sequential dehydrogenation steps of adsorbed hydrazine molecules on the catalyst surface. The resulting intermediates (N_2H_x, $x = 1$–3) undergo proton–electron-coupled processes until molecular nitrogen is released. The kinetics of this pathway strongly depend on the nature of the electrocatalyst, with transition-metal-based materials (e.g., Ni, Co, Fe, and their alloys or heterostructures) demonstrating particularly promising performance [30, 31]. An effective catalyst must not only lower the activation barrier for hydrazine dehydrogenation but also facilitate the rapid desorption of N_2 and H_2 to avoid catalyst poisoning and bubble accumulation.

The main advantages of hydrazine-assisted electrolysis include lower energy consumption, the potential use of hydrazine-containing wastewater as a feedstock, and reduced anodic stress in compact electrolyzer designs [32]. However, limitations persist due to the toxicity and instability of hydrazine and the durability issues of catalysts under sustained HzOR operation. In addition, maintaining long-term catalyst durability under continuous HzOR operation is a major concern, with issues

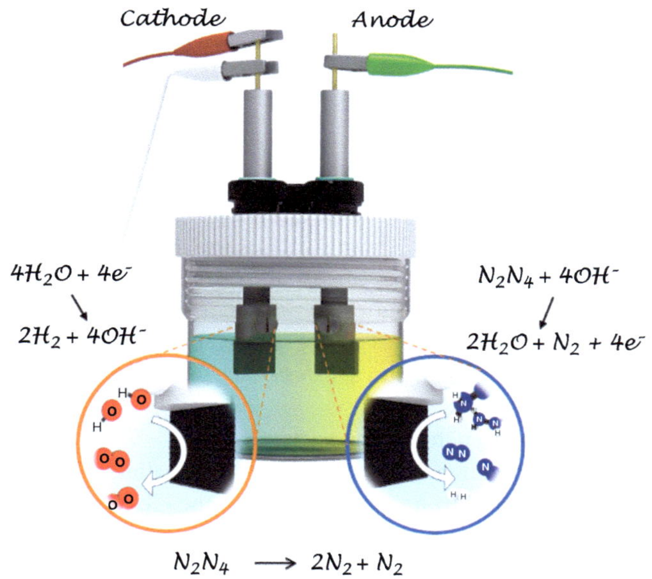

Figure 8.2 Illustration of the hydrazine-assisted water electrolysis setup, highlighting the anodic HzOR and the cathodic HER.

such as corrosion and deactivation. As a result, current research focuses on developing robust, bifunctional electrocatalysts capable of efficiently catalyzing both HzOR and HER while maintaining long-term operational stability [31].

8.3 Transition-metal Phosphides

8.3.1 Structure, Electronic Properties, and Relevance to Electrocatalysis

8.3.1.1 Crystal Structure and Classification of Metal Phosphides

Metal phosphides are a class of compounds that are broadly composed of phosphorus and transition metals such as Fe, Co, Ni, Mo, W, Pt, Ru, etc., in varying stoichiometries and crystal structures. The nature of the metal–phosphorus interactions gives rise to distinct atomic arrangements and electronic characteristics, which are the reason for their improved catalytic behavior. Structurally, many metal phosphides (especially transition-metal phosphides) can be regarded as interstitial compounds, where P atoms occupy interstitial sites in a metal lattice (or vice versa) and the metal–metal and metal–phosphorus bonding network yields a range of phases with metallic or semimetallic character.

In transition metals, the metal-to-phosphorus ratio (M:P) plays a significant role in determining their structure and properties, largely due to the metals' variable valencies. Metal phosphides can be broadly categorized into "metal-rich" phases (such as Ni_2P, Co_2P, Fe_2P, etc.), which typically exhibit metallic conductivity, and "P-rich" phases (such as NiP_2, NiP_3, CoP_3, RuP_2, etc.), which often display semiconducting behavior. In P-rich compositions, the increased presence of P—P bonding and the reduced extent of metal–metal interactions generally give rise to semiconducting characteristics. In contrast, metal-rich phases exhibit stronger M—M bonding networks, which contribute to their metallic conductivity. Beyond this compositional classification, metal phosphides crystallize into several well-established structural families such as the WC-type, NiAs-type, NbAs-type, NaCl-type, MnP-type, NiP-type, Fe_2P-type, and Fe_3P-type structures. These structural arrangements differ in coordination geometry, metal packing arrangement, and bonding topology, all of which influence the electronic structure and catalytic activity of the material. A detailed summary of these crystal structures, along with representative compounds and pictorial illustrations, is provided in Table 8.1.

8.3.1.2 Electronic Structure and Conductivity

The incorporation of phosphorus into metal lattices leads to substantial modifications in the electronic structure, which strongly influence both the electrical properties and catalytic behavior of metal phosphides. One of the significant effects shown with the phosphorus incorporation is the modulation of the d-band center in metal phosphide. Because phosphorus is more electronegative than most transition

Table 8.1 Overview of common crystal structures in metal phosphides.

Structure type	Crystal system	Key structural features	Representative examples
WC-type	Hexagonal	Linear stacking of trigonal prisms; P located inside trigonal prismatic coordination; metal-rich packing	MoP
NiAs-type	Hexagonal	Trigonal prisms laterally displaced by ½ lattice spacing; close-packed M layers	VP
NbAs-type	Tetragonal/ cubic (distorted NiAs-related)	Similar prism units as NiAs but different stacking arrangements	NbP, TaP
MnP-type	Orthorhombic	Distorted NiAs framework; P atoms form chains; mixed M—P and M—M bonding	WP, CrP, MnP, FeP, and CoP
NiP-type	Orthorhombic	Distortion of NiAs; P atoms form P—P pairs instead of chains	NiP
Fe_2P-type	Hexagonal	Two distinct metal sites; trigonal prisms + tetrahedra; mixed M—P and M—M connectivity	Fe_2P, Ni_2P, and Mn_2P
Fe_3P-type	Orthorhombic	More complex polyhedral packing; multiple metal sites with mixed coordination	Fe_3P, Ni_3P, and Mn_3P
NaCl-type	Cubic	Rock-salt metal/phosphorus arrangement; octahedral coordination; typical for early actinide phosphides	UP, ThP, PuP, YP, and ZrP

metals, it withdraws electron density from the metal d-orbitals, thereby shifting the d-band center relative to the Fermi level and altering the density of states in its vicinity. These changes have a direct impact on the adsorption energetics of reaction intermediates and thus on the catalytic performance of transition-metal phosphides.

Another key aspect is the Fermi-level position and charge-carrier distribution. Metal-rich metal phosphides, characterized by extensive metal–metal bonding, typically display metallic or quasi-metallic conductivity due to the presence of delocalized electrons. In contrast, P-rich compositions contain more localized electronic states, reduced M—M interactions, and often exhibit partial semiconducting behavior. This compositional dependence results in significant variability in charge-carrier density and electronic transport. Phosphorus also plays a central role in the modulation of catalytically active sites. Its higher electronegativity enables it to polarize adjacent metal atoms, facilitating the adsorption of protons or other charged intermediates. Simultaneously, the incorporation of P into the metal lattice induces lattice distortions by modifying lattice parameters and M—P bond lengths. These distortions can generate coordinatively unsaturated metal centers, defect-rich surface regions, and higher concentrations of active sites, which simultaneously enhance catalytic reactivity.

A further contribution arises from the bonding and hybridization characteristics of metal phosphides, where the M—P bond typically displays a mixed metallic and covalent nature. This results in altered bond distances and a partial compression of the metal d-bands compared to pure metallic phases. This hybridization increases the density of states near the Fermi level and supports improved electronic conductivity, which is essential for efficient charge transfer during electrocatalytic processes. This further results in high electrical conductivity, active-site, and adsorption–desorption characteristics that are needed for intermediates formation in a wide range of electrocatalytic applications.

8.3.1.3 Relevance to Electrocatalysis

Electrocatalytic reactions, such as the HER, OER, and other substrate transformation reactions, impose stringent requirements on catalytic materials. This requires good electrical conductivity for efficient electron transfer, well-balanced adsorption and desorption of intermediates, stability under operating conditions, and cost-effectiveness for large-scale use. In such scenarios, transition-metal phosphides satisfy many of these criteria and have therefore gained considerable attention as promising electrocatalytic materials. Their appeal comes mainly from their intrinsic metallic or near-metallic conductivity, especially in metal-rich forms, which allows for fast movement of the charges while keeping the structure stable in both alkaline and, in some cases, acidic media. Furthermore, electrolytic application requires a high degree of tunability, which is more favorably offered by transition-metal-composed phosphides. This can be achieved by selecting different metal centers (such as Fe, Co, Ni, Mo, or W) and varying the M:P stoichiometry (e.g., MP, M_2P), which adjust the adsorption energies of key intermediates such as H*, OH*, and OOH* for improved catalytic performance. For HER, achieving a hydrogen adsorption free energy (ΔG_H^*) close to zero is crucial, and the incorporation of phosphorus modulates the metal d-states in a way that allows fine control over this parameter.

Another advantage of metal phosphides (transition-metal-based phosphides) is their versatility across a wide pH range, as many compositions exhibit effective activity in acidic, neutral, and alkaline electrolytes. In addition, the use of earth-abundant first-row transition metals confers a significant cost advantage over noble-metal catalysts like Pt, Ir, and Ru. Transition-metal phosphides also integrate well into advanced nanoarchitectures—such as carbon-supported structures, oxide–phosphide composites, and heterojunction systems—which can further enhance catalytic activity, durability, and the exposure of active sites. As a result, numerous studies have reported transition-metal phosphides achieving performance comparable to or even exceeding that of conventional noble-metal benchmarks for HER and OER.

In the context of hydrazine-assisted water electrolysis, transition-metal phosphides hold particular relevance. Since hydrazine oxidation occurs at a substantially lower potential than the OER, catalysts must remain stable under both reductive and oxidative conditions, a requirement well met by many transition-metal phosphides. The presence of phosphorus can also facilitate the adsorption and activation of hydrazine-derived intermediates through favorable electronic interactions. Moreover, transition-metal phosphides are capable of catalyzing the complementary HER with low overpotentials, which is essential for maximizing the energy efficiency of the overall hydrazine-assisted electrolysis process. Collectively, these attributes make transition-metal phosphides strong contenders for integration into next-generation electrocatalytic systems for sustainable hydrogen production.

8.3.2 Synthesis Strategies for Metal Phosphides

The performance of metal phosphides is strongly dependent on their composition, morphology, nanostructure, phase purity, defect density, and surface exposure. Many methods have been developed to synthesize these materials with specific features for catalytic performances. Below are the common methods that have been focused on so far in the reported studies, corresponding to the synthesis routes illustrated in Figure 8.3a–f.

8.3.2.1 Phosphidation of Metal Hydroxide or Metal Oxide Precursors

Phosphidation of metal oxides and hydroxides is a key strategy for synthesizing metal phosphides, which are widely used in electrocatalysis and energy-storage applications [33, 34]. In this method, a preformed metal or metal oxide/hydroxide precursor is exposed to a phosphorus-containing source under elevated temperatures (Figure 8.3a), leading to the formation of M—P bonds and subsequent conversion into the desired phosphide phase. Typically, the process begins with the preparation of a metal-based precursor such as metal oxide (NiO, CoO_x, Al_2O_3, etc.), metal hydroxide ($Cu(OH)_2$, $Ni(OH)_2$, $Bi(OH)_3$, etc.), or oxyhydroxide (FeOOH, NiOOH, etc.). A phosphorus source, commonly sodium hypophosphite ($NaH_2PO_2 \cdot H_2O$), red phosphorus, or phosphine gas (PH_3), is then introduced. Under an inert or reducing atmosphere (N_2 or Ar), the phosphorus source thermally decomposes to generate reactive phosphorus species (e.g., PH_3 or P vapor), which subsequently reacts with the metal precursor to form the metal phosphide.

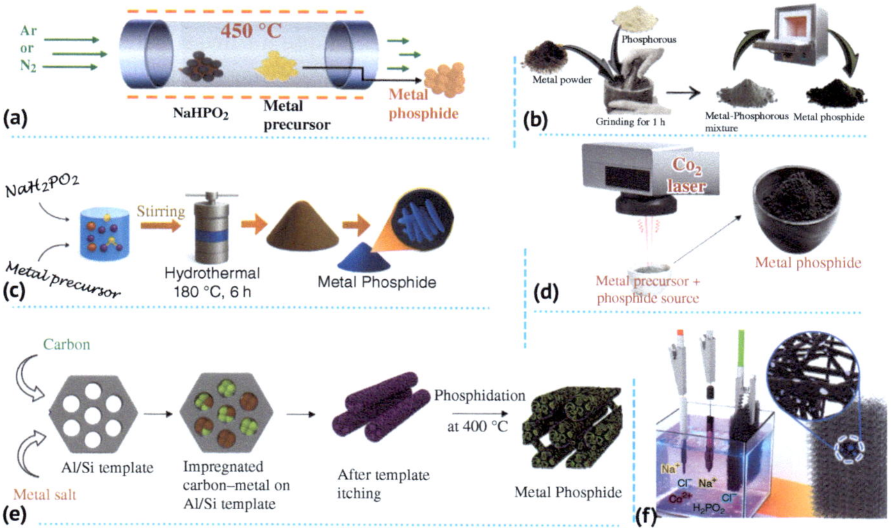

Figure 8.3 Overview of representative synthesis strategies employed for the preparation of metal phosphides: (a) phosphidation of metal hydroxide or metal oxide precursors, (b) direct solid-state reaction, (c) hydrothermal phosphidation, (d) CO_2 laser-assisted method, (e) template or MOF-derived methods, and (f) electrochemical or in situ phosphidation.

Phosphidation is generally carried out at temperatures ranging from 300 to 600 °C, with the final phase composed of metal phosphides like Ni_2P, CoP, CoP_2, or FeP, which is governed by reaction temperature, exposure time, and phosphorus concentration. Posttreatment steps typically involve washing or etching to remove unreacted phosphorus species or residual by-products. Due to its adaptability and reproducibility, this method provides excellent control over stoichiometry, crystallinity, and phase purity, and is therefore widely utilized for the synthesis of metal phosphides.

8.3.2.2 Direct Solid-state Reaction

Another traditional approach for synthesizing metal phosphides is the direct solid-state reaction, as illustrated in Figure 8.3b, where metal and phosphorus sources are physically mixed and subsequently subjected to high-temperature treatment. In this method, metal powders are combined with a solid phosphorus source such as red phosphorus, often using ball milling to ensure intimate mixing and homogenization. The mixture is then annealed under an inert or reducing atmosphere (typically N_2 or Ar), enabling diffusion-driven formation of M—P bonds and conversion to the desired phosphide phase. This route generally yields bulk or microscale phosphide materials rather than nanoscale architectures, owing to the high-temperature, diffusion-controlled nature of the reaction.

Although the direct solid-state method is operationally simple and useful for preparing phase-pure bulk metal phosphide for fundamental structural or electronic studies, it offers limited control over particle size, morphology, and surface characteristics. As a result, it is typically less suitable for producing high-surface-area nanostructured electrocatalysts, which are essential for achieving high catalytic activity. Nonetheless, it remains an important technique for investigating intrinsic material properties and for applications where nanoscale morphology is not a critical requirement.

8.3.2.3 Hydrothermal Phosphidation

The hydrothermal method serves as an excellent initial step in synthesizing catalytic materials, particularly for creating precise metal hydroxide or oxide nanostructures. This technique is highly effective in forming various morphologies, such as nanorods, nanosheets, or nanowires, which are ideal precursors due to their high surface area and controlled shapes. These preformed nanostructures are then subjected to a subsequent phosphidation process, as illustrated in Figure 8.3c. This conversion transforms the metal hydroxides or oxides into metal phosphides, which are often the catalytically active species. Therefore, this two-step methodology allows for precise control over the material's morphology and significantly improves its structural framework. For instance, cobalt phosphide nanorods with well-controlled morphology have been synthesized using a hydrothermal-phosphorization approach, demonstrating exceptional supercapacitive performance [35]. Another example involves the hydrothermal preparation of a Co(OH)F precursor, which is subsequently phosphidated to form a supported cathode [36]. This two-step process can lead to smoother surfaces and larger diameters, which enhance the material's electrochemical activity and improve charge-transport properties.

This strategy is particularly valuable for creating well-defined, high-surface-area architectures, where catalyst structure and its electrochemical behavior are critical for efficient energy conversion.

8.3.2.4 CO_2 Laser-assisted Method

The CO_2 laser-assisted method is a unique rapid-heating strategy for synthesizing metal phosphides with tailored nanostructures. In this technique, a powdered mixture of metal precursors and a phosphorus source is typically placed in a ceramic crucible, which is later irradiated using a high-power CO_2 laser as depicted in Figure 8.3d [37, 38]. Upon laser exposure, rapid thermal activation induces decomposition of the phosphorus source (e.g., NaH_2PO_2, red phosphorus), generating reactive phosphorus vapors that diffuse and react with the metal precursor. This results in the formation of metal–phosphorus bonds and the direct conversion of the precursor mixture into crystalline metal phosphide phases. The extremely fast heating and cooling cycles minimize grain growth, often yielding fine nanoparticles or defect-rich microstructures favorable for electrocatalysis. A major advantage of CO_2 laser processing is the ability to achieve phosphidation within seconds to minutes, in contrast to conventional furnace methods that require extended heating times.

8.3.2.5 Template or MOF-derived Methods

To obtain high-surface-area, nanostructured metal phosphides with tailored architectures—such as nanowires, hollow frameworks, and porous nanosheets—researchers increasingly employ metal–organic frameworks (MOFs) or other sacrificial templates as precursors. In a typical approach as shown in Figure 8.3e, a metal-based MOF (commonly Fe-, Co-, or Ni-based) is first synthesized, often directly grown on a conductive substrate to ensure strong electrical contact. Subsequent thermal treatment or pyrolysis of the MOF leads to its decomposition into a metal, metal–oxide, or metal/carbon composite intermediate while partially retaining the original porous framework. This intermediate is then subjected to phosphidation, during which reactive phosphorus species convert the metal centers into the corresponding phosphide phase. Depending on the MOF composition and pyrolysis conditions, the resulting metal phosphide nanostructure frequently incorporates residual carbon and exhibits well-defined porosity, hierarchical morphology, and high structural stability.

The principal advantage of MOF-derived synthesis lies in its ability to transfer the intrinsic order, porosity, and morphology of the template to the final phosphide, thereby generating materials with large surface areas, abundant exposed active sites, and enhanced electrical conductivity when carbon is retained in the structure. Such templated architectures have been widely explored for electrocatalytic applications. For instance, porous iron phosphide derived from an SBA-15 template, followed by template etching and phosphidation, has exhibited significantly enhanced HER activity. Its ordered mesoporous framework provides a high density of electrochemically accessible active sites and facilitates efficient mass transport, collectively contributing to superior catalytic performance. Such examples show the effectiveness of template-assisted synthetic strategies in producing structurally engineered metal phosphides for advanced electrocatalytic applications.

8.3.2.6 Electrochemical In Situ Phosphidation

Electrochemical or in situ phosphidation represents a more recent class of strategies in which phosphorus species are introduced directly under electrochemical conditions. Figure 8.3f visually depicts the electrochemical phosphidation process, showing the direct deposition of the desired metal phosphide phase onto the substrate. Although less common than conventional thermal phosphidation, these approaches offer unique advantages, including the generation of defect-rich, highly active surfaces and the possibility of producing self-regenerating catalytic interfaces. Recent studies and reviews have highlighted such electrochemically reconstructed metal phosphide surfaces as an emerging pathway for enhancing catalytic stability and activity under operational conditions.

8.4 Characterization Techniques

Accurate characterization of metal phosphide electrocatalysts is essential for establishing clear correlations between their structure, morphology, electronic configuration, and catalytic performance. A combination of complementary techniques is typically employed to probe phase identity, nanoscale architecture, surface chemistry, and electronic properties. The most relevant methods and their significance for metal phosphide systems are outlined below.

8.4.1 X-ray Diffraction

X-ray diffraction (XRD) is one of the most powerful and widely used techniques for verifying the formation of target metal phosphide phases and identifying residual precursors such as metal oxides, hydroxides, or unreacted metals. As shown in Figure 8.4a, the diffraction peaks of Sm-doped nickel phosphides show a progressive peak shift compared to pristine Ni_2P, indicating lattice distortion due to dopant incorporation. A closer look into the (111) plane region (Figure 8.4b) confirms a gradual shift toward higher angles with increasing Sm content, suggesting reduced lattice spacing and strong structural interaction between Sm and Ni_2P [39]. Further, by indexing diffraction peaks against standard reference data cards, researchers can accurately confirm phase purity of respective metal phosphide. Beyond qualitative identification, XRD provides quantitative structural information essential for understanding material behavior. Lattice parameters, for example, can be precisely

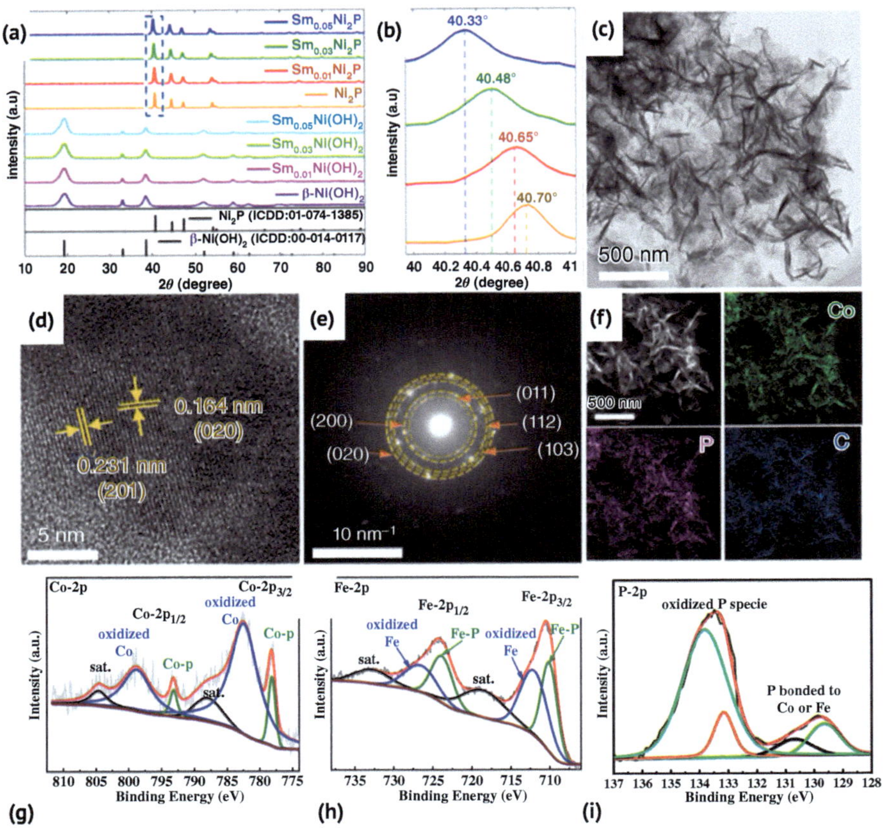

Figure 8.4 (a) XRD patterns of samples; (b) high-resolution XRD patterns of nickel phosphide samples around (111) plane of Ni_2P. *Source*: [39]/Springer Nature. CC BY 4.0. (c–e) TEM image, HRTEM image, and electron diffraction pattern of CoP/rGO. (f) Elemental mapping images of Co, P, and C elements in CoP/rGO. *Source*: [45]/Springer Nature. CC BY 4.0. XPS characterization of the CoFeP/C nanoboxes: (g–i) high-resolution XPS spectra of Co-2p, Fe-2p, and P-2p. *Source*: [48]/with permission of American Chemical Society.

determined and used to track compositional or substitutional changes. Studies on Cr-doped zinc phosphide nanoparticles reported systematic lattice expansion (from $a = b = 8.1129$ Å, $c = 11.1029$ Å to $a = b = 8.1687$ Å, $c = 11.1268$ Å) with increasing Cr content, illustrating how doping alters crystal geometry [40]. Peak shifts toward lower angles generally reflect such lattice expansion, while peak broadening arises from reduced crystallite size or increased microstrain. In NiFeP thin films, XRD analysis showed crystallite size first increased and then decreased with deposition time, while microstrain reduced with longer precursor reaction times [41]. These structural parameters strongly influence functional performance: for instance, Ni_2P embedded in nitrogen–phosphorus-doped carbon exhibits significantly enhanced electrocatalytic oxygen evolution activity, directly linked to its lattice structure and defect distribution revealed by XRD [42]. Thus, XRD-derived structural insights are essential to understand the materials' active structure or phase in catalytic reactions.

8.4.2 Transmission Electron Microscopy

Transmission electron microscopy (TEM) is a key technique for visualizing the morphology and internal structure of metal phosphides. TEM images of nickel phosphide ($Ni_{12}P_5$) nanospheres, for example, reveal a hollow structure with a diameter of around 100 nm and a shell thickness of 15–20 nm [43]. When combined with selected-area electron diffraction (SAED), TEM can also identify crystal phases, and the structures which are the core architecture needed for energy storage performance. High-resolution transmission electron microscopy (HRTEM) provides further details at the atomic level, enabling visualization of lattice fringes and interplanar spacings. HRTEM analysis of CoP nanostructures revealed exposed (111) facets, which turned out to be key to enhancing electrocatalytic activity [44]. As mentioned earlier, SAED further supports these techniques by confirming the local crystallographic structure of materials, like the crystalline phase in $Ni_{12}P_5$ nanospheres [43]. Figure 8.4c–e highlight thin nanosheet-like CoP/rGO structures with highly crystalline lattice fringes and clearly indexed (020) and (201) planes, confirming robust phosphide formation. The ring-like SAED pattern further validates the polycrystalline nature of the CoP nanostructures anchored on reduced graphene oxide, providing an interconnected network beneficial for charge transfer and electrocatalytic activity [45]. These morphological properties like porosity, crystallinity, and defects affect the material's surface area, how it transports molecules, and how easily active sites are accessed, making it ideal for use in electrocatalysis, energy storage, and environmental sensing.

8.4.3 X-ray Photoelectron Spectroscopy

X-ray photoelectron spectroscopy (XPS) is a powerful tool for analyzing the surface composition and oxidation states of metals and phosphorus in metal phosphides. Based on the photoelectric effect, XPS works by using X-ray photons to ionize atoms, causing the emission of core-level electrons. The binding energy of these emitted electrons gives detailed information about the element and its chemical environment.

In metal phosphides, these binding energies are key to understanding bonding and the electronic structure. For example, in binary transition-metal monophosphides like CrP and CoP, shifts in the phosphorus-2p binding energies show how ionic the metal–phosphorus bond is [46]. In metal-rich phosphides like M_2P and M_3P, the phosphorus binding energies decrease as the ionic character increases, signaling the presence of anionic phosphorus [47]. The Co-2p and Fe-2p spectra (Figure 8.4g,h) reveal the presence of both metal–phosphide states and oxidized metal species, confirming the coexistence of catalytically active electronic environments (Figure 8.4f) [48]. Additionally, the P-2p spectrum in Figure 8.4i indicates P bonded to both Co and Fe, alongside surface-oxidized phosphate species, validating strong M–P interactions essential for hydrazine and hydrogen electrocatalysis [48]. Thus, XPS is essential for understanding the charge transfer between metals and phosphorus, especially in bimetallic phosphides, which affects their catalytic and electronic properties.

8.4.4 Inductively Coupled Plasma

Inductively coupled plasma (ICP) techniques, like ICP–OES (optical emission spectrometry) and ICP–MS (mass spectrometry), are key tools for determining the bulk composition of metal phosphides. These methods offer high sensitivity and precise measurements of elemental composition, making it possible to accurately analyze metals, sulfur, and phosphorus in a variety of materials. For example, studies have used ICP–OES to confirm the composition of nanoparticles like Ni_2P, Ni_1Mo_1P, and $Ni_{1.6}Mo_{0.4}P$ supported on silica, with results that closely match the expected values, demonstrating the accuracy of the technique [49]. ICP–MS is particularly useful for detecting trace elements, such as phosphorus in compounds like indium phosphide (InP) and aluminum phosphide (AlP), where phosphorus levels can be measured at very low concentrations ($ng{\cdot}g^{-1}$) [50]. In the past, the phosphine group in compounds like AlP was analyzed by converting phosphine gas into a mercury compound and then using iodometric titration [51]. Today, ICP techniques remain essential for understanding the bulk stoichiometry of metal phosphides, particularly in applications like catalysis and energy conversion, where accurate metal-to-phosphorus ratios are crucial. By providing detailed elemental information, ICP helps link material properties to their crystallographic structure and catalytic performance, guiding the design of metal phosphides with optimized functionality for advanced applications.

8.4.5 Study of Surface Area, Porosity, and Active Sites

Quantifying surface area and porosity is critical for understanding how nanostructure and morphology influence the catalytic performance of metal phosphides. The Brunauer–Emmett–Teller (BET) nitrogen adsorption–desorption method is commonly used to measure the specific surface area ($m^2{\cdot}g^{-1}$) and pore-size distribution of materials, especially those with porous structures or derived from MOFs. Materials with large surface areas and complex porosity provide more accessible active sites, improving mass transport and boosting electrochemical reactions like the OER. For instance, mesoporous cobalt phosphide (meso-CoP), with its high surface area, showed 37 times better electrical conductivity compared to nonporous cobalt phosphide, leading to significantly better performance in electrochemical applications like HER [52]. Similarly, iron phosphide integrated with porous carbon frameworks displayed high catalytic activity due to its large surface area and well-developed porosity [53].

In addition to BET, the electrochemical active surface area (ECSA) is a key parameter for evaluating the accessible catalytic sites on metal phosphide electrodes. ECSA is typically measured by calculating the electrochemical double-layer capacitance (C_{dl}) using cyclic voltammetry (CV) in the non-faradaic region. A higher C_{dl} indicates a greater density of active sites, which directly enhances the catalyst's electrocatalytic performance. For example, aluminum-doped cobalt phosphide showed a fourfold increase in ECSA compared to pure cobalt phosphide, significantly boosting its OER performance [54]. To more accurately compare catalysts with different morphologies, studies often normalize catalytic current density by ECSA instead of geometric area or BET surface area. This method provides a better evaluation of the material's true activity, particularly for complex structures like hollow nanostructures, nanowires, or nanosheets. For example, hollow nanostructured iron phosphide catalysts demonstrated improved OER performance by maximizing the exposure of active sites, enhancing electrochemical efficiency [53].

8.5 Electrochemical Performance Evaluation

To evaluate the electrochemical performance of electrocatalysts in hydrazine-assisted water splitting, a variety of techniques are used to understand the catalyst's activity, efficiency, and stability. These methods include linear sweep voltammetry (LSV), chronopotentiometry, electrochemical impedance spectroscopy (EIS), and Tafel analysis. Together, they provide valuable insights into how well the catalyst performs in reactions like the HER and HzOR.

Many transition-metal phosphides, such as CoP, NiP, and FeP, showed promising activity for HER [55]. For example, ruthenium phosphide (RuP) stands out with its exceptional performance, requiring just 46, 47, and 22 mV at a current density of 10 mA·cm^{-2} in acidic, neutral, and alkaline conditions, respectively [56]. This excellent performance highlights the versatility of RuP across a broad pH range. Similarly, bimetallic phosphides like NiCoP also demonstrate outstanding HER efficiency, requiring only 32 mV to reach 10 mA·cm^{-2}, further emphasizing the effectiveness of these materials for hydrogen production [57]. In addition, catalysts such as NiFeP nanosheets and RhP ultrathin nanosheets exhibit enhanced electrocatalytic activity, reinforcing the strong potential of metal phosphides in high-performance HER systems [57–59]. Such improvements can be clearly observed in Figure 8.5a,b. The $Ni(OH)_2/Ni_2P/NF$ catalyst shows significantly lower HER overpotentials and superior catalytic currents compared to benchmark electrodes. The formation of a Ni_2P core with a surface $Ni(OH)_2$ layer creates optimized catalytic interfaces

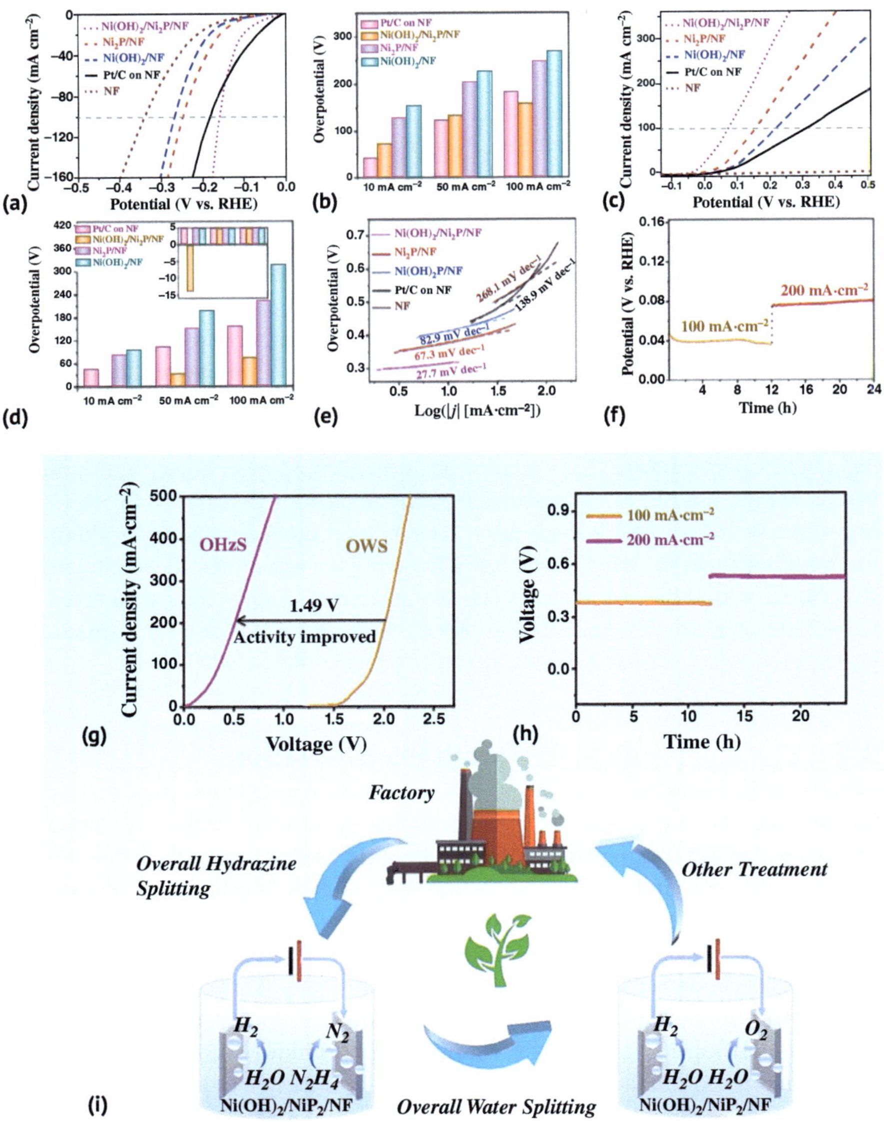

Figure 8.5 Electrocatalytic performance of $Ni(OH)_2/Ni_2P/NF$ for HER and HzOR. (a) LSV curves of $Ni(OH)_2/Ni_2P/NF$ and comparison samples for HER. (b) Overpotentials at 10, 50, and 100 mA·cm^{-2} for HER. (c) LSV curves for HzOR. (d) Overpotentials at 10, 50, and 100 mA·cm^{-2} for HzOR. (e) Tafel plots for HER and HzOR. (f) Long-term stability at 100 and 200 mA·cm^{-2} over 24 hours. (g) Long-term stability at fixed current densities over 24 hours. (h)LSV comparison of Ni(OH)2/Ni2P/NF for HzOR and OER. (i) Schematic of the eco-friendly cycling process. *Source*: [63]/with permission of Elsevier.

that accelerate water dissociation while also supporting fast electron transfer to active sites. One particularly promising example is $NiCo(OH)_x$@NiCoP/NF, which achieves 100 mA·cm^{-2} at an ultralow potential of just 0.04 V, far lower than the voltage typically required for conventional OER. This excellent performance makes

NiCo(OH)$_x$@NiCoP/NF a strong candidate for high-current-density hydrazine-assisted electrolysis for highly efficient hydrogen production [60].

When evaluating electrocatalysts, several performance metrics like overpotential, Tafel slope, current density, and stability are foremost important. Overpotential is the extra voltage needed to drive a reaction beyond its theoretical minimum. A lower overpotential indicates a more efficient catalyst. For instance, CoP–Co_2P@CC-300 showed excellent hydrazine oxidation performance, achieving 10 mA·cm^{-2} at −89 mV and 100 mA·cm^{-2} at −35 mV for HzOR [61]. The Tafel slope is another key factor that gives insight into how fast the reaction occurs; smaller slopes indicate faster charge transfer. For instance, Mo–Ni_2P_v@MNF had a Tafel slope of 48.9 mV·dec^{-1} for HzOR, showing much faster reaction kinetics than many other metal phosphide catalysts [62]. In Figure 8.5c–e, $Ni(OH)_2/Ni_2P$/NF again surpasses the performance of comparison catalysts, delivering higher current densities at much lower applied potentials for HzOR. The favorable kinetics and enhanced adsorption of hydrazine intermediates on this heterostructure contribute to this improved activity. Current density required to measure the rate of electrochemical reactions, is also an important metric for practical applications. For industrial-level hydrogen production, achieving current densities of up to 400 mA·cm^{-2} is crucial. Additionally, stability is one of the critical factors for long-term studies needed while utilizing for practical applications. NiFeP nanosheets on nickel foam, for example, demonstrated stability for over 40 hours of continuous electrolysis, proving their reliability [58]. In agreement, Figure 8.5f,g reveal that $Ni(OH)_2/Ni_2P$/NF remains highly stable at high current densities of 100 and 200 mA·cm^{-2} for extended 24-hour tests. The beneficial impact of replacing OER with HzOR is highlighted in Figure 8.5h, where a major drop in cell voltage is clearly visible when shifting from conventional water splitting to hydrazine-assisted electrolysis. This energy-saving feature is crucial for practical large-scale hydrogen production. Further, a sustainable concept is shown in Figure 8.5i where hydrazine from industrial wastewater can be reused as an anodic fuel, simultaneously producing hydrogen and mitigating environmental hazards. This closed-loop system highlights the dual environmental and energy benefits of this technology.

In addition to these, EIS is used to measure charge-transfer resistance, while chronoamperometry monitors the durability of the catalyst by tracking potential changes at constant current densities. Other parameters like faradaic efficiency, turnover frequency (TOF), and gas-evolution analysis (H_2, N_2) further quantify catalytic efficiency and selectivity. By combining electrochemical performance data with structural and electronic characterization, scientists can correlate the material's composition and structure with its catalytic behavior. This helps them optimize the properties of metal phosphides for better performance in hydrazine-assisted water splitting.

8.6 Post-characterization Techniques

Post-characterization techniques are important to understanding the changes that metal phosphide catalysts go through during hydrazine-assisted water splitting. They offer valuable insights into the stability of the catalyst, how it degrades, and how its active sites evolve during the reaction.

XPS, being a surface-sensitive technique, is especially useful for analyzing the catalyst's surface after electrochemical testing. It can detect shifts in the oxidation states of metal and phosphorus, revealing if the surface has reconstructed or formed new species, such as oxides, hydroxides, or phosphates. For example, after long-term testing of NiCoP@CoFeP in the OER, XPS showed persistent metal oxide peaks and shifts in oxidation states, indicating changes at the surface [14]. XPS can also identify surface adsorbates that might poison the catalyst, offering critical insights into surface degradation or poisoning during the reaction. In conjunction with XPS, XAS-composed techniques like XANES and EXAFS help us understand the local atomic and electronic structures of the catalyst. Ex-situ XAS is valuable for studying structural changes after electrochemical reactions, revealing coordination environments and identifying the active phase that forms during the reaction. While ex-situ XAS is common, in situ or operando XAS—though outside the current scope—provides real-time data on catalyst behavior, capturing transient species and reaction mechanisms under actual operating conditions.

Additionally, XRD is another fundamental technique for identifying crystalline phases in metal phosphides before and after testing. Comparing XRD patterns helps detect phase transformations, crystallization of amorphous regions, or the formation of secondary crystalline products. Changes in peak positions or intensities can indicate lattice strain or shifts in crystallite size, which can affect the catalyst's stability and performance in hydrazine-assisted water splitting. Raman spectroscopy is also useful for examining the molecular structure, crystallinity, and defects of metal phosphides. Ex-situ Raman analysis can reveal how vibrational modes change after the reaction, shedding light on chemical transformations or new compounds formed on the surface. In situ/operando Raman spectroscopy, particularly valuable under working conditions, tracks the dynamic behavior of the catalyst. It helps identify reaction intermediates, monitor the reconstruction of active sites, and even track the breaking of N—N bonds during hydrazine oxidation. This technique provides a clearer picture of the reaction pathways and offers deeper insights into the catalytic mechanisms.

8.7 Conclusion

The future of metal phosphide electrocatalysts for hydrazine-assisted water electrolysis is highly promising, but several important challenges must still be addressed. One of the most important areas for progress is bridging the gap between theoretical predictions and experimental observations. To truly optimize catalytic activity, researchers need a deeper understanding of how the surfaces of transition-metal phosphides evolve during electrolysis. During this process, surface reconstruction, oxidation, and defect formation can dramatically affect long-term catalyst performance. To tackle this, researchers must focus on achieving precise control over catalyst synthesis, ensuring that the materials have well-defined facets, minimal defects, and uniform compositions. Only with such control can we begin to establish reliable structure–activity relationships, linking the physical properties of the catalysts to their electrochemical performance.

Another major challenge lies in understanding the catalyst–electrolyte interface at the molecular level. Traditional ex-situ characterization techniques often fail to capture the dynamic changes that take place under real operating conditions. To address this, the integration of advanced in situ techniques such as in situ Raman spectroscopy, operando XPS, and in situ XAS will be essential. These techniques will allow scientists to capture real-time data on intermediate species, oxidation states, and structural transformations that drive the HER and the HzOR, providing critical insights into how the catalyst behaves under actual operating conditions. By continuously monitoring these changes, we can better understand and fine-tune catalyst performance for improved efficiency and stability.

However, beyond the scientific challenges of improving catalysis, there is also a practical consideration that cannot be overlooked like the safe handling of hydrazine. As a toxic and volatile compound, hydrazine presents serious safety concerns in large-scale applications. For hydrazine-assisted electrolysis to become feasible at industrial scales, it is crucial to ensure safe handling procedures. This includes the use of sealed electrochemical cells, proper ventilation systems, and protective equipment to avoid exposure. Furthermore, the design of practical systems will require automated hydrazine delivery, leak detection sensors, and controlled concentration operation to ensure that the process can run safely and reliably over long periods.

Looking ahead, density functional theory (DFT) will continue to play a pivotal role in the design of next-generation catalysts. By providing atomic-scale insights, DFT will help scientists understand the intricate reaction pathways, the influence of crystal structure, the role of phosphorus, and how heteroatom doping or alloying can improve catalytic performance. The combination of theoretical modeling and in situ experimental data holds immense potential for guiding the design of transition-metal phosphide catalysts that offer improved stability, faster kinetics, and higher efficiency.

In conclusion, the future of metal phosphide electrocatalysts in hydrazine-assisted water electrolysis depends on a multifaceted approach. Progress in understanding surface chemistry, improving structural control, and developing advanced characterization tools will drive the development of next-generation materials for efficient and sustainable hydrogen production. By addressing these scientific, technical, and safety challenges, we can unlock the full potential of metal phosphide catalysts for clean, scalable energy production.

References

1 Cherusseri, J. & Kar, K. K. (2015). Recent progress in nanocomposites based on carbon nanomaterials and electronically conducting polymers. In *Polymer Nanocomposites Based on Inorganic and Organic Nanomaterials* (eds. S. Mohanty, S. K. Nayak, B. S. Kaith, and S. Kalia), 229–256. Hoboken, NJ: John Wiley & Sons. https://doi.org/10.1002/9781119179108.ch8

2 Thomas, S. A., Cherusseri, J., & Rajendran, D. N. (2024). Rapid synthesis of hierarchical tin disulfide (SnS_2) nanostructures by a microwave-assisted hydrothermal method for high-performance supercapatteries. *ACS Appl. Electron. Mater. 6*, 3346–3361.

3 Thomas, S. A., Cherusseri, J., Pallavolu, M. R., Rajendran, D. N., & Kumar, D. (2024). Boron carbon nitride (BCN): An emerging two-dimensional material for rechargeable batteries. *Energy Fuels 38*, 13704–13721.

4 Sundarraj, S., Vadivel, N., Murthy, A. P., Theerthagiri, J., & Choi, M. Y. (2025). MXene electrocatalysts: Transformative approaches in hydrogen production with alternative anode reactions. *Small 21*(2), 2407120.

5 Dufour, J., Serrano, D. P., Gálvez, J. L., Moreno, J., & González, A. (2011). Hydrogen production from fossil fuels: Life cycle assessment of technologies with low greenhouse gas emissions. *Energy Fuels 25*, 2194–2202.

6 Li, Y., Wei, X., Chen, L., & Shi, J. (2021). Electrocatalytic hydrogen production trilogy. *Angew. Chem. Int. Ed. 60*, 19550–19571.

7 Aslam, S., Rani, S., Lal, K., Fatima, M., Hardwick, T., Shirinfar, B., & Ahmed, N. (2023). Electrochemical hydrogen production: Sustainable hydrogen economy. *Green Chem. 25*, 9543–9573.

8 Vos, J. G., Liu, Z., Speck, F. D., Perini, N., Fu, W., Cherevko, S., & Koper, M. T. (2019). Selectivity trends between oxygen evolution and chlorine evolution on iridium-based double perovskites in acidic media. *ACS Catal. 9*, 8561–8574.

9 Plevová, M., Hnát, J., & Bouzek, K. (2021). Electrocatalysts for the oxygen evolution reaction in alkaline and neutral media. A comparative review. *J. Power Sources 507*, 230072. https://doi.org/10.1016/j.jpowsour.2021.230072

10 Park, C. E., Kerkar, R., Arumugam, D., Theerthagiri, J., Ramasamy, S., & Choi, M. Y. (2025). Interface triggered phosphate-functionalized NiCo oxide for boosting water oxidation reaction. *Chem. Eng. J. 524*, 168973.

11 Liu, H., Liu, Y., Li, M., Liu, X., & Luo, J. (2020). Transition-metal-based electrocatalysts for hydrazine-assisted hydrogen production. *Mater. Today Adv. 7*, 100083. https://doi.org/10.1016/j.mtadv.2020.100083.

12 Xu, J., Zhong, M., Yan, S., Chen, X., Li, W., Xu, M., Wang, C., & Lu, X. (2025). Partial oxidation of Rh/Ru nanoparticles within carbon nanofibers for high-efficiency hydrazine oxidation-assisted hydrogen generation. *J. Colloid Interface Sci. 679*, 171–180.

13 Tang, C., Zhang, R., Lu, W., Wang, Z., Liu, D., Hao, S., Du, G., Asiri, A. M., & Sun, X. (2017). Energy-saving electrolytic hydrogen generation: Ni_2P nanoarray as a high-performance non-noble-metal electrocatalyst. *Angew. Chem. Int. Ed. 56*, 842–846.

14 Peng, Y., Huang, M., Yang, Q., Xing, Z., & Lu, Z. H. (2023). Replacing oxygen evolution with hydrazine borane oxidation for energy-saving electrochemical hydrogen production. *Inorg. Chem. 62*, 11056–11063.

15 Cao, M., Li, B., Cao, Y., Li, Y., Tian, R., Shen, Q., Xie, W., & Gu, W. (2025). Co-Fe-Mo phosphides' triphasic heterostructure loaded on nitrogen-doped carbon nanofibers by electrospinning as efficient bifunctional electrocatalysts for overall water splitting. *ACS Appl. Mater. Interfaces 17*, 15259–15273.

16 Kandel, M. R., Pan, U. N., Paudel, D. R., Dhakal, P. P., Kim, N. H., & Lee, J. H. (2022). Hybridized bimetallic phosphides of Ni–Mo, Co–Mo, and Co–Ni in a single ultrathin-3D-nanosheets for efficient HER and OER in alkaline media. *Compos. B: Eng. 239*, 109992.

17 Govindasamy, R., Moon, C. J., Senthil, R. A., Maheskumar, V., Kumar, A., Lee, S., Ubaidullah, M., & Choi, M. Y. (2025). Ultra-low voltage hydrazine splitting with pulsed laser harnessed Ir/CoP for efficient Zn-hydrazine battery systems. *Chem. Eng. J. 522*, 167074.

18 Jin, Y., Fan, X., Cheng, W., Zhou, Y., Xiao, L., & Luo, W. (2024). The role of phosphorus on alkaline hydrogen oxidation electrocatalysis for ruthenium phosphides. *Angew. Chem. 136*, e202406888.

19 Li, Y., Xin, T., Cao, Z., Zheng, W., He, P., & Lee, L. Y. S. (2024). Optimized transition metal phosphides for direct seawater electrolysis: Current trends. *ChemSusChem 17*, e202301926. https://doi.org/10.1002/cssc.202301926.

20 Zeng, L., An, L., Zhang, Z., Zheng, J., Chen, W., Liu, C., Zheng, Y., Dang, D., & Liu, Z. Q. (2024). Heterogeneous strategy constructing a built-in electric field: CoP/FeP bifunctional electrode for overall water splitting. *Chem. Eng. J. 491*, 152084.

21 Xu, S., Zhao, H., Li, T., Liang, J., Lu, S., Chen, G., Gao, S., Asiri, A. M., Wu, Q., & Sun, X. (2020). Iron-based phosphides as electrocatalysts for the hydrogen evolution reaction: Recent advances and future prospects. *J. Mater. Chem. A 8*, 19729–19745.

22 Pan, Y., Lin, Y., Chen, Y., Liu, Y., & Liu, C. (2016). Cobalt phosphide-based electrocatalysts: Synthesis and phase catalytic activity comparison for hydrogen evolution. *J. Mater. Chem. A 4*, 4745–4754.

23 Deng, R., Guo, M., Wang, C., & Zhang, Q. (2024). Recent advances in cobalt phosphide-based materials for electrocatalytic water splitting: From catalytic mechanism and synthesis method to optimization design. *Nano Mater. Sci. 6*, 139–173. https://doi.org/10.1016/j.nanoms.2022.04.003.

24 Sun, F., Qin, J., Wang, Z., Yu, M., Wu, X., Sun, X., & Qiu, J. (2021). Energy-saving hydrogen production by chlorine-free hybrid seawater splitting coupling hydrazine degradation. *Nat. Commun. 12*, 4182.

25 da Silva, A. B., dos Reis, E. A., Ribeiro, C., & Mascaro, L. H. (2025). Overcoming stability hurdles of transition metal phosphides for H_2 evolution: A review. *ChemSusChem 18*, 202501388. https://doi.org/10.1002/cssc.202501388

26 Su, T., Wang, J., & Cai, C. (2024). Ni_2P nanoparticles supported on N, P co-doped carbon as catalyst for α-alkylation of arylacetonitriles with alcohols. *ACS Appl. Nano Mater. 7*, 8914–8925.

27 Gumpu, M. B. & Karvembu, R. (2025). Carbon-supported cobalt–nickel phosphide as a high-performance electrocatalyst for the efficient overall water splitting. *ACS Appl. Energy Mater. 8*, 13264–13277.

28 Pan, Y., Hu, W., Liu, D., Liu, Y., & Liu, C. (2015). Carbon nanotubes decorated with nickel phosphide nanoparticles as efficient nanohybrid electrocatalysts for the hydrogen evolution reaction. *J. Mater. Chem. A 3*, 13087–13094.

29 Wang, J., Li, X., Wei, B., Sun, R., Yu, W., Hoh, H. Y., Xu, H., Li, J., Ge, X., Chen, Z., & Su, C. (2020). Activating basal planes of $NiPS_3$ for hydrogen evolution by nonmetal heteroatom doping. *Adv. Funct. Mater. 30*, 1908708.

30 Wang, H. Y. & Yuan, Z. Y. (2025). Hydrazine-assisted water electrolysis system: Performance enhancement and application expansion. *Mater. Horiz. 12*, 5123–5148. https://doi.org/10.1039/d5mh00118h

31 Wang, H. Y., Wang, L., Ren, J. T., Tian, W. W., Sun, M. L., & Yuan, Z. Y. (2023). Heteroatom-induced accelerated kinetics on nickel selenide for highly efficient hydrazine-assisted water splitting and Zn-hydrazine battery. *Nano-Micro Lett. 15*, 155.

32 Tong, Y. & Chen, P. (2024). Recent progress of advanced electrocatalysts for hydrogen production via hydrazine-assisted water electrolysis. *Inorg. Chem. Front. 11*, 6218–6245. https://doi.org/10.1039/d4qi01789g

33 Dutta, A. & Pradhan, N. (2017). Developments of metal phosphides as efficient OER precatalysts. *J. Phys. Chem. Lett. 8*, 144–152. https://doi.org/10.1021/acs.jpclett.6b02249

34 Zhang, Y., Lee, S., Jeong, S., Son, E., Baik, J. M., Han, Y. K., & Park, H. (2024). Phase-bridged hierarchical catalysts for efficient and stable water electrolysis. *Adv. Funct. Mater. 34*, 2309250.

35 Xiao, G., Chen, H., Zhu, S., Zhao, W., Chen, S., Wen, Y., Chen, J., Ha, R., Hao, Y., Wang, J., & Wen, G. (2025). Rational design of cobalt phosphide nanorods via hydrothermal-phosphorization for high-performance asymmetric supercapacitors. *J. Alloys Compd. 1029*, 180758.

36 Tian, J., Liu, Q., Asiri, A. M., & Sun, X. (2014). Self-supported nanoporous cobalt phosphide nanowire arrays: An efficient 3D hydrogen-evolving cathode over the wide range of pH 0–14. *J. Am. Chem. Soc. 136*, 7587–7590.

37 Xu, Z., Chen, Q., Han, X., Wang, J., Wang, P., Zheng, T., Pang, S. Y., Wang, J., Li, H., Xia, Z., & Hao, J. (2024). Ultrafast synthesis of transition metal phosphides in air via pulsed laser shock. *Nano Lett. 24*, 12254–12262.

38 Guo, W., Ye, G., Huang, L., Li, Z., Song, Y., Su, J., Cao, X., Li, G., Liu, Y., Xin, Y., & Zhang, Q. (2025). Tailoring the catalytic activity of metal catalysts by laser irradiation. *Chem. Eur. J. 31*, e202404378. https://doi.org/10.1002/chem.202404378

39 Shahroudi, A. & Habibzadeh, S. (2024). Enhanced hydrogen evolution reaction activity through samarium-doped nickel phosphide (Ni_2P) electrocatalyst. *Sci. Rep. 14*, 16818.

40 Praveenkumar, N., Rao, N. M., & Madhuri, K. V. (2025). Optical and magnetic properties of Cr-doped zinc phosphide nanoparticles prepared by the solid-state reaction method. *Next Mater. 8*, 100676.

41 Kalaivani, A., Khan, T., Gayathri, V., Balraj, B., & Sebaey, T. A. (2025). Effects of deposition time and phosphorus concentration on the structural, morphological, and magnetic properties of galvanostatically electrodeposited NiFeP thin films. *ACS Omega 10*, 35451–35461.

42 Mahmood, S., Riaz, M. S., Ammar, M., Wang, Z., Iqbal, M. J., Ashraf, G. A., Afshan, N., Hassan, N., Bahadur, A., Iqbal, S., & Saad, M. (2024). Enhancing selective nitrate-to-ammonia electrocatalysis with high-performing Ni_2P

embedded nitrogen phosphide doped carbon (NPC) deposited on CP: Unprecedented performance and stability. *Int. J. Hydrog. Energy 70*, 315–324.

43 Ni, Y., Tao, A., Hu, G., Cao, X., Wei, X., & Yang, Z. (2006). Synthesis, characterization and properties of hollow nickel phosphide nanospheres. *Nanotechnology 17*, 5013–5018.

44 Joo, J., Kim, T., Lee, J., Choi, S.-I., & Lee, K. (2019). Morphology-controlled metal sulfides and phosphides for electrochemical water splitting. *Adv. Mater. 31*, 1806682. https://doi.org/10.1002/adma.201806682

45 Xu, X., Lu, Y., Shi, J., Hao, X., Ma, Z., Yang, K., Zhang, T., Li, C., Zhang, D., Huang, X., & He, Y. (2023). Corrosion-resistant cobalt phosphide electrocatalysts for salinity tolerance hydrogen evolution. *Nat. Commun. 14*, 7708.

46 Grosvenor, A. P., Wik, S. D., Cavell, R. G., & Mar, A. (2005). Examination of the bonding in binary transition-metal monophosphides MP (M = Cr, Mn, Fe, Co) by X-ray photoelectron spectroscopy. *Inorg. Chem. 44*, 8988–8998.

47 Blanchard, P. E. R., Grosvenor, A. P., Cavell, R. G., & Mar, A. (2008). X-ray photoelectron and absorption spectroscopy of metal-rich phosphides M_2P and M_3P (M = Cr-Ni). *Chem. Mater. 20*, 7081–7088.

48 Li, H., Chen, S., Su, F., Li, Z., & Tang, K. W. (2024). N-doped carbon-incorporated cobalt-iron mixed-metal phosphide nanoboxes as efficient bifunctional catalysts for overall water splitting. *ACS Appl. Mater. Interfaces 16*, 69210–69220.

49 García-Pérez, D., Álvarez-Galván, M. C., Capel-Sánchez, M. C., Campos-Martín, J. M., & Habas, S. E. (2024). Metal phosphide nanoparticles generated via a molecular precursor route for hydrotreatment of methyl laurate. *ChemCatChem 16*, e202400169.

50 Santos, E. J. D., Zagonel, G. F., Herrmann, A. B., Fantin, E. B., & Sturgeon, R. E. (2018). Determination of aluminium phosphide (AlP) in fumigants by ICP OES. *Anal. Methods 10*, 5047–5050.

51 Fritz, G. & Poppenburg, G. (1964). Zur quantitativen bestimmung der phosphin-gruppe in hydrolysierbaren element-phosphor-verbindungen. *Z. anorg. allg. Chem. 331*, 147–150.

52 Pramanik, M., Tominaka, S., Wang, Z. L., Takei, T., & Yamauchi, Y. (2017). Mesoporous semimetallic conductors: Structural and electronic properties of cobalt phosphide systems. *Angew. Chem. Int. Ed. 56*, 13508–13512.

53 Mohammadi, S., Gholivand, M. B., Mirzaei, M., & Amiri, M. (2024). Iron phosphide nanoparticles anchored on 3D nitrogen-doped graphene as an efficient electrocatalysts for hydrogen evolution reaction in alkaline media. *Mater. Today Commun. 41*, 110758.

54 Rosser, T. E., Sousa, J. P., Ziouani, Y., Bondarchuk, O., Petrovykh, D. Y., Wei, X. K., Humphrey, J. J., Heggen, M., Kolen'ko, Y. V., & Wain, A. J. (2020). Enhanced oxygen evolution catalysis by aluminium-doped cobalt phosphide through: In situ surface area increase. *Catal. Sci. Technol. 10*, 2398–2406.

55 Coleman, N., Lovander, M. D., Leddy, J., & Gillan, E. G. (2019). Phosphorus-rich metal phosphides: Direct and tin flux-assisted synthesis and evaluation as hydrogen evolution electrocatalysts. *Inorg. Chem. 58*, 5013–5024.

56 Chang, Q., Ma, J., Zhu, Y., Li, Z., Xu, D., Duan, X., Peng, W., Li, Y., Zhang, G., Zhang, F., & Fan, X. (2018). Controllable synthesis of ruthenium phosphides (RuP and RuP_2) for pH-universal hydrogen evolution reaction. *ACS Sustain. Chem. Eng. 6*, 6388–6394.

57 Liang, H., Gandi, A. N., Anjum, D. H., Wang, X., Schwingenschlögl, U., & Alshareef, H. N. (2016). Plasma-assisted synthesis of NiCoP for efficient overall water splitting. *Nano Lett. 16*, 7718–7725.

58 Wang, H. & Tao, S. (2021). Fabrication of a porous NiFeP/Ni electrode for highly efficient hydrazine oxidation boosted H_2 evolution. *Nanoscale Adv. 3*, 2280–2286.

59 Zhao, Y., Jia, N., Wu, X. R., Li, F. M., Chen, P., Jin, P. J., Yin, S., & Chen, Y. (2020). Rhodium phosphide ultrathin nanosheets for hydrazine oxidation boosted electrochemical water splitting. *Appl. Catal. B: Environ. 270*, 118880.

60 Li, M., Zhang, Z., Xiong, H., Wang, L., Zhuang, S., Argyle, M. D., Tang, Y., Yang, X., Chen, Y., Wan, P., & Fan, M. (2020). 0.03 V electrolysis voltage driven hydrazine assisted hydrogen generation on NiCo phosphide nanowires supported NiCo hydroxide nanosheets. *ChemElectroChem 7*, 3089–3097.

61 Guo, R., Gao, L., Zhang, Y., Zhang, X., Ma, M., & Hu, T. (2023). Enhancing hydrazine-assisted hydrogen production by constructing CoP-Co_2P bifunctional catalysts. *Appl. Surf. Sci. 617*, 156602.

62 Chi, J., Guo, L., Mao, J., Cui, T., Zhu, J., Xia, Y., Lai, J., & Wang, L. (2023). Modulation of electron structure and dehydrogenation kinetics of nickel phosphide for hydrazine-assisted self-powered hydrogen production in seawater. *Adv. Funct. Mater. 33*, 2300625.

63 Yang, H. M., Wang, H. Y., Sun, M. L., & Yuan, Z. Y. (2023). Interface engineering of bifunctional nickel hydroxide/nickel phosphide heterostructure for efficient intermittent hydrazine-assisted water splitting. *Chem. Eng. J. 475*, 146134.

9

Metal Carbides and Nitrides Electrocatalysts for Hydrazine-assisted Water Electrolysis

Binaya Kumar Sahu[1,*] *and Pooja Sahoo*[2]

[1]*Department of Physics, Panchayat Degree College, Sambalpur University, Sambalpur, India*
[2]*Department of Physics, Veer Surendra Sai University of Technology, Sambalpur, India*

*Corresponding author: binayamsc12@gmail.com

9.1 Introduction

The immediate need to slow down climate change and reduce our dependence on finite fossil fuels which are significant causes of greenhouse gas emissions and geopolitical instability is what is driving the global energy transition [1–3]. Due to its high energy density, zero emissions when used, and adaptability as a fuel and chemical feedstock, hydrogen has become a major facilitator of a green energy system. However, traditional methods of producing hydrogen, especially steam methane reforming, are energy-intensive and produce a lot of CO_2 [4–6]. A greener option is provided by electrochemical water splitting, particularly when it is driven by renewable energy. However, the slow and energy-intensive oxygen evolution reaction (OER) at the anode, which demands high overpotentials and includes a complicated four-electron process, prevents its widespread use. To get over these restrictions, scientists have looked into using anodic oxidation of small organic molecules like alcohols, urea, amines, and hydrazine (N_2H_4) in place of OER, which lowers energy input and improves system efficiency overall [7, 8]. The hydrazine oxidation reaction (HzOR) is unique among these because of its quick kinetics, pure by-products (water and nitrogen), and extremely favorable thermodynamics (−0.33 V vs. RHE [reversible hydrogen electrode]), which make it both ecologically beneficial and energy-efficient [9–11]. When HzOR is incorporated into water electrolysis systems (i.e., hydrazine-assisted water electrolysis [Hz-WE]), the cell voltage is significantly reduced (by 1–1.5 V), allowing for the simultaneous purification of hydrazine-rich industrial effluent and the manufacture of inexpensive hydrogen [12, 13]. Recent developments in electrocatalyst design have made Hz-WE much more feasible. Materials with exceptional bifunctional activity toward both HzOR and the hydrogen evolution reaction (HER) include noble metals, carbon-based nanostructures, metal oxides, sulfides, phosphides, etc. [10, 11, 14–16]. These materials provide high current densities at ultralow voltages (as low as 0.2–0.4 V) with nearly 100% faradaic efficiency and brilliant durability. However, the creation of reliable, highly active, and reasonably priced catalysts is essential to these systems' functionality. To maximize activity and stability, engineering techniques such surface defect tuning, doping, heterostructure, and nanoarray structuring are essential. In addition, hydrazine-based water catalysis is the core behind using seawater for hydrogen evolution more effectively [17, 18]. All things considered, Hz-WE offers an unprecedented method of producing hydrogen sustainably by fusing economic feasibility, environmental cleanup, and energy efficiency. Further, this idea can be extended to the construction of direct hydrazine fuel cell (DHFC). In a DHFC, hydrazine is directly oxidized at the anode, yielding either water and energy or nitrogen and water as by-products (Figure 9.1) [19, 20]. Hydrazine's high energy density and comparatively low operating temperature make these fuel cells appealing. They are a sustainable energy source since they have the ability to emit no CO_2.

Because of their exceptional activity and conductivity, noble-metal-based catalysts like Pt (platinum), RuO_2, and IrO_2 have historically set the standard for HER and OER [6, 21–23]. However, their actual employment in Hz-WE is limited by their high cost, limited availability, and poisoning susceptibility, particularly when hydrazine and reaction intermediates are present. In this regard, catalysts based on metal carbides and nitrides have shown promise. Because of their remarkable chemical and thermal stability, high intrinsic electrical conductivity, and noble-metal-like d-band electronic structures, transition-metal carbides (TMCs; like MoC, WC [tungsten carbide], and TiC) and nitrides (TMNs; like CoN, VN, TiN, and Ni_3N) are becoming increasingly recognized as viable alternatives to noble metals in electrochemical applications [24–28]. These materials perform exceptionally well in a variety of energy-related applications, such as fuel cells, oxygen reduction reaction

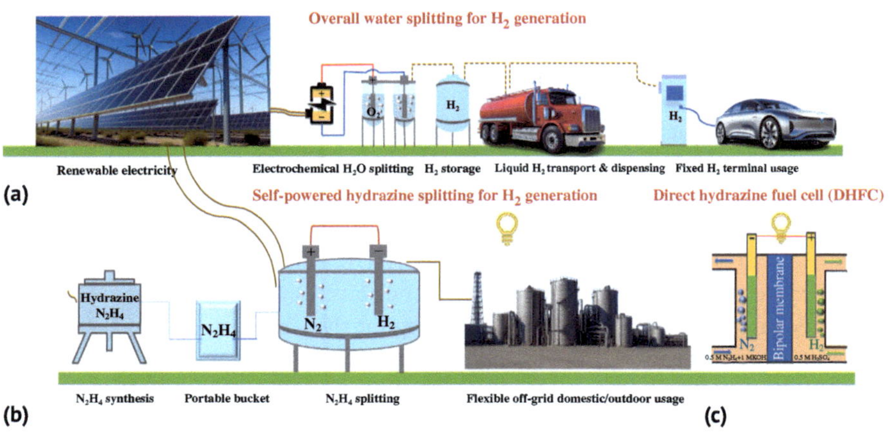

Figure 9.1 Schematics showing (a) overall water splitting for H_2 generation, (b) self-powered hydrazine splitting for H_2 generation, and (c) DHFC.

(ORR), and lithium-ion or lithium-sulfur batteries, in addition to bifunctional catalysis in Hz-WE systems.

Their electronic structural engineering plays a major role in their catalytic performance. Small atoms like carbon or nitrogen occupy the interstitial positions in the metal lattice of metal carbide and nitride, which are both frequently referred to as interstitial alloys. The electronic density of states (DOS) close to the Fermi level is modulated by the addition of interstitial atoms, increasing the material's catalytic potential and reactivity. Specifically, a wider unoccupied d-band that resembles the electronic structure of noble metals like platinum is produced in metal carbides when the metal d-orbitals (usually from groups IV–VI for transition metals) interact with carbon s- and p-orbitals [21, 28–30]. Likewise, the addition of nitrogen atoms to metal nitrides reinforces metal–nitrogen interactions and modifies the d-band center, leading to high stability and adjustable catalytic behavior in a range of electrochemical scenarios. Metal carbides and nitrides exhibit Pt-like behavior in HER as a result of these characteristics, which makes them affordable and efficient substitutes for platinum-based catalysts [24–28, 30]. Additionally, their superior mechanical and thermal robustness, chemical inertness in acidic or alkaline environments, and resistance to poisoning (e.g., by sulfur compounds) further confirm their place in next-generation electrocatalysis. These materials are notable for their ability to exist in several crystalline phases and stoichiometries, which enables the surface composition and coordination environment to be fine-tuned. Further, their ease of tunability ensures engineering the catalysts with minimal energy barriers and optimized active sites for hydrogen/ammonia adsorption and desorption, two essential processes for effective HER and HzOR during water electrocatalysis. Interestingly, nearly all transition metals from groups IV–VI may produce carbides or nitrides, providing a wide compositional playground for catalyst invention, whereas noble metals are typically limited by their rarity and expense. Numerous processes, including carbothermal reduction, nitridation of metal oxides, chemical vapor deposition (CVD), sol–gel procedures, and pulsed laser ablation in liquid (PLAL), can be used to customize the synthesis of metal carbide and nitride electrocatalysts [3, 31–33]. These methods provide control over surface area, phase purity, crystallinity, and nanostructure morphology all of which have a direct impact on catalytic performance. To increase electrocatalytic activity even more, sophisticated modification techniques are used: (i) defect engineering: by adding vacancies (such as carbon or nitrogen vacancies), active sites become denser and the electronic structure is customized, which enhances intermediate adsorption [34–36]. (ii) heteroatom doping: doping with Co, Ni, Fe, or Se lowers the energy barriers for reaction steps and modifies the electronic band structure [37–39]. (iii) single-atom catalysts (SACs): to promote effective HER and HzOR, isolated metal atoms anchored on carbides or nitrides make use of strong metal–support interactions (SMSI) and maximal atomic efficiency [5, 40, 41]. (iv) hierarchical nanostructures: high performance at industrial current densities is made possible by designing 3D architectures with mesoporous channels, which improve mass transport, electrolyte penetration, and gas release [40, 42]. (v) composite engineering: combining with suitable counterpart for enhancing the catalytic behavior [43].

A thorough review of the most recent advancements in the design and development of metal carbide and nitride-based electrocatalysts for Hz-WE is given in this chapter. It discusses the basic ideas of HzOR for HER and DHFC, the benefits of using metal carbides and nitrides as bifunctional catalysts, and the several engineering approaches and synthetic methods utilized to improve their catalytic performance. Surface/interface design, electrical structure modulation, and structure–activity connections are given special attention. The chapter also examines the difficulties and potential for developing Hz-WE as a sustainable method of producing hydrogen and cleaning up the environment, as well as contrasting these materials with traditional electrocatalysts for fuel cell application.

9.2 Structure and Properties of Metal Carbides and Nitrides

When Levy and Boudart showed in 1973 that adding carbon atoms to tungsten (W) changed its electronic structure to the point where the resulting tungsten carbide resembled platinum's catalytic behavior, the catalytic potential of TMCs was first widely acknowledged [44]. This finding paved the way for a great deal of investigation into early TMCs, carbonitrides, and nitrides as less expensive substitutes for noble metals in a range of catalytic processes, such as electrochemical energy conversion. Metal carbides and nitrides are classes of interstitial compounds formed by embedding light nonmetal atoms (carbon or nitrogen, respectively) into the lattice of metals (e.g., Mo, W, Ta, Ti, Nb, V, Ni, Cr, Fe, etc.). This incorporation fundamentally alters the structural, electronic, and catalytic properties of the parent metals, resulting in hybrid materials that combine the attributes of metals, ceramics, and semiconductors [32, 45, 46]. These materials have recently attracted immense attention in electrocatalysis, particularly for hydrazine-assisted water splitting, due to their synergistic bonding, tunable electronic structures, and environmental robustness. Interestingly, orbital hybridization and valence electron concentration control the structural configurations. The native metal lattice can change from body-centered cubic (bcc) to hcp and then to face-centered cubic (fcc) symmetry when the number of sp electrons rises, hence altering the crystal symmetry and affecting electrical behavior (Figure 9.2a,b) [47, 48]. Because of such structural behavior, multiple polymorphs are seen for carbides and nitrides. For instance, molybdenum carbides display multiple polymorphs: η-Mo_3C_2 and β-Mo_2C adopt orthorhombic structures; α-MoC_{1-x} exhibits an fcc structure, favorable for high-density active-site exposure [49–51]. The electronic environments and interatomic distances inside the metal host lattice are altered by the addition of carbon or nitrogen. This modifies the adsorption energies of important chemical intermediates, redistributes the DOS, and modifies the d-band center (Figure 9.2d–f) [46, 52]. Consequently, metal carbides and nitrides can facilitate the desorption of reaction products, which is essential for high catalytic turnover, encourage stepwise dehydrogenation of hydrazine, and lower activation energies for important bond cleavages (such as N—H in HzOR) [11, 39]. When oxygen is partially incorporated

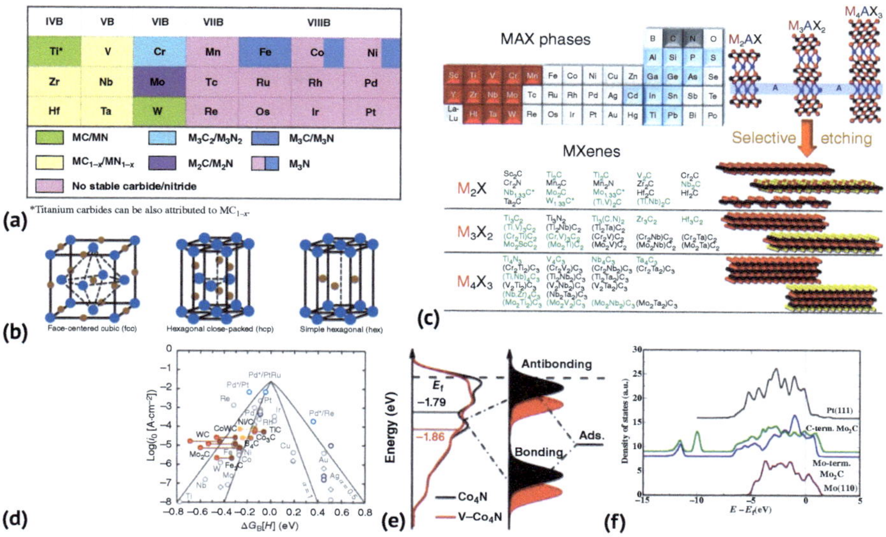

Figure 9.2 (a) Representative transition metals present in TMCs and TMNs. *Source*: [54]/with permission of Elsevier. (b) Common crystal structures of TMCs and TMNs, where blue spheres denote transition-metal atoms and brown spheres denote carbon/nitrogen atoms. *Source*: [48]. John Wiley & Sons. CC BY 4.0. (c) Structural diversity of MAX phases and MXenes: by combining 12 transition metals (red), 12 Group-A elements (blue), and two X elements (black), nearly 100 MAX phases with M_2AX, M_3AX_2, or M_4AX_3 structures have been identified. Transition metals in the M-layers can also form solid solutions and/or double-M-ordered phases, generating complex multielemental compositions. Selective A-layer etching from MAX phases has yielded ~30 experimentally verified MXenes (green) and many more theoretically predicted. Asterisks indicate MXenes with ordered divacancies. *Source*: [55]/with permission of John Wiley & Sons. (d) Volcano plots correlating exchange current density with hydrogen binding energy for representative HER electrocatalysts. *Source*: [24]/with permission of American Chemical Society. (e) Schematic illustration of d-band tuning in CoN through V doping. *Source*: [56]/with permission of John Wiley & Sons. (f) Calculated electronic d-band structures of Pt (111), β-Mo_2C (0001) surface, and Mo (110). *Source*: [57]/with permission of Elsevier.

into metal carbides or nitrides, either deliberately or unintentionally, oxycarbide or nitride composites may occur. By altering surface polarity, charge-transfer properties, and active-site density, these oxygen species can also affect catalytic kinetics [49]. Carbides and nitrides formed with the transition metals are of special importance and categorically called as TMCs and TMNs, respectively [28, 52]. Other than this, MXenes are a rapidly emerging class of two-dimensional (2D) materials derived from layered TMCs, TMNs, and transition-metal carbonitrides. These materials are synthesized via the selective chemical etching of their parent compounds, known as MAX phases, which follow the general formula $M_{n+1}AX_n$ (where $n = 1–4$). In this formula, "M" stands for an early transition metal (e.g., Ti, V, and Mo), "A" is a Group 13–15 element (such as Al or Si), and "X" represents carbon and/or nitrogen atoms. Unlike other 2D materials such as graphene and transition-metal dichalcogenides (TMDs), which rely on weak van der Waals forces between the layers, MAX phases feature a strong metallic M—A bond and a metallic–covalent–ionic M—X bond [43, 53]. During the etching process, typically involving acids like hydrofluoric acid (HF), the A-layer is selectively removed while preserving the M—X framework, resulting in the formation of MXenes with the general formula $M_{n+1}X_nT_x$. Here, T_x denotes surface terminations such as —F, —OH, —O, or —Cl, which arise during etching and play a crucial role in defining the surface chemistry, hydrophilicity, and electrochemical properties of MXenes (Figure 9.2c) [53].

The first MXene, $Ti_3C_2T_x$, was successfully synthesized in 2011 by the research group of Professors M. W. Barsoum and Y. Gogotsi at Drexel University, by etching Ti_3AlC_2 using HF. Since then, a wide variety of MXenes have been developed, including Ti_2CT_x, Nb_2CT_x, Mo_2CT_x, Ti_3CNT_x, $Zr_3C_2T_x$, and multicomponent MXenes such as $(Nb_{0.8}Ti_{0.2})_4C_3T_x$. To date, over 100 MAX phases have been synthesized using different combinations of M, A, and X elements, offering a rich library of precursors for novel MXene compositions. The surface termination groups not only impact MXenes' hydrophilicity and stability, but also play a pivotal role in modulating ion transport, surface charge, and electrical conductivity all of which are critical for applications in energy storage, catalysis, sensors, and beyond. Additionally, MXenes with higher n values tend to exhibit greater structural stability, which makes them particularly attractive for long-term functional use. With their tunable composition, layered architecture, and superior physicochemical properties, MXenes represent a unique and highly versatile platform for the development of next-generation catalysis for hydrazine-assisted water catalysis. Further, several advanced synthesis techniques have been devised to improve the catalytic performance of metal nitrides and carbides.

To enhance bifunctional catalytic activity for both hydrazine oxidation and hydrogen evolution events, heteroatom doping with elements such as Co, Ni, W, or Fe efficiently modifies the electronic structure and creates synergistic interactions [15, 61–63]. Besides, defect engineering speeds up reaction kinetics by facilitating the optimal adsorption and desorption of important intermediates like $N_2H_4^*$, NH_2^*, and H^* (Figure 9.3a,b). This includes the production of nitrogen or carbon vacancies [34, 36]. Furthermore, combining conductive supports such as graphene or carbon nanotubes (CNTs) with TMNs/TMCs improves electron transport while strengthening structural stability in electrochemical settings. Single-atom (SA) catalysis, in which individual metal atoms (such as M—N or M—C coordination sites) are anchored to MC, is another intriguing strategy [62]. This maximizes catalytic selectivity and atom efficiency by making use of SMSI. When combined, these tactics allow catalytic characteristics to be fine-tuned for next-generation energy applications (Figure 9.3) [40, 41, 64].

Phase, stoichiometry, shape, and surface chemistry all have a significant impact on their performance and can be controlled with the right synthesis techniques. Physical techniques like reactive magnetron sputtering, pulsed laser deposition (PLD), and physical vapor deposition (PVD) are very helpful for creating high-purity, binder-free thin films straight onto conductive substrates like Si, FTO, or ITO. Grain orientation, thickness, and crystallinity—parameters essential to electrocatalytic activity—can be precisely controlled thanks to these methods [65, 66]. Because of their high conductivity and strong metal–nonmetal bonding, TiN, Mo_2C,

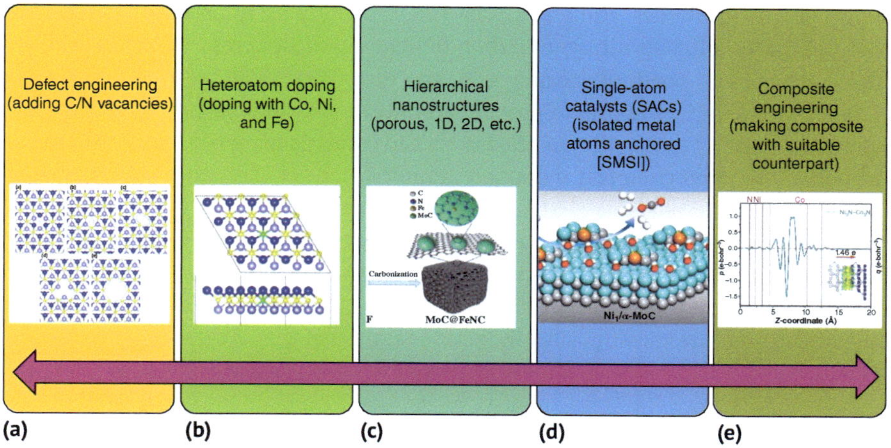

Figure 9.3 Schematic of manipulating metal carbides and nitrides using different tactics. (a) Defect engineering (showing vacancy in MoC). *Source*: [58]/with permission of Elsevier. (b) Heteroatom doping (depicting heteroatom doping in MoC). *Source*: [58]/with permission of Elsevier. (c) Hierarchical nanostructuring (schematic diagram of porous MoC). *Source*: [40]/with permission of American Chemical Society. (d) SA loading (Ni-SA-loaded MoC). *Source*: [59]/with permission of American Chemical Society. (e) Composite engineering (heterojunction between Ni_3N and Co_3N). *Source*: [60]/with permission of John Wiley & Sons.

and WC thin films produced using these methods, for example, have outstanding HER and HzOR activities. Additionally, active-site availability can be improved by adjusting factors such as sputtering power, gas ratio, and substrate temperature to induce defect/vacancy generation. Through laser-induced surface melting and reactivity with reactive nitrogen or carbon sources, pulsed laser irradiation (PLI) allows metal or oxide films to be transformed in-situ into nitrides or carbides. By facilitating vacancy engineering and the inclusion of nitrogen/carbon defects during quick quenching, this technique produces catalytically active sites. For colloidal MNs with a lot of surface vacancies and superior catalytic qualities, such as VN and MoN, PLAL offers a sustainable, surfactant-free production method [67, 68]. Later on, the prepared samples are coated with suitable binder for catalytic property. Techniques for chemical synthesis provide flexibility in regulating composition and shape. MNs/MCs can be synthesized from nanoparticles into porous three-dimensional structures using the sol–gel, coprecipitation, salt-templating, and urea–glass methods [69, 70]. By introducing dopant precursors during synthesis, altering electrical structures, and encouraging bifunctional activity, heteroatom doping (with Fe, Co, and Ni) can be accomplished (Figure 9.3). Similarly, MOFs, salt scaffolds, or hard or soft templates can be used to guide morphology tuning, such as creating hollow structures, nanorods, or nanosheets, to increase surface area and diffusion pathways [71]. Crystalline nitrides and carbides can be formed more easily by high-temperature nitridation/carburization in ammonia, N_2/H_2, or methane atmospheres. By altering hydrogen adsorption characteristics, partial substitution with oxygen (forming oxynitrides or oxycarbides) significantly improves HER kinetics. Enhancing the HzOR activity by improving intermediate desorption, employing nitrogen/carbon-deficient atmospheres, or varying the gas flow rate can also result in vacancy-rich structures [72]. SMSI are used in SA catalysis on MN or MC supports to provide atomically distributed metal centers (such as Pt, Co, and Ni) that optimize utilization and catalytic activity [21, 41]. These can be added via photochemical reduction, atomic-layer deposition, or wet impregnation. Stability and effective charge transfer are guaranteed when SAs are anchored on doped or faulty MN/MC matrices. To improve mechanical integrity and electron mobility, further hybridization with conductive supports like graphene, CNTs, or MXenes is used. In addition to providing hierarchical porosity and preventing agglomeration, these composites increase mass transfer and catalytic efficiency in general.

9.3 Fundamentals of Hydrazine-assisted Water Catalysis for Hydrogen Production

HER at the cathode and HzOR at the anode power hydrazine-assisted hydrogen generation systems. These two-electrode arrangements have been studied in aqueous electrolytes that are neutral, acidic, and more frequently alkaline. Because of their advantageous kinetics and catalyst stability, alkaline media have been the subject of most research among them. HER usually follows the Volmer–Tafel (VT) or Volmer–Heyrovsky (VH) processes, which are two-electron-transfer pathways, in alkaline conditions. A more intricate four-electron-transfer procedure, comprising four successive dehydrogenation processes, is used in HzOR. The following equations describe these chemical pathways [73]:

Cathodic (HER):

$$4H_2O + 4e^- \rightarrow 2H_2 + 4OH^- \tag{9.1}$$

$$\text{Volmer: } H_2O + e^- \rightarrow H^* + OH^- \tag{9.2}$$

$$\text{Tafel: } H^* + H^* \rightarrow H_2 \tag{9.3}$$

$$\text{Heyrovsky: } H^* + H_2O + e^- \rightarrow H_2 + OH^- \tag{9.4}$$

In Hz-WE, HzOR is essential because it offers a substitute anodic reaction for the slow OER. HzOR generates nitrogen gas (N_2) as a harmless by-product and has a lower thermodynamic potential (0.33 V vs. RHE) than OER. Energy-efficient hydrogen synthesis in alkaline or neutral circumstances is made possible by HzOR in conjunction with hydrogen evolution at the cathode. Because of their tunable electronic structures, high electrical conductivity, and chemical stability, TMNs are attractive electrocatalysts for HzOR. Strong hydrazine molecule adsorption and activation are made possible by their special metallic–covalent bonding, which also preserves effective electron-transfer channels:

Overall anodic reaction (HzOR):

$$N_2H_4 \rightarrow N_2 + 4H^+ + 4e^- \ (E = 0.33 \text{ V vs. RHE}) \tag{9.5}$$

In alkaline medium, the protons react with OH^-:

$$N_2H_4 + 4OH^- \rightarrow N_2 + 4H_2O + 4e^- \quad (9.6)$$

During the stepwise mechanism of HzOR on TMNs/TMCs, multiple intermediates are involved in the four-electron oxidation of hydrazine. The following basic steps, which take place on the TMN/TMC surface (shown by the symbol *), can be used to describe the process [11, 74]:

Adsorption of hydrazine:

$$N_2H_4\ (aq) \rightarrow N_2H_4^* \quad (9.7)$$

Dehydrogenation steps:

$$N_2H_4^* \rightarrow N_2H_3^* + H^+ + e^- \quad (9.8)$$

$$N_2H_3^* \rightarrow N_2H_2^* + H^+ + e^- \quad (9.9)$$

$$N_2H_2^* \rightarrow N_2H^* + H^+ + e^- \quad (9.10)$$

$$N_2H^* \rightarrow N_2^* + H^+ + e^- \quad (9.11)$$

Desorption of nitrogen:

$$N_2^* \rightarrow N_2(g) + * \quad (9.12)$$

Metal–nitrogen bonds on the catalyst surface stabilize each intermediate, and the electronic structure and surface energy of the TMNs determine the energy barrier for each dehydrogenation phase. For example, TMNs with nitrogen vacancies and a modest hydrogen binding energy offer the perfect platform for hydrogen desorption and hydrazine activation:

Proposed pathway summary:

$$N_2H_4 \rightarrow N_2H_3 \rightarrow N_2H_2 \rightarrow N_2H \rightarrow N_2(g) \quad (9.13)$$

The coupled HER–HzOR can be written as follows:

When TMNs are used as bifunctional catalysts in a two-electrode setup, the overall reaction becomes:

$$\text{Anode (HzOR): } N_2H_4 + 4OH^- \rightarrow N_2 + 4H_2O + 4e^- \quad (9.14)$$

$$\text{Cathode (HER): } 4H_2O + 4e^- \rightarrow 2H_2 + 4OH^- \quad (9.15)$$

$$\text{Net reaction: } N_2H_4 \rightarrow N_2 + 2H_2 \quad (9.16)$$

This pathway provides a low-voltage route for hydrogen production (typically <0.4 V cell voltage) without involving OER, and with much lower energy input compared to traditional water electrolysis.

Favorable adsorption of N_2H_4: through back-donation mechanisms and lone-pair interactions, the nitrogen-rich surface of TMNs/TMCs can encourage robust adsorption and activation of hydrazine. (i) Optimal intermediate binding: to ensure effective catalysis, TMNs/TMCs provide a balance between binding too weakly, which does not activate hydrazine, and binding too forcefully, which poisons the surface. (ii) Improved electron transport: during the multi-electron oxidation process, TMNs/TMCs' intrinsic metallic conductivity facilitates quick charge transfer. (iii) The function of nitrogen/carbon vacancies: these surface imperfections can serve as active sites to lower the energy barrier for breaking N—H bonds [75].

9.4 Fundamentals of Hydrazine-assisted Water Catalysis for Fuel Cells

Additionally, DHFCs, which aspire to generate power and hydrogen in a single device, are compatible with hydrazine-assisted water electrocatalysis. In this case, hydrazine undergoes direct electrochemical oxidation at the anode to produce protons, electrons, and nitrogen gas. By avoiding the production of greenhouse gases like CO_2 and producing a high theoretical cell voltage of 1.56 V, this technique makes DHFCs more environmentally friendly than hydrocarbon-based systems. The power density and efficiency of DHFCs are further increased by the quick reaction kinetics of hydrazine oxidation and the lack of CO poisoning. A closed-loop, energy-sustainable platform for the production and use of clean fuel can be created by combining hydrazine-assisted electrolysis with DHFC technology to make self-powered hydrogen production systems that use excess electricity from the fuel cell to drive hydrogen evolution from water (Figure 9.4a).

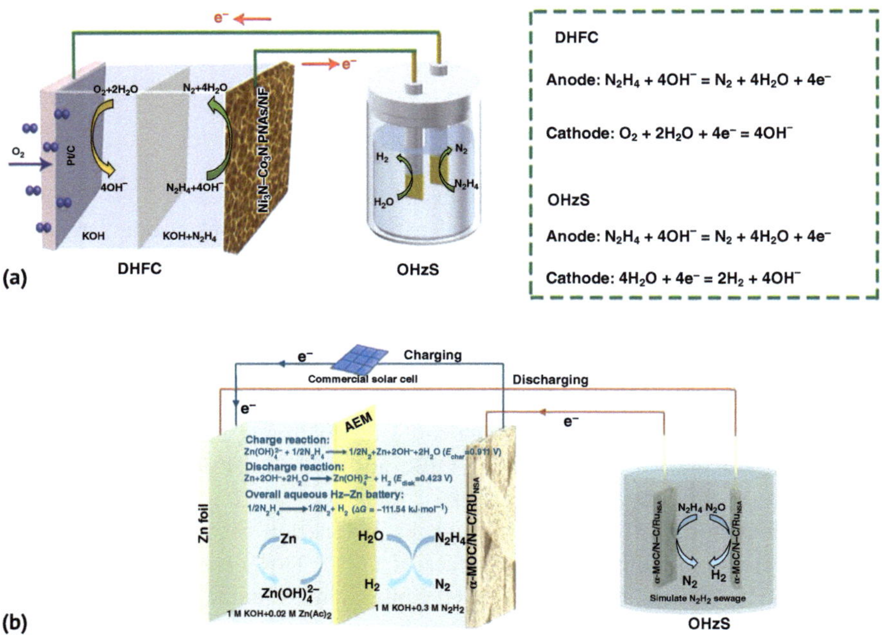

Figure 9.4 (a) Image illustrating a system for simultaneous hydrogen fuel synthesis and power generation using DHFC integrated with a hydrogen evolution system (OHzS). *Source*: [76]/with permission of John Wiley & Sons. (b) Zn-battery-coupled HzOR. *Source*: [9]/with permission of John Wiley & Sons.

DHFC commonly consists of an ORR electrocatalyst on the cathode and HzOR electrocatalyst on the cathode:

$$\text{Anode (HzOR): } N_2H_4 + 4OH^- \rightarrow N_2 + 4H_2O + 4e^- \tag{9.17}$$

$$\text{Cathode (OER): } O_2 + 4H^+ + 4e^- \rightarrow 2H_2O \tag{9.18}$$

$$\text{Net reaction: } N_2H_4 \rightarrow N_2 + 4H^+ + 4e^- \tag{9.19}$$

An inventive hybrid energy system that combines fuel-driven energy conversion and electrochemical storage is the hydrazine–zinc (Hz–Zn) battery. It combines the reduction of zinc ions at the cathode with HzOR at the anode. The system runs at a lower input voltage than traditional water electrolysis because of hydrazine's low theoretical oxidation potential (−0.33 V vs. RHE), which makes it an energy-efficient method for producing hydrogen. Zn^{2+} ions are reduced and deposited as metallic zinc at the cathode during discharge, while hydrazine undergoes electrochemical oxidation at the anode, releasing electrons. This dual capability provides a sustainable method for energy conversion and storage by enabling the simultaneous generation of hydrogen and power (Figure 9.4b).

The whole reaction can be shown as follows:

Hz–Zn battery reactions:

During discharging:

$$\text{Anode: } N_2H_4 + 4OH^- \rightarrow N_2 + 4H_2O + 4e^- \tag{9.20}$$

$$\text{Cathode: } Zn^{2+} + 2e^- \rightarrow Zn \tag{9.21}$$

Overall discharge reaction:

$$N_2H_4 + 2OH^- + Zn^{2+} \rightarrow N_2 + 2H_2O + Zn \tag{9.22}$$

During charging:

$$\text{Anode: } Zn \rightarrow Zn^{2+} + 2e^- \tag{9.23}$$

$$\text{Cathode: } 2H_2O + 2e^- \rightarrow H_2 + 2OH^- \text{ (cathode reaction during charge)} \tag{9.24}$$

Overall discharge reaction:

$$Zn + 2H_2O \rightarrow Zn^{2+} + H_2 + 2OH^- \tag{9.25}$$

$$N_2H_4 + 4OH^- \rightarrow N_2 + 4H_2O + 4e^- \text{ (anode oxidation)} \tag{9.26}$$

$$2H_2O + 2e^- \rightarrow H_2 + 2OH^- \text{ (cathode hydrogen evolution)} \tag{9.27}$$

In addition to theoretical methods, experimental methods such as electrochemical impedance spectroscopy (EIS) and Tafel analysis can be used to assess the reaction kinetics of HER and HzOR. Improved reaction rates, which underpin more effective electrocatalytic processes, are specifically indicated by lower Tafel slopes and decreased charge-transfer resistance (R_{ct}). Additionally, the energy barriers of crucial stages in HER and HzOR are frequently estimated using density functional theory (DFT) calculations, which provide important information about catalytic behavior. DFT calculations showed that lower energy barriers are linked to better catalytic performance and quicker reaction kinetics.

9.5 Measurement Criteria for HzOR-assisted Water Electrocatalysis

9.5.1 Overpotential

Overpotential (η) is a key indicator of electrocatalyst efficiency in HzOR. The typical thermodynamic potential for HzOR, as determined by the Gibbs free energy equation, is 0.33 V vs. RHE at standard conditions (25 °C and 1 atm):

$$\Delta G = -nFE_{\text{eq}} \tag{9.28}$$

Here, n is the number of electrons exchanged (4 in the case of HzOR), F is the Faraday constant, E_{eq} is the standard equilibrium potential, and ΔG is the Gibbs free energy change. However, because of the intricacy of electron–proton-linked transfer stages, HzOR, like other multi-electron processes, suffers kinetic limits despite its advantageous thermodynamics. Therefore, in order for the reaction to proceed at a practical rate, the applied potential (E) must be greater than the equilibrium potential. Two benchmarks that are frequently used to assess HzOR activity are: the potential at which there is a noticeable rise in current density, signifying the start of substantial hydrazine oxidation, is known as the onset overpotential (η_0). In practical applications in hydrogen production systems, the overpotential at 10 mA·cm^{-2} (η_{10}) is the potential needed to provide a current density of 10 mA·cm^{-2}. Cyclic voltammetry (CV) or linear sweep voltammetry (LSV) measurements are used to identify these overpotentials. The applied potentials at various current densities are then determined by analyzing the polarization curves that occur. The overpotential is obtained by subtracting the observed applied potential at a given current density from the equilibrium potential of HzOR (0.33 V vs. RHE). In conclusion, the more effectively a catalyst promotes hydrazine oxidation with the least amount of energy loss, the lower the onset and η_{10} overpotentials. This is a crucial requirement for developing hydrazine-assisted water-splitting technologies.

9.5.2 Tafel Slope

The methods used to assess the properties of electrocatalysts depend heavily on the Tafel slope (b). To have a better understanding of the Tafel slope, we will begin by discussing the exchange current density (i_0). The total current (j) for a specific electrochemical reaction is equal to the sum of the currents in the anode (j_a) and cathode (j_c) (Eq. 9.29). More importantly, the well-known Butler–Volmer (BV) equation (as given in Eq. 9.30) indicates that α_a and α_c represent the anode and cathode transfer coefficients, respectively. While n and F stand for the number of electrons exchanged in the reaction and the Faraday constant, respectively, R and T stand for the universal gas constant and absolute temperature, correspondingly. Furthermore, the electrode area (A) can be divided to determine the exchange current, j_0:

$$j = j_a + j_c \tag{9.29}$$

$$j = j\left[\exp(\alpha_a nF\eta/RT) + \exp(\alpha_c nF\eta/RT)\right] \tag{9.30}$$

$$\frac{j_0}{A} = i_0 \tag{9.31}$$

The BV equation can be simplified to the form of Eq. (9.32), which was originally put forward by Tafel as an empirical formula, in which a refers to the overpotential at a current density of 1 $mA{\cdot}cm^{-2}$ and b is the Tafel slope:

$$\eta = a + b\ \log(j) \tag{9.32}$$

The amount of potential that must be increased in order to attain a unit increase in current density is represented by the Tafel slope (b). Since it reflects quicker reaction kinetics and better current responsiveness at lower overpotentials, a lower Tafel slope in the context of HzOR denotes higher catalytic efficiency.

9.5.3 Electrochemical Active Surface Area

When assessing electrocatalysts, particularly those with porous or nanostructured surfaces frequently employed in HzOR, the electrochemical active surface area (ECSA) is a crucial metric. Double-layer capacitance (C_{dl}) measurement is usually used to estimate ECSA. It is possible to determine that the observed current originates exclusively from the charging and discharging of the electrical double layer by examining the CV curves in the non-faradaic region, which is typically a 0.1 V window close to the open-circuit potential and where no redox reactions take place. Since the charging current (i_c) and scan rate (v) are linearly related, the value of C_{dl} can be found by calculating the slope of the i_c–v plot. ECSA can be computed using the following formula after C_{dl} is known:

$$i_c = vC_{dl} \tag{9.33}$$

$$\text{ECSA} = C_{dl}/C_s \tag{9.34}$$

A larger ECSA indicates more accessible active sites for HzOR, leading to better catalytic performance.

9.5.4 Electrochemical Impedance

A key tool for examining the internal resistance and charge-transfer characteristics of electrocatalysts during HzOR is EIS. Charge-transfer resistance (R_{ct}) can be evaluated by examining the semicircles in Nyquist plots obtained from EIS. Smaller semicircles signify faster kinetics. Additionally, EIS helps in the rectification of polarization curves and displays solution resistance. It is a useful tool for assessing catalyst performance and stability in HzOR systems since it also aids in monitoring impedance changes associated with morphological or structural changes.

9.5.5 Stability

Long-term stability is an important factor for assessing electrocatalysts, particularly for their practical use in HzOR, in addition to activity-related measures. Extended constant voltage/current testing (usually spanning 10 hours) or accelerated CV is the often used method to evaluate stability. Over time, a good electrocatalyst should sustain a constant current with little deterioration. For practical applications, long-term durability is still a major obstacle that must be overcome for all electrocatalyst classes.

9.6 Metal Carbides for Hydrazine-assisted Water Electrocatalysis

The first ever carbide-based electrocatalyst for HzOR was presented in a pioneering work. A multi-doped Fe_2MoC/N-doped carbon hybrid was created by mimicking the enzymatic cooperativity present in nitrogen-cycle enzymes. This catalyst has a high faradaic efficiency (3.6 $e^-\cdot[N_2H_4]^{-1}$), an ultralow onset potential (0.28 V vs. RHE at pH 14), and remarkable stability over 2,000 cycles. At active sites, Fe, Mo, and N dopants cooperate, and its hierarchical porous structure improves mass and charge transmission. This groundbreaking study opened up new avenues for non-precious metal catalysis by establishing carbides as practical and extremely effective substitutes for precious metals in HzOR [61].

Molybdenum carbide–nitrogen-doped carbon (MoC_x–NC) composites were developed in a previous work as effective, noble-metal-free electrocatalysts for DHFCs. The optimized MoC_x–NC (900 °C, 5% Mo), which was synthesized by high-temperature pyrolysis, had exceptional electrocatalytic activity for HzOR, attaining a peak current density of 3.56 $mA\cdot cm^{-2}$ and outstanding cycling stability (86.13% after 500 cycles). The presence of amorphous molybdenum carbide embedded in a nitrogen-doped carbon matrix, which improves conductivity and active-site exposure, was verified by physical characterization. In DHFCs, the electrocatalyst showed long-term operational stability and a high power density of 158.26 $mW\cdot cm^{-2}$. A diffusion-controlled HzOR mechanism with a low activation energy was discovered through kinetic investigations. The carbon matrix's nitrogen species and MoC_x work in concert to make a substantial contribution [62].

Furthermore, using hydrazine-assisted water splitting, Feng et al. offered a novel electrocatalyst, Mo_2C@C/Co@C, with dual-independent active sites for effective hydrogen production. By employing a one-pot synthesis technique, they created Mo_2C@C/Co@C, wherein Mo_2C nanoparticles and Co particles are inserted independently in carbon layers to produce unique HER and HzOR active sites. Co@C is in charge of the stepwise dehydrogenation of hydrazine in HzOR, whereas Mo_2C@C promotes HER by water dissociation. At 10 $mA\cdot cm^{-2}$, the electrocatalyst attains low overpotentials of 71 mV for HER and −83 mV for HzOR. The stability and bifunctional efficiency of the catalyst are confirmed by theoretical (DFT) and electrochemical investigations. The system reaches 300 $mA\cdot cm^{-2}$ at just 0.49 V in a symmetric hydrazine-splitting electrolyzer (OHzS [overall hydrazine splitting]) and maintains stability for more than 50 hours. Furthermore, an Hz–Zn battery with the same catalyst shows long-term cycling durability (300 cycles) and 94% energy efficiency. In addition to providing a low-energy, carbon-free approach for scalable hydrogen synthesis, this dual-site design is a major leap in bifunctional electrocatalysis and has great potential for real-world energy applications [77].

Inspired by DHFCs, Li et al. suggested a self-powered hydrogen production system that runs on a hydrazine–water galvanic cell without the need for oxygen or outside energy. A bifunctional catalyst—low-valent Ru SAs stabilized on Mo_2C nanobelts via a vacancy-trapping technique—is at the heart of this breakthrough. With HER and HzOR, the Ru SA/v-Mo_2C catalyst exhibits exceptional bifunctionality, ultralow overpotentials, and outstanding kinetics. The system reaches a record-breaking H_2 production rate of 18 mol·h^{-1}·m^{-2} (403.2 L·h^{-1}·m^{-2}) and an open-circuit voltage of 1.37 V. An Ru–Mo dual-site synergy that promotes *H-spillover and maximizes adsorption/desorption energetics for both HER and HzOR is confirmed by DFT calculations and in-situ spectroscopy. The galvanic cell powers a linked hydrazine electrolyzer by using asymmetric pH electrolytes to optimize voltage output. In both alkaline and acidic HER, the catalyst has greater activity than Pt/C, and it performs noticeably better in HzOR. With a high energy density (358 Wh·g^{-1} N_2H_4) and exceptional endurance, the system runs constantly. This concept is a prime example of a revolutionary approach to hydrogen production that combines electrochemical energy collection, catalyst engineering, and reaction coupling [9].

In the same direction to energy-intensive water electrolysis, a self-powered hydrogen synthesis method inspired by hydrazine–water galvanic cells was proposed. Using a vacancy-trapping technique, a low-valent ruthenium SA catalyst supported on Mo_2C nanobelts (Ru SA/v-Mo_2C) was engineered. This catalyst outperforms Pt-based benchmarks with great stability and exhibits exceptional bifunctional activity for both the HzOR and the HER. The system produced a record hydrogen generation rate of 18 mol·h^{-1}·m^{-2} ($\approx$ 403 L·h^{-1}·m^{-2}), an open-circuit voltage of 1.37 V, and an energy density of 358 Wh·g^{-1} N_2H_4 by linking Ru SA/v-Mo_2C with a pH-decoupled hydrazine–H_3O battery. Ru–Mo dual-site synergy was demonstrated by theoretical and in-situ spectroscopic investigations, where Ru sites promote OH adsorption and water dissociation (Figure 9.5a) [78].

Further, a recent study investigated the controversial role of iron carbide (Fe_3C) in Fe–N–C electrocatalysts used for ORR and HzOR in alkaline media. Although Fe_3C is commonly found in pyrolyzed Fe–N–C materials, its catalytic significance remained unclear. Researchers synthesized five Fe_3C-rich carbon composites with varying particle sizes and environments, followed by selective Fe_3C removal using mild acid leaching. Surprisingly, these Fe_3C-rich samples showed poor catalytic activity, while performance improved significantly after Fe_3C removal, even after accounting for increased electrochemical surface area. This indicates that Fe_3C obstructs access to catalytically active Fe–N_x, nitrogen, or carbon sites rather than enhancing catalysis. DFT calculations revealed that Fe_3C binds intermediates too strongly, impeding catalytic turnover. Spectroscopic and poisoning experiments further confirmed Fe–N_x as the primary active site. The study concluded that Fe_3C is an inactive spectator and highlighted the need to focus on Fe–N_x and carbon–nitrogen functionalities for designing efficient, platinum-free electrocatalysts [79].

Mo_2C MXene was investigated theoretically as an electrocatalyst for HzOR, a substitute for OER in the synthesis of hydrogen. Three surface types—O/OH-terminated, nitrogen-rich, and SA centers—were examined using DFT simulations. Although it featured less active sites, the nitrogen-modified surface (three O, four OH, and one N) displayed the maximum activity according to the G_{max} descriptor. Although SA centers offered additional sites, their intrinsic activity was decreased by their excessively high hydrazine binding. The paper highlighted a trade-off between site density and activity, providing guidance for creating effective catalysts based on MXene for sustainable hydrogen production (Figure 9.5b) [80].

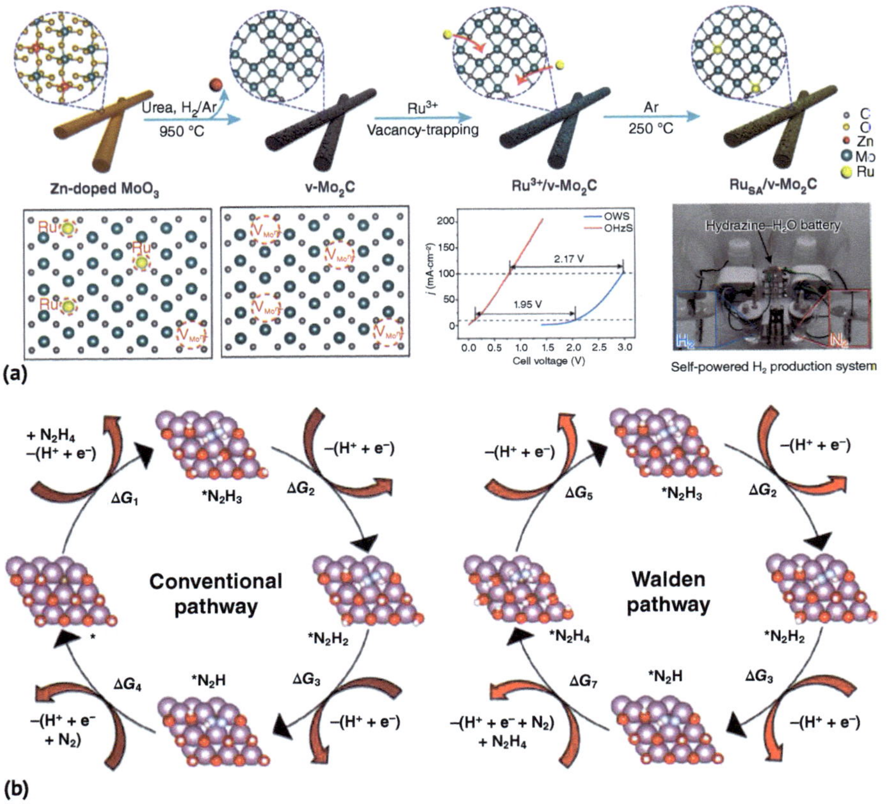

Figure 9.5 (a) Synthesis of vacancy-trapped Ru-SA-decorated MoC and the corresponding application for HzOR. *Source*: [78]/with permission of John Wiley & Sons. (b) Schematic showing two different pathways of hydrazine oxidation. *Source*: [80]/with permission of American Chemical Society.

In a separate work, $Ru_{0.91}Ni_{0.09}$–N/O–Ti_3C_2, a new electrocatalyst for effective hydrogen generation aided by hydrazine, was introduced. Ruthenium–nickel nanoclusters were embedded into nitrogen/oxygen-functionalized MXene (Ti_3C_2) frameworks, resulting in improved performance for HzOR and HER. The d-band center of Ru is modulated by Ni, which maximizes charge transfer and intermediate adsorption. The catalyst produced industrial-level current densities (1 A·cm^{-2} at 0.92 V) and a very low overpotential (29.3 mV for HER, −29.9 mV for HzOR). Robust interfacial contacts and advantageous electronic structures were confirmed by structural characterization and DFT simulations. The system is highly durable, uses little energy (0.35 kWh·m^{-3} H_2), and functions effectively in a membrane-free cell. This work established a scalable and cost-effective route for green hydrogen production, leveraging the synergistic properties of engineered Ru–Ni clusters and MXene supports [64].

A new $BaMnO_3$ nanoparticle-doped $Ti_3C_2X_2$ MXene catalyst for hydrazine electrooxidation was synthesized and evaluated in a previous work. $Ti_3C_2X_2$ MXene was made by hydrothermally depositing $BaMnO_3$ nanoparticles after Ti_3AlC_2 was etched with HF. The effective synthesis was validated by thorough characterization using XRD, SEM, TEM, EDX, and XPS. $BaMnO_3$/MXene showed dramatically improved performance in electrochemical tests (CV, CA, and EIS), with a high mass activity of 731.7 $mA{\cdot}mg^{-1}$, more than twice that of pristine MXene. Because of the synergistic actions of $BaMnO_3$ and MXene, the catalyst demonstrated enhanced stability, a lower onset potential (0.211 V vs. 0.443 V), and a decreased charge-transfer resistance. Strong metal anchoring and improved electron transport were made possible by MXene's defects and functional groups. The $BaMnO_3$/MXene demonstrated stability and competitive activity in comparison to other non-noble metal catalysts, making it a viable low-cost electrocatalyst for small organic fuel electrooxidation and DHFCs [81].

For HzOR, vacancy-engineered $Ti_3C_2T_x$ MXene-supported Ni SACs were introduced. Stable Ni—C coordination is formed by immobilizing Ni atoms by a "self-reduction" process that makes use of intrinsic Ti vacancies and surface terminations of $Ti_3C_2T_x$. With an ultralow onset potential (−0.03 V vs. RHE) and remarkable stability over 24,000 seconds, the resultant Ni SACs/$Ti_3C_2T_x$ demonstrated exceptional catalytic activity. Advanced characterization methods such as XAS, EXAFS, and HAADF–STEM validated the bonding environment and atomic dispersion of Ni. According to DFT simulations, Ni atoms alter the electronic DOS, improving hydrazine adsorption and lowering dehydrogenation energy barriers. The SACs exhibited better durability, quicker kinetics, and a lower charge-transfer resistance than Ni nanoparticles (Figure 9.6a) [64].

A high-performance nanocatalyst system made of ultrafine NiRh nanoparticles immobilized on 2D vanadium carbide (V_2C), a form of MXene, was shown in a previous study. The NiRh/V_2C catalyst, which is intended to selectively break down hydrous hydrazine ($N_2H_4{\cdot}H_2O$) in order to produce hydrogen under mild circumstances, achieved 100% hydrogen selectivity and an impressive turnover frequency (TOF) of 987.5 h^{-1} at 323 K. The strong electronic metal–support interactions (EMSI) that V_2C substrate provides, along with the synergistic effect of Ni and Rh, are responsible for the remarkable catalytic activity. The many surface functional groups on V_2C facilitate charge transfer, inhibit aggregation, and encourage the dispersion of 2.1-nm NiRh nanoparticles. The superiority of V_2C was validated by comparative research with various supports and compositions. The catalyst exhibited outstanding recyclability and thermal stability. This work supported the development of hydrogen energy systems by presenting an economical method for creating reliable, effective nanocatalysts for hydrogen generation [82].

Further, Feng et al. discussed the role of additional HzORs in rechargeable Hz–Zn batteries. A Hz–Zn battery using a bifunctional 3D Mo_2C/Ni@C/CS electrocatalyst allows for decoupled HER and HzOR during discharge and charge, respectively. This system achieved high energy efficiency (96%) with ultralow input voltage (< 0.4 V), and excellent durability across 600 cycles, revealing a realistic route for energy-saving, on-demand hydrogen generation (Figure 9.6b) [83].

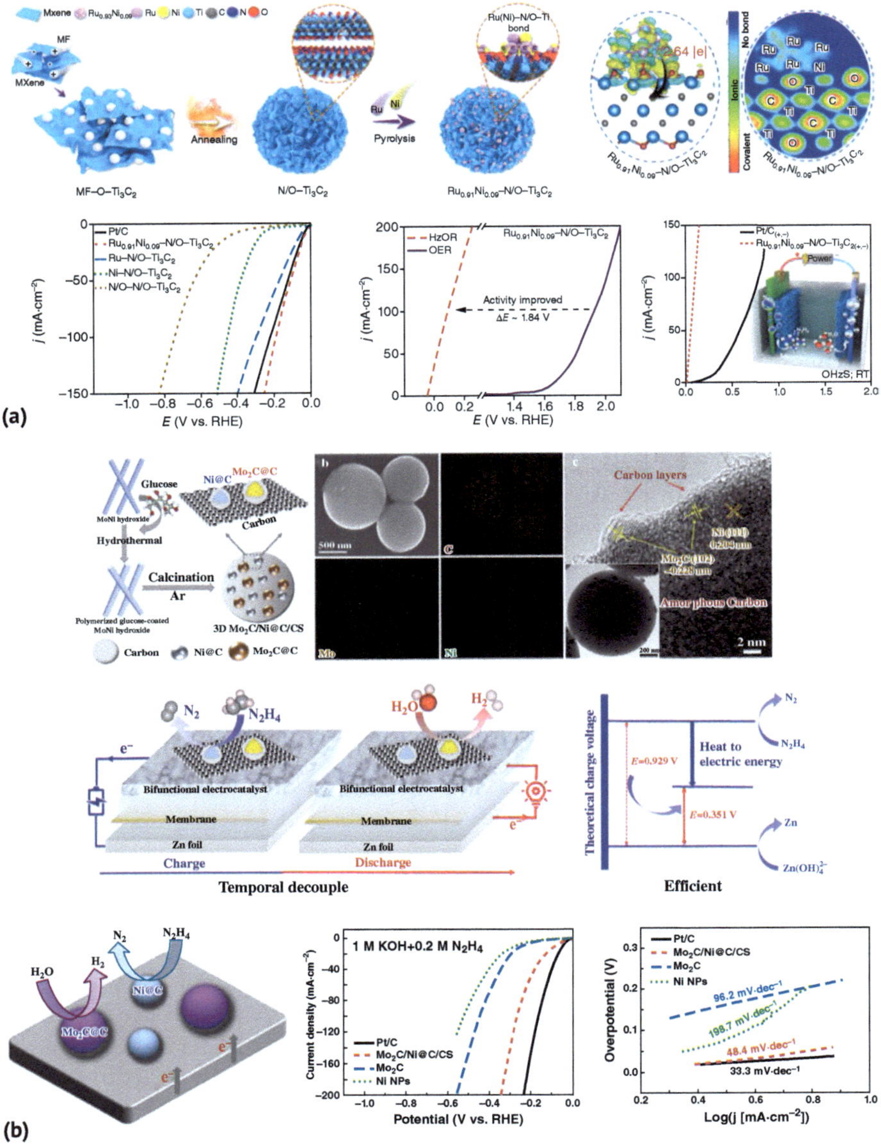

Figure 9.6 (a) Ni SA immobilization on vacancy-rich MXene for HzOR. *Source*: [64]/with permission of John Wiley & Sons. (b) Incorporation of Ni@C and $Mo_2C@C$ for HzOR-based Zn battery. *Source*: [83]. Springer Nature. CC BY 4.0.

9.7 Metal Nitrides for Hydrazine-assisted Water Electrocatalysis

In another work, a cerium-doped nickel nitride catalyst was used to couple the HER with the HzOR, presenting a unique method for energy-efficient hydrogen synthesis. Porous Ce-doped Ni_3N nanosheet arrays on nickel foam (Ce–Ni_3N/NF), which the researchers synthesized, demonstrated exceptional bifunctional catalytic performance for both HER and HzOR. The system ran steadily for more than 100 hours at 400 mA·cm^{-2} and required an exceptionally low cell voltage of only 0.156 V at 10 mA·cm^{-2}. Low charge-transfer resistance, rapid diffusion kinetics, and improved active-site exposure were revealed by structural characterizations. Ce doping increases the inherent catalytic activity by optimizing the hydrogen adsorption free energy, according to DFT calculations [39].

Further, a new method for producing hydrogen with less energy by growing Mo-doped Ni_3N/Ni heterostructure nanosheets on Ni foam (Mo–Ni_3N/Ni/NF) was introduced in another study. This catalyst achieved outstanding bifunctionality for both HER and HzOR in alkaline circumstances by skillfully combining interface engineering and chemical substitution. At 10 mA·cm^{-2}, it provided a low working potential of −0.3 mV for HzOR and 45 mV overpotential for HER. It is impressive that an OHzS system employing this material has a near 100% faradaic efficiency, strong durability, and only needs 55 mV to operate. Its usefulness was further demonstrated by proof-of-concept experiments utilizing the built DHFC, solar cell, and discarded AAA battery. Mo substitution improves H* adsorption behavior and reduces the energy barrier for HzOR, according to calculations using DFT. This study presented a viable and economical approach to producing high-performance hydrogen [84].

According to a study, copper–nickel nitride (Cu_1Ni_2–N) nanosheets are effective bifunctional electrocatalysts for the generation of hydrogen with hydrazine assistance. The layered hydroxide precursors on carbon fiber fabric were thermally ammonolyzed to create the nanosheets. With low overpotentials of 71.4 mV for the HER and 0.5 mV for the HzOR at 10 mA·cm^{-2}, Cu_1Ni_2–N demonstrated outstanding catalytic activity for both reactions. For more than 75 hours, a two-electrode system using Cu_1Ni_2–N as the anode and cathode produced hydrogen at a low voltage of 0.24 V with excellent stability. The material's strong conductivity, synergistic Cu_4N/Ni_3N interface, and porous nanostructure all contribute to this performance. Through Hz-WE, the study identified Cu_1Ni_2–N as a potential, inexpensive bifunctional catalyst for energy-efficient hydrogen synthesis [85].

A vanadium-doped nickel nitride (V–Ni_3N) nanosheet catalyst developed on nickel foam was presented for effective hydrogen production via HER and HzOR. The electronic structure of Ni_3N was modulated by V doping, which improved hydrazine dehydrogenation and hydrogen adsorption/desorption. At 10 mA·cm^{-2}, the V–Ni_3N demonstrated exceptional bifunctional catalytic activity, using just 2 mV for HzOR and 70 mV for HER. This catalyst produced an ultralow voltage of 94 mV for total hydrazine splitting in a two-electrode system. DFT computations and structural and electrochemical investigations validated the contribution of V to enhancing reaction kinetics and reducing energy barriers [86].

High-valent tungsten doping of Ni_3N improves water and hydrazine activation by lowering the d-band center, introducing compressive strain, and optimizing intermediate binding. Additionally, this strain facilitates HER and HzOR by encouraging electron transfer to antibonding orbitals. Low overpotentials (46 mV for HER and 81 mV for HzOR) and outstanding long-term stability (≈ 450 hours) were attained by the W–Ni_3N catalyst. Spectroscopic studies and DFT computations showed that strain and electronic modulation work together to lower energy barriers and increase catalytic activity. Notably, utilizing hydrazine results in significant energy and cost savings, since the dual-electrode electrolyzer employing this catalyst only needs 0.185 V at 50 $mA{\cdot}cm^{-2}$ (Figure 9.7a) [63].

An effective bifunctional electrocatalyst for energy-efficient hydrogen production by hydrazine-assisted seawater splitting over a broad pH range (7–14) was shown using Iron-doped cobalt nitride (Fe–Co_3N). Using a two-step hydrothermal-nitridation procedure, the material was directly synthesized on nickel foam to create porous microspheres with a protective amorphous iron oxide surface layer. High catalytic activity and stability are encouraged by this design for both HERs and HzORs. The Fe–Co_3N system stabilized at 500 $mA{\cdot}cm^{-2}$ for 95 hours and reached an industrial-scale current density of 1,000 $mA{\cdot}cm^{-2}$ at just 0.65 V in alkaline seawater. At 0.75 V, it maintained 200 $mA{\cdot}cm^{-2}$ even in neutral seawater (Figure 9.7b) [38].

According to a further study, a new dual-doped Co_3N nanowire array electrode (PW–Co_3N NWA/NF) co-doped with tungsten and phosphorus (P) exhibited remarkable electrocatalytic activity for HER and HzOR. Low overpotentials of −55 mV for HzOR and −41 mV for HER were attained by this bifunctional catalyst at 10 $mA{\cdot}cm^{-2}$. In a two-electrode system, total hydrazine splitting (OHzS) may be driven at the same current density with a record-low cell voltage of only 28 mV. The doping optimizes the dehydrogenation kinetics and hydrogen adsorption energy, according to DFT studies. Interestingly, by combining a DHFC with the hydrazine electrolyzer, a self-powered hydrogen generation system was further demonstrated, with a hydrogen production rate of 1.25 $mmol{\cdot}h^{-1}$ at normal temperature. This study demonstrated the promise of hydrazine-based systems and dual-doping techniques for portable, energy-efficient hydrogen synthesis [42].

Mn@Ni_3N–Co_3N/NF, a bifunctional electrocatalyst with lattice-matched Ni_3N and Co_3N phases improved by Mn doping, was designed. High activity for the HzOR and HER was made possible by the designed interface, which facilitated effective electron redistribution. The catalyst used 53.3% less energy than traditional water electrolysis while achieving industrial-scale current densities (500 $mA{\cdot}cm^{-2}$) at only 0.49 V. It also demonstrated exceptional reusability and durability. Sustainable hydrogen production and the cleanup of hydrazine wastewater are two advantages of this strategy, which also offered insights into the design of new metal nitride catalysts for low-energy green hydrogen technologies [60].

The synthesis of Ni_3N–Co_3N porous nanosheet arrays (PNAs) on Ni foam, a bifunctional electrocatalyst intended to improve the HzOR and HER, was made. These heterostructured nanosheets have exceptional electrocatalytic activity, achieving 10 $mA{\cdot}cm^{-2}$ current at low overpotentials of 43 mV (HER) and −88 mV (HzOR). They

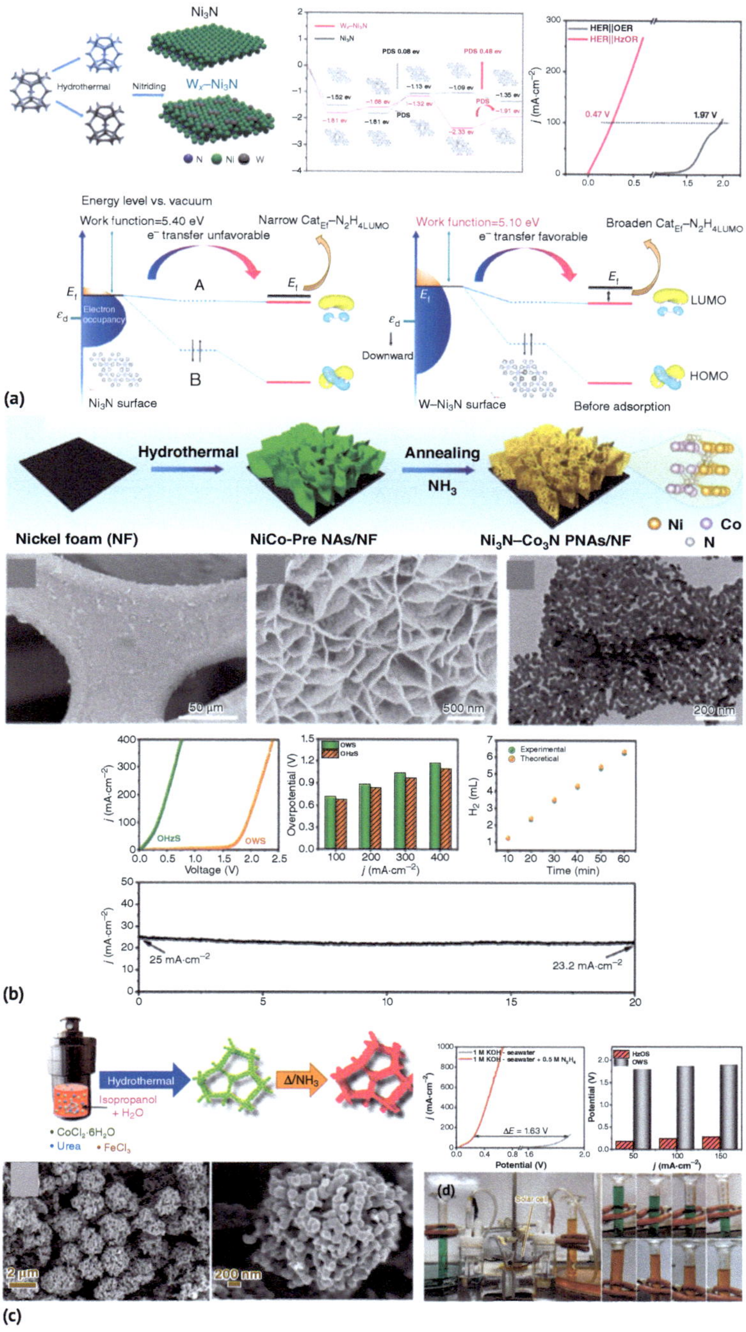

Figure 9.7 (a) Facile synthesis of electronic structure-engineered W–Ni_3N for HzOR-assisted H_2 production. *Source*: [63]/with permission of John Wiley & Sons. (b) Depiction of Fe-doped Co_3N for HzOR-aided seawater splitting for reaching industry-level current density. *Source*: [38]/with permission of Elsevier. (c) Interfacing Ni_3N and Co_3N for efficient HzOR catalysis. *Source*: [76]/with permission of John Wiley & Sons.

were created using a straightforward hydrothermal-nitridation process. Record-low cell voltages of 0.071 and 0.76 V were demonstrated by an integrated hydrazine electrolyzer to drive 10 and 400 $mA{\cdot}cm^{-2}$, respectively. When combined with a solar cell or DHFC, the technique also makes it possible to produce hydrogen on its own. Interfacial charge transfer in the Ni_3N–Co_3N structure enhances hydrogen adsorption and hydrazine dehydrogenation kinetics, according to DFT studies (Figure 9.7c) [76].

In order to improve rechargeable Zn–CO_2 batteries (ZCBs), a study proposed a novel method that used hydrazine oxidation to lower charge voltage. Isolated FeN_3 sites were embedded in hierarchical porous carbon nanoboxes (Fe-HPCNs), which were made from ferrocene and ZIF-8 covered with polydopamine. These FeN_3 sites enabled bifunctional electrocatalysis for both CO_2 reduction (ECO_2R) and HzOR, leading to noteworthy improvements: a 96% CO Faraday efficiency (FE_{CO}), reduced overpotential for HzOR, and ultralow charge voltage (~1.84 V). The isolated FeN_3 structure was validated by structural characterization, and low energy barriers and favorable intermediate formation were found by in-situ FTIR and DFT calculations. Fe-HPCNs showed excellent discharge selectivity and 89% energy efficiency when used in ZCBs [87].

Using hydrothermal synthesis and nitridation, a study provided a hierarchically nanostructured NiFeN electrocatalyst based on nickel foam. The catalyst, which was created for the HzOR in DHFCs, performed exceptionally well. It showed a high current density of 1,017 $mA{\cdot}cm^{-2}$ at 0.3 V vs. RHE, good stability (90.8% retention after 1,000 cycles), and almost 100% faradaic selectivity. Its high intrinsic activity, excellent conductivity, and 3D nanoporous design are the reasons for its outstanding activity. Crucially, it inhibited the breakdown of non-faradaic hydrazine. The potential of TMNs including iron in energy conversion applications was demonstrated by this work [88].

In a different work, 2D Ti_2NT_x MXene materials made from Ti_2AlN MAX phases by selective chemical etching were synthesized and characterized. XRD, XPS, BET, and HRTEM investigations verified that the resultant Ti_2NT_x had an accordion-like lamellar shape with enlarged surface area (2.0867 $m^2{\cdot}g^{-1}$ vs. 0.5375 $m^2{\cdot}g^{-1}$ for Ti_2AlN) and distinctive functional groups (—OH, —F, and —O). Comparing Ti_2NT_x MXenes to Ti_2AlN, electrochemical investigations employing CV, chronoamperometry, and EIS demonstrated that the former exhibited noticeably better electrocatalytic activity for hydrazine electrooxidation. The Ti_2NT_x catalyst demonstrated long-term stability, minimal charge-transfer resistance, and a specific activity of 2.739 $mA{\cdot}cm^{-2}$ and a mass activity of 44.1 $mA{\cdot}mg^{-1}$. The presence of active surface terminations and an increase in electrochemically active surface area are responsible for the enhanced performance. The potential of Ti_2NT_x MXene as a very effective and reliable catalyst for hydrazine-based fuel cell applications was highlighted in this paper [89].

9.8 Conclusion

Hz-WE offers a promising alternative to conventional water splitting by replacing the sluggish OER with the HzOR, which is thermodynamically more favorable and kinetically faster. This approach significantly reduces the required cell voltage, enhances energy efficiency, and opens avenues for integrated hydrogen production and environmental remediation. TMCs and TMNs have emerged as highly effective bifunctional electrocatalysts for Hz-WE due to their noble-metal-like electronic structures, high electrical conductivity, and excellent chemical stability. These materials offer significant advantages over traditional noble-metal catalysts, including lower cost, structural tunability, and enhanced tolerance to harsh electrochemical environments. This chapter reviewed the structural features, synthesis strategies, and electrocatalytic properties of TMCs and TMNs, with a focus on their role in HzOR and HER. Techniques such as carbothermal reduction, sputtering, PLD, and sol–gel processes enable precise control over material morphology and surface properties. Additionally, performance can be significantly improved through defect engineering, heteroatom doping, SAC design, and the development of hierarchical nanostructures. Recent experimental studies demonstrated that these materials can achieve ultralow overpotentials, high faradaic efficiencies, and stable operation at industrial-scale current densities. Their integration into energy systems such as DHFCs and Hz–Zn hybrid batteries further highlights their versatility and real-world applicability. DFT calculations have provided deep mechanistic insights, guiding the rational design of active sites and optimized reaction pathways. In summary, TMCs and TMNs are powerful candidates for next-generation electrocatalysis, particularly in the context of energy-efficient hydrogen production via Hz-WE. Their tunable structures, favorable reaction kinetics, and broad design space make them key materials in advancing low-cost, high-performance hydrogen generation technologies. Future research should focus on scalable synthesis, long-term durability, and the integration of these materials into practical devices to fully realize their potential in clean energy applications.

Acknowledgments

One of the authors (BKS) acknowledges extramural research funding under MRIP-2023-Physics (23EM/PH/123) for support during the preparation of this book chapter.

References

1 Shanmugam, V., Neuberg, S., Zapf, R., Pennemann, H., & Kolb, G. (2020). Hydrogen production over highly active Pt based catalyst coatings by steam reforming of methanol: Effect of support and co-support. *Int. J. Hydrog. Energy* *45*(3), 1658–1670.

2 McKendry, I. G., Mohamad, L. J., Thenuwara, A. C., Marshall, T., Borguet, E., Strongin, D. R., & Zdilla, M. J. (2018). Synergistic in-layer cobalt doping and interlayer iron intercalation into layered MnO_2 produces an efficient water oxidation electrocatalyst. *ACS Energy Lett.* *3*(9), 2280–2285.

3 Sahu, B. K. & Das, A. (2019). Significance of oxygen defects in SnO_2 quantum dots as hybrid electrochemical capacitors. *J. Phys. Chem. Solids* *129*, 293–297.

4 Jeong, Y., Theerthagiri, J., Lee, S. J., Chitumalla, R. K., Moon, C. J., Min, A., Kheawhom, S., Jang, J., & Choi, M. Y. (2024). Unique production strategy of Pt/C electrocatalysts via pulsed laser for hydrogen generation: Insights screening by DFT calculations. *Electrochim. Acta 488*, 144218.

5 Wang, W., Wu, Y., Lin, Y., Yao, J., Wu, X., Wu, C., Zuo, X., Yang, Q., Ge, B., Yang, L., & Li, G. (2022). Confining zero-valent platinum single atoms in α-MoC_{1-x} for pH-universal hydrogen evolution reaction. *Adv. Funct. Mater. 32*(12), 2108464.

6 Park, J., Sahu, B. K., Moon, C. J., Theerthagiri, J., Min, A., Yun, S., & Choi, M. Y. (2024). Probing hydrogen evolution on pulsed laser-crafted Pt-infused oxygen-deficient black TiO_2 in real-time using Raman spectroscopy. *J. Phys. Chem. C 128*, 14987–14906.

7 Wu, Z., Yang, P., Li, Q., Xiao, W., Li, Z., Xu, G., Liu, F., Jia, B., Ma, T., Feng, S., & Wang, L. (2023). Microwave synthesis of Pt clusters on black TiO_2 with abundant oxygen vacancies for efficient acidic electrocatalytic hydrogen evolution. *Angew. Chem. 135*(14), e202300406.

8 Tang, Y., Zheng, S., Cao, S., Xue, H., & Pang, H. (2020). Advances in the application of manganese dioxide and its composites as electrocatalysts for the oxygen evolution reaction. *J. Mater. Chem. A 8*(36), 18492–18514.

9 Li, Y., Niu, S., Liu, P., Pan, R., Zhang, H., Ahmad, N., Shi, Y., Liang, X., Cheng, M., Chen, S., & Du, J. (2024). Ruthenium nanoclusters and single atoms on α-MoC/N-doped carbon achieves low-input/input-free hydrogen evolution via decoupled/coupled hydrazine oxidation. *Angew. Chem. 136*(30), e202316755.

10 Liu, H., Liu, Y., Li, M., Liu, X., & Luo, J. (2020). Transition-metal-based electrocatalysts for hydrazine-assisted hydrogen production. *Mater. Today Adv. 7*, 100083.

11 Wang, H.-Y. & Yuan, Z.-Y. (2025). Hydrazine-assisted water electrolysis system: Performance enhancement and application expansion. *Mater. Horiz. 12*, 5123–5148.

12 Zhou, M., Yu, Z., Yu, G., Fu, R., Wang, S., Yang, W., Liao, X., Zhao, Y., & Wang, Z. (2025). Constructing of n-type semiconductor heterostructures for efficient hydrazine-assisted hydrogen production. *Adv. Funct. Mater. 35*(6), 2415058.

13 Yang, Y., Li, X., Liu, G., Liu, H., Shi, Y., Ye, C., Fang, Z., Ye, M., & Shen, J. (2024). Hierarchical ohmic contact interface engineering for efficient hydrazine-assisted hydrogen evolution reaction. *Adv. Funct. Mater. 36*(7), 2307979.

14 Roh, H., Choi, H., Kim, J.-S., Shin, H., Park, T., & Yong, K. (2022). Low power consumed PV-electrolysis with CoFeP nanowires for hydrazine-assisted hydrogen production. *Appl. Surf. Sci. 606*, 154951.

15 Sahoo, P., Mishra, S. B., Debata, S., Sharma, A., Sahu, B. K., Padhan, S., Thangavel, R., & Lee, B. K. (2024). Hydrothermally grown halogen-doped ZnO nanorods for photoelectrochemical water oxidation: Experimental and DFT insights. *J. Phys. Chem. C 128*(44), 18711–18723.

16 Sahu, B. K. & Das, A. (2020). Graphene oxide surface chemistry regulated growth of SnO_2 nanoparticles for electrochemical application. *J. Alloys Compd. 834*, 154901.

17 Moon, C. J., Maheskumar, V., Min, A., Kumar, A., Lee, S., Senthil, R. A., Ubaidullah, M., & Choi, M. Y. (2025). Laser-regulated iridium-diffused

nitrogen–carbon sites for enhanced hydrazine-assisted alkaline seawater splitting and zinc–hydrazine batteries. *Small 21*(8), 2408569.

18 Li, R.-Q., Guo, S., Wang, X., Wan, X., Xie, S., Liu, Y., Wang, C., Zhang, G., Cao, J., Dai, J., & Ge, M. (2024). Dual-strategy engineered nickel phosphide for achieving efficient hydrazine-assisted hydrogen production in seawater. *Chem. Sci. 15*(26), 10084–10091.

19 Khalafallah, D., Zhi, M., & Hong, Z. J. C. (2021). Development trends on nickel-based electrocatalysts for direct hydrazine fuel cells. *Small 13*(1), 81–110.

20 Oshchepkov, A. G., Braesch, G., Bonnefont, A., Savinova, E. R., & Chatenet, M. (2020). Recent advances in the understanding of nickel-based catalysts for the oxidation of hydrogen-containing fuels in alkaline media. *ACS Catal. 10*(13), 7043–7068.

21 Liu, W., Wang, A., Zhang, J., Yu, S., Wang, M., Tian, S., Tang, H., Zhao, Z., Ren, X., Guo, Y., & Ma, D. (2025). Pt/α-MoC catalyst boosting pH-universal hydrogen evolution reaction at high current densities. *ACS Nano 19*(10), 10038–10047.

22 Li, R., Shang, J., Wang, F., Lu, Q., Yan, H., Tuo, Y., Liu, Y., Feng, X., Chen, X., Chen, D., & Yang, C. (2025). Quantification and optimization of platinum–molybdenum carbide interfacial sites to enhance low-temperature water-gas shift reaction. *Nat. Commun. 16*(1), 1098.

23 Over, H. (2021). Fundamental studies of planar single-crystalline oxide model electrodes (RuO_2, IrO_2) for acidic water splitting. *ACS Catal. 11*(14), 8848–8871.

24 Michalsky, R., Zhang, Y.-J., & Peterson, A. (2014). Trends in the hydrogen evolution activity of metal carbide catalysts. *ACS Catal. 4*(5), 1274–1278.

25 Zhang, X., Shi, C., Chen, B., Kuhn, A. N., Ma, D., & Yang, H. (2018). Progress in hydrogen production over transition metal carbide catalysts: Challenges and opportunities. *Curr. Opin. Chem. Eng. 20*, 68–77.

26 Gao, Q., Zhang, W., Shi, Z., Yang, L., & Tang, Y. (2019). Structural design and electronic modulation of transition-metal-carbide electrocatalysts toward efficient hydrogen evolution. *Adv. Mater. 31*(2), 1802880.

27 Yang, Y., Ahmed, A. T., Patel, P., Al-Hetty, H. R. A. K., Ramachandran, T., Shankhyan, A., Mann, V. S., Gupta, D., Kariem, M., & Nafea, M. A. (2025). Metal nitrides vs. metal carbides: A comparative analysis of catalysts for the hydrogen evolution reaction. *Braz. J. Chem. Eng.* https://doi.org/10.1007/s43153-025-00552-5

28 Theerthagiri, J., Lee, S. J., Murthy, A. P., Madhavan, J., & Choi, M. (2020). Fundamental aspects and recent advances in transition metal nitrides as electrocatalysts for hydrogen evolution reaction: A review. *Curr. Opin. Solid State Mater. Sci. 24*(1), 100805.

29 Chen, M., Hu, L., Xu, L., Wei, J., Wu, P., Guan, G., Wang, T., & Ma, Y. (2023). Synergistically tuning surface states of hierarchical MoC by Pt-N dual-doping engineering for optimizing hydrogen evolution activity. *Small 7*(9), 2300308.

30 Gupta, M., Pandey, N., Reddy, V., Phase, D., Schlage, K., Wille, H.-C., & Gupta, A. (2019). In-situ growth of iron mononitride thin films studied using X-ray absorption spectroscopy and nuclear resonant scattering. *Hyperfine Interact. 240*(1), 99.

31 Bonu, V., Sahu, B. K., Das, A., Amirthapandian, S., Dhara, S., & Barshilia, H. C. (2019). Sub-wavelength waveguide properties of 1D and surface-functionalized SnO_2 nanostructures of various morphologies. *Beilstein J. Nanotechnol. 10*(1), 379–388.

32 Yao, W., Makowski, P., Giordano, C., & Goettmann, F. (2009). Synthesis of early-transition-metal carbide and nitride nanoparticles through the urea route and their use as alkylation catalysts. *Chem. Eur. J. 15*(44), 11999–12004.

33 Haemers, J., Gusmão, R., & Sofer, Z. (2020). Synthesis protocols of the most common layered carbide and nitride MAX phases. *Small Methods 4*(3), 1900780.

34 Xie, J.-Z. & Jiang, H. (2023). Revealing carbon vacancy distribution on α-MoC_{1-x} surfaces by machine-learning force-field-aided cluster expansion approach. *J. Phys. Chem. C 127*(27), 13228–13237.

35 Råsander, M., Hugosson, H. W., & Delin, A. (2017). Density functional study of carbon vacancies in titanium carbide. *J. Phys.: Condens. Matter 30*(1), 015702.

36 Ye, T.-N., Park, S.-W., Lu, Y., Li, J., Sasase, M., Kitano, M., & Hosono, H. (2020). Contribution of nitrogen vacancies to ammonia synthesis over metal nitride catalysts. *J. Am. Chem. Soc. 142*(33), 14374–14383.

37 Wan, C. & Leonard, B. (2015). Iron-doped molybdenum carbide catalyst with high activity and stability for the hydrogen evolution reaction. *Chem. Mater. 27*(12), 4281–4288.

38 Sinha, N., Das, C., & Roy, P. (2024). Iron-doped cobalt nitride as an efficient electrocatalyst towards energy saving hydrazine assisted seawater splitting in near neutral to highly alkaline pH achieving industry level current density. *Int. J. Hydrog. Energy 51*, 1011–1021.

39 Li, R.-Q., Zeng, S., Sang, B., Xue, C., Qu, K., Zhang, Y., Zhang, W., Zhang, G., Liu, X., Deng, J., & Fontaine, O. (2023). Regulating electronic structure of porous nickel nitride nanosheet arrays by cerium doping for energy-saving hydrogen production coupling hydrazine oxidation. *Nano Res. 16*(2), 2543–2550.

40 Li, Q., Zhao, J., Li, P., Xu, Z., Feng, J., Chen, B., & Liu, R. (2024). Coembedding Fe single atom–coupled MoC nanoparticles in N-doped hierarchically porous carbon cubes for oxygen electroreduction. *ACS Nano 18*(33), 21975–21984.

41 Tang, N., Liu, D., Chen, S., Wang, Z., Ma, Y., Li, Q., Li, Y., Xu, G., Wu, C., Kang, L., & Luo, W. (2024). Pt atom-substituted MoC single-atom catalyst for enhancing H_2 production. *ACS Catal. 14*(19), 14297–14307.

42 Xiong, D., He, X., Liu, X., Gong, S., Xu, C., Tu, Z., Wu, D., Wang, J., & Chen, Z. (2024). 1D/3D heterogeneous assembling body of cobalt nitrides for highly efficient overall hydrazine splitting and supercapacitors. *Small 20*(8), 2306100.

43 Lim, K. R. G., Handoko, A. D., Nemani, S. K., Wyatt, B., Jiang, H.-Y., Tang, J., Anasori, B., & Seh, Z. W. (2020). Rational design of two-dimensional transition metal carbide/nitride (MXene) hybrids and nanocomposites for catalytic energy storage and conversion. *ACS Nano 14*(9), 10834–10864.

44 Levy, R. & Boudart, M. (1973). Platinum-like behavior of tungsten carbide in surface catalysis. *Science 181*(4099), 547–549.

45 Xie, J. & Xie, Y. (2016). Transition metal nitrides for electrocatalytic energy conversion: opportunities and challenges. *Chem. Eur. J. 22*(11), 3588–3598.

46 Tian, D., Denny, S. R., Li, K., Wang, H., Kattel, S., & Chen, J. (2021). Density functional theory studies of transition metal carbides and nitrides as electrocatalysts. *Chem. Soc. Rev. 50*(22), 12338–12376.

47 Yu, X.-X., Weinberger, C. R., & Thompson, G. (2016). Ab initio investigations of the phase stability in group IVB and VB transition metal carbides. *Comput. Mater. Sci. 112*, 318–326.

48 Zhong, Y., Xia, X., Shi, F., Zhan, J., Tu, J., & Fan, H. (2016). Transition metal carbides and nitrides in energy storage and conversion. *Adv. Sci. 3*(5), 1500286.

49 Gautam, A. K., Yu, S., He, H., Yang, H., & Mironenko, A. (2025). Role of surface oxygen in α-MoC catalyst stability and activity under electrooxidation conditions. *ACS Catal. 15*(8), 6640–6651.

50 Zhang, X., Wang, J., Guo, T., Liu, T., Wu, Z., Cavallo, L., Cao, Z., & Wang, D. (2019). Structure and phase regulation in Mo_xC (α-$MoC_{1-}x$/β-Mo_2C) to enhance hydrogen evolution. *Appl. Catal. B: Environ. 247*, 78–85.

51 Wan, C., Regmi, Y. N., & Leonard, B. (2014). Multiple phases of molybdenum carbide as electrocatalysts for the hydrogen evolution reaction. *Angew. Chem. 126*(25), 6525–6528.

52 Shah, M. S. A. S., Jang, G. Y., Zhang, K., & Park, J. H. (2023). Transition metal carbide-based nanostructures for electrochemical hydrogen and oxygen evolution reactions. *EcoEnergy 1*(2), 344–374.

53 Sundarraj, S., Vadivel, N., Murthy, A. P., Theerthagiri, J., & Choi, M. (2025). MXene electrocatalysts: transformative approaches in hydrogen production with alternative anode reactions. *Small 21*(2), 2407120.

54 Shaikh, N. S., Ubale, S. B., Mane, V. J., Shaikh, J. S., Lokhande, V. C., Praserthdam, S., Lokhande, C. D., & Kanjanaboos, P. (2022). Novel electrodes for supercapacitor: Conducting polymers, metal oxides, chalcogenides, carbides, nitrides, MXenes, and their composites with graphene. *J. Alloys Compd. 893*, 161998.

55 Hantanasirisakul, K. & Gogotsi, Y. (2023). Electronic and optical properties of 2D transition metal carbides and nitrides (MXenes). *Adv. Mater. 30*(52), 1804779.

56 Chen, Z., Song, Y., Cai, J., Zheng, X., Han, D., Wu, Y., Zang, Y., Niu, S., Liu, Y., Zhu, J., & Liu, X. (2018). Tailoring the d-band centers enables Co_4N nanosheets to be highly active for hydrogen evolution catalysis. *Angew. Chem. Int. Ed. 57*(18), 5076–5080.

57 Kitchin, J. R., Nørskov, J. K., Barteau, M. A., & Chen, J. (2005). Trends in the chemical properties of early transition metal carbide surfaces: A density functional study. *Catal. Today 105*(1), 66–73.

58 Cao, X., Bai, H., Wu, W., Bao, H., & Li, Y. (2022). Improving the catalytic activity of two-dimensional Mo_2C for hydrogen evolution reaction by doping and vacancy defects. *Int. J. Hydrog. Energy 47*(91), 38517–38523.

59 Lin, L., Yu, Q., Peng, M., Li, A., Yao, S., Tian, S., Liu, X., Li, A., Jiang, Z., Gao, R., & Han, X. (2020). Atomically dispersed Ni/α-MoC catalyst for hydrogen production from methanol/water. *J. Am. Chem. Soc. 143*(1), 309–317.

60 Wang, X., Hu, H., Yan, X., Zhang, Z., & Yang, M. (2024). Activating interfacial electron redistribution in lattice-matched biphasic Ni_3N-Co_3N for energy-efficient electrocatalytic hydrogen production via coupled hydrazine degradation. *Angew. Chem. 136*(19), e202401364.

61 Ojha, K., Farber, E. M., Burshtein, T. Y., & Eisenberg, D. (2018). A multi-doped electrocatalyst for efficient hydrazine oxidation. *Angew. Chem. Int. Ed. 57*(52), 17168–17172.

62 Deng, J., Li, X., Imhanria, S., Chen, K., Deng, X., & Wang, W. (2021). Molybdenum carbide-nitrogen doped carbon composites as effective non-precious electrocatalyst for direct hydrazine fuel cell. *Electrochim. Acta 384*, 138417.

63 Qin, H., Lin, G., Zhang, J., Cao, X., Xia, W., Yang, H., Yuan, K., Jin, T., Wang, Q., & Jiao, L. (2025). Enhanced cooperative generalized compressive strain and electronic structure engineering in W-Ni_3N for efficient hydrazine oxidation facilitating H_2 production. *Adv. Mater. 37*(8), 2417593.

64 Nguyen, T. H., Tran, D. T., Malhotra, D., Tran, P. K. L., Dinh, V. A., Ta, T. T. N., Dong, C. L., Kim, N. H., & Lee, J. H. (2025). D-orbital-modulated ruthenium embedded within functionalized hollow MXene networks for enhanced hydrazine-assisted hydrogen production. *Small 21*(22), 2502553.

65 Yang, D., Su, Z., Chen, Y., Srinivas, K., Gao, J., Zhang, W., Wang, Z., & Lin, H. (2021). Electronic modulation of hierarchical spongy nanosheets toward efficient and stable water electrolysis. *Small 17*(4), 2006881.

66 Wang, J., Liu, S., Wang, Y., Wang, T., Shang, S., & Ren, W. (2020). Magnetron-sputtering deposited molybdenum carbide MXene thin films as a saturable absorber for passively Q-switched lasers. *J. Mater. Chem. C 8*(5), 1608–1613.

67 Oh, Y., Theerthagiri, J., Min, A., Moon, C. J., Yu, Y., & Choi, M. (2024). Pulsed laser interference patterning of transition-metal carbides for stable alkaline water electrolysis kinetics. *Carbon Energy 6*(5), e448.

68 Davodi, F., Mühlhausen, E., Settipani, D., Rautama, E.-L., Honkanen, A.-P., Huotari, S., Marzun, G., Taskinen, P., & Kallio, T. (2019). Comprehensive study to design advanced metal-carbide@graphene and metal-carbide@iron oxide nanoparticles with tunable structure by the laser ablation in liquid. *J. Colloid Interface Sci. 556*, 180–192.

69 Jamil, R., Ali, R., Loomba, S., Xian, J., Yousaf, M., Khan, K., Shabbir, B., McConville, C. F., Mahmood, A., & Mahmood, N. (2021). The role of nitrogen in transition-metal nitrides in electrochemical water splitting. *Chem Catal. 1*(4), 802–854.

70 Yang, W., Rehman, S., Chu, X., Hou, Y., & Gao, S. (2015). Transition metal (Fe, Co and Ni) carbide and nitride nanomaterials: Structure, chemical synthesis and applications. *ChemNanoMat 1*(6), 376–398.

71 Ashraf, I., Rizwan, S., & Iqbal, M. (2020). A comprehensive review on the synthesis and energy applications of nano-structured metal nitrides. *Front. Mater. 7*, 181.

72 Dong, J., Fu, Q., Jiang, Z., Mei, B., & Bao, X. (2018). Carbide-supported Au catalysts for water–gas shift reactions: A new territory for the strong metal–support interaction effect. *J. Am. Chem. Soc. 140*(42), 13808–13816.

73 Yu, Y., Lee, S. J., Theerthagiri, J., Lee, Y., & Choi, M. (2022). Architecting the AuPt alloys for hydrazine oxidation as an anolyte in fuel cell: Comparative analysis of hydrazine splitting and water splitting for energy-saving H_2 generation. *Appl. Catal. B: Environ. 316*, 121603.

74 Park, C. E., Theerthagiri, J., Maheskumar, V., Kumar, A., Jeong, G. H., & Choi, M. (2025). Cocktail effect of 4*d*/5*d* band twisted high-entropy alloys on carbon nanotube for hydrazine splitting. *Small 21*(29), 2410304.

75 Burshtein, T. Y., Yasman, Y., Munoz-Moene, L., Zagal, J. H., & Eisenberg, D. (2024). Hydrazine oxidation electrocatalysis. *ACS Catal. 14*(4), 2264–2283.

76 Qian, Q., Zhang, J., Li, J., Li, Y., Jin, X., Zhu, Y., Liu, Y., Li, Z., El-Harairy, A., Xiao, C., & Zhang, G. (2021). Artificial heterointerfaces achieve delicate reaction kinetics towards hydrogen evolution and hydrazine oxidation catalysis. *Angew. Chem. 133*(11), 6049–6058.

77 Feng, Y., Guan, Y., Wen, L., & Liu, Y. (2024). Dual-independent active sites for efficient hydrogen production. *J. Mater. Chem. A 12*(31), 20300–20306.

78 Zhang, L., Li, M., Sun, C., Wang, H. T., Xiao, Y., Ma, K., Cai, Y., Lee, C. Y., Shao, Y. C., Wang, C. H., & Zhao, S. (2025). A hydrazine–water galvanic cell-inspired self-powered high-rate hydrogen production via overall hydrazine electrosplitting. *Adv. Funct. Mater. 35*(19), 2420163.

79 Burshtein, T. Y., Aias, D., Wang, J., Sananis, M., Farber, E. M., Gazit, O. M., Grinberg, I., & Eisenberg, D. (2021). Fe–N–C electrocatalysts in the oxygen and nitrogen cycles in alkaline media: The role of iron carbide. *Phys. Chem. Chem. Phys. 23*(47), 26674–26679.

80 Singh, D. & Exner, K. S. (2025). Atomic-scale insights into electrochemical hydrazine oxidation on Mo_2C MXene. *Energy Fuels 39*(28), 13801–13810

81 Ulas, B., Çetin, T., Kaya, Ş., Akinay, Y., & Kivrak, H. (2024). Novel $Ti_3C_2X_2$ MXene supported $BaMnO_3$ nanoparticles as hydrazine electrooxidation catalysts. *Int. J. Hydrog. Energy 58*, 726–736.

82 Tang, S., Zhang, Z., Xu, L., Qin, H., Dong, J., Lv, Q., Han, J., & Song, F. (2024). Ultrafine nickel-rhodium nanoparticles anchored on two-dimensional vanadium carbide for high performance hydrous hydrazine decomposition at mild conditions. *J. Colloid Interface Sci. 669*, 228–235.

83 Feng, Y., Shi, Q., Lin, J., Chai, E., Zhang, X., Liu, Z., Jiao, L., & Wang, Y. (2022). Decoupled electrochemical hydrazine "splitting" via a rechargeable Zn–hydrazine battery. *Adv. Mater. 34*(51), 2207747.

84 Liu, Y., Zhang, J., Li, Y., Qian, Q., Li, Z., & Zhang, G. (2021). Realizing the synergy of interface engineering and chemical substitution for Ni_3N enables its bifunctionality toward hydrazine oxidation assisted energy-saving hydrogen production. *Adv. Funct. Mater. 31*(35), 2103673.

85 Wang, Z., Xu, L., Huang, F., Qu, L., Li, J., Owusu, K. A., Liu, Z., Lin, Z., Xiang, B., Liu, X., & Zhao, K. (2019). Copper–nickel nitride nanosheets as efficient bifunctional catalysts for hydrazine-assisted electrolytic hydrogen production. *Adv. Funct. Mater. 9*(21), 1900390.

86 Zhang, J., Liu, Y., Li, J., Jin, X., Li, Y., Qian, Q., Wang, Y., El-Harairy, A., Li, Z., Zhu, Y., & Zhang, H. (2021). Vanadium substitution steering reaction kinetics acceleration for Ni_3N nanosheets endows exceptionally energy-saving hydrogen evolution coupled with hydrazine oxidation. *ACS Appl. Mater. Interfaces 13*(3), 3881–3890.

87 Gao, S., Li, H., Lu, Z., Meng, S., Zhao, X., Wang, X., Liu, X., & Hu, G. (2025). Isolated FeN_3 sites anchored hierarchical porous carbon nanoboxes for hydrazine-assisted rechargeable Zn-CO_2 batteries with ultralow charge voltage. *Carbon Energy* 7(1), e637.

88 Tang, P., Wen, H., & Wang, P. (2022). Hierarchically nanostructured Ni_2Fe_2N as an efficient electrocatalyst for hydrazine oxidation reaction. *Chem. Eng. J. 431*, 134123.

89 Ulas, B., Cetin, T., Topuz, M., & Akinay, Y. (2024). Ti_2NT_x MXene materials derived from Ti_2AlN MAX phases: Their characterization and electrocatalytic activity toward hydrazine electrooxidation. *Int. J. Hydrog. Energy 82*, 892–900.

10

Metal Alloy Electrocatalysts for Hydrazine-assisted Water Electrolysis

Shreyanka Shankar Naik

National Institute for Scientific Research–Energy, Materials, and Telecommunications Center, Varennes, QC, Canada

Corresponding author: shreyanka.shankar@inrs.ca

In recent years, hydrogen (H_2) has emerged as a clean energy carrier and alternative energy source due to its extremely high gravimetric energy density. So far, there are several traditional methods involved in the production of H_2, such as steam methane reforming (SMR), coal gasification, and partial oxidation of hydrocarbons. Nonetheless, these routes have significant environmental impacts. These processes are highly carbon-intensive, releasing large amounts of carbon dioxide, carbon monoxide, and other greenhouse gases that contribute to global warming. One of the proven alternatives to these conventional methods is electrolysis of water, which generates clean hydrogen with zero carbon emission. Therefore, most studies have focused on developing electrocatalysts for the cathode, while comparatively less attention has been given to anode materials. As discussed in the earlier chapters, the kinetically sluggish oxygen evolution reaction (OER) involves a four-electron-transfer process and requires a high thermodynamic potential of 1.23 V. Therefore, the key restraint for the conversion efficiency of water to hydrogen is that it is limited by the sluggish OER process. There have been investigations on several path-breaking strategies for hydrogen production to replace kinetically sluggish OER with oxidation of small organic molecules, such as alcohols, aldehydes, carbohydrates, urea, and hydrazine hydrate. Coupling of hydrogen evolution reaction (HER) with a kinetically favorable anodic reaction, with low theoretical potentials, is highly necessary to lower the overpotential. Among these alternatives, the hydrazine (N_2H_4) oxidation reaction (HzOR) draws attention due to its small theoretical potential (−0.33 V vs. RHE) and high power density. The success of achieving low overpotential mainly relies on the electrocatalyst material. Moreover, N_2H_4 is commonly used in the agrochemical and pharmaceutical industries. The chemical waste containing hydrazine is carcinogenic and toxic when it is released into the environment. Hence, it is advantageous to perform a hydrazine-assisted water-splitting reaction where a conversion of N_2H_4 results in the elimination of toxic chemicals, while generating clean hydrogen and inert nitrogen gas. It also saves more than 50% electricity and eliminates the hazardous H_2/O_2 process [1, 2].

Among the various known electrocatalysts, noble-metal- and transition-metal-based systems have demonstrated significant catalytic activity and efficiency. Alloying these metals offers several advantages, including the creation of additional active sites, improved charge-transfer properties, and high configurational entropy. Fundamentally, the concept of alloying aims to regulate the available active sites and the electronic structure of surface atoms, thereby optimizing the adsorption–desorption behavior of reactants. Therefore, the electrochemical performances of catalysts have shown significant improvement when alloyed, compared to their monometallic counterparts. In this chapter, we will discuss the impact of alloying and the role of individual elements in enhancing HzOR performance. So far, bimetallic, multimetallic alloys, and high-entropy alloys (HEAs) have proven effective toward the HzOR. During the evolution of hybrid structures, bimetallic alloys stood out as highly effective and promising materials in the field of catalysis.

As an example, a class of noble-metal-based AuPt alloys with various compositions were developed by Yu et al., using a unique pulsed laser technique. In this study, the authors managed to reduce the Au and Pt ions dissolved in the mixture of methanol–water into a face-centered-cubic (fcc) AuPt nanoalloy structure through laser irradiation, coupling it with a sonicator, which limits the coagulation of particles and allows the particles to grow uniformly. Hence, the authors have developed various combinations of AuPt alloys, namely, Au8Pt1, Au4Pt1, Au1Pt1, Au1Pt4, and Au1Pt8, respectively. Among these alloys, Au1Pt8 has outperformed other electrocatalysts during HER, due to the availability of a large concentration of Pt atomic percentage, which promotes HER at the lowest overpotential of 0.26 V to reach 10 $mA\cdot cm^{-2}$. Moreover, AuPt alloys show a high chemisorption tendency, particularly between the alloys and the lone pair of nitrogen from the hydrazine structure. Thereby, a coordination bond is established that directs a charge transfer between the nitrogen and empty orbitals of the Pt atom. Thus, HzOR was performed using 0.5 M N_2H_4 in 1 M KOH solution. It can be seen that Au1Pt8 alloy substantially improved HzOR, allowing the smallest overpotential of 0.50 V to attain 10 $mA\cdot cm^{-2}$. This study revealed that the appropriate proportion of Au:Pt alloy efficiently optimizes catalytic reaction energy barriers, as a result, these catalysts can be employed as a trifunctional material (HzOR, HER, and OER). Furthermore, when Au1Pt8 alloys were used as anodic and cathodic electrodes (Au1Pt8/C||Au1Pt8/C) in an electrolyzer setup, the overall cell potential of the reaction significantly reduced to 0.172 V to deliver 10 $mA\cdot cm^{-2}$, which is substantially lower than 1.773 V required by the electrolysis of water [3].

Similarly, Fu et al. synthesized wavy $RhRu_{0.5}$ alloy nanowires with fcc crystal phase via a solvothermal/polyol method [4]. Among the well-established noble metals, Rh and Ru have the lowest half-wave potential for HzOR. In addition, Ru typically exhibits the lowest onset for HzOR but is unable to achieve the complete electrooxidation reaction. Therefore, alloying Ru with Rh-based electrocatalysts provides synergistic benefits to lower potential while simultaneously achieving total oxidation of N_2H_4. The authors have demonstrated that ultrafine wavy $RhRu_{0.5}$ alloy has outperformed bare Rh nanowires and state-of-the-art Pt/C toward HER and HzOR. During

HER, $RhRu_{0.5}$-alloy nanowires exhibited a mass activity of 1.89 $A{\cdot}mg^{-1}$, which outperforms the Rh nanowires (1.11 $A{\cdot}mg^{-1}$) and Pt/GC (0.57 $A{\cdot}mg^{-1}$) at −0.07 V vs. RHE. DFT calculations of the projected density of states (PDOS) showed that the d-band center of Rh atoms in the $RhRu_{0.5}$ alloy is downshifted from the Fermi level by 0.26 eV relative to bare Rh. This shift weakens the interaction between the catalyst surface and adsorbates, promoting N_2 desorption during the HzOR process and thereby enhancing catalytic activity. Further, $RhRu_{0.5}$ alloy was used as a cathode and an anode material in a hydrazine electrolyzer cell. At a very low mass loading of 0.2 $mg{\cdot}cm^{-2}$, $RhRu_{0.5}$ electrodes generate a high current density of 100 $mA{\cdot}cm^{-2}$ involving an ultralow voltage of 54 mV. Finally, this study contemplated the pathway to convert hazardous hydrazine into eco-friendly products [5].

So far, we have discussed the role of noble metals and their performances in the complex oxidation of N_2H_4. However, their limited availability and high cost make them less suitable as electrocatalysts for a sustainable future. Nonetheless, we cannot ignore the fact that they deliver high current densities with minimal catalyst loading at lower overpotentials. Therefore, Ao et al. fabricated a carbon nanofiber (CNF) embedded with PdCo alloy NPs using the carbonization method following the electrospinning technique. The morphological analysis technique revealed that the PdCo NPs have been uniformly distributed over the surface of CNF. Through electrochemical reactions, it has been observed that the Pd_1Co_1—CNFs catalyst showed an extraordinary activity toward the HER process. Moreover, it could reach a current density of 10 $mA{\cdot}cm^{-2}$ at 258 mV, which is significantly better than Pd_1Co_2—CNF, Pd_2Co_1—CNF, Pd—CNF, and Co—CNF catalysts. Further, the catalytic performances were investigated for HzOR replacing the well-known OER in a mixture of 3 M KOH and 0.2 M hydrazine solution. Surprisingly, the current response quickly increased in the presence of hydrazine, which clearly shows the selectivity of the catalyst for hydrazine. During this process, again Pd_1Co_1—CNFs catalyst exhibits a small working potential of 0.258 V to drive a current density of 50 $mA{\cdot}cm^{-2}$. For establishment in commercial applications, Pd_1Co_1—CNF has the potential to deliver a current density of 200 $mA{\cdot}cm^{-2}$ at 0.363 V. Moreover, Pd_1Co_1—CNF has been identified as a bifunctional catalyst. Hence, the overall water splitting (OWS) reaction was performed using two-electrode setup (Pd_1Co_1—CNF||Pd_1Co_1—CNF) and it requires large cell potential of 1.936 V to drive a current density of 10 $mA{\cdot}cm^{-2}$, while the overall hydrazine-splitting (OHzS) device assembled with Pd_1Co_1—CNF electrodes only delivers a small voltage of 0.440 V. The superior activity of Pd_1Co_1—CNFs electrode is due to the synergistic effect caused by the Pd and Co that promotes the redistribution of charges between their interfaces, which can tune their atomic electronic structure and improve intermediate adsorption during HER and HzOR. Also, the unique 1D CNF with encapsulated PdCo alloy provides strong structural stability, exhibits large electrochemically available active sites for the reaction, and promotes better electron transfer. These characteristics of the hybrid structure contribute significantly to the high electrochemical performance of PdCo—CNFs electrode, endowing it with promising application in energy-saving hydrogen generation [6]. Thereafter, Jeong et al. synthesized a transition-metal–noble-metal bimetallic CoPd alloy nanoparticles using a novel pulsed laser ablation (PLA) technique. Herein, the Co target was placed in a methanolic solution containing Pd^{2+} ions.

During the PLA of the Co target, metal ions are generated from the Co surface, and the free Co^{2+} ions interact with the nearby Pd^{2+}, and they gradually nucleate and grow into a CoPd alloy structure. The composition ratio of the Co:Pd atoms in an alloy is controlled by the Pd ion concentration in the solution. XRD analysis revealed that CoPd alloy has a cubic fcc phase. From the ICP–OES, the elemental atomic mass proportions in the alloy samples were estimated, and the results showed that the ratios of Co:Pd concentration in the alloy structures are Co7Pd3, Co6Pd4, Co2Pd8, and Co1Pd9. The electrocatalytic performances of these electrodes were also compared with the bare metallic Co and Pd nanoparticles. It is interesting to note that Co1Pd9 exhibits better performance with the smallest overpotentials (0.22 V at 10 $mA{\cdot}cm^{-2}$) among all the alloy samples, and it outperforms even the bare Pd sample (0.32 V at 10 $mA{\cdot}cm^{-2}$). Therefore, a clear positive synergetic effect was observed in the Co1Pd9 alloy structure during HER. Also, the HER kinetics study revealed that Co1Pd9 exhibits the smallest Tafel slope value of 128 $mV{\cdot}dec^{-1}$, following a two-electron–two-proton Volmer–Heyrovsky mechanism. Furthermore, the authors have replaced a kinetically sluggish OER and performed HzOR in 1 M KOH/0.5 M N_2H_4 solution. The proposed general mechanism of HzOR over a metal catalyst in alkaline medium is shown in Eqs. (10.1) and (10.2):

$$N_2H_4 + 4OH^- \longrightarrow N_2 + 4H_2O + 4e^- \tag{10.1}$$

$$N_2H_4 + xOH^- \longrightarrow N_2 + (4-x)/2H_2 + xH_2O + xe^- \quad -1 \le x \le 3 \tag{10.2}$$

Authors have shown that the polarization curves obtained for the HzOR over the Co1Pd9 alloy exhibit superior activity with an ultralow potential of −0.1 mV to reach 10 $mA{\cdot}cm^{-2}$, which is closer to the theoretical potential of HzOR. The synthesized Co1Pd9 alloy combination has remarkable HzOR activity and exceeds those of many recently reported electrocatalysts. Also, Co1Pd9 has a large TOF (s^{-1}), followed by Co2Pd8 > Co6Pd4 > Co and Co7Pd3. Finally, the enhancement in the activities of Co1Pd9 might be due to faster electron mobility between Co and Pd and the strain effect. When the OHzS electrolyzer was established using di-electrode Co1Pd9||Co1Pd9, the cell potential drastically lowered from 1.837 V of OWS to 0.228 V (OHzS). Hence, a significant drop in the cell potential could be due to the effective N_2H_4 chemisorption facilitated by a dative bond between the lone pair of electrons of N and the empty 5s-orbital of Pd and 3d-orbital of low-spin Co [7, 8]. Further, N—H bond dissociation and radical formation during the oxidation process produce N_2 and generate charges, which facilitate hydrogen production [9]. Likewise, Jeong et al. have successfully developed CuPd alloy nanostructures using a similar pulsed laser technique as discussed previously. Among the various alloy compositions, Cu1Pd3/C demonstrates remarkable HER performance with a low overpotential of 0.315 V at 10 $mA{\cdot}cm^{-2}$. Also, the smallest Tafel slope of 45.3 $mV{\cdot}dec^{-1}$ in 1 M KOH electrolyte shows the favorable reaction kinetics toward hydrogen production. Moreover, Cu1Pd3/C demonstrates an ultralow overpotential of 0.560 V at 10 $mA{\cdot}cm^{-2}$ during anodic oxidation of hydrazine. Finally, the constructed Cu1Pd3/C||Cu1Pd3/C electrolyzer generates a cell bias of 0.505 V at 10 $mA{\cdot}cm^{-2}$ for OHzS with long stability of 5 hours, prevailing over the OWS significantly (1.960 V) [10].

Another approach to enhance the activity of the electrocatalyst is through modulating the electronic structure by changing the widely accepted d–d hybridization in d-block elements to p–d orbital hybridization (between p- and d-block elements). Therefore, Jin and coworkers synthesized a novel amorphous RhPb nanoflower (a-RhPb NF) alloy structure through one-pot hydrothermal reaction. The amorphous nanoflower-like structure of the sample was achieved by the CO produced by the decomposition of formaldehyde at high temperature, which acts as a structure-directing agent and reducing agent. In addition, the authors developed crystalline RhPb alloys using the same reaction conditions but with an elevated temperature to investigate the influence of crystallinity on electrochemical performances. The HER performances of the a-RhPb NFs and c-RhPb NFs were investigated in a three-electrode system under 1 M KOH solution. The obtained overpotential of a-RhPb NFs is −36 mV, which is much smaller than c-RhPb NFs (−47 mV), Rh NFs (−58 mV), and Pt/C (−51 mV) at 10 $mA{\cdot}cm^{-2}$. The authors further investigated the HzOR performance of the electrocatalyst in 1 M KOH with 0.1, 0.3, and 0.5 M concentrations of N_2H_4. Surprisingly, the anodic current density has a prodigious increment after the addition of hydrazine solutions, indicating that a-RhPb NFs are active and sensitive for HzOR. Among a-RhPb NFs, c-RhPb NFs, Rh NFs, and Pt/C in 1 M KOH/0.1 M N_2H_4 electrolyte, the a-RhPb NFs demonstrated extremely low potentials of 0.018 mV to attain oxidative current densities of 10 $mA{\cdot}cm^{-2}$. The better performance of a-RhPb is due to the unique 3D nanoflower structure that provides a larger specific surface area and abundant active sites. In order to verify the performance of a-RhPb NFs in a two-electrode hydrazine electrolyzer system for OHzS, the authors fabricated anode and cathode electrodes with a catalyst loading of 0.4 $mg{\cdot}cm^{-2}$. Further, LSV was performed in the presence of hydrazine, requiring ultrasmall voltages of 0.095 and 0.321 V to achieve 10 and 100 $mA{\cdot}cm^{-2}$, respectively, which are much lower than those for the water-splitting system (1.535 and 1.764 V). Later, they prepared a unique RuSb nanobranch with p–d-orbital hybridization using the hydrothermal method. The authors examined their efficiency for HER and HzOR in alkaline seawater electrolytic conditions. Surprisingly, $Ru_{50}Sb_{50}$ can attain a current density of 10 $mA{\cdot}cm^{-2}$ at low overpotentials of 39 and 252 mV for HER and HzOR, respectively. Due to their impressive redox performance, an OHzS electrolyzer was designed with RuSb||RuSb as the cathode||anode couple electrodes. During the OHzS reaction, an exceptionally low cell voltage of 35 mV was involved to drive 10 $mA{\cdot}cm^{-2}$, which outperformed the Pt/C||Pt/C couple in the same two-electrode electrolyzer. Lastly, DFT calculations unveiled that Sb doping majorly helps in hydrogen adsorption more thermoneutrally and also lowers the energy barrier for dehydrogenation from $^*N_2H_3$ to $^*N_2H_2$. Hence, the RuSb catalyst can be considered as a model candidate to explore the other electrocatalysts for OHzS tuned by p-block metals or metalloids [11].

In the previous discussion, we have clearly observed a combined synergetic effect in tuning the electronic structure in bimetallic alloys toward HzOR-assisted hydrogen production when using platinum group metals (PGMs). To overcome the dependency on PGMs, the involvement of non-precious metals in designing alloys is an ultimate pathway to improve catalytic performance and reduce electrolysis costs.

For instance, hydrogen evolution assisted with a wet-chemical reaction was performed to synthesize trimetallic PdNiRu alloys with nanochain-like network (NCN) structures. It has been observed that NCN architectures possess large ECSA ($84.32\ m^2 \cdot g_{Pd}^{-1}$), which greatly promotes the interfacial catalytic reaction toward HzOR. Moreover, it outperformed PdNi NPs, PdRu NPs, NiRu NPs, and Pd black catalysts. The CV curves revealed that the PdNiRu can generate the maximum anodic current density of 30.86 $mA{\cdot}cm^{-2}$ in the presence of 0.5 M KOH and 10 mM hydrazine solution, which is slightly higher than those of PdNi NPs (30.84 $mA{\cdot}cm^{-2}$), PdRu NPs (6.96 $mA{\cdot}cm^{-2}$), NiRu NPs (1.02 $mA{\cdot}cm^{-2}$), and Pd black (6.66 $mA{\cdot}cm^{-2}$). These results revealed that the NCN structure of PdNiRu alloys with a large number of active sites assists the enhanced catalytic oxidation of hydrazine molecules. Overall, hydrazine can be easily oxidized to N_2 without forming any toxic intermediate species and thereby outperforms alcohol-based oxidation reactions as a clean energy alternative [12].

Recently, HEAs have emerged as a new class of alloys and have gained huge attention in the field of catalysis. They are monophasic solid compounds containing multiple primary metallic elements. They exhibit unique physicochemical characteristics, such as the ability to tune the electronic properties [13] and d-band centers [14]. Nonetheless, the superlative performance of the HEAs has not been sufficiently investigated. In catalysis, HEAs, which possess unique and tunable properties, showed extraordinary stability and performance in various applications, especially during hydrazine-assisted water electrolysis reactions. So far, varieties of HEAs have been developed and employed in various applications. Katiyar et al. have developed a highly stable AgAuPtPdCu alloy nanocatalyst structure comprising five elements for the oxidation of hydrazine molecules. Interestingly, the preparation of HEAs involved casting-cum-cryomilling method, which is unique in nature. During the HzOR reaction, it was observed that AgAuPtPdCu alloy NPs showed excellent activity by producing N_2 and O_2, lowering the overpotential even in buffer medium. Moreover, production of oxygen at 0.3 V is due to the OER, which is activated at potentials as low as 0.3 V (Ag/AgCl) or 1.13 V (RHE), closer to the thermodynamic potential range. A linear current response during the addition of hydrazine suggests that the hydrazine oxidation process is controlled by the diffusion process or mass-transfer kinetics. Theoretical calculation using DFT revealed a favorable HzOR with a parallel endothermicity of 0.65 eV at the Pt—Pt sites. Thus, it could be reduced by the assistance of the oxygen evolution process, where the electron-donating effect of H_2O stabilizes the reactant $^*NHNH_2$ less than the product *NH—NH in the rate-determining elementary step. Also, the electron acceptance of $^*NHNH_2$ from the surface decreases the energy gap between the Fermi level of HEA and the highest occupied molecular orbital (HOMO) of H_2O that helps in the favorable OER [15]. In a similar work by Feng et al., an idea was proposed to develop high-entropy alloy nanoclusters (HEANCs) having a mixture of transition-metal and noble-metal (Pt:Ru:Ni:Co:Mo) elements was proposed, delivering fully active sites and super-active sites for the electrochemical reactions. The prepared single-phasic HEANCs have an average particle size of 1.48 nm with only seven atomic layers and are distributed uniformly over carbon supports. The selected combination of elements in HEA is ideal for both HER and HzOR reactions. During HER, HEANC/C electrocatalyst requires only small overpotentials of 9.5 mV to achieve a current density of 10 $mA{\cdot}cm^{-2}$, while commercial Pt/C requires 32 mV. Also, HEANC/C exhibits mass activity of 12.85 $A{\cdot}mg^{-1}$ Pt + Ru, which is 17.36-fold that of commercial Pt/C (0.74 $A{\cdot}mg^{-1}$ Pt). Similarly, HEANC/C has been ultra-active during the oxidation of 0.1 M hydrazine in 1 M KOH solution. The performance of the catalyst was estimated by the catalyst's loading. Here, the obtained mass activity of HEANC/C is 3.26 $A{\cdot}mg^{-1}$ at 0.05 V vs. RHE, which is 13.04 times larger than Pt/C (0.25 $A{\cdot}mg^{-1}$). Lastly, when considering HEANC/C as both anode and cathode electrocatalysts for an OHzS process, the HEANC/C catalyst requires ultralow cell voltages of 0.025 and 0.181 V to reach 10 and 100 $mA{\cdot}cm^{-2}$, respectively, significantly outperforming the potentials of OWS. The authors postulated, through theoretical calculations, that the ultralow working potential of the OHzS reaction is endorsed by the hyperactive sites of all the elements in the alloy structure. This work appeared to be a pathway for developing a bifunctional HEA catalyst for other electrooxidation and reduction reactions. This work not only highlighted HEANCs as sufficiently effective bifunctional catalysts in OHzS systems but also supplied a guiding line in the design of high-activity catalysts in energy-related fields [16].

In conclusion, the electrooxidation of N_2H_4 is a highly anticipated reaction mainly in the field of fuel cells and electrolyzers. Despite being a multi-proton and multi-electron process, HzOR has been significantly efficient in lowering the overall operational potential of the reaction, in turn assisting hydrogen evolution more efficiently. It has been proven that alloy structures have the ability to efficiently oxidize hydrazine during electrochemical reaction under alkaline medium. However, a clear correlation between the superficial structure and molecular interactions is still unclear in the multimetallic alloy structure. Hence, there is still room for the investigation of surface behavior and partial electronic states of alloy states. So far, most of the research has been focused on PGMs, however, cost and abundance are always key factors while developing a catalyst for industrial-scale production. Therefore, a combination of non-PGM alloy structures with PGM single-atom-doped catalysts can be an alternative technique for the development of low-cost catalysts with high efficiency, which has been least investigated currently.

References

1 Liu, Y., Zhang, J., Li, Y., Qian, Q., Li, Z., Zhu, Y., & Zhang, G. (2020). Manipulating dehydrogenation kinetics through dual-doping Co_3N electrode enables highly efficient hydrazine oxidation assisting self-powered H_2 production. *Nat. Commun.* *11*(1), 1853.

2 Wang, T., Cao, X., & Jiao, L. (2022). Progress in hydrogen production coupled with electrochemical oxidation of small molecules. *Angew. Chem.* *134*(51), e202213328.

3 Yu, Y., Lee, S. J., Theerthagiri, J., Lee, Y., & Choi, M. Y. (2022). Architecting the AuPt alloys for hydrazine oxidation as an anolyte in fuel cell: Comparative

analysis of hydrazine splitting and water splitting for energy-saving H_2 generation. *Appl. Catal. B: Environ. 316*, 121603. https://doi.org/10.1016/j.apcatb.2022.121603

4 Fu, X., Zhao, Z., Wan, C., Wang, Y., Fan, Z., Song, F., Cao, B., Li, M., Xue, W., Huang, Y., & Duan, X. (2019). Ultrathin wavy Rh nanowires as highly effective electrocatalysts for methanol oxidation reaction with ultrahigh ECSA. *Nano Res. 12*(1), 211–215. https://doi.org/10.1007/s12274-018-2204-8

5 Fu, X., Cheng, D., Wan, C., Kumari, S., Zhang, H., Zhang, A., Huyan, H., Zhou, J., Ren, H., & Wang, S. (2023). Bifunctional ultrathin $RhRu_{0.5}$-alloy nanowire electrocatalysts for hydrazine-assisted water splitting. *Adv. Mater. 35*(23), 2301533.

6 Ao, Y., Chen, S., Wang, C., & Lu, X. (2021). Palladium cobalt alloy encapsulated in carbon nanofibers as bifunctional electrocatalyst for high-efficiency overall hydrazine splitting. *J. Colloid Interface Sci. 601*, 495–504. https://doi.org/10.1016/j.jcis.2021.05.119

7 Miao, R. & Compton, R. G. (2021). Mechanism of hydrazine oxidation at palladium electrodes: Long-lived radical di-cation formation. *Electrochim. Acta 388*, 138655. https://doi.org/10.1016/j.electacta.2021.138655

8 Miao, R. & Compton, R. G. (2021). The electro-oxidation of hydrazine: A self-inhibiting reaction. *J. Phys. Chem. Lett. 12*(6), 1601–1605. https://doi.org/10.1021/acs.jpclett.1c00070

9 Jeong, Y., Naik, S. S., Theerthagiri, J., Moon, C. J., Min, A., Aruna Kumari, M. L., & Choi, M. Y. (2023). Manifolding surface sites of compositional CoPd alloys via pulsed laser for hydrazine oxidation-assisted energy-saving electrolyzer: Activity origin and mechanism discovery. *Chem. Eng. J. 470*, 144034. https://doi.org/10.1016/j.cej.2023.144034

10 Jeong, Y., Naik, S. S., Yu, Y., Theerthagiri, J., Lee, S. J., Show, P. L., Choi, H. C., & Choi, M. Y. (2023). Ligand-free monophasic CuPd alloys endow boosted reaction kinetics toward energy-efficient hydrogen fuel production paired with hydrazine oxidation. *J. Mater. Sci. Technol. 143*, 20–29. https://doi.org/10.1016/j.jmst.2022.09.043

11 Liu, X., Wang, T., Chen, Y., Wang, J., Xie, W., Wu, R., Xu, X., Pang, L., Zhang, X., Lv, Y., Wang, G., Yamauchi, Y., & Jin, T. (2023). Distinctive p-d orbital hybridization in RuSb nanobranches for simultaneously enhanced hydrogen evolution and hydrazine oxidation in alkaline seawater. *Appl. Catal. B: Environ. 333*, 122771. https://doi.org/10.1016/j.apcatb.2023.122771

12 Yuan, T., Wang, A., Fang, K., Wang, Z., & Feng, J. (2017). Hydrogen evolution-assisted one-pot aqueous synthesis of hierarchical trimetallic PdNiRu nanochains for hydrazine oxidation reaction. *J. Energy Chem. 26*(6), 1231–1237. https://doi.org/10.1016/j.jechem.2017.08.003

13 Feng, G., Ning, F., Song, J., Shang, H., Zhang, K., Ding, Z., Gao, P., Chu, W., & Xia, D. (2021). Sub-2 nm ultrasmall high-entropy alloy nanoparticles for extremely superior electrocatalytic hydrogen evolution. *J. Am. Chem. Soc. 143*(41), 17117–17127.

14 Feng, G., Ning, F., Pan, Y., Chen, T., Song, J., Wang, Y., Zou, R., Su, D., & Xia, D. (2023). Engineering structurally ordered high-entropy intermetallic nanoparticles with high-activity facets for oxygen reduction in practical fuel cells. *J. Am. Chem. Soc.* *145*(20), 11140–11150.

15 Katiyar, N. K., Dhakar, S., Parui, A., Gakhad, P., Singh, A. K., Biswas, K., Tiwary, C. S., & Sharma, S. (2021). Electrooxidation of hydrazine utilizing high-entropy alloys: Assisting the oxygen evolution reaction at the thermodynamic voltage. *ACS Catal.* *11*(22), 14000–14007. https://doi.org/10.1021/acscatal.1c03571

16 Feng, G., Pan, Y., Su, D., & Xia, D. (2024). Constructing fully-active and ultra-active sites in high-entropy alloy nanoclusters for hydrazine oxidation-assisted electrolytic hydrogen production. *Adv. Mater.* *36*(13), 2309715. https://doi.org/10.1002/adma.202309715

11

Metal Borides and Selenides for Hydrazine-assisted Water Electrolysis

Soumya Ranjan Mishra[1,], Shalu Rawat[1], Vishwajit Chavda[2], and B. M. Nagaraja[1,*]*

[1]*Centre for Nano and Material Sciences, Jain University, Jain Global Campus, Bengaluru, India*
[2]*Department of Chemistry, Institute of Science, Humanities and Liberal Studies, Indus University, Ahmedabad, India*

*Corresponding authors: soumyamikan1998@gmail.com; bm.nagaraja@jainuniversity.ac.in

11.1 Introduction

For the production of hydrogen (H_2), hydrazine-assisted water electrolysis (Hz-WE) has been a suitable option because of its low energy consumption compared to conventional water electrolysis [1, 2]. The higher energy usage in the case of the traditional method is due to its sluggish kinetics, high overpotential requirement, and high oxidation potential (+1.23 V vs. SHE) for the oxygen evolution reaction (OER) to occur at the anode [3]. However, in the case of Hz-WE, the cell voltage essential for electrolysis is significantly reduced as the hydrazine oxidation reaction (HzOR) at the anode requires a lower thermodynamic oxidation potential of −0.33 V vs. SHE [4, 5]. This also leads to synergistic H_2 generation at the cathode, decreasing the operational costs and enhancing the system integration along with overall efficiency. In comparison to OER, HzOR is less energy-intensive and simpler, as it is a fast process requiring less overpotential [6]. The decrease in total energy demand has resulted in a shift to HzOR from OER as observed in recent studies, leading to sustainable and scalable hydrogen production [7].

The fabrication of a cost-effective, efficient, and robust electrocatalyst is an important aspect to determine the full potential of Hz-WE [8]. Precious metals like Pt, Ru, and Ir have dominated the field of water electrolysis because of their high catalytic activity; however, the high cost and scarcity have limited the research on a large scale [9]. Hence, recent studies have shifted their attention to non-precious catalysts as intense research is being carried out on an emerging class of electrocatalysts, such as metal borides and selenides [10]. Due to their high chemical stability, electrical conductivity, and tunable electronic structures, metal borides (such as MoB_2, Co–Ni–B, and Ni_2B) are considered to be excellent candidates for Hz-WE. Metal borides also show consistent performance even in a wide pH range due to rapid charge-transfer characteristics. A study carried out by Lee and Fokwa [11] has reported that adding boron to molybdenum borides improved the hydrogen evolution reaction (HER) activity. In some cases, specific boride phases even competed with the performance of Pt catalysts. Moreover, due to their high electron densities at the Fermi level, metal-rich diboride nanomaterials have shown excellent HER activity. Along with metal borides, metal selenides (such as $NiSe_2$, CoSe, and FeSe) show excellent stability and catalytic activity for OER and HER in alkaline medium [12]. The high performance of selenides is attributed to their structural and compositional tunability, which helps in optimizing the catalytic sites and enhancing the intrinsic activity. Masud et al. [13] developed a molecular nickel complex with a metal selenide core, which achieved excellent HER and OER results with low overpotentials of 310 and 200 mV, respectively, at 10 $mA\cdot cm^{-2}$.

Moreover, there has also been significant progress in other classes of non-precious catalysts, such as metal sulfides, phosphides, carbides, nitrides, and alloys [14–16]. Despite the advances, these metal classes show significant scientific gaps, such as poor conductivity, surface oxidation, corrosion, complex synthesis methods, and durability. The metal borides and selenides cover these gaps, and due to their excellent conductivity, intrinsic catalytic activity, and stability, they show superior Hz-WE activity compared to the above classes of non-precious catalysts. The ability to tailor the active sites via compositional engineering and the unique electronic structure enhances the potential for Hz-WE, as the HER and HzOR are efficiently catalyzed at minimal overpotentials [17]. This chapter focuses on the structure, activity relationship, and catalytic activity of metal borides and selenides for Hz-WE. The fundamentals and mechanistic insights of Hz-WE will also be elucidated. By critically analyzing the latest research, the chapter will identify future directions for the development of next-generation non-precious electrocatalysts for sustainable hydrogen production.

11.2 Fundamentals of Hz-WE

Compared to conventional water electrocatalysis, Hz-WE is an advanced process as it significantly reduces the energy input required for H_2 generation [18]. In addition, Hz-WE shows better thermodynamic and kinetic properties as it replaces the OER at the anode with HzOR (Eq. 11.1), while the cathode exhibits the HER (Eq. 11.2) [19]. The reactions presented below generally occur in the alkaline media:

$$N_2H_4 + 4OH^- \rightarrow N_2 + 4H_2O + 4e^- \quad (11.1)$$

$$2H_2O + 2e^- \rightarrow H_2 + 2OH^- \quad (11.2)$$

The overall cell reaction can be represented as shown in Eq. (11.3),

$$N_2H_4 + 2H_2O \rightarrow N_2 + 2H_2 + 2OH^- \quad (11.3)$$

Moreover, the thermodynamic benefits of the hydrazine-assisted process are higher than those of traditional electrocatalysis. The thermodynamic oxidation potential of HzOR is −0.33 V vs. SHE, which is much lower compared to +1.23 V vs. SHE of OER. It introduces a novel electrochemical setup for hydrazine-assisted hydrogen production, combining HER/HzOR, referred to as overall hydrazine splitting (OHzS). This approach holds important potential for advancing sustainable energy storage and conversion technologies (Figure 11.1) [20]. This can be observed in a recent work carried out by Hu et al. [21], where hydrogen generation was observed at a very low cell voltage of 0.34 V at 10 $mA{\cdot}cm^{-2}$. Sun et al. [22] also claimed that the lower voltage requirement of HzOR makes it 48% more energy efficient compared to commercial OER. Additionally, the kinetics of HzOR is more favorable than OER as hydrazine (N_2H_4) undergoes facile dehydrogenation due to its four available protons, which leads to faster electron transfer and maximum current densities at low overpotential. Furthermore, computational and in-situ spectroscopic studies done by Zhu et al. [23] revealed that an optimized catalyst surface eases the electron transfer by HzOR, thus decreasing the adsorption energy barriers for both hydrogen and hydrazine intermediates. Katiyar et al. [24] conducted density functional theory (DFT) calculations and observed that the electron-accepting nature of hydrazine intermediates enhances favorable orbital overlap with the catalyst Fermi level, whereas the donating effect of water molecules to the catalyst surface reduces the endothermicity of HzOR.

Although the benefits of hydrazine in enhancing the kinetics of water electrolysis reactions and saving energy are quite substantial, the toxicity and ecological

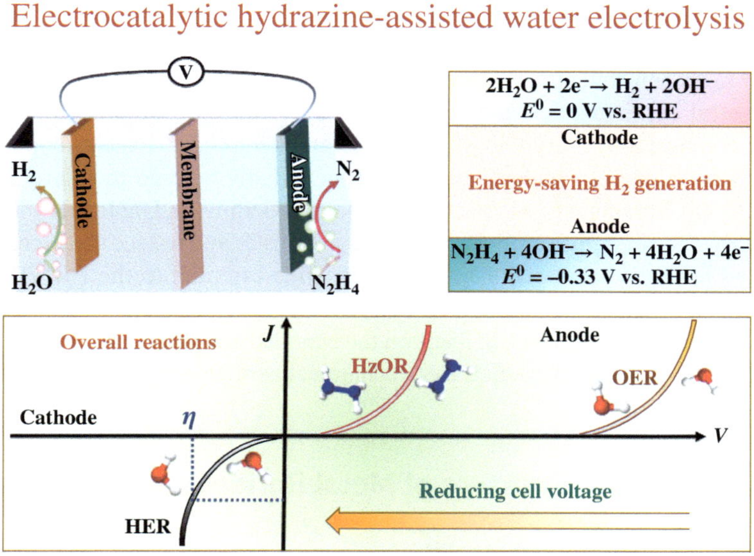

Figure 11.1 Graphical illustration of the hydrazine-assisted water electrocatalysis process. *Source*: [20]/with permission of The Royal Society of Chemistry.

concerns connected with hydrazine must be dealt with. Hydrazine's inhalation, ingestion, or dermal contact makes it one of the most toxic and probable cancer-causing substances with great health effects [25]. Central nervous system, liver, and kidney damage are some of the issues to be dealt with from chronic exposure to it. Agencies such as the US Environmental Protection Agency (EPA) and European Chemicals Agency (ECHA) are dealing with these issues as it is considered a hazardous air pollutant. Because of its volatile nature paired with acute toxic behaviors, occupational exposure limits have to be set far below the safe levels, such as 0.01–0.03 ppm [26].

HzOR can utilize hydrazine as an anodic fuel while simultaneously removing it electrochemically from wastewater streams. Through the process of HzOR, hydrazine is completely oxidized to water and nitrogen gas, which is harmless to the environment, for removal, and aids in the generation of hydrogen fuel [17]. With such attributes, Hz-WE technology has become fascinating for sustainable energy generation from hydrazine-contaminated wastewater remnants found in aerospace, agriculture, and pharmaceutical industries [27]. The effectiveness of hydrazine removal depends a lot on how active and selective the catalyst is; this is because partial oxidation can lead to the creation of dangerous intermediates like ammonia or nitrogen oxides [28]. So, it is important to make strong, selective catalysts that will not only get the most hydrogen out of hydrazine but also make sure that it is completely detoxified.

Practical application of energy-efficient Hz-WE for hydrogen production hinges on developing bifunctional catalysts that can effortlessly facilitate HzOR at the anode and HER at the cathode with minimal overpotentials [29, 30]. Such catalysts must have high electrical conductivity for rapid charge transport, defined active sites to stabilize and activate hydrazine for solid-state electrolysis, and effective proton reduction for hydrogen generation. Moreover, long-lasting durability requires resistance to chemical destabilization and deactivation under alkaline conditions [31]. The catalyst composition, morphology, electronic architecture, and their cooperative function often need to achieve such multifunctionality tailored to a specific design [32]. Recent work has shown that certain heterostructures and materials with abundant interfaces can significantly improve charge transfer and reduce the energy barrier needed for both half-reactions. This approach is essential to the Hz-WE's thermodynamic and kinetic advantages to take full effect [33]. Based on the above discussion, the upcoming sections will focus on the chemistry and structure of metal borides and selenides and analyze their electrochemical performance for Hz-WE.

11.3 Chemistry and Structure of Metal Borides

11.3.1 Structural Features and Stability

Metal borides include both transition-metal-rich phases and boron-rich phases, and are known to have great structural or electronic diversity. NiB, CoB, FeB, MoB, and

TiB_2 are some of the most common metal borides and have different crystal chemistries [34]. Boron-rich crystalline phases like TiB_2 and MoB tend to have highly ordered structures where boron atoms create covalent networks in the form of chains, sheets, or three-dimensional structures embedded in a metal matrix [35]. These frameworks exhibit remarkable mechanical hardness, high melting points, exceptional resistance to chemical breakdown, and are stabilized by strong covalent B—B bonds and metallic M—B interactions [36].

Amorphous borides, on the other hand, can be classified as low-temperature-synthesized, under-catalyzed, or unstructured products lacking long-range order while still retaining some metal–boron coordination [11]. Such materials manifest a large density of active sites and a large surface area, beneficial for electrochemical catalysis. Moreover, the exceptional charge mobility in both crystalline and amorphous borides is due to the hybridization of metallic and covalent bonding, leading to delocalized electron density. This results in rapid electron transfer with minimal resistive loss during operation, which is very advantageous for electrocatalysis [37]. In addition, in the presence of strong acids and bases, the robust metal–boron framework helps to maintain the structural integrity and does not affect the catalytic activity [38].

11.3.2 Synthesis of Compositional Control

There are currently a number of physical and chemical techniques that go into developing metal borides. Each one has its advantages for controlling particle form, surface properties, and phase [39]. Among them, the chemical reduction approach is the widely used method as it is quite effective in fabricating nanocrystalline and amorphous borides such as NiB and CoB. In this method, the metal salts are reduced with the help of sodium borohydride ($NaBH_4$) or potassium borohydride (KBH_4) in aqueous or organic solvents, which leads to the formation of high-surface-area and tunable-particle-size borides. The morphology and surface functionalization of borides via this method are dependent on a few reaction parameters such as reducing agent concentration, surfactants, and temperature. Moreover, the electrodeposition method is also used to allow the direct development of boride films or nanostructures on conductive substrates. This method is very useful in developing core–shell structures and other specific morphologies, as accurate control can be gained on the microstructure, thickness, and composition by tuning the bath composition and deposition potential.

For the fabrication of high-crystalline borides such as TiB_2 and FeB, solid-state high-temperature synthesis is often carried out, which includes a conventional ceramic pathway and self-propagating solid-state metathesis (SSM) reactions [40]. These SSM reactions are important for the scalable production of borides as the process is rapid, highly energy efficient, and gives high yield and excellent phase purity. Meanwhile, the fabrication of well-defined nanostructures at relatively low temperatures is carried out by a hybrid technique that combines ultrasonication with hydrothermal processing. Defect densities within boride nanorods and nanoparticles, as well as core–shell structures, enable their synthesis through these methods. The possibility to optimize the geometry is very important because nanostructuring can improve the catalytic efficiency by augmenting the active surface area and the utilization of the catalytic sites. Furthermore, surface modification either during the synthesis of the material or after can increase the dispersibility and the chemical stability of the material, including active sites, thus increasing the overall efficiency of the catalyst.

From these considerations, it is evident that the distinct crystal chemistry, electronic structure of metal borides, along with the recent advances in synthesis techniques, explain their rapid adoption as reliable and high-performance electrochemical catalysts [41]. The possibility to design boride materials with custom phase compositions, morphology, and surface characteristics enhances the strategy toward optimizing their catalytic activities, particularly in the newly developed applications like Hz-WE [35].

11.3.3 Electrocatalytic Behavior in Hz-WE

Enhancing the energy efficiency of large-scale high-purity H_2 synthesis using metal borides can be achieved by combining the thermodynamically advantageous HzOR with HER in a hybrid water electrolyzer. IrB NSs/NF have been effectively synthesized by Duan et al. [42] using a two-step process. The as-obtained IrB NSs/NF demonstrated enough optimal electronic structure, exposed active sites, and thus exceptional activity for both HzOR and HER, which were attributed to the highly curved nanosheet shape and lattice strain following B doping. Significantly, in N_2H_4-assisted water splitting, the two-electrode system consisting of IrB NSs/NF performed exceptionally well, achieving current densities of 10 and 100 $mA{\cdot}cm^{-2}$ at surprisingly low voltages of 0.033 and 0.142 V, respectively. However, compared to the conventional water splitting, i.e., 1.41 V, the equivalent operating voltage is much lower.

Meanwhile, HER and HzOR-based sustainable H_2 generation via heterostructured nanocomposites has garnered considerable interest, as it uses less energy than whole water electrolysis. To enable sustainable hydrogen production and hydrazine removal in alkaline conditions, Wang et al. [43] focused on designing and constructing heterostructured CoPB@NiFe–OH as efficient bifunctional catalysts. At a current density of 10 $mA{\cdot}cm^{-2}$, the CoPB@NiFe–OH heterointerface demonstrated a notably low HzOR potential of −135 mV with a phosphorus-to-boron (P/B) atom ratio of 0.2, and HER potential of −32 mV when the P/B ratio was 0.5. This surpasses the majority of documented catalysts by producing hydrogen without an external voltage and with a modest current density output of 25 $mA{\cdot}cm^{-2}$. Moreover, DFT calculations revealed that electron transfer from CoPB to NiFe–OH at the heterointerface enables the Ni sites in NiFe–OH to optimize the adsorption energy of $N_2H_4^*$, while the Co sites in CoPB promote H^* adsorption. This synergistic effect significantly enhances the efficiency of Hz-WE by coupling HER with HzOR.

11.4 Chemistry and Structure of Metal Selenides

11.4.1 Structural Features and Stability

Transition-metal selenides like ZnSe, $FeSe_2$, HgSe, CuSe, Cu_3Se_2, CdSe, and $MoSe_2$ have attracted significant interest due to their versatile industrial applications, including in semiconductors, photovoltaics, catalysis, energy storage, and photocatalysis [44]. They exhibit high metallicity, enhancing electron transport and catalytic reactions, making them ideal for electrochemical water splitting [12]. Transition-metal selenides share similar properties with sulfides since selenium and sulfur belong to the same group. Due to selenium's electrical conductivity and higher density, selenides often offer better volumetric energy density and rate capability as electrode materials [45]. Selenium offers advantages over sulfur due to its higher nucleophilicity and reactivity with oxygen species, and also outperforms metal oxides and phosphides as electrode materials by providing lower cost, higher capacity, richer redox chemistry, better conductivity, and greater stability [46]. Metal selenides are classified into layered and non-layered types. Layered metal selenides attract more attention due to their unique structure, where metal atoms covalently bond with two adjacent selenide layers, forming a sandwich-like Se—M—Se structure [47]. Selenium's electron configuration is $4s^2 4p^4$, and its vacant 3d-orbitals can participate in bonding, especially with transition metals, by gaining electrons to form Se^{2-} or sharing electrons to create covalent bonds. For diatomic transition-metal selenides (MSe type), molecular orbitals form from the metal's nd- and $(n+1)$ s-orbitals combined with selenium's 4s- and 4p-orbitals. Strong hybridization between metal and selenium orbitals enables electron wave functions to delocalize across atoms, dispersing charge density over metal–ligand units. This electron delocalization alleviates charge imbalances caused by divalent ion insertion, enhancing ion storage stability and ion diffusion [48].

Metal selenides form different crystals or structural forms or different phases while keeping their chemical composition the same, showing polymorphism. It is well documented in the case of $MoSe_2$, which possess a semiconducting 2H phase and the metallic 1T phase depending on Se atom arrangement [49]. The 2H phase has a trigonal prismatic coordination and is stable, while the 1T phase, with octahedral coordination, is metastable but often shows higher electrical conductivity and catalytic activity. Controlling the phase of metal selenides allows tailoring their electronic structure and surface properties to optimize performance [50].

11.4.2 Synthesis and Compositional Control

Development of metal selenides can be achieved by different methods. Hydrothermal/solvothermal synthesis is the most common, which uses high temperature (>100 °C) and pressure to produce uniform nanostructures, with product quality influenced by pressure, temperature, and water/solvent other than water. Solvothermal synthesis allows better control over particle agglomeration, morphology, and phase [51].

Other methods include chemical bath deposition (CBD) and chemical vapor deposition (CVD). CBD relies on slow precursor reactions to form nanocrystalline films without particle aggregation, it efficiently produces transition-metal selenide thin films with advantages like low cost, simple setup, low temperature, eco-friendliness, large area coverage, and good crystallite size control [52]. CVD, including metal-organic CVD, exposes substrates to volatile precursors that react or decompose to form thin films. CVD is widely used for high-performance transition-metal selenide films [53]. Besides them, metal selenides are also efficiently synthesized by pulsed laser deposition (PLD), which uses high-power laser pulses to ablate material from a target, forming a plasma plume that deposits thin films on substrates [45]. And in ultrasonication, ultrasonic radiation generates extreme local heat and pressure, enabling fast, eco-friendly synthesis of nanomaterials without harsh conditions [54]. Electrodeposition is one of the most appealing techniques for the synthesis of metal selenides. It is a fast, low-cost, and scalable method for growing thick, uniform metal selenide films on large or curved surfaces at room temperature and ambient pressure without expensive equipment. However, its crystalline quality and microstructure are generally inferior to those achieved by techniques like MBE, PLD, or CVD [55].

Pristine transition-metal selenides face challenges in electrocatalysis due to poor stability and lower catalytic activity compared to noble metals. To overcome these drawbacks, extensive research has focused on various modification strategies such as defect engineering and element doping [56]. Heteroatom doping with elements like N, P, O, and B is known to increase the electrochemical performance of electrode materials. However, achieving multi-heteroatom-doped composites typically requires multiple synthetic steps, making the process more complex [57]. Heteroatom doping is an effective strategy to improve selenides by introducing defects and modifying their electronic structure, enhancing conductivity and rate performance. It is categorized into nonmetallic doping, which boosts reaction kinetics for high-power battery applications, and metallic doping, which adds electron conduction channels and offers high-temperature stability [58].

11.4.3 Electrocatalytic Behavior in Hz-WE

Metal selenides are crucial for HER because they are a high-efficiency alternative to noble-metal catalysts, offering low cost. Their electronic properties, structural tunability, and stability make them highly valuable in developing sustainable hydrogen technologies, both electrochemical and photocatalytic. Several metal selenides of different morphologies were synthesized by different techniques. Extensive studies have shown that binary and ternary metal selenides, synthesized via various techniques and morphologies, offer improved performance in HER and other anodic reactions such as urea oxidation reaction (UOR), OER, HzOR, and methanol oxidation reaction (MOR). For instance, Ni-based selenide catalysts (NiSe-3, NiSe-2, and NiSe-1) were electrodeposited on nickel foam using varying $NiCl_2{\cdot}6H_2O$-to-SeO_2 ratios. Among them, NiSe-2, with a nanoporous structure, showed the best bifunctional performance for both OER and HzOR. It achieved a low overpotential of 252 mV for OER (10 $mA{\cdot}cm^{-2}$) and an exceptionally high current density of 318 $mA{\cdot}cm^{-2}$ at 0.4 V for HzOR, along with long-term stability. The enhanced activity is attributed to its high conductivity, large electrochemical surface area, and superaerophobic nanoporous surface [59]. Similarly, $MWSe_x$ (M = Ni, Co, Fe, and Mn) catalysts were prepared via a simple hot-injection method and evaluated for HER at soft interfaces. Incorporation of Ni and Co into WSe_x significantly improved catalytic activity, while Fe and Mn showed minimal effect. HER performance followed the order of: Ni > Co ≫ Fe ≈ Mn. Lower activity in Fe/Mn-based amorphous structures was attributed to weaker binding affinities and lower electronegativities, as confirmed by TEM [60]. Another work presented a bifunctional tubular cobalt perselenide ($CoSe_2$) nanosheet electrode for hydrogen production via electrolysis in HzOR. The $CoSe_2$ electrode delivers 10 $mA{\cdot}cm^{-2}$ at −84 mV for HER and −17 mV for HzOR, demonstrating excellent activity and durability. Remarkably, stable water electrolysis is achieved at an ultralow cell voltage of 164 mV over 14 hours, significantly lower than the theoretical 1.23 V [61].

In addition, defect-enriched 2D nickel copper selenide (D/Ni–Cu–Se) was developed by Khalafallah et al. [62] by growing on nickel foam, as a bifunctional electrocatalyst for efficient HER and UOR, through a hydrothermal and defect-engineering process. The self-supported electrode achieves 10 $mA{\cdot}cm^{-2}$ at 87.7 mV (HER) and 1.335 V (UOR) in alkaline electrolyte. Urea and hydrazine-assisted water electrolyzers using D/Ni–Cu–Se/NF showed significantly reduced cell voltages (~1.395 and 0.268 V) compared to conventional water electrolysis (1.572 V), with stable operation over 48 hours. Aslan et al. [60] synthesized ternary amorphous metal selenides ($MWSe_x$; M = Fe, Co, Ni, and Mn) via a hot-injection method and tested them as HER catalysts in a water/1,2-dichloroethane biphasic system. Catalytic performance followed the trend of: Fe ≈ Mn ≪ Co < Ni. $FeWSe_x$ and $MnWSe_x$ showed little improvement due to their amorphous nature, while $CoWSe_x$ and $NiWSe_x$ significantly enhanced HER activity compared to WSe_x, highlighting the beneficial effect of Co and Ni incorporation. Multimetallic and MOF-derived selenides also enable alternative anodic reactions, such as UOR and MOR, achieving reduced energy demands and higher hydrogen production rates. A Ce/Co/Ni multimetal selenide composite on nickel foam ($CeSe/Co_3Se_4$@NiSe–NF) was developed as a bifunctional electrocatalyst for both MOR and HER. It showed excellent catalytic performance, with high current density (135.6 $mA{\cdot}cm^{-2}$ at 1.45 V vs. RHE) for MOR and a low overpotential (121 mV at 10 $mA{\cdot}cm^{-2}$) for HER. In a methanol-assisted water electrolysis cell, it delivered 10 $mA{\cdot}cm^{-2}$ at just 1.45 V with stable operation for over 20 hours [63]. Meshesha et al. [64] prepared an MOF-derived Ni and Co-doped iron selenide (NiCoFeSe) hexagonal nanorod catalyst, with excellent electrocatalytic performance for HER, OER, and UOR with low overpotentials (as low as 210 mV at 50 $mA{\cdot}cm^{-2}$).

The mechanism behind HER-based metal selenides involves several aspects, and heterojunctions and cocatalysts are known to enhance the activity of metal selenides. For instance, Wang and Jin [65] reported that under visible light irradiation, CdS nanorods undergo bandgap excitation, generating electron–hole (e^-/h^+) pairs. Due to the excellent conductivity of $NiSe_2$ and its intimate contact with CdS, the photogenerated electrons in the conduction band (CB) of CdS transfer efficiently to the surface of $NiSe_2$ nanoparticles, where they reduce water to produce H_2. This synergy increases the separation of e^-/h^+ pairs, improves electron transfer, and provides abundant active sites, thereby significantly boosting the photocatalytic hydrogen evolution efficiency of the $NiSe_2$/CdS composite. Du et al. [66] also reported enhanced activity of $CdS_{0.95}Se_{0.05}$ nanowires when Cu_2Se was used as a cocatalyst due to lowering of e^-/h^+ recombination. The performance improved as Cu_2Se loading increased up to 20% and about 7.1 times higher HER activity was achieved than pure $CdS_{0.95}Se_{0.05}$. The mechanism involved is shown in Figure 11.2 [66]. The study highlighted the necessity of intimate contact and heterojunction formation, as close contact facilitates efficient charge transfer, suppressing e^-/h^+ recombination and boosting photocatalytic hydrogen generation. Similarly, TMCoSe/CdSSe composite (illustrated using FeCoSe/CdSSe) was also reported to have improved activity due to the formation of a Schottky junction at the interface. The work function (WF) of CdSSe (4.45 eV) is lower than that of FeCoSe (4.84 eV), and electrons transfer from CdSSe to FeCoSe upon intimate contact. This causes the CB and valence band (VB) of CdSSe to bend upward, creating a Schottky barrier that traps photogenerated electrons, preventing recombination. The highly conductive FeCoSe cocatalyst efficiently transfers these trapped electrons to its surface, where they catalyze the reduction of water to produce hydrogen gas with high activity [67]. Further, selenium vacancies also play an important role in HER, as shown in the case of molybdenum selenides. $MoSe_2$ nanosheets with selenium vacancies exhibit enhanced HER performance, featuring low overpotential and Tafel slope due to abundant Se vacancy defects that serve as extra active sites, increased surface area, and improved conductivity. These Se vacancies accelerate electrocatalytic kinetics by reinforcing active sites [68, 69].

These advances highlight the potential of metal selenides as efficient, scalable, and environmentally friendly electrocatalysts for next-generation hydrogen production technologies.

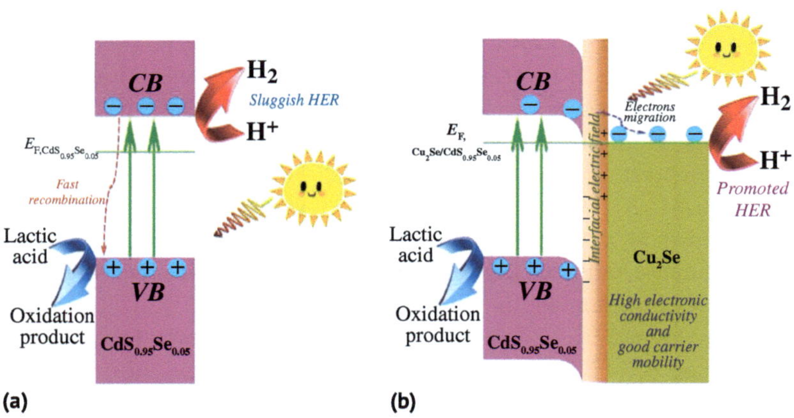

Figure 11.2 Representation of the mechanism proposed for photocatalytic HER by (a) pure $CdS_{0.95}Se_{0.05}$ nanowires and (b) y-Cu_2Se/$CdS_{0.95}Se_{0.05}$ heterojunction. *Source*: [66]/with permission of Elsevier.

11.5 Challenges and Future Prospects

Hz-WE is showing tremendous promise in the field of H_2 generation; however, there are substantial critical challenges that must be addressed before its large-scale adoption. The first gap that must be taken care of is long-term operational stability, which arises due to material degradation and catalyst leaching. In some cases, metal selenides and borides exhibit a lack of activity and contamination of the electrolyte even in highly alkaline or oxidative circumstances due to surface oxidation, dissolution of active components, or phase transformation. At high current densities, some studies had reported that metal borides underwent boron leaching or partial oxidation during prolonged electrolysis. In addition, selenides also exhibited leaching, affecting both performance and the environment under specific electrochemical conditions. The second gap that must be looked at is the safety protocols for handling hydrazine and its optimum use due to its carcinogenicity, toxicity, and volatility. WHO has set a strict industrial exposure limit of 0.01–0.03 ppm, as continuous exposure can lead to respiratory and neurological damage. Meanwhile, for any large-scale Hz-WE system, it is critical to implement real-time monitoring and immediate neutralization strategies to avoid fatal hydrazine exposure. In addition, by-products such as ammonia and nitrogen oxides must be avoided in Hz-WE systems, as it is crucial to avoid complete electrooxidation of hydrazine to nitrogen.

New strategies in advanced materials are being developed to address existing challenges. A boride or selenide-based single-atom catalyst (SAC) can maximize atomic economy and have tunable electronic environments, which enhance its catalytic activity and selectivity. Janus-type dual-function composites are gaining interest, which utilize separate regions with distinct anodic and cathodic catalytic actions on a single nanostructure. Recent studies showed that metal centers anchored on boron supports have durable catalysis due to leaching resistance, providing exceptionally active sites to catalyze HzOR and HER reactions. Achievements like these enable the design of structures that can synergistically improve charge transport while spatially decoupling reaction pathways to improve selectivity. Furthermore, solar-sustained reaction systems, photothermal Hz-WE hybrids that utilize solar energy to selectively heat the catalyst, increase reaction kinetics, and allow off-grid hydrogen production, would significantly reduce energy input.

The primary focus for the comprehension of Hz-WE technology's potential revolves around material synthesis, cell engineering, and scalable production. For commercial viability, self-propagating high-temperature synthesis or template-assisted electrodeposition synthesis should be considered for the fabrication of metal selenides and borides. In addition, for industrial deployment, the electrochemical system should be designed in such a way as to maintain efficient hydrazine delivery and removal systems, incorporate safety features, carry out real-time monitoring of catalyst stability, and maintain effluent composition. It is crucial to mention that in spite of the positive characteristics that metal borides possess, there are not many documented studies focusing on their use in Hz-WE. Most works done so far concentrate on HER and OER, with a very small number of works showcasing borides' bifunctional activity on hydrazine oxidation and hydrogen evolution. This reveals a considerable scientific void and stresses the importance of comprehensive studies on boride catalysts for Hz-WE, including works on mechanisms of reaction, long-term performance, and life cycle analysis.

In short, Hz-WE is a fascinating technique to make hydrogen in a way that is good for the environment while also cleaning up pollutants. However, it will be important to address the interconnected problems of catalyst lifetime, hydrazine safety, and system integration at a realistic scale. For Hz-WE to be employed in industry and for environmental objectives, there has to be further progress in catalyst research, notably with boride and selenide compounds, as well as better cell designs and safety rules.

11.6 Conclusion

The use of hydrazine in water electrolysis systems dramatically enhances hydrogen production by making the process more energy efficient while simultaneously mitigating hydrazine pollution. This technology is particularly attractive because it combines rigorous thermodynamics with fast kinetics simultaneously for energy and environmental applications. Among non-precious catalysts, metal borides and selenides stand out because of their enhanced chemical stability and electrical conductivity, as well as their ability to be chemically tailored. Implementation of new designs, including nanostructured, heterostructured, and SACs, has improved the selectivity and efficiency of Hz-WE systems. To date, the progress made in developing Hz-WE systems remains limited, primarily due to challenges involving catalyst leaching and degradation, complete and safe oxidation of hydrazine, robust protocols for handling hydrazine, and the development of comprehensive safety measures. Furthermore, the scarcity of literature focused on boride-based catalysts for Hz-WE draws attention to the urgent need for targeted investigation into their multifunctional behavior, mechanistic pathways, and operational longevity. Future works will focus on interdisciplinary efforts in materials science, electrochemical engineering, and environmental safety, resulting in scalable synthesis, advanced cell design, and life cycle assessment. Hence, addressing these challenges will help in incorporating Hz-WE on a large scale, making energy production green and cost-effective.

Acknowledgments

The authors would like to acknowledge CNMS, JAIN (Deemed-to-be University) and Nano Mission, DST, Government of India, for financial support SR/NM/NS-20/2014, minor research project (No. JU/MRP/CNMS/11/2022), and postdoctoral fellowship grants JUCRTAPDF006 and JUIAPPICRTA/2024/786.

References

1 Li, Y., Zhao, Y., Li, F. M., Dang, Z., & Gao, P. (2021). Ultrathin NiSe nanosheets on Ni foam for efficient and durable hydrazine-assisted electrolytic hydrogen production. *ACS Appl. Mater. Interfaces 13*(29), 34457–34467.

2 Darband, G. B., Lotfi, N., Aliabadi, A., Hyun, S., & Shanmugam, S. (2021). Hydrazine-assisted electrochemical hydrogen production by efficient and self-supported electrodeposited Ni-Cu-P@Ni-Cu nano-micro dendrite catalyst. *Electrochim. Acta 382*, 138335.

3 Li, K., He, J. F., Guan, X., Tong, Y., Ye, Y., Chen, L., & Chen, P. (2023). Phosphorus-modified amorphous high-entropy CoFeNiCrMn compound as high-performance electrocatalyst for hydrazine-assisted water electrolysis. *Small 19*(42), 2302130.

4 Wang, H. Y. & Yuan, Z. Y. (2025). Hydrazine-assisted water electrolysis system: Performance enhancement and application expansion. *Mater. Horiz. 12*, 5123–5148.

5 Darband, G. B., Maleki, M., Toghraei, A., & Shanmugam, S. (2023). Electrodeposition of self-supported transition metal phosphides nanosheets as efficient hydrazine-assisted electrolytic hydrogen production catalyst. *Int. J. Hydrog. Energy 48*(11), 4253–4263.

6 Tian, W., Zhang, X., Wang, Z., Cui, L., Li, M., Xu, Y., Li, X., Wang, L., & Wang, H. (2022). Amorphization activated RhPb nanflowers for energy-saving hydrogen production by hydrazine-assisted water electrolysis. *Chem. Eng. J. 440*, 135848.

7 Ji, K., Wang, S., Yao, S., Ji, Y., Wang, G., Ren, W., Wang, J., Zhang, F., Xie, J., Yang, Z., & Yan, Y. M. (2025). Modeling carbon-free energy conversion systems: enhanced hydrazine-assisted hydrogen production with dual-electric-field effect on needle-like Ru/CoP catalysts. *Energy Environ. Sci. 18*(10), 4764–4774.

8 Wang, H. Y., Yan, F., Wang, H., Zhai, S., Ren, J. T., Wang, L., Sun, M., & Yuan, Z. Y. (2024). The combination of electronic structure and lattice strain engineering for multi-powered hydrazine-assisted seawater electrolysis system at high current densities. *Adv. Energy Mater. 14*(42), 2402611.

9 Fu, X., Cheng, D., Wan, C., Kumari, S., Zhang, H., Zhang, A., Huyan, H., Zhou, J., Ren, H., Wang, S., Zhao, Z., Zhao, X., Chen, J., Pan, X., Sautet, P., Huang, Y., & Duan, X. (2023). Bifunctional ultrathin $RhRu_{0.5}$-alloy nanowire electrocatalysts for hydrazine-assisted water splitting. *Adv. Mater. 35*(23), 2301533.

10 Wang, H. Y., Wang, L., Ren, J. T., Tian, W. W., Sun, M. L., & Yuan, Z. Y. (2023). Heteroatom-induced accelerated kinetics on nickel selenide for highly efficient hydrazine-assisted water splitting and Zn-hydrazine battery. *Nano-Micro Lett. 15*(1), 155.

11 Lee, E. & Fokwa, B. P. T. (2022). Nonprecious metal borides: Emerging electrocatalysts for hydrogen production. *Acc. Chem. Res. 55*(1), 56–64.

12 Xia, X., Wang, L., Sui, N., Colvin, V. L., & Yu, W. W. (2020). Recent progress in transition metal selenide electrocatalysts for water splitting. *Nanoscale 12*(23), 12249–12262.

13 Masud, J., Ioannou, P. C., Levesanos, N., Kyritsis, P., & Nath, M. (2016). A molecular Ni-complex containing tetrahedral nickel selenide core as highly efficient electrocatalyst for water oxidation. *ChemSusChem 9*(22), 3128–3132.

14 Hou, J., Mei, K., Jiang, T., Yu, X., & Wu, M. (2024). NiFeP nanosheets for efficient and durable hydrazine-assisted electrolytic hydrogen production. *Dalton Trans. 53*(10), 4574–4579.

15 Yang, H. M., Wang, H. Y., Zhai, S., Ren, J. T., & Yuan, Z. Y. (2024). Aluminum-induced electron structure on nickel phosphide for effective hydrazine-assisted water splitting at large current density. *Chem. Eng. J. 489*, 151236.

16 Yan, F., Song, X., Wang, H., Wang, H., & Yuan, Z.-Y. (2025). Heteroatom-doped carbon coated CoP/MoP heterostructure for highly efficient hydrazine-assisted water electrolysis at high current densities. *J. Colloid Interface Sci. 700*, 138340.

17 Yang, H. M., Wang, H. Y., Sun, M. L., & Yuan, Z. Y. (2023). Interface engineering of bifunctional nickel hydroxide/nickel phosphide heterostructure for efficient intermittent hydrazine-assisted water splitting. *Chem. Eng. J. 475*, 146134.

18 Liu, Y., Zhang, J., Li, Y., Qian, Q., Li, Z., Zhu, Y., & Zhang, G. (2020). Manipulating dehydrogenation kinetics through dual-doping Co_3N electrode enables highly efficient hydrazine oxidation assisting self-powered H_2 production. *Nat. Commun. 11*(1), 1853.

19 Zhao, Y., Sun, Y., Li, H., Zeng, S., Li, R., Yao, Q., Chen, H., Zheng, Y., & Qu, K. (2023). Highly enhanced hydrazine oxidation on bifunctional Ni tailored by alloying for energy-efficient hydrogen production. *J. Colloid Interface Sci. 652*, 1848–1856.

20 Tong, Y. & Chen, P. (2024). Recent progress of advanced electrocatalysts for hydrogen production via hydrazine-assisted water electrolysis. *Inorg. Chem. Front. 11*(19), 6218–6245.

21 Hu, J., Wang, X., Zhou, Y., Liu, M., Wang, C., Li, M., Liu, H., Li, H., Tang, Y., & Fu, G. (2025). Asymmetric Rh–O–Co bridge sites enable superior bifunctional catalysis for hydrazine-assisted hydrogen production. *Chem. Sci. 16*(4), 1837–1848.

22 Sun, F., Qin, J., Wang, Z., Yu, M., Wu, X., Sun, X., & Qiu, J. (2021). Energy-saving hydrogen production by chlorine-free hybrid seawater splitting coupling hydrazine degradation. *Nat. Commun. 12*(1), 4182.

23 Zhu, L., Huang, J., Meng, G., Wu, T., Chen, C., Tian, H., Chen, Y., Kong, F., Chang, Z., Cui, X., & Shi, J. (2023). Active site recovery and N-N bond breakage during hydrazine oxidation boosting the electrochemical hydrogen production. *Nat. Commun. 14*(1), 1997.

24 Katiyar, N. K., Dhakar, S., Parui, A., Gakhad, P., Singh, A. K., Biswas, K., Tiwary, C. S., & Sharma, S. (2021). Electrooxidation of hydrazine utilizing high-entropy alloys: Assisting the oxygen evolution reaction at the thermodynamic voltage. *ACS Catal. 11*(22), 14000–14007.

25 Roh, H., Choi, H., Kim, J. S., Shin, H., Park, T., & Yong, K. (2022). Low power consumed PV-electrolysis with CoFeP nanowires for hydrazine-assisted hydrogen production. *Appl. Surf. Sci. 606*, 154951.

26 Wang, H. Y., Zhai, S., Wang, H., Yan, F., Ren, J. T., Wang, L., Sun, M., & Yuan, Z. Y. (2024). Taking advantage of potential coincidence region: Insights into gas production behavior in advanced self-activated hydrazine-assisted alkaline seawater electrolysis. *ACS Nano 18*(30), 19682–19693.

27 Guo, R., Zhang, Y., Zhang, X., Yang, H., & Hu, T. (2024). Synthesis of $Co_{0.52}Cu_{0.48}$/Cu@S–C arrays for the enhanced performance of hydrazine-assisted hydrogen production. *Int. J. Hydrog. Energy 68*, 472–480.

28 Xin, Y., Shen, K., Guo, T., Chen, L., & Li, Y. (2023). Coupling hydrazine oxidation with seawater electrolysis for energy-saving hydrogen production over bifunctional CoNC nanoarray electrocatalysts. *Small 19*(21), 2300019.

29 Feng, C., Lv, M., Shao, J., Wu, H., Zhou, W., Qi, S., Deng, C., Chai, X., Yang, H., Hu, Q., & He, C. (2023). Lattice strain engineering of Ni_2P enables efficient catalytic hydrazine oxidation-assisted hydrogen production. *Adv. Mater. 35*(42), 2305598.

30 Yang, H., Wang, H., Wang, L., Sun, M., Xu, F., Ye, H., Ren, J., & Yuan, Z. Y. (2025). Optimized electronic structure induced by cobalt metaphosphate/phosphide for highly efficient hydrazine-assisted water splitting at high current densities. *J. Colloid Interface Sci. 695*, 137765.

31 Wang, Z., Xu, L., Huang, F., Qu, L., Li, J., Owusu, K. A., Liu, Z., Lin, Z., Xiang, B., Liu, X., Zhao, K., Liao, X., Yang, W., Cheng, Y. B., & Mai, L. (2019). Copper–nickel nitride nanosheets as efficient bifunctional catalysts for hydrazine-assisted electrolytic hydrogen production. *Adv. Energy Mater. 9*(21), 1900390.

32 Zhou, M., Yu, Z., Yu, G., Fu, R., Wang, S., Yang, W., Liao, X., Zhao, Y., & Wang, Z. (2025). Constructing of *n*-type semiconductor heterostructures for efficient hydrazine-assisted hydrogen production. *Adv. Funct. Mater. 35*(6), 2415058.

33 Ren, X. & Tong, Y. (2024). The self-reconstruction of Co-modified bimetallic hydroxysulfide nanosheet arrays for efficient hydrazine assisted water splitting. *Int. J. Hydrog. Energy 49*, 489–497.

34 Wang, F. G., Liu, X., Lv, Q. X., Liu, B., Chai, Y. M., & Dong, B. (2022). Transition metal boride-based materials for electrocatalytic water splitting. *Chin. J. Struct. Chem. 41*(9), 2209008–2209044.

35 Chen, H. & Zou, X. (2020). Intermetallic borides: Structures, synthesis and applications in electrocatalysis. *Inorg. Chem. Front. 7*(11), 2248–2264.

36 Pu, Z., Liu, T., Zhang, G., Liu, X., Gauthier, M. A., Chen, Z., & Sun, S. (2021). Nanostructured metal borides for energy-related electrocatalysis: Recent progress, challenges, and perspectives. *Small Methods 5*(10), 2100699.

37 Du, Y., He, X., Yan, C., Hu, Q., Zhang, J., & Yang, F. (2025). Promoted surface reconstruction of amorphous nickel boride electrocatalysts by boron dissolution for boosting the oxygen evolution reaction. *J. Mater. Chem. A 13*(11), 7962–7972.

38 Sitler, S. J., Raja, K. S., & Charit, I. (2016). Metal-rich transition metal diborides as electrocatalysts for hydrogen evolution reactions in a wide range of pH. *J. Electrochem. Soc. 163*(13), H1069–H1075.

39 Yao, Y., Zhang, Z., & Jiao, L. (2022). Development strategies in transition metal borides for electrochemical water splitting. *Energy Environ. Mater. 5*(2), 470–485.

40 Abeysinghe, J. P., Kölln, A. F., & Gillan, E. G. (2022). Rapid and energetic solid-state metathesis reactions for iron, cobalt, and nickel boride formation and their investigation as bifunctional water splitting electrocatalysts. *ACS Mater. Au 2*(4), 489–504.

41 Gouget, G., Beaunier, P., Portehault, D., & Sanchez, C. (2016). New route toward nanosized crystalline metal borides with tuneable stoichiometry and variable morphologies. *Faraday Discuss. 191*, 511–525.

42 Duan, Z., Ren, T., Xu, Y., Wang, Z., Yu, H., Deng, K., Wang, L., & Wang, H. (2023). Boron-doped iridium nanosheets array for energy-saving hydrogen production by hydrazine-assisted water electrolysis. *Int. J. Hydrog. Energy 48*(95), 37045–37052.

43 Wang, J., Qing, S., Tong, X., Xiang, T., Luo, G., Zhang, K., & Xu, L. (2024). Heterointerface of nickel-ferric hydroxide/cobalt-phosphide-boride boosting hydrazine-assisted water electrolysis. *Green Energy Environ. 10*(10), 1957–1967.

44 Altaf, S., Ajaz, H., Imran, M., Ul-Hamid, A., Naz, M., Aqeel, M., Shahzadi, A., Shahbaz, A., & Ikram, M. (2020). Synthesis and characterization of binary selenides of transition metals to investigate its photocatalytic, antimicrobial and anticancer efficacy. *Appl. Nanosci. 10*(7), 2113–2127.

45 Lu, T., Dong, S., Zhang, C., Zhang, L., & Cui, G. (2017). Fabrication of transition metal selenides and their applications in energy storage. *Coord. Chem. Rev. 332*, 75–99.

46 Hussain, I., Sahoo, S., Lamiel, C., Nguyen, T. T., Ahmed, M., Xi, C., Iqbal, S., Ali, A., Abbas, N., Javed, M. S., & Zhang, K. (2022). Research progress and future aspects: Metal selenides as effective electrodes. *Energy Storage Mater. 47*, 13–43.

47 Luo, M., Yu, H., Hu, F., Liu, T., Cheng, X., Zheng, R., Bai, Y., Shui, M., & Shu, J. (2020). Metal selenides for high performance sodium ion batteries. *Chem. Eng. J. 380*, 122557.

48 Chen, D., Zhao, Z., Chen, G., Li, T., Chen, J., Ye, Z., & Lu, J. (2023). Metal selenides for energy storage and conversion: A comprehensive review. *Coord. Chem. Rev. 479*, 214984.

49 Wang, Y., Xiao, X., Ding, T., & Lu, M. (2021). Mixed 1T-2H phase $MoSe_2$ as interfacial charge-transfer-bridge to boosting photocatalytic activity of dual Z-scheme AgI/1T-2H $MoSe_2/Bi_4O_5Br_2$ heterojunction. *J. Alloys Compd. 875*, 160092.

50 Deng, S., Zhong, Y., Zeng, Y., Wang, Y., Yao, Z., Yang, F., Lin, S., Wang, X., Lu, X., Xia, X., & Tu, J. (2017). Directional construction of vertical nitrogen-doped 1T-2H $MoSe_2$/graphene shell/core nanoflake arrays for efficient hydrogen evolution reaction. *Adv. Mater. 29*(21), 1700748.

51 Feng, W., Pang, W., Xu, Y., Guo, A., Gao, X., Qiu, X., & Chen, W. (2020). Transition metal selenides for electrocatalytic hydrogen evolution reaction. *ChemElectroChem 7*(1), 31–54.

52 Vanderhyde, C. A., Sartale, S. D., Patil, J. M., Ghoderao, K. P., Sawant, J. P., & Kale, R. B. (2015). Room temperature chemical bath deposition of cadmium selenide, cadmium sulfide and cadmium sulfoselenide thin films with novel nanostructures. *Solid State Sci. 48*, 186–192.

53 Konar, R. & Nessim, G. D. (2022). A mini-review focusing on ambient-pressure chemical vapor deposition (AP-CVD) based synthesis of layered transition metal selenides for energy storage applications. *Mater. Adv. 3*(11), 4471–4488.

54 Monjezi, F., Jamali-Sheini, F., & Yousefi, R. (2019). Ultrasound-assisted electrodeposition of Cu_3Se_2 nanosheets and efficient solar cell performance. *J. Alloys Compd. 780*, 626–633.

55 Piperno, L., Celentano, G., & Sotgiu, G. (2023). Electrodeposition of iron selenide: A review. *Coatings 13*(11), 1905.

56 Deng, Y., Xiao, S., Zheng, Y., Rong, X., Bai, M., Tang, Y., Ma, T., Cheng, C., & Zhao, C. (2023). Emerging electrocatalytic activities in transition metal selenides: Synthesis, electronic modulation, and structure-performance correlations. *Chem. Eng. J. 451*, 138514.

57 Shi, X., Fang, L., Peng, H., Deng, X., & Sun, Z. (2022). Metal-organic framework-derived NiSe embedded into a porous multi-heteroatom self-doped carbon matrix as a promising anode for sodium-ion battery. *Nanomaterials 12*(19), 3345.

58 Li, L., Wang, S., Peng, J., Lai, J., Zhang, H., Yang, J., Bella, F., Li, L., Wang, S., Peng, J., Lai, J., Zhang, H., & Yang, J. (2024). Transition metal selenide-based anodes for advanced sodium-ion batteries: Electronic structure manipulation and heterojunction construction aspect. *Molecules 29*(13), 3083.

59 Feng, Z., Zhang, H., Wang, L., Gao, B., Lu, P., & Xing, P. (2020). Nanoporous nickel-selenide as high-active bifunctional electrocatalyst for oxygen evolution and hydrazine oxidation. *J. Electroanal. Chem. 876*, 114740.

60 Aslan, E., Sarilmaz, A., Yanalak, G., Ozel, S. S., Ozel, F., & Patir, I. H. (2020). Transition metal–incorporated tungsten-based ternary refractory metal selenides ($MWSe_x$; M = Fe, Co, Ni, and Mn) as hydrogen evolution catalysts at soft interfaces. *Mater. Today Energy 18*, 100510.

61 Zhang, J. Y., Wang, H., Tian, Y., Yan, Y., Xue, Q., He, T., Liu, H., Wang, C., Chen, Y., & Xia, B. Y. (2018). Anodic hydrazine oxidation assists energy-efficient hydrogen evolution over a bifunctional cobalt perselenide nanosheet electrode. *Angew. Chem. Int. Ed. 57*(26), 7649–7653.

62 Khalafallah, D., Zhang, Y., & Zhang, Q. (2024). Maximizing urea-/hydrazine-assisted electrolytic hydrogen production by defective nickel copper selenide nanostructures. *FlatChem 43*, 100602.

63 Jalali, F., Sheikhi, S., & Hassani, N. (2024). Novel selenide composite as an effective bifunctional electrocatalyst for energy-saving hydrogen production through methanol-assisted water electrolysis. *Energy Fuels 39*(1), 992–1004.

64 Meshesha, M. M., Chanda, D., Balu, R., Jang, S. G., Ahmed, S., & Yang, B. L. (2024). Efficient green hydrogen production through metal–organic framework-derived Ni and Co mediated iron selenide hexagonal nanorods and wireless coupled with photovoltaics for urea and alkaline water electrolysis. *Appl. Catal. B: Environ. Energy 344*, 123635.

65 Wang, G. & Jin, Z. (2019). Function of $NiSe_2$ over CdS nanorods for enhancement of photocatalytic hydrogen production—From preparation to mechanism. *Appl. Surf. Sci. 467–468*, 1239–1248.

66 Du, S., Li, G., Lin, X., Zhang, S., Xu, H., & Fang, P. (2021). Highly efficient H_2 generation over Cu_2Se decorated $CdS_{0.95}Se_{0.05}$ nanowires by photocatalytic water reduction. *Chem. Eng. J. 409*, 128157.

67 Du, S., Lin, S., Ren, K., Li, C., & Zhang, F. (2023). Revealing the effects of transition metal doping on CoSe cocatalyst for enhancing photocatalytic H_2 production. *Appl. Catal. B: Environ. 328*, 122503.

68 Zhang, K., Li, Y., Deng, S., Shen, S., Zhang, Y., Pan, G., Xiong, Q., Liu, Q., Xia, X., Wang, X., & Tu, J. (2019). Molybdenum selenide electrocatalysts for electrochemical hydrogen evolution reaction. *ChemElectroChem 6*(14), 3530–3548.

69 Zhou, X., Jiang, J., Ding, T., Zhang, J., Pan, B., Zuo, J., & Yang, Q. (2014). Fast colloidal synthesis of scalable Mo-rich hierarchical ultrathin $MoSe_{2-x}$ nanosheets for high-performance hydrogen evolution. *Nanoscale 6*(19), 11046–11051.

12

Single-atom Electrocatalysts for Hydrazine-assisted Water Electrolysis

Talshyn Begildayeva and Juan A. Lopez-Ruiz*

Institute for Integrated Catalysis, Pacific Northwest National Laboratory, Richland, WA, USA

*Corresponding author: talshyn.begildayeva@pnnl.gov

12.1 Introduction

Water electrolysis (WE) was estimated to hold 4–5% of global hydrogen (H_2) production in 2020 (4.5 million metric tons [MMT] per year) according to the BloombergNEF Hydrogen Economy Outlook from 2020 and the 2021 Hydrogen Insights Report [1, 2]. According to reports, by 2030, WE could increase to 25 MMT·y^{-1}, assuming electrical infrastructure could support the high energy demand (1,250 TWh at 50 kWh [kgH_2]$^{-1}$) [3]. According to a report, as of 2022, alkaline water electrolyzers (AWEs) accounted for approximately 60% of the market share, followed by proton-exchange-membrane water electrolyzers (PEMWEs) with around 30%, while other electrolyzer technologies comprised the remaining portion [4]. However, the deployment of electrolyzers remains challenging mainly due to high engineering, procurement, and construction (EPC) capacity. This aspect is listed as a reason for the small share of WE in global H_2 production in a report from the United Nations Industrial Development Organization [5]. An International Energy Agency report further emphasized that energy efficiency—often bottlenecked by the oxygen evolution reaction (OER)—is central to cost reduction and scalability [6]. Although it lacked specific quantification of OER-related EPC contributions, the inefficiencies in energy requirements inevitably influence the size and design of the plant. Fundamentally, the thermodynamic potential required for water splitting is 1.23 V, in which the OER accounts for most of the energy input in the reaction [7]. In real-world applications, however, the operating potentials of both AWE and PEMWE often exceed 2 V, with energy losses of 0.57–0.97 V attributed primarily to OER overpotential due to kinetic barriers for breaking strong oxygen bonds and sluggish four-electron-transfer reactions [8]. Indeed, while the OER supplies electrons and protons necessary for the hydrogen evolution reaction (HER), it generates molecular oxygen (O_2), which, although valuable for niche applications such as medical uses, has far less industrial demand compared to H_2 [9]. As a result, advanced electrolyzer and electrocatalyst designs must compensate for (i) enhanced electrocatalyst performance, (ii) increased system complexity, and (iii) higher capital costs, which can exceed $1,000 per kilowatt (kW) for PEMWE systems [10]. One strategy to address these challenges is to replace the traditional OER with more favorable oxidation reactions—such as glycerol oxidation reaction (GEOR), alcohol oxidation reaction (AOR), or hydrazine oxidation reaction (HzOR)—that operate at lower onset potentials (e.g., 0.3–0.6 V vs. RHE for GEOR and ~0.23 V vs. RHE for HzOR, compared to >1.5 V vs. RHE for OER) (Figure 12.1) [11–14].

These oxidation reactions not only decrease the energy input required per mole of H_2 produced, but also could generate valuable coproducts—such as aldehydes, ketones, carboxylic acids, and olefins (via non-Kolbe electrolysis of carboxylic acids) from alcohols—thereby improving the overall techno-economic profile of the electrocatalytic process [13, 15, 16]. While OER substitution represents only one aspect of the electrolyzer optimization, it remains a focal point, with extensive efforts dedicated to the development of electrocatalysts and integrated system designs (i.e., coupling H_2 production with HzOR) that collectively aim to enable cost-effective and scalable H_2 production [17, 18]. The effectiveness of this approach depends on electrocatalyst properties—particularly active-site accessibility, molecular selectivity, surface area, and overpotential [19, 20]. Electrocatalysts must not only adsorb and activate organic substrates efficiently but must also suppress competing side reactions such as OER. In this context, nanostructured materials such as single-atom catalysts (SACs) have shown great potential toward improving HzOR over OER due to their high surface-to-volume ratio, tunable coordination environments, and ability to provide isolated active sites for selective oxidation of organic compounds [21, 22].

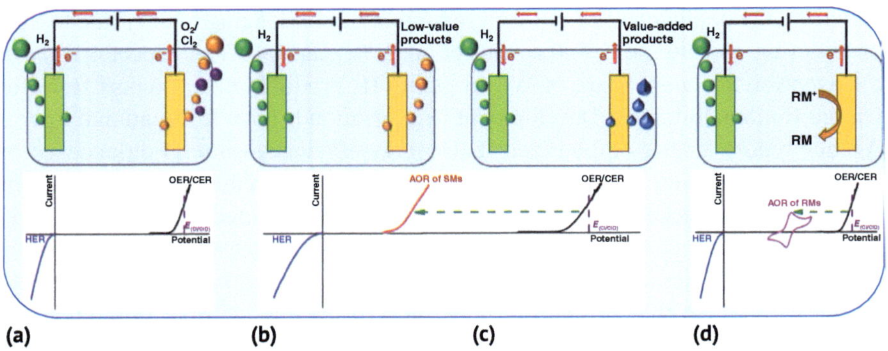

Figure 12.1 Current landscape of water electrolyzers and conceptual approaches emphasizing emerging strategies to replace the OER with alternative oxidation reactions, highlighting potential benefits for improved efficiency and value cogeneration: (a) conventional HER coupled with OER; (b) OER replacement with small-molecule oxidation reactions that yield products of little value (e.g., N_2 or CO_2); (c) high-value-added products; and (d) OER replacement with reactions through redox mediators (RMs). *Source*: [14]/with permission of John Wiley & Sons.

This chapter focuses specifically on the application of SACs for HzOR—a strategy that lowers the overall electrolyzer voltage for H_2 production from >1.9 V (typical for WE) to as low as ~0.8 V at an operating current density of 200 mA cm^{-2} [23]. By exploring the structure–activity relationships, synthetic strategies, and performance metrics of SACs for HzOR, we aim to provide a comprehensive understanding of their prospect to overcome fundamental limitations for electrocatalytic performance.

12.2 SACs: What They Are, How They Work, and Why They Matter

To address the kinetic and energetic limitations of OER, electrocatalyst design has increasingly focused on tailoring the active-site environment at the atomic level to maximize atomic utilization and tunable electronic structures. Jirkovský et al. proposed the concept of SACs in 2011 [24], in which isolated metal sites that drive electrocatalytic reactions are the active centers on SACs. Considering the achievements in SACs synthesis since their discovery, it is now possible to deposit most common metal elements onto conductive supports such as nitrogen-doped graphene or metal–organic frameworks (Figure 12.2a and b).

A review from Gloag et al. provided fundamental understanding of cocatalytic metal–support interactions in SACs and identified prospects for developing high-performance electrocatalytic systems. The authors suggested that reactions proceed via two distinct pathways, depending on whether the substrate of SAC (i) is catalytically active and participates directly in the reaction mechanism or (ii) remains inert under the given conditions [25]. Thus, SACs offer a unique combination of maximum atom utilization, well-defined active sites, and tunable coordination environments that collectively distinguish them from traditional catalyst platforms such as nanomaterials (i.e., alloys, oxides, high-entropy materials, etc.). In nanoparticle (NP) systems, only a fraction of the metal atoms are surface-accessible, depending on the NP size, and are catalytically active due to the high proportion of atoms buried within the particle core [26, 27]. In contrast, SACs can achieve almost complete atomic utilization because every metal atom is individually dispersed and exposed on the support surface (Figure 12.3a) [28–30]. For example, the CeO_2 (100) facet has a greater number of oxygen vacancies than CeO_2 (111) and can trap Pd atoms, thus facilitating atomic dispersion. As a result, atomically dispersed Pd_1/CeO_2 (100) demonstrated 2.8 times higher selectivity (>90%) for the catalytic N-alkylation reaction at 140 °C in benzyl alcohol (7 mL) containing *n*-dodecane (20 μL) under Ar atmosphere and using aniline (50 μmol) compared to clusters of Pd/CeO_2 (111) [31, 32]. Also, defective NiO supports with Ni vacancies were found to stabilize Au single atoms (SAs), i.e., the Au_1/NiO SAC exhibited over one order of magnitude higher CO oxidation than comparable Au NP/NiO catalysts at 120 °C over 120 hours (TOF 0.41 s^{-1} for Au_1/NiO and 0.033 s^{-1} for Au NPs/NiO) [33]. Liu et al. showed that the surface energy of single Pt sites on CeO_2 substrate is two times higher than that of

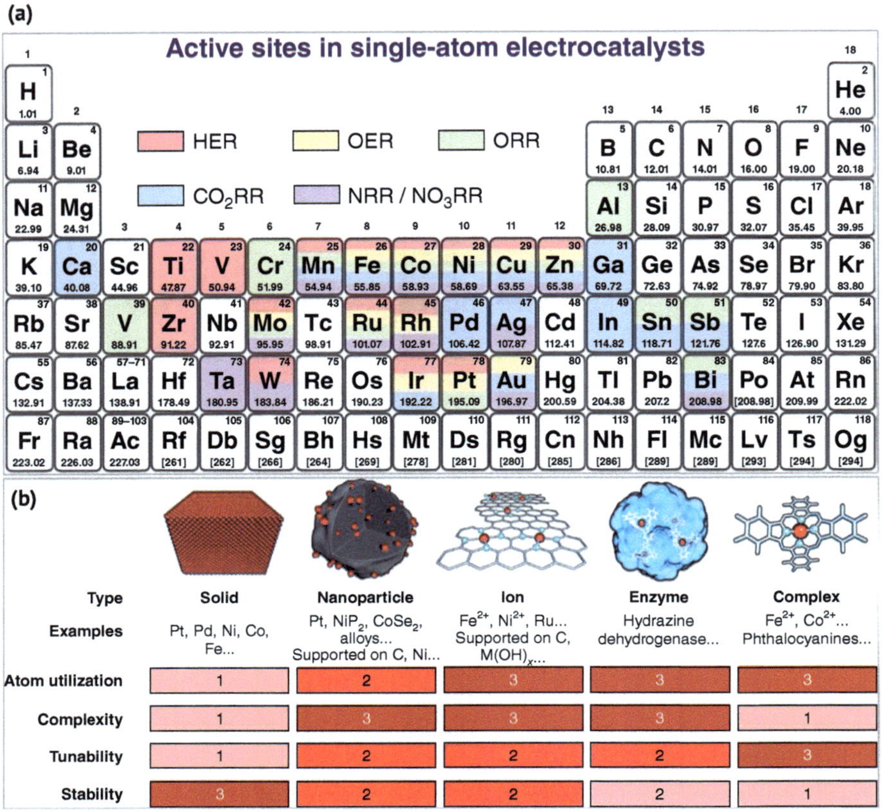

Figure 12.2 Summary of catalysts by: (a) elemental composition, with highlighted metals reported as active sites in SACs tested for various electrocatalytic reactions such as HER, OER, ORR, CO_2RR, N_2 reduction reaction (NRR), and nitrate (NO_3^-) reduction reaction (NO_3RR). *Source*: [36]/with permission of John Wiley & Sons; and (b) type, tested for HzOR. *Source*: [37]/with permission of American Chemical Society.

Pt NPs (0.190 eV Å^{-2} for Pt–substrate interaction and 0.09 eV Å^{-2} for Pt–Pt interaction, respectively) and related to higher catalytic activity [34]. SACs further offer precisely defined local coordination environments (e.g., M–N_4 or M–O_4) that enable controlled adsorption and reaction pathways. Zhou et al. [35] performed a comparative analysis of single copper (Cu) atoms supported on nitrogen-doped carbon nanosheets (Cu/NCNSs) and Cu NPs for the electrooxidation of biomass-derived molecule, 5-hydroxymethylfurfural (HMF). Zhou et al. [35] also showed that Cu SACs had over two orders of magnitude higher specific density than Cu NPs, 653.2 and 442.5 mA mg^{-1} compared to 0.23 and 2.06 mA mg^{-1}, respectively, in 0.1 M potassium hydroxide (KOH) + 10 mM 5-HMF electrolyte at 1.5 V vs. RHE. Cu/NCNSs exhibited a peak specific current density of 2,354 mA mg^{-1} during glucose oxidation at 1.6 V vs. RHE, which was over 600 times higher than that of Cu NPs [35]. The same Cu/NCNSs demonstrated lower onset potentials (1.13, 1.18, 1.23, and

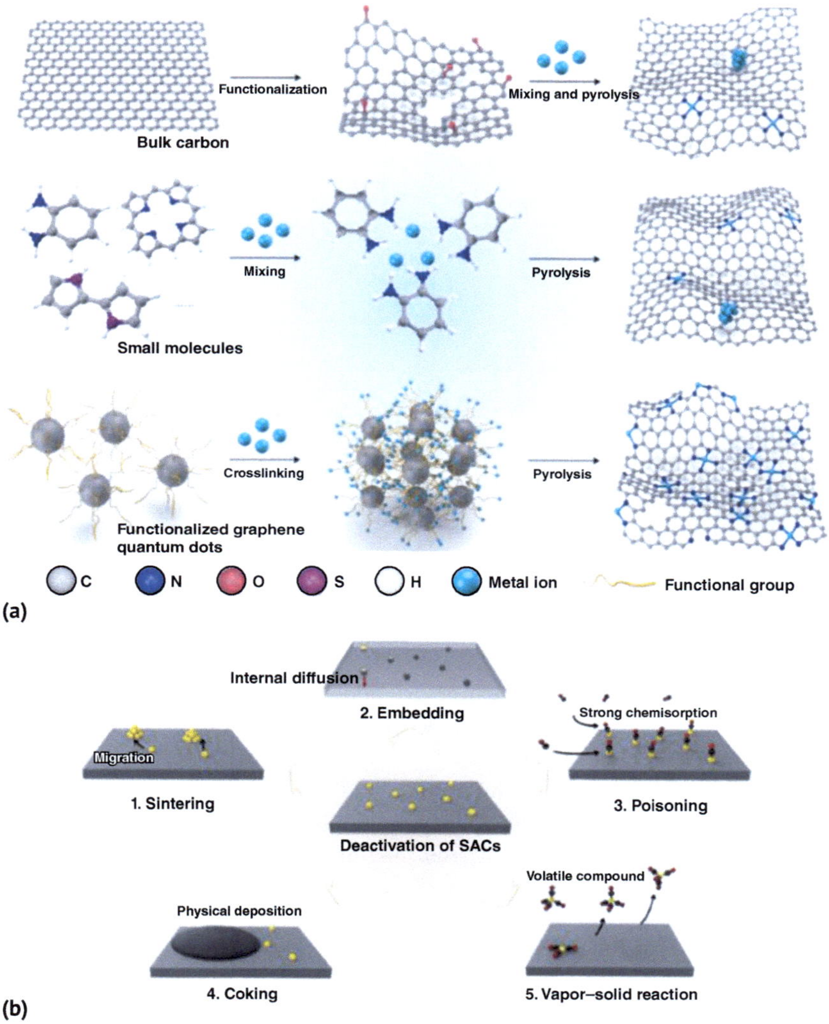

Figure 12.3 General SAC (a) synthetic strategies. *Source*: [39]/with permission of Springer Nature; and (b) representative deactivation mechanisms. *Source*: [40]/Springer Nature/CC BY 4.0.

1.20 V vs. RHE) compared to Cu NPs (1.32, 1.20, 1.35, and 1.39 V vs. RHE) for the electrooxidation of glucose, glycerol, 5-HMF, and benzyl alcohol, respectively [35].

The role of local coordination was further highlighted by Lei et al., where OER over Ir SAC on $Ni_{(3-x)}Fe_xS_2$ (Ir_1/NFS) and Ir SAC on nickel–iron hydroxide (Ir_1/NFH) substrates was compared. Ir_1/NFS exhibited improved kinetic performance by generating 10 mA cm^{-2} at 20 mV lower overpotential than Ir_1/NFH (170 and 190 mV, respectively). Furthermore, Ir_1/NFH showed a nearly one order of magnitude higher TOF at 300 mV overpotential (9.85 and 1.15 s^{-1}, respectively) for OER in 1 M KOH. Lei et al. [38] attributed the high performance to the chemical

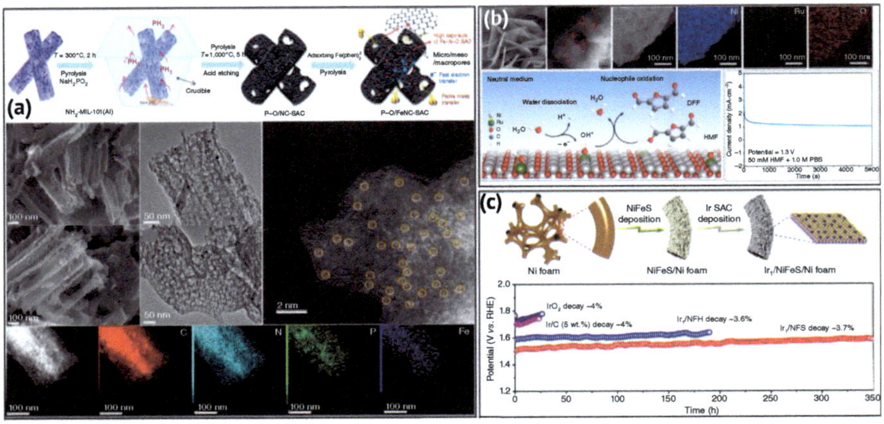

Figure 12.4 Illustration of SACs with various compositions and structures tested for (a) ORR on Fe—N—C. *Source*: [43]/with permission of American Chemical Society; (b) biomass-derived alcohol oxidation on Ru_1–NiO. *Source*: [11]/with permission of John Wiley & Sons; and (c) OER on Ir_1/NFS. *Source*: [38]/Springer Nature/CC BY 4.0.

environment of Ir in Ir—S—M (M=Fe or Ni) bonds that aids formation of surface OOH (*OOH) from surface O (*O) based on density functional theory (DFT) calculations (Figure 12.4c) [38].

This coordination can favor reduction reactions such as HER and the oxygen reduction reaction (ORR), as was shown by Chen et al. [41], Peng et al. [42] and Liu et al. [43]. For example, Liu et al. showed that Fe—N_4 centers supported on N-doped carbon (NC) material improved with P—O bonds (P—O/Fe—N—C-SAC) and outperformed 20 wt.% Pt/C in ORR. Specifically, the half-cell potential was 0.912 V vs. RHE compared to 0.864 V vs. RHE for Pt/C, and the limiting current density at 0.2 V vs. RHE was −5.48 mA cm^{-2} for P—O/Fe—N—C-SAC vs. −4.91 mA cm^{-2} for Pt/C [43]. Favorable ORR with this SAC was further supported by DFT results, which showed that introduction of P—O bonds to the support matrix increased the electron density of Fe, lowering the reaction energy barrier for *OOH (Figure 12.4a). Similar SACs with Co centers integrated within an unsymmetric Co–N_5S_1 moiety, along with NC complexes embedded in amberlite-derived N, P, and S co-doped carbon frameworks (Co_{SA+NC}/NPSC), exhibited ORR/OER bifunctional performance in 1 M KOH. The half-wave potential of Co_{SA+NC}/NPSC was measured at 0.827 V vs. RHE, which is slightly lower than that reported for the Fe–N_4 SAC, and an overpotential of 350 mV at 10 mA cm^{-2} current density for OER [42].

In addition, a tungsten-based SAC supported on metal–organic framework-derived NC (W-SAC) with W sites coordinated in a $W_1N_1C_3$ moiety required only 85 mV of overpotential to reach 10 mA cm^{-2} current density with a small Tafel slope of 53 mV dec^{-1} during HER in a 0.1 M KOH alkaline electrolyte, which is the same as state-of-the-art 20 wt.% Pt/C [41]. The implementation of M–O_4 coordination can be seen in Ru_1–NiO SAC with single Ru atoms doped into the lattice site of Ni atoms of NiO. This catalyst outperformed bulk RuO_2 and NiO and exhibited a low HMFOR onset potential of 1.068 V vs. RHE when tested under neutral conditions in 1 M PBS with 50 mM 5-HMF. Unlike SACs reported before, the Ru center in Ru_1–NiO indirectly participated in the reaction—the Ru site acted as an active center for water dissociation into adsorbed hydroxyl intermediates (*OH) through coupled electron–proton transfer. These *OH species, rather than Ru atoms directly, drove the oxidation reaction (Figure 12.4b) [11]. These findings highlight the critical role of atomic-level control in electrocatalyst design to develop more energy-efficient electrochemical systems.

While SACs have demonstrated improved performance under diverse reaction conditions, elucidating the deactivation mechanism is essential to enhance their long-term activity and guide the development of next-generation electrocatalysts. A review paper by Chae et al. summarized the deactivation mechanisms for SACs in thermocatalysis, which could be extended to electro- and photocatalytic processes (Figure 12.3b) [40]. The most common deactivation mechanisms for SACs were SA (i) sintering into NPs, (ii) embedding into the matrix of the support material, (iii) leaching into solution, and (iv) active-site poisoning. The latter is of critical importance when working with real feedstocks with fluctuating compositions or harsh pH conditions. The duration of stability tests reported in the literature ranges from 10 minutes to 100 hours (Table 12.1); thus, the long-term stability of SACs and possible deactivation mechanisms remain to be explored. For example, one of the promising Ru_1–NiO SACs for HMFOR in neutral media (1 M PBS) was evaluated for only about 83 minutes at 1.3 V vs. RHE (~1 mA cm^{-2}) and a low concentration of 5-HMF (50 mM), which may be a limiting factor for continuous tests. Co_{SA+NC}/NPSC and Fe–N_4 SACs were tested for 400 and 833 minutes at 0.7 and 0.85 V vs. RHE and demonstrated ~9% and 15% losses of current density (with respect to an initial current density of 10 mA cm^{-2} for both cases), respectively. To further demonstrate the reliability of SACs, zinc–air batteries were assembled using these two electrocatalysts (Co_{SA+NC}/NPSC and Fe–N_4) with power densities of 62.18 and 166.5 mW cm^{-2} that delivered a smaller charge–discharge voltage gap and better durability over 105 (~0 °C) and 300 hours, respectively [42, 43]. Co_{SA+NC}/NPSC additionally performed well with no visible deactivation for 60 hours of continuous OER at constant 10 mA cm^{-2} (~1.59 V). A study using single-molecule fluorescence microscopy revealed that SACs in general, and Pt SAC particularly, possess higher resistance to the so-called "catalysis-induced surface restructuring" compared to Pt NPs. HAADF–STEM analysis results after 10 hours of fluorogenic reaction revealed almost no variation on Pt SAC sites supported on CeO_2, while observable crystallographic restructuring was detected in Pt NP structures. The authors reported that the interaction energy between a single Pt atom and the CeO_2 support is approximately twice as favorable compared to the binding energy of a Pt atom coordinated with neighboring Pt atoms in Pt NP, indicating a stronger metal–support binding in the SA configuration (estimated values from DFT calculations are −12.6 eV for Pt SAC on CeO_2 vs. −6.97 eV for Pt NP) [34]. Another SAC, Ir_1/NFS, demonstrated about 350 hours of stability at 100 mA cm^{-2} and ~1.5 V vs. RHE in 1 M KOH with only ~3.7% of activity decay, which was better than each Ir/C (5 wt.%) and IrO_2 bulk catalysts with 4% decay after 25 hours (Figure 12.4c) [38]. In thermocatalytic conversion of cellulose to ethylene glycol at 245 °C and 60 bar H_2, the Ni–N–C SAC maintained stable activity over seven cycles, with each cycle lasting 105 minutes, whereas the Ni/AC NP catalyst lost 50% of its initial activity by the second cycle due to NP agglomeration [44].

Currently, a key limitation of SACs is their inability to perform multi-electron transfers, which is most likely due to the isolation of active sites. For example, Ni SACs on nitrogen-doped carbon nanotubes (NiSAC@N-CNTs), Ni SAs on the surface of the carbon nanotubes (Ni_{SA}/CNTs), metal–nitrogen sites doped within defective carbon nanotubes (Fe, Co, or Ni–CNTs-MW), and SACs used for the CO_2 electrocatalytic reduction (CO_2RR) typically yield only CO and are unable to promote further multi-electron transfer steps toward more reduced products [32, 45, 46]. Liu et al. highlighted the importance of the Langmuir–Hinshelwood (LH) mechanism and neighboring active sites in facilitating the further reduction of *CO on the Cu surface using DFT calculations. The authors showed that the pristine metal–nitrogen–carbon (M–N–C) SACs always fail to achieve CO_2RR due to the lack of active sites near the MN_4 centers that can place the * H species decomposed from H_2O (a key step in the LH mechanism), given that the graphene skeleton is intrinsically inert. They suggested a regulation strategy to enable CO_2RR on the SACs by doping the graphene skeleton with a heteroatom (B, P, and S) to facilitate * H formation. The authors extended the study to X–FeN_4 and X–MnN_4 systems (X = B, P, and S) and observed electron density redistribution on X–FeN_4, which makes these systems dynamically more favorable than the primitive FeN_4. However, they exhibited higher kinetic barriers for the LH step [47]. This insight enables the rational design of SACs to enhance their catalytic performance in multi-electron transfer reactions.

12.3 HzOR as a Promising Alternative to the OER in Hybrid WE

Hydrazine (N_2H_4) possesses several favorable properties: it is a liquid at ambient conditions (boiling point of 114 °C), is completely miscible with water, and offers a high theoretical energy density of 5 kWh kg^{-1} of N_2H_4—comparable to that of methanol (6.1 kWh kg^{-1}) [48]. Additionally, its complete oxidation produces only nitrogen gas (N_2), avoiding CO_2 emissions during its utilization (Figure 12.1) [14]. Therefore, HzOR has gained attention as an attractive alternative due to its lower thermodynamic oxidation potential ($E^0 = -0.38$ V vs. RHE), which is nearly 1.2 V lower than that of OER ($E^0 = 1.23$ V vs. RHE) [49, 50]. This allows hybrid WE systems utilizing HzOR as the anodic process to operate at lower full electrolyzer voltages, often <0.6 V at an operating current density of 200 mA cm^{-2} (~13.58 kWh $[kgH_2]^{-1}$), thereby lowering energy consumption by up to 80% compared to commercial AWE and PEMWE (~55.65 kWh $[kgH_2]^{-1}$) [14, 51]. From a mechanistic standpoint, HzOR is a four-electron process ($N_2H_4 \rightarrow N_2 + 4H^+ + 4e^-$) with relatively low activation barriers (0.20–0.40 eV vs. 0.45–0.70 eV for OER), potentially promoting reaction kinetics under acidic, neutral, and alkaline conditions [52, 53]. There is a

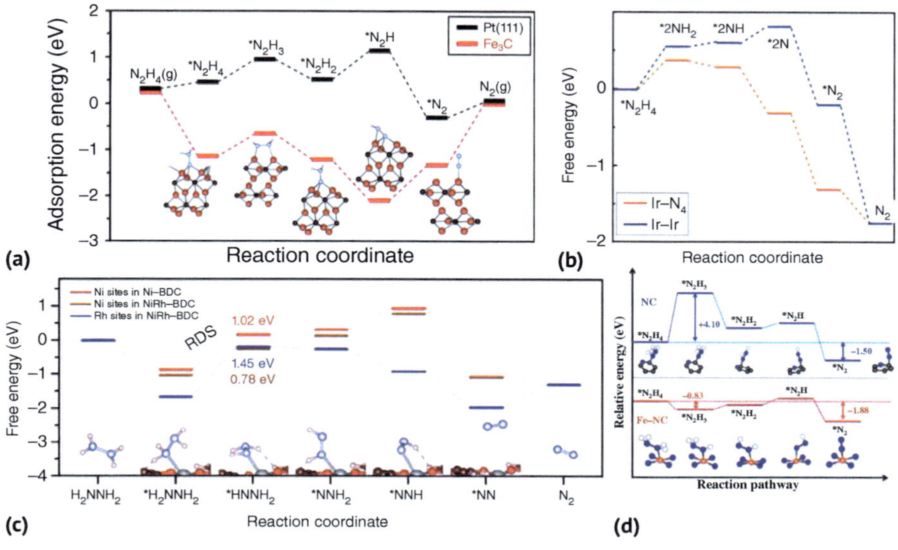

Figure 12.5 HzOR reaction pathways calculated via DFT methods for SACs with (a) Fe_3C and Pt (111), (b) Ir−N_4 and Ir−Ir, (c) Rh site in NiRh−BDC, and (d) NC and Fe—N—C. Panel (a): *Source*: [57]/with permission of The Royal Society of Chemistry. Panel (b): *Source*: [54]/John Wiley & Sons/CC BY 4.0. Panel (c): *Source*: [55]/with permission of American Chemical Society. Panel (d): *Source*: [56]/with permission of American Chemical Society.

commonly accepted reaction mechanism for HzOR to proceed in an alkaline electrolyte, i.e., $N_2H_4 \rightarrow {}^*N_2H_4 \rightarrow {}^*N_2H_3 \rightarrow {}^*N_2H_2 \rightarrow {}^*N_2H \rightarrow {}^*N_2 \rightarrow N_2$. In an acidic electrolyte, however, the reaction may start with undesired dissociation of the N—N bond, i.e., $N_2H_4 \rightarrow {}^*N_2H_4 \rightarrow {}^*2NH_2 \rightarrow {}^*2NH \rightarrow 2N \rightarrow {}^*N_2 \rightarrow N_2$ (Figure 12.5) [54–56].

Overpotential is a key parameter in electrochemical reactions, and HzOR is no exception. For over more than 50 years of research, electrocatalysts for HzOR have demonstrated that the overpotential required to initiate and sustain the reaction governs its efficiency. According to a review by Burshtein et al., HzOR active electrocatalysts can be categorized into five distinct categories, each exhibiting specific advantages and associated challenges (Figure 12.2b) [37].

Among these electrocatalyst categories, SACs are included under the classification of "ion catalysts." As previously mentioned, SACs exhibit high atomic utilization, making them ideal catalysts for mechanistic HzOR studies [57–59]. However, their synthesis often involves complex procedures, including precise control over coordination environments, which represents the main challenge associated with SACs (Figure 12.3a). Notably, SACs also demonstrated promising stability during HzOR, positioning them as competitive candidates among the various electrocatalyst categories. One of the earliest applications of SACs for alkaline HzOR was reported by Wang et al., where atomically dispersed Fe-N_x sites supported by hierarchically porous carbon membranes (FeN_4/HPCM) had an onset potential of 0.2 V vs. RHE, which was 0.4 V lower than that of a pure carbon sample (0.6 V vs. RHE) in

1 M KOH with 100 mM of N_2H_4. DFT calculations revealed that the activation of $N_2H_4^*$ to yield $N_2H_3^*$ ($^*N_2H_4 \rightarrow {}^*N_2H_3 + H^+ + e^-$) was the rate-determining step during HzOR [59]. Among the three M–N_4–CC SACs (M=Fe, Co, and Ni; CC = carbon cloth) tested in HzOR in 1 M KOH with 100 mM N_2H_4, the Fe-based SAC followed the four-electron reaction pathway to produce N_2 and demonstrated the lowest onset potential (0.303 V vs. RHE) compared to the Co- and Ni-based SACs (0.366 and 0.415 V vs. RHE, respectively). The lower onset potential of the Fe-based SAC was associated with a higher electronic charge of Fe in the M–N_4–C configuration and a lower barrier for $^*N_2H_4 \rightarrow {}^*N_2H_3$ calculated from DFT methods (0.77 eV) compared to the Co- and Ni-based SACs (0.96 and 1.14 eV, respectively) [60]. Burshtein et al. [57] further supported the importance of Fe–N_x coordination in SAC by showing that Fe_3C centers in Fe–N–C materials had negligible activity for HzOR, with onset potentials ranging from 0.35 to 0.7 V vs. RHE and low current densities (not specified). DFT calculations also revealed that Fe_3C centers act as active sites that strongly bind nitrogen-based intermediates (Figure 12.5a) [57]. Xu et al. reported that 1,4-benzenedicarboxylate ligands with Ni atoms (Ni–BDC) partially substituted by Rh atoms ($NiRh_{0.016}$–BDC) can act as promising bifunctional electrocatalysts for both HER and HzOR in a coupled system (Figure 12.5c) [55]. Compared to Ni–BDC (421 mV), $NiRh_{0.016}$–BDC required a lower overpotential (17 mV) at the anode and 0.06 V of full-cell potential to reach 10 mA cm^{-2} during HzOR in 1 M KOH with 0.3 M N_2H_4. DFT calculations suggested that Rh atoms actively participate in charge transfer with neighboring Ni atoms in $NiRh_{0.016}$–BDC, thus lowering the energy barrier for the N_2H_4 dehydrogenation step on Ni active sites from 1.45 eV in Ni–BDC to 0.78 eV in $NiRh_{0.016}$–BDC.

Ni centers on MXenes (Ni SACs/$Ti_3C_2T_x$) showed an onset potential of −0.03 V vs. RHE for HzOR in 1 M KOH with 100 mM N_2H_4, which was enabled by the strong interaction of Ni^{z+} ions (z from 0 to 2) with defects in the MXene support [61]. Jiao et al. systematically explored single transition-metal atoms anchored on nitrogenated holey graphene (TM@C_2N, TM = Ti, Mn, Fe, Co, Ni, Cu, Mo, Rh, Ru, Pd, Pt, Au, Ag, and W) as potential catalysts for HzOR using DFT [62]. Among all systems, Ru-, Mo-, Ti-, Co-, and Fe-based SACs on C_2N can be considered as promising HzOR catalysts due to their low limiting potential (U_L), which was derived from the free energy change of the potential-determining step (ΔG_{PDS}) along the lowest-energy pathway. While Ru-, Co-, and Fe-based SACs have already been experimentally studied, there remain gaps in the synthesis and HzOR application of Mo- and Ti-based SACs [51, 63].

The various works evaluating SACs for HzOR described in this chapter are summarized in Table 12.1, highlighting that the HzOR conditions were different for each test reported. Therefore, we recommend caution when correlating performance with SAC categories because the electrode areas, catalyst loadings, and reaction conditions were different for all the tests.

While various studies have focused on elucidating the activity and mechanistic behavior of SACs for HzOR, some more advanced investigations have progressed beyond fundamental characterization. For example, FeN_4/HPCM,

Table 12.1 Summary of works reporting SACs for HzOR (NR = not reported).

SACs	Electrode area (cm^2)	Catalyst loading ($mg\ cm^{-2}$)	Electrolyte	Hz conc. (mM)	OCV (V)	Onset potential (V)	Potential (V vs. RHE)	Current density ($mA\ cm^{-2}$)	Cell voltage (V)	Length of stability test (h)	Ref.
Cobalt/nitrogen co-doped carbon nanosheet arrays on CC (CC@CoNC)	1	NR	1 M KOH + seawater	500	NR	NR	−0.042	10	0.032	45	[51]
CC@WS_2/Ru	1	5	1 M KOH	500	NR	NR	−0.074	10	0.015	100	[64]
Fe—N—C	0.196	0.408	1 M KOH	100	1.6	0.27	NR	10	1.10	NR	[56]
Fe_3C	0.196	1	1 M KOH	20	NR	0.35	NR	NR	NR	NR	[57]
FeN_4/HPCM	0.50	0.60	1 M KOH	100	NR	0.2	0.6	70	NR	1	[59]
Fe_{SA}/CNTs	NR	2	1 M KOH	100	0.83	0.303	0.77	4	NR	0.1	[60]
Co^{2+} ions on the nitrile group of cyanographene ((CN)–Co)	0.071	NR	0.1 M PBS	50	NR	0.16	0.34	1	NR	2.5	[65]

(Continued)

Table 12.1 (Continued)

SACs	Electrode area (cm^2)	Catalyst loading ($mg\ cm^{-2}$)	Electrolyte	Hz conc. (mM)	OCV (V)	Onset potential (V)	Potential (V vs. RHE)	Current density ($mA\ cm^{-2}$)	Cell voltage (V)	Length of stability test (h)	Ref.
Ir-SA/NC	0.071	0.285	0.5 M H_2SO_4	330	NR	NR	0.39	10	NR	10	[54]
Lanthanoid coordination compounds with Fe–N–C	0.196		1 M KOH	100	NR	0.32	NR	NR	NR	NR	[58]
Ni SACs/ $Ti_3C_2T_x$	NR	NR	1 M KOH	100	NR	−0.03	0.1	70		6.5	[61]
$NiRh_{0.016}$–BDC	1	NR	1 M KOH	300	NR	NR	0.017	10	0.060	40	[55]
Pd-SAs/NiFe	1	1	1 M KOH	200	NR	0.3	0.55	10	NR	NR	[66]
Ru/MPNC	NR	1	1 M KOH	500	1.77		−0.039	100	0.600	10	[67]
Ru/PNC	NR	0.5	1 M KOH	500	1.75	NR	−0.02	100	0.700	10	[63]

Fe/N-doped carbon (Fe–NC), and Fe SAs supported on carbon nanotubes (Fe_{SA}/CNTs) anodes were tested in direct hydrazine fuel cells in 6 M KOH + 1 M N_2H_4, 4 M KOH + 3.12 M N_2H_4, and 1 M KOH + 1.5 M N_2H_4, resulting in peak power densities of 241, 392, and 48.6 mW $(cm^2\ mg_{cat})^{-1}$, respectively. The open-circuit potential values for Fe–NC and Fe_{SA}/CNTs were recorded as 1.6 and 0.83 V at 80 °C, respectively, while not being determined for FeN_4/HPCM [56, 59, 60]. Power generation in a self-driven direct hydrazine–H_2O_2 fuel cell was demonstrated with SACs, as well as Ru/MPNC (Ru particles in mesoporous NC) and Ru/PNC (Ru/porous NC composite) anodes, at 100 mA cm^{-2}, corresponding to 0.6 and 0.7 V cell voltages and peak power densities of 97.6 and 76.5 mW $(cm^2\ mg_{cat})^{-1}$, respectively. This validates SAC performance under close to real operating conditions and highlights their potential for energy conversion technologies [63, 67].

12.4 Challenges for WE Coupled With Hydrazine Oxidation on SACs

The integration of HzOR in electrolyzers represents a promising strategy for lowering the overall energy input required for H_2 production. However, several key challenges must be addressed before large-scale implementation can be realized.

- Safety and environmental risks of hydrazine: unlike water, N_2H_4 is a highly reactive compound classified as toxic, flammable, and potentially carcinogenic. Its use in electrochemical systems for H_2 production raises safety concerns during storage, handling, and operation. For example, the Occupational Safety and Health Administration (OSHA) limits exposure to 1 ppm over an 8-hour time-weighted average, reflecting its hazardous nature [68]. Thus, future research must prioritize the development of containment strategies, robust system sealing, and sensor-integrated safety protocols to mitigate these risks while maintaining electrocatalytic performance.
- Catalyst design and synthesis constraints: while SACs offer nearly 100% atom utilization and distinct coordination environments, their synthesis remains complex. Techniques such as atomic-layer deposition, wet impregnation followed by thermal activation, and MOF-derived methods require strict control over temperature, atmosphere, and precursor dispersion. These procedures are often labor-intensive, sensitive to contamination, difficult to automate, and hinder industrial transition. SACs further face intrinsic limitations when used in reactions requiring multi-electron transfers, such as HzOR. The isolated active sites in SACs typically lack the cooperative behavior seen in NPs or bulk catalysts, leading to insufficient adsorption or activation of complex intermediates. Advancements in dual-site SACs, heteroatomic coordination environments, or SAC-based composites may offer routes to overcome this bottleneck.
- Membrane crossover and efficiency losses: the crossover of nitrogen-containing species (e.g., unreacted N_2H_4) through membranes can lead to

lower electrolyzer efficiency and possible cathode poisoning [69]. Addressing this issue requires advances in membrane design, such as ion-selective barrier layers or dual-membrane architectures, to minimize crossover while maintaining ionic conductivity.

- Consistent electrocatalytic testing: as shown in Table 12.1, the electrocatalytic performance of SACs has not been consistently evaluated because electrochemical reaction conditions (e.g., half-cell potential, *iR* compensation, electrode surface area, concentration of reagents, pH, and test length) and SAC formulations (e.g., metal, wt.% loading, synthesis method, dopants, and support) were different. Furthermore, SACs are not consistently benchmarked against their non-SAC counterparts (i.e., electrocatalysts with metal NPs containing identical metal weight loading as the SACs and on the same support). Consequently, it is difficult to establish meaningful property–activity relationships or to determine whether SACs exhibit better long-term performance than NP-based catalysts. Standardizing reaction conditions and systematically comparing SACs with the corresponding NP-based counterparts will accelerate progress in the field.
- Full techno-economic analysis and life cycle assessment: while it is true that H_2 production from N_2H_4 requires lower full-cell potential compared to water, the energy costs of producing N_2H_4 cannot be ignored. Future studies evaluating the energy efficiency of the HzOR for H_2 (or energy) production should therefore assess the entire cycle, including N_2H_4 production, storage, and distribution. The resulting energy-efficiency, cost, and safety metrics must be benchmarked against conventional baselines such as WE for H_2 production or H_2 fuel cells (supplied by electrolytic H_2) for power generation to determine whether the N_2H_4 cycle offers a genuine advantage and to identify possible fields of application.

Despite these obstacles, HzOR remains an attractive anodic process with a theoretical oxidation potential of 0.23 V vs. RHE, offering a pathway to reduce the full-cell voltage of electrolyzers to below <1 V, compared to >1.5 V with traditional OER.

References

1 BloombergNEF (2020). Hydrogen economy outlook: Key messages. https://data.bloomberglp.com/professional/sites/24/BNEF-Hydrogen-Economy-Outlook-Key-Messages-30-Mar-2020.pdf
2 Hydrogen Council (2021). Hydrogen insights 2021. https://hydrogencouncil.com/
3 IEA (2021). Net zero by 2050: A roadmap for the global energy sector. https://www.iea.org/reports/net-zero-by-2050
4 Hydrogen Council (2023). Hydrogen insights 2023. https://hydrogencouncil.com/en/
5 UNIDO (2020). Global trends in hydrogen production and utilization. https://www.unido.org

6 IEA (2022). Global hydrogen review 2022. https://www.iea.org/reports/global-hydrogen-review-2022

7 Emam, A. S., Hamdan, M. O., Abu-Nabah, B. A., & Elnajjar, E. (2024). A review on recent trends, challenges, and innovations in alkaline water electrolysis. *Int. J. Hydrog. Energy 64*, 599–625. https://doi.org/10.1016/j.ijhydene.2024.03.238

8 Staffell, I., Scamman, D., Abad, A. V., Balcombe, P., Dodds, P. E., Ekins, P., Shah, N., & Ward, K. R. (2019). The role of hydrogen and fuel cells in the global energy system. *Energy Environ. Sci. 12*(2), 463–491. https://doi.org/10.1039/C8EE01157E

9 Lagadec, M. F. & Grimaud, A. (2020). Water electrolysers with closed and open electrochemical systems. *Nat. Mater. 19*(11), 1140–1150. https://doi.org/10.1038/s41563-020-0788-3

10 IRENA (2020). Green hydrogen cost reduction: Scaling up electrolysers to meet the 1.5 °C climate goal. https://www.irena.org/publications

11 Ge, R., Wang, Y., Li, Z., Xu, M., Xu, S.-M., Zhou, H., Ji, K., Chen, F., Zhou, J., & Duan, H. (2022). Selective electrooxidation of biomass-derived alcohols to aldehydes in a neutral medium: Promoted water dissociation over a nickel-oxide-supported ruthenium single-atom catalyst. *Angew. Chem. Int. Ed. 61*(19), e202200211. https://doi.org/10.1002/anie.202200211

12 Ren, J.-T., Chen, L., Wang, H.-Y., Tian, W.-W., & Yuan, Z.-Y. (2024). Water electrolysis for hydrogen production: from hybrid systems to self-powered/catalyzed devices. *Energy Environ. Sci. 17*(1), 49–113. https://doi.org/10.1039/D3EE02467A

13 Terekhina, I., White, J., Cornell, A., & Johnsson, M. (2023). Electrocatalytic oxidation of glycerol to value-added compounds on Pd nanocrystals. *ACS Appl. Nano Mater. 6*(13), 11211–11220. https://doi.org/10.1021/acsanm.3c01236

14 Yu, Z. & Liu, L. (2024). Recent advances in hybrid seawater electrolysis for hydrogen production. *Adv. Mater. 36*(13), 2308647. https://doi.org/10.1002/adma.202308647

15 Qiu, Y., Lopez-Ruiz, J. A., Sanyal, U., Andrews, E., Gutiérrez, O. Y., & Holladay, J. D. (2020). Anodic electrocatalytic conversion of carboxylic acids on thin films of RuO_2, IrO_2, and Pt. *Appl. Catal. B: Environ. 277*, 119277. https://doi.org/10.1016/j.apcatb.2020.119277

16 Wang, D., Wang, P., Wang, S., Chen, Y.-H., Zhang, H., & Lei, A. (2019). Direct electrochemical oxidation of alcohols with hydrogen evolution in continuous-flow reactor. *Nat. Commun. 10*(1), 2796. https://doi.org/10.1038/s41467-019-10928-0

17 Seh, Z. W., Kibsgaard, J., Dickens, C. F., Chorkendorff, I., Nørskov, J. K., & Jaramillo, T. F. (2017). Combining theory and experiment in electrocatalysis: Insights into materials design. *Science 355*(6321), eaad4998. https://doi.org/10.1126/science.aad4998

18 Shi, Y., Pan, H., Xia, J., Li, C., Gong, Y., Niu, L., Liu, X., Sun, C. Q., & Xu, S. (2021). Designing of highly efficient oxygen evolution reaction electrocatalysts utilizing a correlation factor: Theory and experiment. *ACS Appl. Mater. Interfaces 13*(26), 30533–30541. https://doi.org/10.1021/acsami.1c04829

19 Linnemann, J., Kanokkanchana, K., & Tschulik, K. (2021). Design strategies for electrocatalysts from an electrochemist's perspective. *ACS Catal. 11*(9), 5318–5346. https://doi.org/10.1021/acscatal.0c04118

20 Zhu, C., Fu, S., Shi, Q., Du, D., & Lin, Y. (2017). Single-atom electrocatalysts. *Angew. Chem. Int. Ed. 56*(45), 13944–13960. https://doi.org/10.1002/anie.201703864

21 Sun, T., Mitchell, S., Li, J., Lyu, P., Wu, X., Pérez-Ramírez, J., & Lu, J. (2021). Design of local atomic environments in single-atom electrocatalysts for renewable energy conversions. *Adv. Mater. 33*(5), 2003075. https://doi.org/10.1002/adma.202003075

22 Zhang, Q. & Guan, J. (2020). Single-atom catalysts for electrocatalytic applications. *Adv. Funct. Mater. 30*(31), 2000768. https://doi.org/10.1002/adfm.202000768

23 Zhu, L., Huang, J., Meng, G., Wu, T., Chen, C., Tian, H., Chen, Y., Kong, F., Chang, Z., Cui, X., & Shi, J. (2023). Active site recovery and N—N bond breakage during hydrazine oxidation boosting the electrochemical hydrogen production. *Nat. Commun. 14*(1), 1997. https://doi.org/10.1038/s41467-023-37618-2

24 Jirkovský, J. S., Panas, I., Ahlberg, E., Halasa, M., Romani, S., & Schiffrin, D. J. (2011). Single atom hot-spots at Au–Pd nanoalloys for electrocatalytic H_2O_2 production. *J. Am. Chem. Soc. 133*(48), 19432–19441. https://doi.org/10.1021/ja206477z

25 Gloag, L., Somerville, S. V., Gooding, J. J., & Tilley, R. D. (2024). Co-catalytic metal–support interactions in single-atom electrocatalysts. *Nat. Rev. Mater. 9*(3), 173–189. https://doi.org/10.1038/s41578-023-00633-2

26 Bell, A. T. (2003). The impact of nanoscience on heterogeneous catalysis. *Science 299*(5613), 1688–1691. https://doi.org/10.1126/science.1083671

27 Wang, A., Li, J., & Zhang, T. (2018). Heterogeneous single-atom catalysis. *Nat. Rev. Chem. 2*(6), 65–81. https://doi.org/10.1038/s41570-018-0010-1

28 Bergeret, G. & Gallezot, P. (2008). Particle size and dispersion measurements. In *Handbook of Heterogeneous Catalysis* (eds. G. Ertl, H. Knözinger, F. Schüth, and J. Weitkamp), 738–765. Hoboken, NJ: John Wiley & Sons, Ltd. https://doi.org/10.1002/9783527610044.hetcat0038

29 Luo, J., Waterhouse, G. I. N., Peng, L., & Chen, Q. (2023). Recent progress in high-loading single-atom catalysts and their applications. *Ind. Chem. Mater. 1*(4), 486–500. https://doi.org/10.1039/D3IM00062A

30 Qiao, B., Wang, A., Yang, X., Allard, L. F., Jiang, Z., Cui, Y., Liu, J., Li, J., & Zhang, T. (2011). Single-atom catalysis of CO oxidation using Pt_1/FeO_x. Nat. Chem. *3*(8), 634–641. https://doi.org/10.1038/nchem.1095

31 Hu, B., Sun, K., Zhuang, Z., Chen, Z., Liu, S., Cheong, W.-C., Chen, C., Hu, M., Cao, X., Ma, J., Tu, R., Zheng, X., Xiao, H., Chen, X., Cui, Y., Peng, Q., Chen, C., & Li, Y. (2022). Distinct crystal-facet-dependent behaviors for single-atom palladium-on-ceria catalysts: Enhanced stabilization and catalytic properties. *Adv. Mater. 34*(16), 2107721. https://doi.org/10.1002/adma.202107721

32 Zhang, W., Mehmood, A., Ali, G., Liu, H., Chai, L., Wu, J., & Liu, M. (2025). Nickel nanocluster-stabilized unsaturated Ni–N_3 atomic sites for efficient CO_2-to-CO electrolysis at industrial-level current. *Angew. Chem. Int. Ed. 64*(13), e202424552. https://doi.org/10.1002/anie.202424552

33 Mochizuki, C., Inomata, Y., Yasumura, S., Lin, M., Taketoshi, A., Honma, T., Sakaguchi, N., Haruta, M., Shimizu, K.-i., Ishida, T., & Murayama, T. (2022).

Defective NiO as a stabilizer for Au single-atom catalysts. *ACS Catal.* *12*(10), 6149–6158. https://doi.org/10.1021/acscatal.2c00108

34 Liu, X., Ge, X., Cao, J., Xiao, Y., Wang, Y., Zhang, W., Song, P., & Xu, W. (2022). Revealing the catalytic kinetics and dynamics of individual Pt atoms at the single-molecule level. *Proc. Natl. Acad. Sci. USA* *119*(14), e2114639119. https://doi.org/10.1073/pnas.2114639119

35 Zhou, Y., Slater, T. J. A., Luo, X., & Shen, Y. (2023). A versatile single-copper-atom electrocatalyst for biomass valorization. *Appl. Catal. B: Environ.* *324*, 122218. https://doi.org/10.1016/j.apcatb.2022.122218

36 Jia, C., Sun, Q., Liu, R., Mao, G., Maschmeyer, T., Gooding, J. J., Zhang, T., Dai, L., & Zhao, C. (2024). Challenges and opportunities for single-atom electrocatalysts: From lab-scale research to potential industry-level applications. *Adv. Mater.* *36*(42), 2404659. https://doi.org/10.1002/adma.202404659

37 Burshtein, T. Y., Yasman, Y., Muñoz-Moene, L., Zagal, J. H., & Eisenberg, D. (2024). Hydrazine oxidation electrocatalysis. *ACS Catal.* *14*(4), 2264–2283. https://doi.org/10.1021/acscatal.3c05657

38 Lei, Z., Cai, W., Rao, Y., Wang, K., Jiang, Y., Liu, Y., Jin, X., Li, J., Lv, Z., Jiao, S., Zhang, W., Yan, P., Zhang, S., & Cao, R. (2022). Coordination modulation of iridium single-atom catalyst maximizing water oxidation activity. *Nat. Commun.* *13*(1), 24. https://doi.org/10.1038/s41467-021-27664-z

39 Xia, C., Qiu, Y., Xia, Y., Zhu, P., King, G., Zhang, X., Wu, Z., Kim, J. Y., Cullen, D. A., Zheng, D., Li, P., Shakouri, M., Heredia, E., Cui, P., Alshareef, H. N., Hu, Y., & Wang, H. (2021). General synthesis of single-atom catalysts with high metal loading using graphene quantum dots. *Nat. Chem.* *13*(9), 887–894. https://doi.org/10.1038/s41557-021-00734-x

40 Chae, S., Chen, Y., Yang, J., & Su, J. (2024). The stability of single-atom catalysts in thermocatalysis. *Surf. Sci. Technol.* *2*(1), 18. https://doi.org/10.1007/s44251-024-00049-2

41 Chen, W., Pei, J., He, C.-T., Wan, J., Ren, H., Wang, Y., Dong, J., Wu, K., Cheong, W.-C., Mao, J., Zheng, X., Yan, W., Zhuang, Z., Chen, C., Peng, Q., Wang, D., & Li, Y. (2018). Single tungsten atoms supported on MOF-derived N-doped carbon for robust electrochemical hydrogen evolution. *Adv. Mater.* *30*(30), 1800396. https://doi.org/10.1002/adma.201800396

42 Peng, J., Xue, T., Li, Z., Shi, J., Wang, X., & Xu, B. (2025). Modulating electronic structure of Co-N_5S_1 sites in Co single atom catalysts via phosphorus incorporation and nanoclusters to promote oxygen electrocatalytic activity. *Energy Mater.* *5*(6), 500060. https://doi.org/10.20517/energymater.2024.247

43 Liu, M., Sun, T., Peng, T., Wu, J., Li, J., Chen, S., Zhang, L., Li, S., Zhang, J., & Sun, S. (2023). Fe-NC single-atom catalyst with hierarchical porous structure and P—O bond coordination for oxygen reduction. *ACS Energy Lett.* *8*(11), 4531–4539. https://doi.org/10.1021/acsenergylett.3c01998

44 Liu, W., Chen, Y., Qi, H., Zhang, L., Yan, W., Liu, X., Yang, X., Miao, S., Wang, W., Liu, C., Wang, A., Li, J., & Zhang, T. (2018). A durable nickel single-atom catalyst for hydrogenation reactions and cellulose valorization under harsh conditions. *Angew. Chem. Int. Ed.* *57*(24), 7071–7075. https://doi.org/10.1002/anie.201802231

45 Sun, J. W., Wu, X., Liu, P. F., Chen, J., Liu, Y., Lou, Z. X., Zhao, J. Y., Yuan, H. Y., Chen, A., Wang, X. L., Zhu, M., Dai, S., & Yang, H. G. (2023). Scalable synthesis of coordinatively unsaturated metal-nitrogen sites for large-scale CO_2 electrolysis. *Nat. Commun. 14*(1), 1599. https://doi.org/10.1038/s41467-023-36688-6

46 Yadav, D. K., Singh, D. K., & Ganesan, V. (2020). Recent strategy(ies) for the electrocatalytic reduction of CO_2: Ni single-atom catalysts for the selective electrochemical formation of CO in aqueous electrolytes. *Curr. Opin. Electrochem. 22*, 87–93. https://doi.org/10.1016/j.coelec.2020.04.008

47 Liu, T., Wang, Y., & Li, Y. (2023). Can metal–nitrogen–carbon single-atom catalysts boost the electroreduction of carbon monoxide? *JACS Au 3*(3), 943–952. https://doi.org/10.1021/jacsau.3c00026

48 Hwang, H., Hong, S., Kim, D.-H., Kang, M.-S., Park, J.-S., Uhm, S., & Lee, J. (2020). Optimistic performance of carbon-free hydrazine fuel cells based on controlled electrode structure and water management. *J. Energy Chem. 51*, 175–181. https://doi.org/10.1016/j.jechem.2020.03.081

49 Li, K., Zhou, G., Tong, Y., Ye, Y., & Chen, P. (2023). Interface engineering of a hierarchical P-modified Co/Ni_3P heterostructure for highly efficient water electrolysis coupled with hydrazine degradation. *ACS Sustain. Chem. Eng. 11*(38), 14186–14196. https://doi.org/10.1021/acssuschemeng.3c03909

50 Rosca, V., Duca, M., de Groot, M. T., & Koper, M. T. M. (2009). Nitrogen cycle electrocatalysis. *Chem. Rev. 109*(6), 2209–2244. https://doi.org/10.1021/cr8003696

51 Xin, Y., Shen, K., Guo, T., Chen, L., & Li, Y. (2023). Coupling hydrazine oxidation with seawater electrolysis for energy-saving hydrogen production over bifunctional CoNC nanoarray electrocatalysts. *Small 19*(21), 2300019. https://doi.org/10.1002/smll.202300019

52 Liu, F., Jiang, X., Wang, H.-H., Chen, C., Yang, Y.-H., Sheng, T., Wei, Y.-S., Zhao, X.-S., & Wei, L. (2022). Boosting electrocatalytic hydrazine oxidation reaction on high-index faceted Au concave trioctahedral nanocrystals. *ACS Sustain. Chem. Eng. 10*(2), 696–702. https://doi.org/10.1021/acssuschemeng.1c07700

53 Shi, J., Sun, Q., Chen, J., Zhu, W., Cheng, T., Ma, M., Fan, Z., Yang, H., Liao, F., Shao, M., & Kang, Z. (2024). Nitrogen contained rhodium nanosheet catalysts for efficient hydrazine oxidation reaction. *Appl. Catal. B: Environ. 343*, 123561. https://doi.org/10.1016/j.apcatb.2023.123561

54 Luo, F., Pan, S., Xie, Y., Li, C., Yu, Y., Bao, H., & Yang, Z. (2023). Hydrazine-assisted acidic water splitting driven by iridium single atoms. *Adv. Sci. 10*(32), 2305058. https://doi.org/10.1002/advs.202305058

55 Xu, X., Chen, H.-C., Li, L., Humayun, M., Zhang, X., Sun, H., Debecker, D. P., Zhang, W., Dai, L., & Wang, C. (2023). Leveraging metal nodes in metal–organic frameworks for advanced anodic hydrazine oxidation assisted seawater splitting. *ACS Nano 17*(11), 10906–10917. https://doi.org/10.1021/acsnano.3c02749

56 Zheng, Y., He, F., Chen, M., Zhang, J., Hu, G., Ma, D., Guo, J., Fan, H., Li, W., & Hu, X. (2020). Mimicking hydrazine dehydrogenase for efficient electrocatalytic oxidation of N_2H_4 by Fe–NC. *ACS Appl. Mater. Interfaces 12*(34), 38183–38191. https://doi.org/10.1021/acsami.0c10637

57 Burshtein, T. Y., Aias, D., Wang, J., Sananis, M., Farber, E. M., Gazit, O. M., Grinberg, I., & Eisenberg, D. (2021). Fe–N–C electrocatalysts in the oxygen and nitrogen cycles in alkaline media: The role of iron carbide. *Phys. Chem. Chem. Phys. 23*(47), 26674–26679. https://doi.org/10.1039/D1CP03650E

58 Salton, I., Ioffe, K., Burshtein, T. Y., Farber, E. M., Seraphim, N. M., Segal, N., & Eisenberg, D. (2022). Lanthanoid coordination compounds as diverse self-templating agents towards hierarchically porous Fe–N–C electrocatalysts. *Mater. Adv. 3*(21), 7937–7945. https://doi.org/10.1039/D2MA00596D

59 Wang, Y.-C., Wan, L.-Y., Cui, P.-X., Tong, L., Ke, Y.-Q., Sheng, T., Zhang, M., Sun, S.-H., Liang, H.-W., Wang, Y.-S., Zaghib, K., Wang, H., Zhou, Z.-Y., & Yuan, J. (2020). Porous carbon membrane-supported atomically dispersed pyrrole-type Fe-N_4 as active sites for electrochemical hydrazine oxidation reaction. *Small 16*(31), 2002203. https://doi.org/10.1002/smll.202002203

60 Zhang, J., Wang, Y., Yang, C., Chen, S., Li, Z., Cheng, Y., Wang, H., Xiang, Y., Lu, S., & Wang, S. (2021). Elucidating the electro-catalytic oxidation of hydrazine over carbon nanotube-based transition metal single atom catalysts. *Nano Res. 14*(12), 4650–4657. https://doi.org/10.1007/s12274-021-3397-9

61 Zhou, S., Zhao, Y., Shi, R., Wang, Y., Ashok, A., Héraly, F., Zhang, T., & Yuan, J. (2022). Vacancy-rich MXene-immobilized Ni single atoms as a high-performance electrocatalyst for the hydrazine oxidation reaction. *Adv. Mater. 34*(36), 2204388. https://doi.org/10.1002/adma.202204388

62 Jiao, D., Tian, Y., Wang, H., Cai, Q., & Zhao, J. (2020). Single transition metal atoms anchored on a C_2N monolayer as efficient catalysts for hydrazine electrooxidation. *Phys. Chem. Chem. Phys. 22*(29), 16691–16700. https://doi.org/10.1039/D0CP02930K

63 Guan, X., Wu, Q., Li, H., Zeng, S., Yao, Q., Li, R., Chen, H., Zheng, Y., & Qu, K. (2023). Identifying the roles of Ru single atoms and nanoclusters for energy-efficient hydrogen production assisted by electrocatalytic hydrazine oxidation. *Appl. Catal. B: Environ. 323*, 122145. https://doi.org/10.1016/j.apcatb.2022.122145

64 Li, J., Li, Y., Wang, J., Zhang, C., Ma, H., Zhu, C., Fan, D., Guo, Z., Xu, M., Wang, Y., & Ma, H. (2022). Elucidating the critical role of ruthenium single atom sites in water dissociation and dehydrogenation behaviors for robust hydrazine oxidation-boosted alkaline hydrogen evolution. *Adv. Funct. Mater. 32*(16), 2109439. https://doi.org/10.1002/adfm.202109439

65 Kadam, R. G., Zhang, T., Zaoralová, D., Medved', M., Bakandritsos, A., Tomanec, O., Petr, M., Chen, J. Z., Miller, J. T., Otyepka, M., Zbořil, R., Asefa, T., & Gawande, M. B. (2021). Single Co-atoms as electrocatalysts for efficient hydrazine oxidation reaction. *Small 17*(16), 2006477. https://doi.org/10.1002/smll.202006477

66 Liu, G., Nie, T., Wang, H., Shen, T., Sun, X., Bai, S., Zheng, L., & Song, Y.-F. (2022). Size sensitivity of supported palladium species on layered double hydroxides for the electro-oxidation dehydrogenation of hydrazine: From nanoparticles to nanoclusters and single atoms. *ACS Catal. 12*(17), 10711–10717. https://doi.org/10.1021/acscatal.2c02628

67 Wang, J., Guan, X., Li, H., Zeng, S., Li, R., Yao, Q., Chen, H., Zheng, Y., & Qu, K. (2022). Robust Ru-N metal-support interaction to promote self-powered H_2 production assisted by hydrazine oxidation. *Nano Energy 100*, 107467. https://doi.org/10.1016/j.nanoen.2022.107467

68 The National Institute for Occupational Safety and Health (NIOSH) (2019). Hydrazine. https://www.cdc.gov/niosh/npg/npgd0329.html

69 Rees, N. V. & Compton, R. G. (2011). Carbon-free energy: A review of ammonia- and hydrazine-based electrochemical fuel cells. *Energy Environ. Sci. 4*(4), 1255–1260. https://doi.org/10.1039/C0EE00809E

Part IV

Urea and Furfural-assisted Water Electrolysis

13

Metal Sulfides and Phosphides Electrocatalysts for Urea-assisted Water Electrolysis

Marciélli Karoline Rodrigues de Souza, Eduardo dos Santos Freitas Cardoso, and Marcos Roberto de Vasconcelos Lanza**

São Carlos Institute of Chemistry, University of São Paulo, São Carlos, Brazil

*Corresponding authors: marciellisouza@outlook.com.br; marcoslanza@usp.br

13.1 Introduction

The accelerated growing demand for energy worldwide has spurred the continuous reliance on fossil fuels, including coal, petroleum, and natural gas, and this has contributed toward exacerbating climate change and environmental degradation [1, 2]. In response to these critical policy and environmental challenges, the development of renewable and sustainable energy technologies has become, essentially, an imperative task. Among the wide range of alternatives to fossil fuels, hydrogen (H_2) is considered an outstanding alternative primarily because of its high energy density and its unique combustion by-product—water, which makes it entirely free of harmful emissions [3, 4]. The outstanding advantages and properties of hydrogen have spurred extensive research into diverse pathways for its production; these pathways include biomass conversion, thermochemical processes, and photocatalytic or electrocatalytic water-splitting techniques [5–7].

Among the wide range of approaches to hydrogen production, water electrolysis stands out because of its high efficiency and scalability for industrial applications. The overall process comprises two essential electrochemical half-reactions: (i) hydrogen evolution reaction (HER), which occurs at the cathode; and (ii) oxygen evolution reaction (OER), which takes place at the anode [4]. The use of OER in hydrogen production is constrained primarily by the following underlying factors: (i) its inherent need for high overpotential, and (ii) slow reaction kinetics; these factors hinder the energy efficiency and economic viability of water electrolysis technologies [8].

To address the challenges posed by the sluggish kinetics and high overpotential of OER, two main approaches have been widely explored: (i) the design of efficient and long-lasting electrocatalysts specifically for OER, and (ii) the substitution of

OER by alternative anodic reactions which involve low-oxidation-potential compounds such as small organic molecules [8, 9]. In this context, the electrochemical oxidation of urea has emerged as a promising route with great benefits.

Urea oxidation reaction (UOR) is regarded as an outstanding electrochemical oxidation technique due to its considerably lower thermodynamic onset potential of 0.37 V vs. the standard hydrogen electrode (SHE), which substantially decreases the amount of energy required for hydrogen production [10, 11]. Moreover, urea is an abundant nitrogenous compound which is commonly found in industrial effluents and municipal wastewater and whose electrochemical oxidation simultaneously facilitates wastewater remediation in addition to hydrogen production [8]. The physicochemical properties of urea, which include high water solubility, thermal stability, nonflammability, and ease of handling and storage, provide further reason and rationale for its use as an anodic feedstock [1, 12].

The advances made in the urea-assisted water-splitting technique are closely linked to the development of both active and stable electrocatalysts that are highly efficient for the simultaneous promotion of UOR and HER in alkaline media. Due to the high costs and limited durability of noble-metal catalysts, research efforts have increasingly been directed toward the development of alternative catalysts using transition-metal-based compounds—particularly transition-metal sulfides (TMSs) and transition-metal phosphides (TMPs), which are composed of earth-abundant elements [13]. These TMPs and TMSs exhibit superior electrical conductivity, high intrinsic catalytic activity, and better structural stability [1, 14]. Additionally, the ability to modulate the structural features—such as surface morphology, crystallinity, and composition—to obtain the needed or desirable properties essentially helps to regulate the catalytic performance and efficiency [15–18].

This chapter presents a comprehensive analysis of recent advances in the design and application of metal sulfides and phosphides-based electrocatalysts for urea-assisted electrolysis. The chapter investigates the correlations between structural features and catalytic activity, analyzes, in a thorough fashion, the synthesis and post-synthesis modification strategies, and sheds light on the mechanistic pathways that lead to UOR and HER under the application of metal sulfides and phosphides-based electrocatalysts. Additionally, the chapter offers a forward-looking perspective on the practical implementation of these electrocatalysts in clean hydrogen production and wastewater remediation.

13.2 The Mechanism Involving UOR in Alkaline Media

Urea dissociation in alkaline media is driven by electrical energy and water electrolysis, which lead to hydrogen production. The significantly lower theoretical cell potential (0.37 V vs. SHE) required in urea-assisted hydrogen production compared to conventional water electrolysis (1.23 V vs. SHE) highlights its energy-saving advantage [19, 20]. Both UOR and HER involve the transfer of six electrons, with hydroxide ions playing a fundamental role in maintaining the alkaline medium [1].

Hydrogen generation and urea oxidation are governed by two half-reactions: at the cathode, hydrogen is produced through HER; at the anode, urea oxidation produces nitrogen (N_2) and carbon dioxide (CO_2) [21]. As a beneficial by-product, purified water is obtained from both UOR and HER [1]. For a much clearer understanding of the electrochemical pathways involved, the fundamental half-reactions governing urea oxidation and hydrogen evolution are outlined below [1, 22]:

At the anode (UOR) in alkaline media:

$$CO(NH_2)_2\ (aq) + 6OH^- \leftrightarrows 5H_2O(l) + N_2(g) + CO_2(g) + 6e^-,\\ E^\theta = -0.46 \text{ vs. SHE} \tag{13.1}$$

At the cathode (HER) in alkaline media:

$$6H_2O\ (l) + 6e^- \leftrightarrows 3H_2(g) + 6OH^-,\ E^\theta = -0.83 \text{ V vs. SHE} \tag{13.2}$$

Overall catalytic reaction (UOR/HER):

$$CO(NH_2)_2 + H_2O \leftrightarrows N_2 + CO_2 + 3H_2,\ E^\theta = 0.37 \text{ V vs. SHE} \tag{13.3}$$

As shown in Eqs. (13.1)–(13.3), the oxidation of one mole of urea with six moles of hydroxide ions yields CO_2, H_2O, and N_2, with a theoretical potential of −0.46 V vs. SHE. On the cathodic side, the generation of molecular hydrogen requires a potential of −0.83 V vs. SHE. Taken together, these half-reactions indicate that hydrogen production via urea electrolysis can be energetically more favorable, as the overall theoretical cell voltage (Eq. 13.3) is notably lower than the standard 1.23 V vs. SHE required for the traditional water-splitting technique [1, 21, 22].

The UOR is characterized by slow reaction kinetics, which is primarily attributed to its intricate multistep pathway; this pathway involves a six-electron-transfer process [9]. Despite this limitation (slow reaction kinetics), the UOR offers a significant advantage: it can be carried out under alkaline conditions without the need for noble-metal catalysts, and this reduces costs and allows its application on a large scale [22, 23]. Bearing that in mind, numerous catalyst engineering approaches have been explored with a view to enhancing the performance of HER, aiming at improving the system efficiency and economic competitiveness [9].

Optimizing catalytic performance has been the central focus in materials research, and this has played a pivotal role in the development of metal sulfides and phosphides-based compounds. Owing to their abundance in nature, low production costs, and versatile physicochemical characteristics, these materials have attracted considerable attention among researchers and have been increasingly employed as efficient bifunctional catalysts for both HER and UOR in alkaline media [16]. The ongoing efforts focused on tailoring the catalytic behaviors of metal sulfides and phosphides-based compounds through compositional tuning and electronic structure engineering remain essential when it comes to fully exploiting the benefits of urea-assisted electrolysis, as this contributes to the advancement of sustainable and energy-efficient hydrogen generation technologies [24].

13.3 Synthesis of Sulfides and Phosphides-based Electrocatalysts

The synthesis of efficient metal sulfides and phosphides-based electrocatalysts for UOR and urea electrolysis relies heavily on the strategic application of wet-chemical methods that offer precise control over morphology, surface properties, and crystallinity [9, 16, 25]. To enhance the catalytic efficiency, combined techniques are often employed. Among these techniques, hydrothermal, solvothermal, and electrodeposition techniques have gained prominence due to their adaptability, scalability, and ability to produce nanostructured materials with enhanced catalytic behavior [26–28].

13.3.1 Integrated Techniques for the Synthesis of Sulfides and Phosphides-based Electrocatalysts

13.3.1.1 Hydrothermal, Solvothermal, and Electrochemical Synthesis Techniques

Hydrothermal and solvothermal methods involve the action of water and organic solvents, respectively, under high temperature and pressure in a closed system, typically using Teflon-lined vessels placed inside a stainless-steel autoclave [27, 28]. These techniques provide an optimal environment for controlled crystal growth, allowing precise regulation of morphology and particle size by adjusting time, temperature, pressure, and pH. Hydrothermal and solvothermal techniques are highly efficient in the sense that their application leads to high yields, excellent purity, and superior homogeneity of the final products [27].

In a recent study conducted by Yang et al., [18], the authors reported using hydrothermal treatment and electrodeposition for the synthesis and formation of crystalline/amorphous structure, NiFeS@NiMoP, grown in situ on nickel foam (NiFeS@NiMoP/NF); according to the authors, the hydrothermal method played a crucial role in the formation of NiFeS nanosheets, ensuring precise regulation of morphology and crystallinity before electrodeposition. The reaction mechanism involved the integration of crystalline and amorphous phases and the synergistic interaction between TMSs and TMPs, which contributed significantly to the improvement of the reaction kinetics. The structural configuration adopted in the aforementioned study helped enhance the adsorption of key intermediates, and this resulted in superior bifunctional catalytic performance.

Similarly, solvothermal techniques have enabled the fabrication of catalysts like Co–Ni–S@NF, where the morphology and surface chemistry are influenced by the solvent environment. The inclusion of cobalt has been shown to enhance charge-transfer dynamics and facilitate the formation of NiOOH species, which is crucial for urea oxidation (using only 1.35 V at 100 $mA{\cdot}cm^{-2}$, with 100 hours of stability) [29].

Electrodeposition can be used as a complementary technique, often in tandem with hydrothermal synthesis; this technique requires the immersion of two or three electrodes in a specific electrolytic solution, where an external potential is applied in order to promote metal deposition at the cathode. The composition and thickness of the films can be precisely controlled by adjusting the applied potential, electrode current density, electrolyte concentration, and deposition time. This technique offers significant advantages, including the controlled growth of standardized nanostructured crystals and the guarantee of high purity and excellent conductivity [26]. Additionally, the electrodeposition technique is easy to execute and is highly adaptable for customization, making it a viable and efficient approach for electrocatalyst synthesis [30].

Another notable example is the $CoP@Ni_3S_2$ catalyst which is synthesized by interfacial engineering through hydrothermal synthesis and subsequent electrodeposition [16]. The electrodeposition functionalization of Ni_3S_2 elevated the surface oxidation state of CoP (increasing Co^{3+}, crucial for the formation of active CoOOH in UOR), while CoP modulated the electronic structure of sulfur in Ni_3S_2, enhancing the HER activity. This hierarchical architecture yielded excellent performance metrics, including low Tafel slopes (27.7 $mV{\cdot}dec^{-1}$ for UOR and 115.4 $mV{\cdot}dec^{-1}$ for HER) and stability of 27 hours at 50 $mA{\cdot}cm^{-2}$.

The synergy between these fabrication techniques provides a powerful pathway for the construction of electrocatalyst with hierarchical architecture, optimized electronic structures, and abundant active sites, designed to increase the efficiency of urea-assisted hydrogen production.

13.4 Structural and Catalytic Performances of Metal Sulfides and Phosphides

TMSs and TMPs have been used for the construction of a new generation of non-noble-metal-based bifunctional electrocatalysts. The unique electronic structure, surface redox properties, and structural versatility of these TMSs and TMPs-based electrocatalysts have contributed to the significant improvements observed in urea-assisted hydrogen production technologies [9].

Electrocatalysts derived from transition metals, such as Co, Ni, Mo, Fe, and Cu, in the form of sulfides or phosphides, have demonstrated outstanding electrical conductivity, high catalytic performance, and notable redox behavior; in addition to that, the metals used in the construction of these catalysts are composed of elements that are earth abundant [22, 31–34]. During electrolysis, these electrocatalytic materials often undergo in-situ reconstruction, forming active oxyhydroxide species like CoOOH and NiOOH. These dynamic surface phases are known to facilitate the adsorption of intermediate reactions and electron-transfer processes, which are key factors when it comes to enhancing catalytic kinetics and boosting the overall efficiency of HER [1, 35].

The ability of the materials to host multiple oxidation states, combined with their high electrical conductivity and surface adaptability, creates a dynamic environment that supports both HER and UOR under alkaline conditions [11, 14, 17, 32]. These properties make sulfides and phosphides-based materials ideal platforms for the

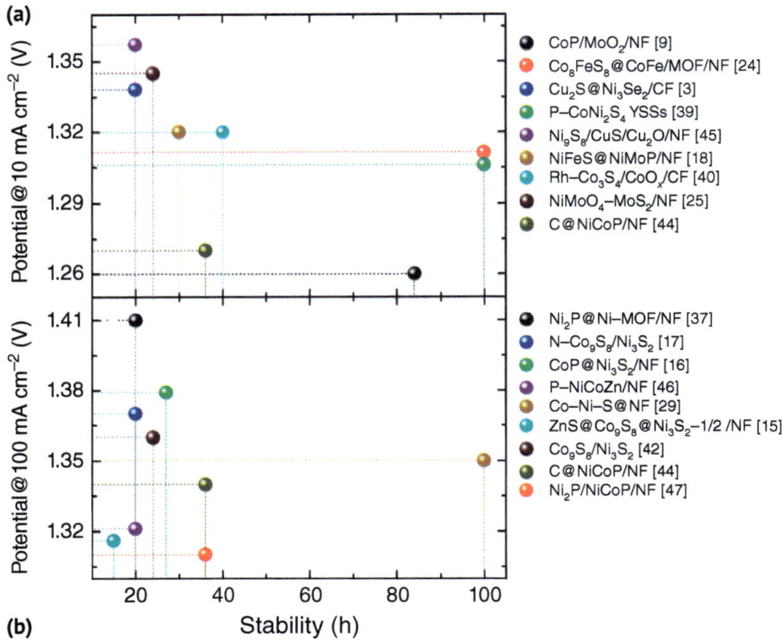

Figure 13.1 Comparative analysis of stability in UOR under varying potentials at (a) 10 $mA{\cdot}cm^{-2}$ and (b) 100 $mA{\cdot}cm^{-2}$ using recently reported UOR reference electrocatalysts.

rational design of highly efficient and cost-effective catalysts. As shown in Figure 13.1, when applied for UOR, sulfides and phosphides-based electrocatalysts have been found to exhibit excellent electrochemical stability under low potential values.

This section presents a comprehensive overview of the roles that structural, morphological, compositional, and hybrid design strategies play in modulating catalytic performance under alkaline conditions.

13.4.1 Morphological and Interface Engineering

The efficiency and performance of electrocatalysts are intricately linked to the control and modulation of their morphology. The use of conductive structures, such as nickel foam (NF), copper foam (CF), and carbon cloth—which allows the uniform growth of the catalysts, in the design of electrocatalysts provides us with a robust electrical pathway for enhancing the catalytic performance [3, 23, 36]. A number of studies have demonstrated the advantages of morphological engineering.

13.4.1.1 Heterostructures Engineering

Heterostructured catalysts create distinct active sites with complementary properties; furthermore, they enhance the overall performance of the system by forming stable nucleophilic and electrophilic regions that optimize molecular adsorption, dissociation, and reaction kinetics [37]. These principles have been exemplified in recent bifunctional catalysts designed and applied successfully in urea-assisted electrolysis.

Zhu et al. [25] developed and characterized a hexagonal nanosheet heterostructure composed of $NiMoO_4$ and MoS_2, directly grown on NF, which they referred to as $NiMoO_4$–MoS_2/NF. The catalyst was synthesized via a two-step hydrothermal method followed by thermal annealing, and this gave rise to a well-defined hexagonal morphology, as confirmed by scanning electron microscopy (SEM) and energy-dispersive X-ray spectroscopy (EDS) mapping, which revealed a homogeneous distribution of Ni, Mo, O, and S—see Figure 13.2a–c. The high-resolution transmission electron microscopy (HRTEM) images (Figure 13.2d,e) exhibited 0.216, 0.234, and 0.236 nm lattice spacings, which corresponded to the coexistence and effective interfacial interaction of the crystalline $NiMoO_4$ and MoS_2 phases. Complementary X-ray photoelectron spectroscopy (XPS) analyses pointed to significant electronic modulation. The Ni-2p spectrum (Figure 13.2f) exhibited positive binding energy shifts after hybridization, indicating enhanced d-orbital vacancy and stronger interaction with OH* intermediates. The Mo-3d spectrum (Figure 13.2g) identified both Mo^{6+} and Mo^{4+} species, while the additional S-2p peak (Figure 13.2h) around

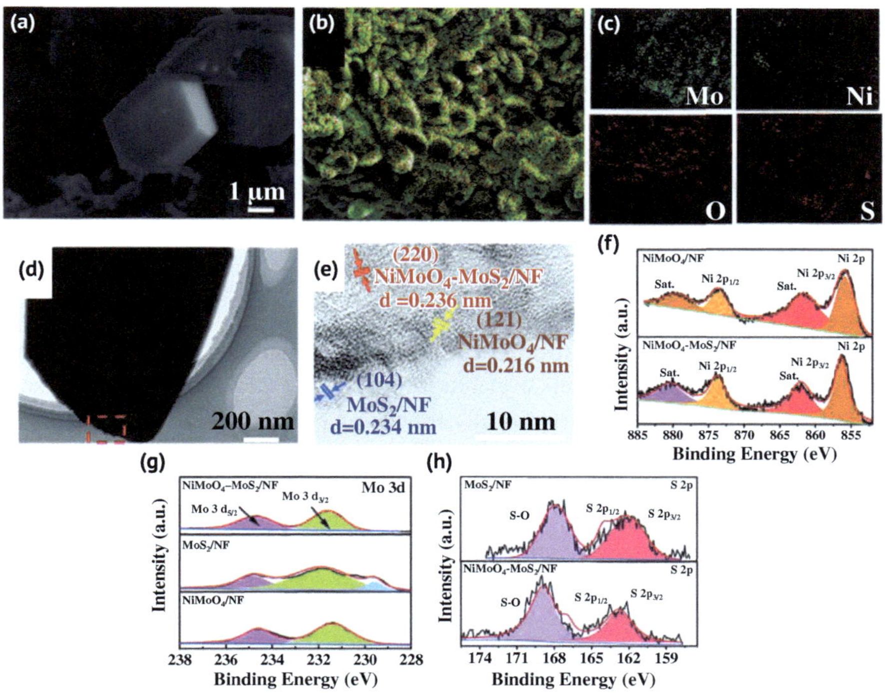

Figure 13.2 (a) SEM images of $NiMoO_4$–MoS_2/NF; (b, c) elemental mapping images of Mo, Ni, O, and S; (d, e) HRTEM images; and (f–h) XPS spectra of Ni-2p, Mo-3d, and S-2p. *Source:* [25]/with permission of The Royal Society of Chemistry.

161.7–162.9 eV, associated with MoS_2, pointed to an increase in electron density on S^{2-}, which contributed toward enhancing H* adsorption. Electrocatalytic tests conducted in an electrolyte containing 1 M KOH and 0.5 M urea pointed to excellent catalytic activity, with a remarkably low overpotential of 23 mV for HER and 1.345 V (vs. RHE) for UOR at a current density of 10 $mA \cdot cm^{-2}$. The system also presented nonnegligible advantages with Tafel slopes of 36.5 $mV \cdot dec^{-1}$ for HER and 77.8 $mV \cdot dec^{-1}$ for UOR, low charge-transfer resistance, and high double-layer capacitance (C_{dl} = 29.1 $mF \cdot cm^{-2}$ for HER; 34.4 $mF \cdot cm^{-2}$ for UOR). When applied in a two-electrode setup for urea-assisted water splitting, a cell voltage of only 1.445 V was sufficient to reach 10 $mA \cdot cm^{-2}$; also, the system exhibited excellent electrochemical stability over 24 hours with minimal current loss. The authors attributed the outstanding electrocatalytic performance to the formation of a hexagonal nanosheet architecture, which offers an extended surface area and facilitates effective charge transfer. In addition, the synergistic interface between the oxide and sulfide phases enhances the intrinsic reactivity of the material and exposes more active sites. These combined structural and electronic advantages show that $NiMoO_4$–MoS_2/NF is a highly efficient bifunctional catalyst which can be used simultaneously for hydrogen generation and the treatment of nitrogen-rich wastewater.

Using a structure similar to $NiMoO_4/MoS_2$, Fereja et al. [38] developed a bifunctional electrocatalyst of a three-dimensional hybrid heterostructure of tungsten-doped NiS_2/MoO_2 (W–NiS_2/MoO_2@CC) grown on carbon cloth. The W–NiS_2/MoO_2@CC catalyst was synthesized via a hydrothermal process followed by sulfurization annealing. Regarding the electrochemical parameters, the catalyst yielded an overpotential of 1.3 V (vs. RHE) at 10 $mA \cdot cm^{-2}$ for UOR, with a Tafel slope of 24.1 $mV \cdot dec^{-1}$, and an overpotential of 52 mV at 10 $mA \cdot cm^{-2}$ for HER, with a Tafel slope of 49 $mV \cdot dec^{-1}$. The heterojunction between NiS_2 and MoO_2, along with defect sites, helped improve the charge transfer and electronic redistribution. Moreover, tungsten doping and oxygen vacancies effectively modulated the electronic structure, reducing the activation energy for C—N bond cleavage and optimizing hydrogen adsorption free energy (ΔG_{H^*}).

Liu et al. [9] synthesized a strongly coupled heterostructure catalyst (CoP/MoO_2/NF), where the MoO_2 domains preferentially adsorbed OH^-, while the CoP sites facilitated H^* adsorption, synergistically enhancing HER efficiency (11 mV overpotential at 10 $mA \cdot cm^{-2}$, surpassing Pt/C); the catalyst recorded low operating potential in both water (1.46 V) and urea-assisted (1.32 V) electrolysis, maintaining an 84-hour stability (1 M KOH + 0.5 M urea). Regarding OER, the CoP/MoO_2/NF catalyst exhibited lower overpotential (282 mV) and Tafel slope (36.2 $mV \cdot dec^{-1}$) compared to RuO_2/NF. Compared to the individual CoP/NP and MoO_2 components, the synergy between CoP and MoO_2 in the CoP/MoO_2/NF catalyst helps overcome the sluggish transfer of six electrons in the UOR, aided by greater electrochemical active surface area (ECSA) and site density. Posttest FTIR/HRTEM analysis helped confirm the in-situ formation of CoOOH (replacing Co—P bonds) and MoO_2, which enabled surface restructuring without degradation. In this mechanism, CoOOH favors NH_2 adsorption/N_2 formation, while MoO_2 catalyzes CO_2 evolution from C=O, which is, essentially, a reflection of complementary nucleophilic/electrophilic interactions.

13.4.1.2 Incorporation of Conductivity Phases and Controlled Morphology

The incorporation of conductive phases into the catalyst architecture plays a key role in enhancing the structural integrity and the catalytic performance. Transition-metal oxides (TMOs) help accelerate charge transfer and promote the release of reaction intermediates, ultimately contributing to improved electrocatalytic activity [9]. Similarly, the application of nanotubes, yolk–shell spheres, and porous nanosheets into the catalyst architecture also helps increase the ECSA, in addition to promoting better electrolyte penetration and accelerating bubble release [36, 39, 40]. The three-dimensional structure of nanotubes enables the occurrence of rapid electrocatalytic reactions due to their open architecture and the abundant availability of nanotubes with effective channels [40, 41]. The application of this structural configuration not only reduces the amount of precursor metals employed but also significantly enhances the electronic conductivity of transition metals.

Nguyen et al. [40] developed advanced multifunctional electrocatalysts based on cobalt sulfide/oxide nanotubes (Co_3S_4/CoO_x NTs) decorated with single atoms and nanoclusters of rhodium (Rh), referred to as Rh–Co_3S_4/CoO_x NTs. Synthesized via a template-assisted wet-chemical approach, these nanotubes (Rh–Co_3S_4/CoO_x NTs), composed of Co_3S_4/CoO_x nanosheets, were anchored on a three-dimensional CF (3D-CF) substrate. The Rh–Co_3S_4/CoO_x NTs catalyst exhibited outstanding electrocatalytic performance, with low overpotentials of 56.1 mV for HER and 1.35 V for urea electrolysis at 10 $mA{\cdot}cm^{-2}$ when applied in 1 M KOH. The use of the three-dimensional nanotube structure, combined with nanosheets, helped increase the active surface area, in addition to promoting efficient mass transport and improving the electron transfer. The superhydrophilic surface, driven by polar Rh—O bonds and metal oxides, facilitated rapid electrolyte diffusion and efficient gas release.

Another related study conducted by Lu et al. [39] investigated the fabrication and electrochemical behavior of phosphorized yolk–shell spherical structures of $CoNi_2S_4$ (P–$CoNi_2S_4$ YSSs) when applied for hydrogen generation through both water and urea electrolysis in 1 M KOH. Transmission electron microscopy (TEM) images confirmed the successful formation of an ultrathin and porous yolk–shell structure (Figure 13.3a), which contributed toward enhancing the accessibility of catalytically active regions. HRTEM images, combined with inverse fast Fourier transform (IFFT) analysis, helped identify (400) and (311) crystalline planes corresponding to $CoNi_2S_4$ (Figure 13.3b, h1 and h2). High-angle annular dark-field scanning transmission electron microscopy (HAADF-STEM) and subsequent elemental mapping analyses pointed to a homogeneous distribution of Co, Ni, S, and P within the structure, confirming the effective incorporation of phosphorus into the $CoNi_2S_4$ structure (Figure 13.3c).

The P–$CoNi_2S_4$ YSS catalyst demonstrated outstanding performance when applied for UOR. The catalyst required only 1.306 V vs. RHE at 10 $mA{\cdot}cm^{-2}$, along with a Tafel slope of 55 $mV{\cdot}dec^{-1}$, for UOR to occur successfully; this is clearly indicative of favorable reaction kinetics. When evaluated in a dual-electrode configuration, the P–$CoNi_2S_4$ pair required only 1.544 V at 10 $mA{\cdot}cm^{-2}$ to effectively drive the water electrolysis process, and the overpotential required was further reduced to 1.402 V

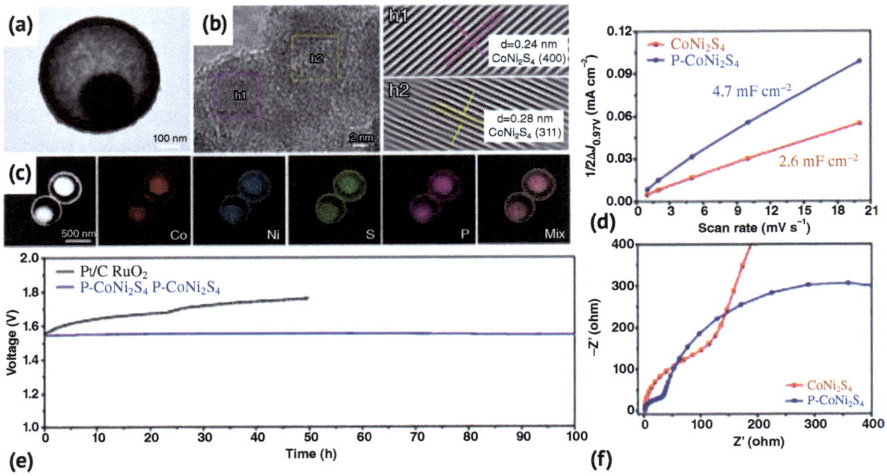

Figure 13.3 (a) TEM images of P-$CoNi_2S_4$ YSSs; (b) HRTEM image (h1 and h2 IFFT images); (c) elemental mapping images; (d) half of the capacitive current density difference at 0.97 V vs. RHE as a function of the scan rate; (e) chronoamperometry curves at 10 mA·cm^{-2}; and (f) EIS spectra of $CoNi_2S_4$ YSSs and P-$CoNi_2S_4$. *Source*: [39]/with permission of John Wiley & Sons.

in urea-assisted electrolysis. This enhanced performance of the P–$CoNi_2S_4$ YSS catalyst is attributed to phosphorus doping, which boosts and creates new active sites while facilitating water dissociation. Electrochemical impedance spectroscopy (EIS) analysis revealed a lower charge-transfer resistance (R_{ct}) (Figure 13.3f) and a higher ECSA of 4.7 mF·cm^{-2} (Figure 13.3d), which may have stemmed from electron delocalization promoted by phosphorus orbitals integrated into the sulfide structure. Importantly, the P–$CoNi_2S_4$ YSS catalyst exhibited superior operational stability, maintaining consistent performance for over 100 hours of continuous operation in both water and urea electrolysis (Figure 13.3e).

These findings underscore the great application potential of this class of materials as efficient, stable, and low-cost bifunctional electrocatalysts. The combination of conductive components with precisely engineered morphological features plays a key role in boosting catalytic efficiency, positioning these materials as strong contenders when it comes to the choice of environmentally sustainable technologies that are suitable for green hydrogen generation and the remediation of wastewater streams, among other applications.

13.4.2 Compositional Modulation and Synergistic Effects

13.4.2.1 Bimetallic Integration and Controlled Doping

When it comes to the design of bimetallic catalysts, choosing metal combinations strategically is crucial to maximizing the electrochemical activities of TMSs and TMPs, which are well known for their outstanding catalytic activities in a wide range of electrocatalytic processes [42].

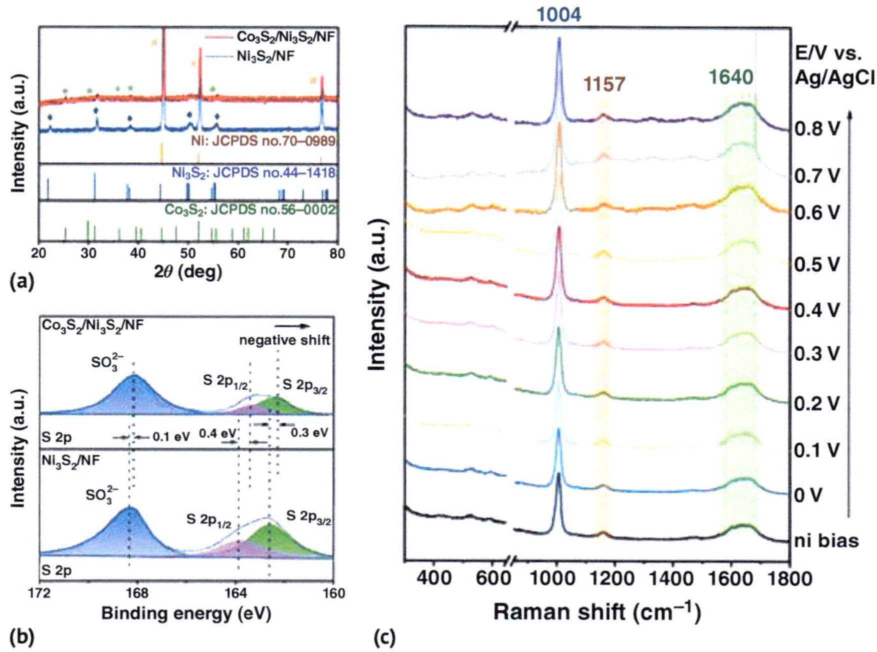

Figure 13.4 (a) XRD pattern of Co_9S_8/Ni_9S_2/NF and Ni_3S_2/NF; (b) XPS spectra of S-2p; and (c) in-situ Raman analysis of Co_9S_8/Ni_3S_2/NF during UOR electrocatalysis. *Source*: [42]/with permission of The American Chemical Society.

Bao et al. [42] synthesized and characterized bimetallic Co_9S_8/Ni_3S_2 nanostructures supported on NF via a one-step hydrothermal process, which resulted in a uniformly distributed nanosheets. SEM and TEM analyses revealed that the Co_9S_8/Ni_3S_2/NF nanostructures exhibited a morphology that favored electrolyte diffusion and gas release, while XRD results confirmed the presence of the corresponding crystalline phases (Figure 13.4a). The XPS data presented in Figure 13.4b show a shift toward lower binding energies for the Co_9S_8/Ni_3S_2/NF heterostructure compared to Ni_3S_2/NF; this points to the existence of a higher electron density around the sulfur atoms, which is caused by the presence of Co. The enhanced charge transfer within the system facilitates the weakening of Ni—S bonds, thereby lowering the adsorption energy barriers for key intermediates such as *OH, *O, and *OOH. Complementary in-situ Raman spectroscopy (Figure 13.4c) analysis revealed the characteristic features of the Co_9S_8/Ni_3S_2/NF catalyst. The symmetric C—N stretching vibration of urea at 1,004 cm^{-1}, along with additional peaks at 1,157 and 1,640 cm^{-1} corresponding to S—O bonds in the metal sulfides, showed that the participation of sulfide sites and the overall architecture contribute toward promoting the catalytic process, as corroborated by the electrochemical results. For UOR, the catalyst required an overpotential of 1.36 V vs. RHE at 100 mA· cm^{-2}, while it required an overpotential of 136 mV at 10 mA·cm^{-2} for HER. In a bifunctional configuration for urea-assisted hydrogen production, the Co_9S_8/Ni_3S_2/NF system operated at 1.57 V (10 mA·cm^{-2}) with approximately 100% faradaic efficiency toward H_2 evolution.

The electrocatalyst also exhibited remarkable stability: 24 hours at 50 mA·cm^{-2} for UOR, 24 hours at 10 mA·cm^{-2} for HER, and 93% activity retention after 24 hours of overall electrolysis, with morphological integrity maintained post-reaction.

The electronic structure of Co_9S_8/Ni_9S_2/NF can be adjusted through doping with specific metals or elements. Doping with metallic cations contributes to reducing the impedance of these materials and enhancing catalytic activity by accelerating the Volmer step and $OH_{adsorption}$ [23]. In turn, nitrogen dopants can modulate the adsorption energy of intermediates, optimizing the catalytic performance. Additionally, this approach lowers the potential barrier for electron transfer coupled with protons, facilitating greater urea adsorption and reducing the energy required for hydrogen production [43].

Based on this approach and using a structure similar to the one described by Bao et al. [42], Xie et al. [17] developed the N–Co_9S_8/Ni_3S_2/NF hybrid catalysts, synthesized via electrodeposition of amorphous CoS_x/NiOH, grown on NF, and subjected to doping with nitrogen, followed by ammonization (Figure 13.5a). The mechanism applied resulted in the conversion of the amorphous phases into crystalline Co_9S_8 and Ni_3S_2 (Figure 13.5b), where the original morphology of the nanosheets was maintained and a highly porous surface was formed (Figure 13.5c–f), which was found to be ideal for electrolyte diffusion and the efficient release of gaseous products. HRTEM analyses of the N–Co_9S_8/Ni_3S_2/NF revealed lattice spacings of 0.30 and 0.41 nm corresponding to the (311) and (101) planes of Co_9S_8 and Ni_3S_2, respectively (Figure 13.5g), which are in line with the X-ray diffraction results. EDS mapping pointed to the homogeneous distribution of N on the nanosheets, in addition to the elements Co, Ni, and S (Figure 13.5h), which, essentially, confirmed the effective incorporation of dopants. DFT calculations indicated that nitrogen doping and the Co_9S_8/Ni_3S_2 heterointerface optimize H^* and urea adsorption, accelerating the kinetics for UOR and HER.

Electrochemical analysis showed that the N–Co_9S_8/Ni_3S_2/NF catalyst reached a current density of 400 mA·cm^{-2} at 1.47 V vs. RHE in UOR and presented an overpotential of 111 mV at 10 mA·cm^{-2} in HER. In bifunctional configurations, the N–Co_9S_8/Ni_3S_2/NF system required 1.40 V for urea electrolysis at 10 mA·cm^{-2}, which is clearly lower than the 1.55 V required for conventional water splitting in both reactions. The material exhibited stability in 20 hours of chronoamperometric testing. These results highlight the crucial influence of dopant incorporation when it comes to tailoring and ensuring the stability of electrocatalysts for energy-efficient applications.

The Co_9S_8/Ni_3S_2 systems, with and without N doping, exhibited synergistic interaction between metal centers, and this played a key role in enhancing the bifunctionality and catalytic stability. N doping in these systems optimized the adsorption of the urea *H intermediate, reducing the HER overpotential from 136 mV for the undoped structure (Co_9S_8/Ni_3S_2) to 111 mV for the N-doped structure (N–Co_9S_8/Ni_3S_2)—both electrochemically in 1 M KOH and 0.5 M urea.

13.4.3 Hybrid Architectures and MOF-derived Platforms

Over the past few decades, metal–organic frameworks (MOFs) have attracted considerable interest among researchers due to their high surface area, pore size, and

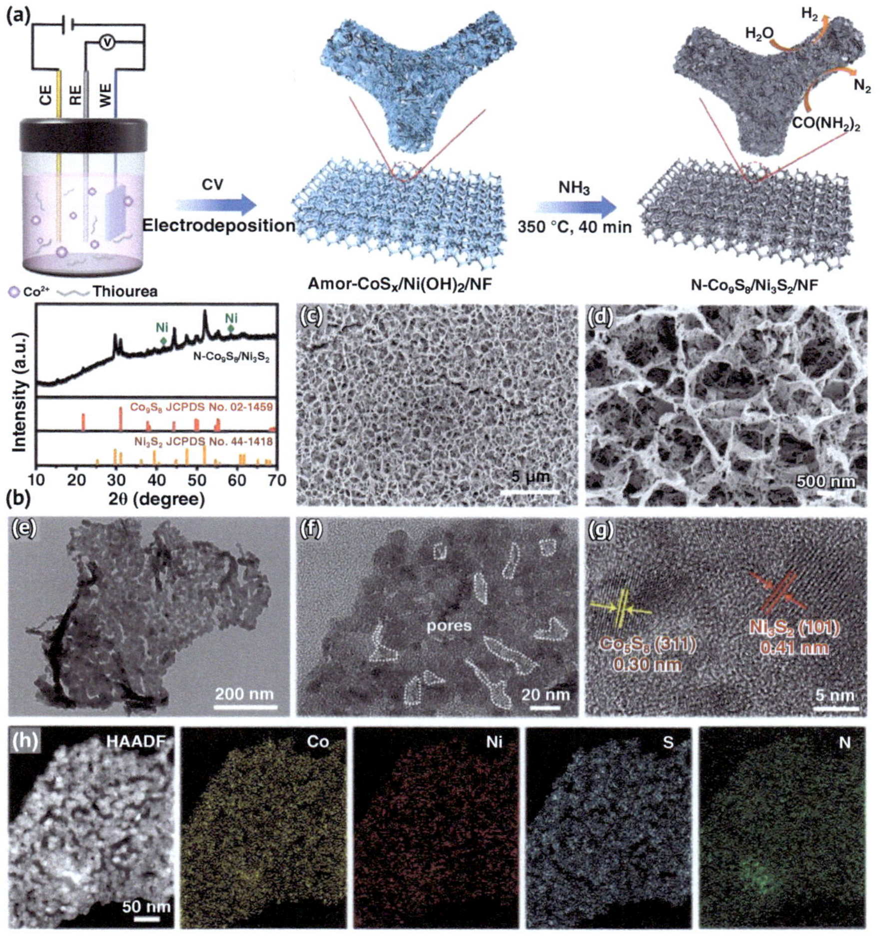

Figure 13.5 (a) Schematic illustration of the synthesis of the N−Co_9S_8/Ni_3S_2/NF electrode. (b) XRD pattern. (c, d) SEM and (e, f) TEM images. (g) HRTEM analysis. (h) HAADF-STEM image and corresponding elemental mapping images. *Source*: [17]/ with permission of John Wiley & Sons.

tunable chemical composition. Although their stand-alone application is not highly efficient, coupling these structures with metal sulfides and phosphides has given rise to highly efficient catalysts with enhanced electrochemical activity for hydrogen evolution. This improvement is attributed to increased conductivity, better adsorption, and nanoparticles tuning [37].

Wang et al. [37] synthesized Ni_2P@Ni-MOF nanosheets on NF through the phosphidation of Ni-MOF sheets. The resulting catalyst, Ni_2P@Ni-MOF/NF, was employed in urea electrolysis, where cell potentials of 1.65 and 1.91 V were recorded for water electrolysis at 100 mA·cm^{-2} (Figure 13.6a). When applied as anode for UOR, the Ni_2P@Ni-MOF/NF catalyst recorded an overpotential of 1.41 V vs. RHE

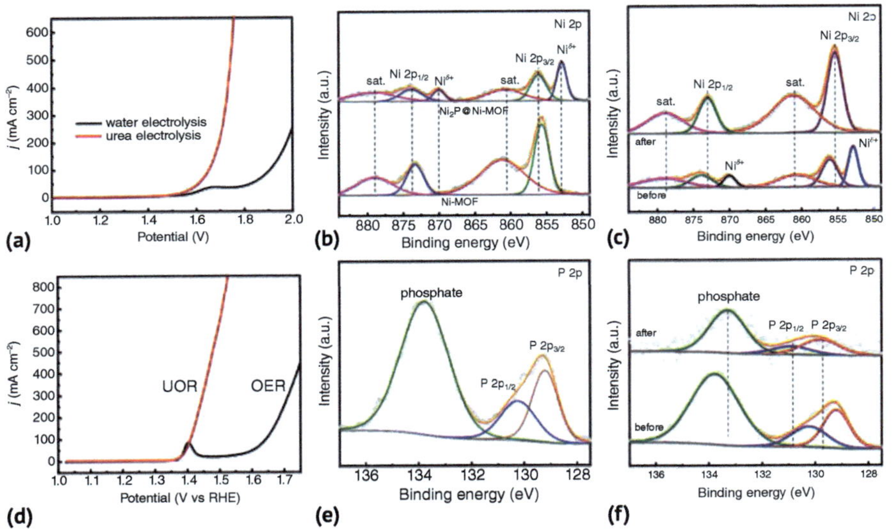

Figure 13.6 (a, d) LSV curves of Ni_2P@Ni-MOF/NF applied in 1 M NaOH with or without 0.33 M urea; and high-resolution XPS spectra of (b) Ni-2p, (c) P-2p, and (e, f) after stability test. *Source*: [37]/Springer Nature/CC BY 4.0.

at 100 $mA{\cdot}cm^{-2}$; on the other hand, when the catalyst was employed as cathode for HER, it yielded a low overpotential of 66 mV at 10 $mA{\cdot}cm^{-2}$. The authors attributed the catalytic performance to the enhanced conductivity of the Ni_2P@Ni-MOF/NF catalyst, where part of the Ni atoms in the Ni-MOF was transformed into exposed Ni_2P, facilitating urea oxidation to Ni^{3+} species. Figure 13.6b shows, in detail, the HRXPS spectrum related to Ni-2p of the Ni-MOF; the spectrum exhibited four main peaks corresponding to Ni-$2p_{3/2}$, Ni-$2p_{1/2}$, and their satellite. Upon subjecting the Ni_2P@Ni-MOF to phosphating, two new peaks associated with reducing $Ni^{\delta+}$ emerged, and the original peaks shifted to higher energies, indicating charge transfer to phosphorous. The P-2p spectrum (Figure 13.6c) exhibited typical peaks of P–Ni species and oxidized phosphate.

This finding shows that UOR proceeds more rapidly than the competing OER, where NiOOH acts as the predominant catalytic species. The Ni_2P@Ni-MOF/NF electrocatalyst exhibited outstanding performance, with impressive current densities up to 850 $mA{\cdot}cm^{-2}$ for UOR (Figure 13.6d); the current density was normalized by C_{dl} (94.3 $mF{\cdot}cm^{-2}$ for Ni_2P@Ni-MOF/NF), with Tafel slopes indicating superior kinetics (43.8 $mV{\cdot}dec^{-1}$) compared to the other catalysts investigated. Additional electrochemical parameters, such as R_{ct} (1.23 Ω) and R_s (0.15 Ω), showed that high-temperature carbonization and phosphidation effectively improved the conductivity of the material. The electrolytic cell maintained stable operation for up to 20 hours at 10 $mA{\cdot}cm^{-2}$, and post-stability XPS (Figure 13.6e,f) confirmed the oxidation of Ni species to higher-valent states. Interestingly, the observed shift of the energy state to a higher-energy state pointed to a charge transfer from P atoms to neighboring atoms, as can be clearly seen in Figure 13.6e.

Shi et al. [44] developed a carbon-coated bimetallic phosphide electrocatalyst (C@NiCoP/NF), derived from an MOF and grown on NF. Synthesized via the phosphorization of ZIF-67 precursors, the C@NiCoP/NF catalyst retained its layered structure, with a ~15 nm carbon coating providing effective corrosion protection. XPS analysis confirmed the formation of Ni—P and Co—P bonds and the favoring of electronic redistribution.

The high performance exhibited by the catalyst is attributed to the following: (i) the in-situ formation of NiOOH/$Co(OH)_2$ during the UOR, (ii) the conductive and protective role of the MOF-derived carbon layer, and (iii) the increased density of active sites. In alkaline medium and in the presence of urea, the C@NiCoP/NF required an overpotential of only 1.27 V vs. RHE to reach the current density of 10 $mA{\cdot}cm^{-2}$ (Tafel slope: 49.66 $mV{\cdot}dec^{-1}$). For the HER, the C@NiCoP/NF catalyst exhibited overpotentials of 91 mV at 10 $mA{\cdot}cm^{-2}$ and 168 mV at 100 $mA{\cdot}cm^{-2}$, with a Tafel slope of 75.25 $mV{\cdot}dec^{-1}$.

In a two-electrode configuration, the catalyst recorded a current density of 100 $mA{\cdot}cm^{-2}$ at 1.51 V, maintaining its stability for 120 hours. These results point to the suitability of the C@NiCoP/NF as an efficient and durable electrocatalyst for urea-assisted water electrolysis.

13.4.4 Summary of Structure–Performance Relationship

A unifying theme across the electrochemical strategies presented in this chapter is the close interplay between structure, composition, and catalytic behavior. Effective electrocatalysts tend to share the following common traits:

- extended interfacial contact between distinct phases;
- optimized electronic environments which help reduce activation barriers;
- hierarchically porous architecture helps ensure mass transport and electrolyte access;
- adaptive surfaces capable of transforming into catalytically active oxides/hydroxides in situ.

These observations are quantitatively supported by metrics such as overpotential (< 1.41 V for UOR), high ECSA (C_{dl} > 20 $mF{\cdot}cm^{-2}$), and stability over 15 hours at industrial current densities (Figure 13.1). Together, they point to rational design principles for developing the next generation of scalable, cost-effective, and high-performance bifunctional catalysts for application in integrated energy and environmental systems.

Table 13.1 presents an overview of the main TMSs and TMPs-based bifunctional electrocatalysts currently employed for water splitting. To make further progress in this technique, more efforts need to be made in the design of new, highly efficient composites.

Table 13.1 A summary of bifunctional sulfates and phosphates-based electrocatalysts for general water splitting.

Catalyst	Substrate	Electrolyte	UOR potential (V) vs. RHE/current density ($mA \cdot cm^{-2}$)	Electrolysis cell potential (V)/current density ($mA \cdot cm^{-2}$)	Overpotential for HER (mV) at 10 $mA \cdot cm^{-2}$	N Tafel slope ($mV \cdot dec^{-1}$)	References
$CoP/MoO_2/NF$	NF	1 M KOH + 0.5 M urea	1.26 V/10 $mA \cdot cm^{-2}$	1.32 V for urea-assisted electrolysis; 1.46 V for total water splitting/ 10 $mA \cdot cm^{-2}$	11	83.6	[9]
Ni_2P@ Ni-MOF/NF	NF	1 M NaOH + 0.33 M urea	1.41 V/100 $mA \cdot cm^{-2}$	1.91 V for water splitting; 1.65 V for urea-assisted electrolysis/ 100 $mA \cdot cm^{-2}$	66	80	[37]
Co_8FeS_8@ CoFe-MOF/NF	NF	1 M KOH + 0.33 M urea	1.311 V/10 $mA \cdot cm^{-2}$	1.62 V for total water electrolysis; 1.55 V for urea-assisted electrolysis/ 10 $mA \cdot cm^{-2}$	191	17.72	[24]
Cu_2S@Ni_3Se_2	CF	1 M KOH + 0.5 M urea	1.338 V/10 $mA \cdot cm^{-2}$	1.70 V for water splitting; 1.48 V for urea-assisted electrolysis/10 $mA \cdot cm^{-2}$	106	85	[3]
P–$CoNi_2S_4$ YSSs	—	1 M KOH + 0.5 M urea	1.306 V/10 $mA \cdot cm^{-2}$	1.544 V for water splitting; 1.402 V for urea-assisted electrolysis/10 $mA \cdot cm^{-2}$	135	65	[39]
$Ni_9S_8/CuS/Cu_2O$	NF	1 M KOH + 0.33 M urea	1.357 V/10 $mA \cdot cm^{-2}$	1.66 V for water splitting; 1.47 V for urea-assisted electrolysis/10 $mA \cdot cm^{-2}$	146	163	[45]
N–Co_9S_8/Ni_3S_2	NF	1 M KOH + 0.5 M urea	1.35, 1.37, 1.41, and 1.44 V/50, 100, 200, and 300 $mA \cdot cm^{-2}$, respectively	1.55 V for water electrolysis; 1.40 V for urea-assisted electrolysis/10 $mA \cdot cm^{-2}$	111	88.6	[17]
NiFeS@NiMoP	NF	1 M KOH + 0.5 M urea	1.32 V/10 $mA \cdot cm^{-2}$	1.40 V for urea-assisted electrolysis/ 10 $mA \cdot cm^{-2}$	56.4	99.9	[18]
Rh–Co_3S_4/CoO_x	CF	1 M KOH + 0.5 M urea	1.32 V/10 $mA \cdot cm^{-2}$	1.45 V for water electrolysis; 1.35 V for urea-assisted electrolysis/10 $mA \cdot cm^{-2}$	56.1	54.3	[40]

CoP@Ni_3S_2	NF	1 M KOH + 0.5 M urea	1.379 V/100 mA·cm^{-2}	1.883 V for water splitting; 1.574 V for urea-assisted electrolysis/100 mA·cm^{-2}	68.9	115.4	[16]
P–NiCoZn or P-LDH/NF-15% or P–NiCoZn- LDH	NF	1 M KOH + 0.5 M urea	1.321 V/100 mA·cm^{-2}	1.644 V for water electrolysis; 1.479 V for urea-assisted electrolysis/100 mA·cm^{-2}	271 at 100 mA·cm^{-2}	121	[46]
Co–Ni–S@NF	NF	1 M KOH + 0.33 M urea	1.35 V/100 mA·cm^{-2}	—	285 at 100 mA·cm^{-2}	29.6	[29]
ZnS@Co_9S_8@Ni_3S_2-1/2	NF	1 M KOH + 0.5 M urea	1.316 V/100 mA·cm^{-2}	Urea-assisted electrolysis: 1.314 V/10 mA·cm^{-2}; 1.506 V/100 mA·cm^{-2} Water electrolysis: 1.522 V/10 mA·cm^{-2}; 1.721 V/100 mA·cm^{-2}	97 at 10 mA·cm^{-2}; 215 at 100 mA·cm^{-2}	42.24	[15]
$NiMoO_4$–MoS_2	NF	1 M KOH containing 0.5 M urea	1.345 V/10 mA·cm^{-2}	1.445 V for urea-assisted electrolysis/10 mA·cm^{-2}	23	77.8	[25]
C@NiCoP	NF	1 M KOH containing 1 M urea	1.27 V/10 mA·cm^{-2} and 1.34 V/100 mA·cm^{-2}	1.51 V for urea-assisted electrolysis/100 mA·cm^{-2}	91 at 10 mA·cm^{-2} and 168 at 100 mA·cm^{-2}	75.25	[44]
Co_9S_8/Ni_3S_2	NF	1 M KOH + 0.5 M urea	1.36 V/100 mA·cm^{-2}	1.57 V for urea-assisted electrolysis; 1.59 V for water electrolysis/10 mA·cm^{-2}	136	135	[42]
Ni_2P/NiCoP	NF	1 M KOH + 0.5 M urea	1.31 V/100 mA·cm^{-2}	1.47 V/10 mA·cm^{-2} and 1.63 V/100 mA·cm^{-2} for urea-assisted electrolysis; 1.82 V for water electrolysis/100 mA·cm^{-2}	—	62.5	[47]

13.5 Conclusions

The incorporation of urea into water electrolysis has emerged as a promising approach for sustainable hydrogen generation, combining lower energy requirements with the added benefit of wastewater treatment. This chapter comprehensively outlined the pivotal role of TMSs and TMPs as bifunctional electrocatalysts that exhibit exceptional catalytic activity and stability for both HER and UOR in alkaline media. The chapter thoroughly elucidated key structural and compositional design principles related to the development of TMSs and TMPs-based electrocatalysts with a view to enhancing charge transfer and maximizing catalytic efficiency. The key design principles that mainly affect the performance and efficiency of these electrocatalysts include heterointerface engineering, morphological control, such as hierarchical 3D architectures, nanotubes, and yolk–shell structures, and doping strategies. The appropriate modulation and tuning of the aforementioned principles have been shown to optimize the intermediate adsorption energetics while increasing the exposure of active sites.

Notably, the in-situ surface reconstruction into catalytically active oxyhydroxides and the synergistic interplay between sulfide and phosphide phases underpin the remarkable performance exhibited by the TMSs and TMPs-based electrocatalysts; this is evidenced by the low overpotentials—such as those observed for UOR (<1.41 V at 100 $mA{\cdot}cm^{-2}$), minimal Tafel slopes, and extended operational stability (>15 hours) of the catalysts.

Despite these advances in the design of TMSs and TMPs-based electrocatalysts, several challenges remain. Future research must address issues related to scalability, long-term durability under industrial conditions, and the integration of these catalysts into modular electrolyzer systems. Furthermore, the development of in-situ/operated characterization techniques and machine learning-guided catalyst design will be crucial for unraveling complex reaction mechanisms and accelerating the discovery of next-generation electrocatalysts. Ultimately, the convergence of material innovation with system-level engineering will be essential to transitioning urea electrolysis from a promising laboratory concept to a scalable solution, consolidating its pivotal role in advancing a green hydrogen-based economy.

References

1 Kumar, S., Bhanuse, G. B., & Fu, Y. P. (2024). Phosphide-based electrocatalysts for urea electrolysis: Recent trends and progress. *ChemPhysChem 25*(8), e202300924. https://doi.org/10.1002/cphc.202300924

2 Staffell, I., Scamman, D., Abad, A. V., Balcombe, P., Dodds, P. E., Ekins, P., Shah, N., & Ward, K. R. (2019). The role of hydrogen and fuel cells in the global energy system. *Energy Environ. Sci. 12*(2), 463–491. https://doi.org/10.1039/c8ee01157e

3 Lv, L., Li, Z., Wan, H., & Wang, C. (2021). Achieving low-energy consumption water-to-hydrogen conversion via urea electrolysis over a bifunctional electrode of

hierarchical cuprous sulfide@nickel selenide nanoarrays. *J. Colloid Interface Sci. 592*, 13–21. https://doi.org/10.1016/j.jcis.2021.02.038

4 Wang, J., Cui, W., Liu, Q., Xing, Z., Asiri, A. M., & Sun, X. (2016). Recent progress in cobalt-based heterogeneous catalysts for electrochemical water splitting. *Adv. Mater. 28*(2), 215–230. https://doi.org/10.1002/adma.201502696

5 Dou, B., Zhang, H., Song, Y., Zhao, L., Jiang, B., He, M., Ruan, C., Chen, H., & Xu, Y. (2019). Hydrogen production from the thermochemical conversion of biomass: Issues and challenges. *Sustain. Energy Fuels 3*(2), 314–342. https://doi.org/10.1039/c8se00535d

6 Martini, B. K. & Maia, G. (2021). Using a combination of Co, Mo, and Pt oxides along with graphene nanoribbon and $MoSe_2$ as efficient catalysts for OER and HER. *Electrochim. Acta 391*, 138907. https://doi.org/10.1016/j.electacta.2021.138907

7 Sun, B., Zhou, W., Li, H., Ren, L., Qiao, P., Li, W., & Fu, H. (2018). Synthesis of particulate hierarchical tandem heterojunctions toward optimized photocatalytic hydrogen production. *Adv. Mater. 30*(43), 1804282. https://doi.org/10.1002/adma.201804282

8 Suen, N. T., Hung, S. F., Quan, Q., Zhang, N., Xu, Y. J., & Chen, H. M. (2017). Electrocatalysis for the oxygen evolution reaction: Recent development and future perspectives. *Chem. Soc. Rev. 46*(2), 337–365. https://doi.org/10.1039/c6cs00328a

9 Liu, Z., Lu, Z., Cao, Y., Xie, J., Hu, J., & Hao, A. (2024). Strongly coupled heterostructured CoP/MoO_2 as an advanced electrocatalyst for urea-assisted water electrolysis. *Inorg. Chem. 63*(5), 2803–2813. https://doi.org/10.1021/acs.inorgchem.3c04342

10 Li, R. Q., Li, S., Lu, M., Shi, Y., Qu, K., & Zhu, Y. (2020). Energy-efficient hydrogen production over a high-performance bifunctional NiMo-based nanorods electrode. *J. Colloid Interface Sci. 571*, 48–54. https://doi.org/10.1016/j.jcis.2020.03.023

11 Sun, X. & Ding, R. (2020). Recent progress with electrocatalysts for urea electrolysis in alkaline media for energy-saving hydrogen production. *Catal. Sci. Technol. 10*(6), 1567–1581. https://doi.org/10.1039/c9cy02618e

12 Paygozar, S., Aghdam, A. S. R., Hassanizadeh, E., Andaveh, R., & Darband, G. B. (2023). Recent progress in non-noble metal-based electrocatalysts for urea-assisted electrochemical hydrogen production. *Int. J. Hydrog. Energy 48*(20), 7219–7259. https://doi.org/10.1016/j.ijhydene.2022.11.087

13 Yu, L., Zhu, Q., Song, S., McElhenny, B., Wang, D., Wu, C., Qin, Z., Bao, J., Yu, Y., Chen, S., & Ren, Z. (2019). Non-noble metal-nitride based electrocatalysts for high-performance alkaline seawater electrolysis. *Nat. Commun. 10*(1), 5106. https://doi.org/10.1038/s41467-019-13092-7

14 Wang, M., Zhang, L., He, Y., & Zhu, H. (2021). Recent advances in transition-metal-sulfide-based bifunctional electrocatalysts for overall water splitting. *J. Mater. Chem. A 9*(9), 5320–5363. https://doi.org/10.1039/d0ta12152e

15 Du, X., Ding, Y., & Zhang, X. (2023). MOF-derived Zn–Co–Ni sulfides with hollow nanosword arrays for high-efficiency overall water and urea electrolysis. *Green Energy Environ. 8*(3), 798–811. https://doi.org/10.1016/j.gee.2021.09.007

16 Wen, D., Ye, M., Xia, Y., Mai, W., Zhang, Z., Zhang, W., Peng, W., Hu, W., & Tian, L. (2022). CoP nanoplates dotted with porous Ni_3S_2 nanospheres for the collaborative enhancement of hydrogen production via urea-water electrolysis. *Appl. Surf. Sci. 586*, 152736. https://doi.org/10.1016/j.apsusc.2022.152736

17 Xie, H., Feng, Y., He, X., Zhu, Y., Li, Z., Liu, H., Zeng, S., Qian, Q., & Zhang, G. (2023). Construction of nitrogen-doped biphasic transition-metal sulfide nanosheet electrode for energy-efficient hydrogen production via urea electrolysis. *Small 19*(17), 2207425. https://doi.org/10.1002/smll.202207425

18 Yang, S., Wen, T., & Gong, Y. (2024). Optimized crystalline/amorphous NiFeS@ NiMoP on nickel foam as dual-functional electrocatalysts for urea-water splitting. *Int. J. Hydrog. Energy 68*, 834–842. https://doi.org/10.1016/j.ijhydene.2024.04.252

19 Shang, J., Wei, T., Yan, X., Fang, Z., Du, L., Shi, J., Sultana, F., Li, T., & Li, R. (2025). CeO_2-enhanced surface reconstruction of Ni_3S_2 nanosheets for improved urea-assisted water splitting performance. *Sustain. Energy Fuels 9*, 1183–1195. https://doi.org/10.1039/d4se01713g

20 Wang, L., Wang, P., Xue, X., Wang, D., Shang, H., Zhao, Y., & Zhang, B. (2024). Interface engineering of three-phase nickel–cobalt sulfide/nickel phosphide/iron phosphide heterostructure for enhanced water splitting and urea electrolysis. *J. Colloid Interface Sci. 665*, 88–99. https://doi.org/10.1016/j.jcis.2024.03.109

21 He, M., Feng, C., Liao, T., Hu, S., Wu, H., & Sun, Z. (2020). Low-cost $Ni_2P/Ni_{0.96}S$ heterostructured bifunctional electrocatalyst toward highly efficient overall urea-water electrolysis. *ACS Appl. Mater. Interfaces 12*(2), 2225–2233. https://doi.org/10.1021/acsami.9b14350

22 Gómez-Sacedón, C., López-Fernández, E., González-Elipe, A. R., Espinós, J. P., Yubero, F., Gil-Rostra, J., & de Lucas-Consuegra, A. (2024). NiFeO/NiFe bilayer electrocatalyst for an efficient urea assisted water electrolysis. *Int. J. Hydrog. Energy 59*, 604–613. https://doi.org/10.1016/j.ijhydene.2024.02.079

23 Wang, Y., Li, X., Du, X., & Zhang, X. (2024). Modulation of the electronic structure of Co_2P by Mo, Fe co-doping for efficient urea electrolysis. *Int. J. Hydrog. Energy 51*, 448–461. https://doi.org/10.1016/j.ijhydene.2023.06.240

24 Zhao, T., Zhong, D., Tian, L., Hao, G., Liu, G., Li, J., & Zhao, Q. (2023). Constructing abundant phase interfaces of the sulfides/metal-organic frameworks p-p heterojunction array for efficient overall water splitting and urea electrolysis. *J. Colloid Interface Sci. 634*, 630–641. https://doi.org/10.1016/j.jcis.2022.11.149

25 Zhu, J., Lv, W., Yang, Y., Huang, L., Yu, W., Wang, X., Han, Q., & Dong, X. (2022). Hexagonal $NiMoO_4$-MoS_2 nanosheet heterostructure as a bifunctional electrocatalyst for urea oxidation assisted overall water electrolysis. *New J. Chem. 46*(21), 10280–10288. https://doi.org/10.1039/d2nj01547a

26 Jiang, L., Pan, Y., Zhang, J., Chen, X., Ye, X., Li, Z., Li, C., & Sun, Q. (2022). Mo propellant boosting the activity of Ni-P for efficient urea-assisted water electrolysis of hydrogen evolution. *J. Colloid Interface Sci. 622*, 192–201. https://doi.org/10.1016/j.jcis.2022.04.050

27 Byrappa, K. & Adschiri, T. (2007). Hydrothermal technology for nanotechnology. *Prog. Cryst. Growth Charact. Mater. 53*(2), 117–166. https://doi.org/10.1016/j.pcrysgrow.2007.04.001

28 Macedo, N. G., Alvim, J. C., Soares, L. C., Souza da Costa, L., Galante, M. T., Lima, V. S., & Longo, C. (2024). Strategies for the synthesis of complex oxides for application as light-responsive electrodes in photoelectrochemical cells—a review. *Mater. Adv. 5*(11), 4541–4562. https://doi.org/10.1039/d3ma01013a

29 Xu, Z., Chen, Q., Chen, Q., Wang, P., Wang, J., Guo, C., Qiu, X., Han, X., & Hao, J. (2022). Interface enables faster surface reconstruction in a heterostructured Co-Ni-S electrocatalyst towards efficient urea oxidation. *J. Mater. Chem. A 16*, 24137–24146. https://doi.org/10.1039/d2ta05494a

30 Gugtapeh, H. S. & Rezaei, M. (2022). Hierarchical flower-like NiFeSbP electrocatalyst toward efficient and stable urea-assisted electrolytic hydrogen production. *ACS Appl. Energy Mater. 5*(12), 15689–15700. https://doi.org/10.1021/acsaem.2c03286

31 Ge, Z., Wang, F., Guo, J., Ma, J., Yu, C., Zhong, A., & Xie, Y. (2021). Low-cost and multi-level structured NiFeMn alloy@NiFeMn oxyhydroxide electrocatalysts for highly-efficient overall water splitting. *Inorg. Chem. Front. 8*, 2713–2724.

32 He, M., Hu, S., Feng, C., Wu, H., Liu, H., & Mei, H. (2020). Interlaced rosette-like MoS_2/Ni_3S_2/NiFe-LDH grown on nickel foam: A bifunctional electrocatalyst for hydrogen production by urea-assisted electrolysis. *Int. J. Hydrog. Energy 45*(1), 23–35. https://doi.org/10.1016/j.ijhydene.2019.10.229

33 Kim, C., Kim, S. H., Lee, S., Kwon, I., Kim, S., Seok, C., Park, Y. S., & Kim, Y. (2022). Boosting overall water splitting by incorporating sulfur into NiFe (oxy)hydroxide. *J. Energy Chem. 64*, 364–371. https://doi.org/10.1016/j.jechem.2021.04.067

34 Niu, Z., Fan, S., Li, X., Liu, Z., Wang, J., Duan, J., Tadé, M. O., & Liu, S. (2022). Facile tailoring of the electronic structure and the d-band center of copper-doped cobaltate for efficient nitrate electrochemical hydrogenation. *ACS Appl. Mater. Interfaces 14*(31), 35477–35484. https://doi.org/10.1021/acsami.2c04789

35 Liu, J., Wang, Y., Liao, Y., Wu, C., Yan, Y., Xie, H., & Chen, Y. (2021). Heterostructured Ni_3S_2-Ni_3P/NF as a bifunctional catalyst for overall urea-water electrolysis for hydrogen generation. *ACS Appl. Mater. Interfaces 13*(23), 26948–26959. https://doi.org/10.1021/acsami.1c04325

36 Zhu, W., Ren, M., Hu, N., Zhang, W., Luo, Z., Wang, R., Weng, J., Huang, L., Suo, Y., & Wang, J. (2018). Traditional $NiCo_2S_4$ phase with porous nanosheets array topology on carbon cloth: A flexible, versatile and fabulous electrocatalyst for overall water and urea electrolysis. *ACS Sustain. Chem. Eng. 6*(4), 5011–5020. https://doi.org/10.1021/acssuschemeng.7b04663

37 Wang, H., Zou, H., Liu, Y., Liu, Z., Sun, W., Lin, K. A., Li, T., & Luo, S. (2021). Ni_2P nanocrystals embedded Ni-MOF nanosheets supported on nickel foam as bifunctional electrocatalyst for urea electrolysis. *Sci. Rep. 11*(1), 21414. https://doi.org/10.1038/s41598-021-00776-8

38 Fereja, S. L., Li, P., Zhang, Z., Guo, J., Fang, Z., Li, Z., He, S., & Chen, W. (2022). W-doping induced abundant active sites in a 3D NiS_2/MoO_2 heterostructure as an efficient electrocatalyst for urea oxidation and hydrogen evolution reaction. *Chem. Eng. J. 432*, 134274. https://doi.org/10.1016/j.cej.2021.134274

39 Lu, X. F., Zhang, S. L., Sim, W. L., Gao, S., & Lou, X. W. (2021). Phosphorized $CoNi_2S_4$ yolk-shell spheres for highly efficient hydrogen production via water and urea electrolysis. *Angew. Chem. Int. Ed. 60*(42), 22885–22891. https://doi.org/10.1002/anie.202108563

40 Nguyen, D. C., Doan, T. L. L., Prabhakaran, S., Kim, D. H., Kim, N. H., & Lee, J. H. (2022). Rh single atoms/clusters confined in metal sulfide/oxide nanotubes as advanced multifunctional catalysts for green and energy-saving hydrogen productions. *Appl. Catal. B: Environ. 313*, 121430. https://doi.org/10.1016/j.apcatb.2022.121430

41 Zhang, Q., Ma, S., Xie, Y., Pan, S., Miao, Z., Wang, J., & Yang, Z. (2024). Cobalt incorporation promotes CO_2 desorption from nickel active sites encapsulated by nitrogen-doped carbon nanotubes in urea-assisted water electrolysis. *Langmuir 40*(49), 26212–26220. https://doi.org/10.1021/acs.langmuir.4c03711

42 Bao, Y., Yang, Y., Jiang, J., Gao, J., He, Y., Wang, S., Bai, X., Dong, X., & Tao, H. (2025). Superhydrophilic bimetallic sulfide nanosheets for hydrogen production using urea-assisted electrocatalytic water splitting. *ACS Appl. Nano Mater. 8*(13), 6530–6540. https://doi.org/10.1021/acsanm.5c00212

43 Song, M., Yang, X., Guo, C., Zhang, S., Ma, J., & Gao, H. (2025). Nanoflower $Mn_xNi_{2-x}P$ as efficient bifunctional catalyst for hydrogen production with urea-assisted energy-saving in alkaline freshwater and seawater. *EcoEnergy 3*(2), 470–481. https://doi.org/10.1002/ece2.90

44 Shi, M., Sultana, F., Qin, X., Zhang, P., Qian, K., Wei, T., Duan, Y., Li, T., Bai, J., & Li, R. (2025). In situ evolution of MOF-derived C@NiCoP/NF promotes urea-assisted electrocatalytic hydrogen production. *Appl. Catal. B: Environ. 371*, 125210. https://doi.org/10.1016/j.apcatb.2025.125210

45 Wei, D., Tang, W., & Wang, Y. (2021). Hairy sphere-like Ni_9S_8/CuS/Cu_2O composites grown on nickel foam as bifunctional electrocatalysts for hydrogen evolution and urea electrooxidation. *Int. J. Hydrog. Energy 46*(40), 20950–20960. https://doi.org/10.1016/j.ijhydene.2021.03.206

46 Yang, Z., Zhang, Y., Feng, C., Wu, H., Ding, Y., & Mei, H. (2021). P doped NiCoZn LDH growth on nickel foam as an highly efficient bifunctional electrocatalyst for overall urea-water electrolysis. *Int. J. Hydrog. Energy 46*(50), 25321–25331. https://doi.org/10.1016/j.ijhydene.2021.05.076

47 Zhang, P., Sultana, F., Li, T., Qin, X., Shi, M., Wei, T., Qian, K., Tan, M., Li, Z., Bai, J., & Li, R. (2025). Regulating Ni_2P electronic structure and morphology with cobalt: a one-step route to enhanced electrocatalytic urea oxidation and water splitting. *Catal. Sci. Technol. 15*, 2733–2744. https://doi.org/10.1039/d5cy00008d

14

Metal Carbides and Nitrides Electrocatalysts for Urea and Furfural-assisted Water Electrolysis

Soorya S. Raj[1,], M. L. Aruna Kumari[1], P. Keerthana[2]*

[1]*Department of Chemistry, The Oxford College of Science, Bengaluru, India*
[2]*Department of Science, Alliance University, Bengaluru, India*

*Corresponding author: sooryasraj111@gmail.com

14.1 Introduction

The critical need for sustainable and clean energy sources has established hydrogen (H_2) as a top energy carrier for the future hydrogen economy [1]. Electrochemical water splitting, or water electrolysis, is a very effective technology for producing high-purity H_2 from renewable electricity. Electrolysis occurs in two half-reactions—the hydrogen evolution reaction (HER) at the cathode and the oxygen evolution reaction (OER) at the anode:

$$\text{Cathode (HER)}: 2H_2O + 2e^- \rightarrow H_2 + 2OH^- \text{ (alkaline)}$$
$$\text{Anode (OER)}: 4OH^- \rightarrow O_2 + 2H_2O + 4e^- \text{ (alkaline)}$$

The thermodynamic cell voltage for water electrolysis is 1.23 V; however, commercial water electrolyzers operate at a considerably higher voltage (generally >1.8 V), which enables them to achieve practical current densities [2]. The high energy input is due to the sluggish kinetics of the anodic OER, a complex four-proton/four-electron-transfer process that involves O—O bond formation, requiring considerable overpotential [3]. The kinetic limitation contributes to an increase of overall energy consumption and operational costs of H_2 production, thus limiting the wide-scale commercialization of green H_2 [4].

To address the energy barriers of OER, a new strategy called hybrid water electrolysis or value-added electrolysis has been established. In this approach, the thermodynamically difficult OER is replaced with a thermodynamically easier oxidation reaction (i.e., lower anodic potential) while H_2 is produced at the cathode. This modification reduces the cell voltage needed to produce H_2 with a remarkable decrease in energy consumption [5]. Notably, the choice of anodic reactions typically involves the conversion of waste or biomass-derived chemicals into chemical products, with dual advantages: the production of H_2 using reasonable energy and chemical production with added value [6]. In this chapter, we will delve into two such examples. Urea oxidation reaction (UOR): a process to produce H_2 from wastewater containing urea (e.g., agricultural runoff or human/animal urine), and the theoretical potential for UOR mechanism is 0.37 V, which is much less than OER [7]. Furfural oxidation reaction (FOR): a process that could better convert a platform molecule generated from lignocellulosic biomass (furfural) into high-value chemicals (maleic acid, furoic acid) while producing H_2 at the cathode [8].

The creation of efficient and affordable electrocatalysts is key for operational hybrid electrolysis systems. For many decades, noble metals such as platinum (Pt) have been the leading HER catalysts, but their high price and limited availability prohibit their usage at scale [9]. This has led to a wealth of research into electrocatalysts that are based on non-precious metals. Some of the most exciting candidates that present more promise are transition-metal carbides (TMCs) and transition-metal nitrides (TMNs) [10]. These interstitial compounds display a novel set of properties that can make them exceptional electrocatalysts. Pt-like electronic structure: the addition of carbon or nitrogen atoms to the transition-metal lattice alters the d-band electronic structure, and produces catalytic properties similar to those of noble metals. High stability and conductivity: TMCs and TMNs display high thermal stability and high electrical conductivity, which are important characteristics for long-lasting electrochemical use. Earth abundance and low cost: constructed from earth-abundant transition metals (e.g., Ni [nickel], Mo [molybdenum], and W [tungsten]), TMCs and TMNs provide a price-competitive option to Pt group metals. This chapter will explore the recent state of the art of metal carbides and nitrides (MCNs) as highly active, non-precious electrocatalysts for urea and furfural-assisted water electrolysis. The chapter addresses the underlying catalytic mechanisms, the role of varied exposures in compositional and structural engineering approaches, and barriers that must be overcome to deploy these energy-efficient H_2 production and value-added molecular production processes from lab to scale.

14.2 Fundamentals of Urea and Furfural-assisted Water Electrolysis

14.2.1 Thermodynamics and Kinetics of UOR

UOR is a different anodic process that has an energy pathway that is lower than the typical OER in electrolyzing water. The overall UOR is:

$$CO(NH_2)_2 + 6OH^- \rightarrow N_2 + CO_2 + 5H_2O + 6e^- (E^0 = 0.37\ \text{V vs. RHE})$$

UOR thermodynamically happens as an overall reaction at a significantly lower potential than the OER (E^0 = 1.23 V vs. RHE), thereby lowering the overall cell voltage necessary for hydrogen production. UOR is particularly appealing to researchers as it would have applications involving urea-rich wastewater streams, providing a simultaneous energy generation and environmental remediation opportunity.

The kinetics of UOR are mediated through adsorption, as well as subsequent dehydrogenation of urea molecules on the catalyst [12]. Ni-containing catalysts, particularly the $Ni(OH)_2$/NiOOH redox couple, are reported to be very active since they can mediate this process via sequential proton-coupled and electron-transfer steps. However, stabilizing the catalyst can be a challenge since it is often poisoned through CO_2 adsorption or upon interaction with intermediate species such as CO or CN^-. The resonant structure of urea with NH_2 and CO groups gives its C—N bonds with partial double-bond character [13]. However, this resonance can be blocked by the binding of Ni catalyst through NH_2 ends, leading to weakening of the C—N bond. This cleavage leads to the formation of undesirable product instead of selective N_2 product. So, to achieve selective N_2 production, we need to avoid C—N cleavage (Figure 14.1) [14]. Approaches to mediate reaction stability and kinetics in the UOR include Fe, Co (cobalt), or Cu doping or the more complex design strategies such as nanostructured interfaces.

14.2.2 FOR: Pathways and Value-added Products

FOR has developed into a viable anodic alternative for the electrochemical production of hydrogen. Furfural is a biomass-derived aldehyde obtained from lignocellulosic feedstocks, resulting in high-value chemicals such as 2-furoic acid, furanone, and maleic acid based on the reaction conditions and the catalyst utilized (Figure 14.2) [15].

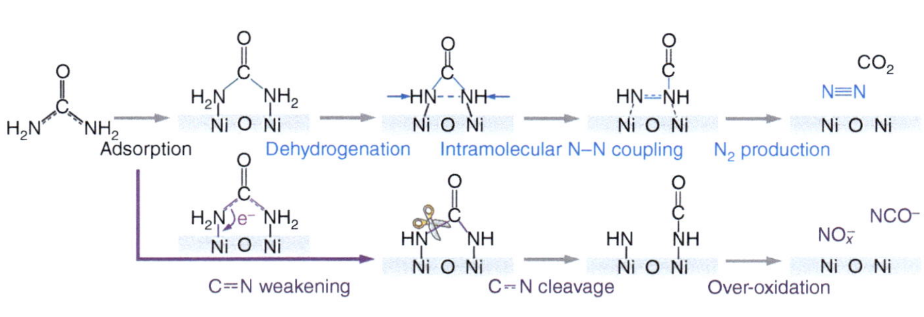

Figure 14.1 The UOR pathways for N_2/CO_2 or NO_x^-/NCO^- production on connected symmetric Ni—O—Ni sites. *Source*: [11]. Springer Nature. CC BY 4.0.

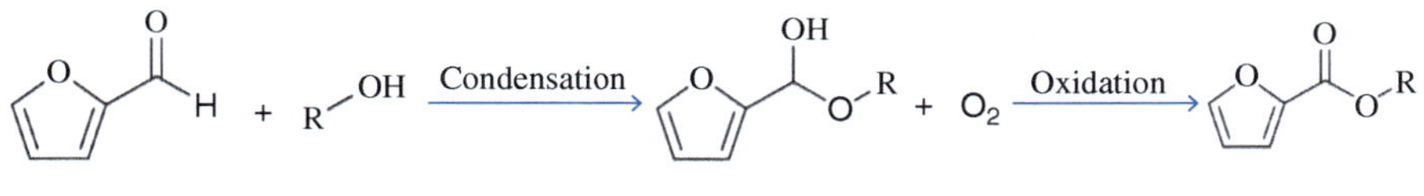

Figure 14.2 Oxidation of furfural. *Source*: [15]/MDPI/CC BY 4.0.

The typical reaction pathways are:

1. Furfural → 2-Furoic acid (partial-oxidation reaction)
2. Furfural → Maleic acid/furanone (deep-oxidation reaction)

The reaction becomes a catalytic process where furfural must be adsorbed on to the catalyst surface then undergo oxidative dehydrogenation. Catalysts based on Au, Pd, and Ni have been investigated and show high selectivity toward 2-furoic acid under alkaline conditions. Notably, oxidation of furfural occurs at potentials well below OER, enabling subsequent hydrogen evolution at the cathode leading to a more efficient overall energy process [16].

14.2.3 Comparison with OER: Potential Reduction and Efficiency Gains

Both the UOR and the FOR can be considered energy-efficient anodic alternatives to the OER for water electrolysis. The OER is intrinsically slow in nature and takes place through a four-electron transfer that requires high overpotentials (>300 mV), whereas UOR and FOR are simpler reaction mechanisms requiring lower activation energy. In terms of thermodynamic potentials, the potential for UOR is 0.37 V vs. RHE and it is 0.45–0.60 V vs. RHE for FOR, while the OER thermodynamic potential is 1.23 V. Thus, substituting UOR or FOR for the OER would reduce the total cell voltage by ≈0.8 V, representing 30–40% savings in total energy efficiency in the process [7]. In addition to their energy efficiency and cost-effectiveness, UOR and FOR result in the formation of coproducts of potential economic value (N_2, CO_2, or furoic acid) with oxygen, thus improving the economics of electrolytic hydrogen production.

14.2.4 Dual Benefits: Sustainable Hydrogen Generation + Chemical Coproduction

The combination of UOR and FOR in the process of water electrolysis signifies a major advancement in the generation of green hydrogen. These reactions allow hydrogen to be sustainably produced at the cathode, and some simultaneous chemical valorization, or wastewater treatment, at the anode [17]. The dual benefit of production and value-adding is supportive of a circular carbon economy, where waste or biomass-derived substrates are converted into high-value chemicals or additives while producing clean energy. The UOR is thermodynamically favored over the OER, such that less energy is needed in order to achieve the electron transfer required for simultaneous cathodic HER (Figure 14.3) [11].

Urea-assisted electrolysis, for example, tackles the wastewater treatment problem while utilizing less energy, while furfural-assisted electrolysis converts renewable biomass into furoic acid, a chemical precursor to the production of pharmaceuticals and resins. Collectively, these systems support the ideals of sustainable chemistry, minimizing the carbon footprint while maximizing process economics

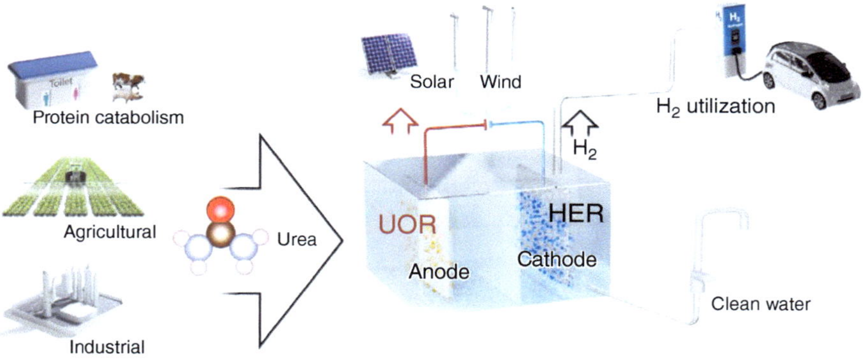

Figure 14.3 Schematic of UOR toward clean water and H_2 utilization. *Source*: [11]/ Springer Nature/CC BY 4.0.

14.3 MCNs as Electrocatalysts

14.3.1 Unique Physicochemical Properties

MCNs have emerged as incredibly remarkable materials for use as electrocatalysts due to their outstanding physicochemical features. They have metallic-like electrical conductance, high melting points, and are mechanically stable, making them the perfect materials for use in electrochemical systems applied to energy conversion systems, including water electrolysis, fuel cells, and CO_2 reduction.

Another advantageous feature of these materials is their tunable electronic structure. When carbon or nitrogen atoms are incorporated into metallic lattices, they modify the d-band center of the resulting metal, which, in turn, modifies the energy of adsorption to intermediates during the reactions; this is referred to as d-band tuning [18]. It is also important to note that the catalyst must balance the adsorption strength between the catalyst surface and the reactants in order to optimize catalytic activity. For example, tungsten carbide (WC) and molybdenum carbide (Mo_2C) have similar electronic structures to platinum, therefore making them excellent non-precious materials for the HER and the oxygen reduction reaction (ORR) [19]. TMNs, such as VN, TiN, and MoN, have excellent electrical properties and catalytic activity, due in part to the strong metal–nitrogen covalency that strongly modifies the charge-transfer properties [20].

14.3.2 Synergy with Transition Metals for Enhanced Activity

The use of metal carbides or nitrides in conjunction with transition metals can provide a synergistic way to increase electrocatalytic activity. This synergy comes from electronic interactions at the interface that can facilitate charge transfer and change the energetics of adsorbing key intermediates. An example is coupling nickel or cobalt with molybdenum carbide to form heterostructures with enhanced HER and OER activity, which results from an advantageous balance between hydrogen adsorption (on Ni/Co sites) and water dissociation (on carbide surfaces) [21]. These hybrid systems create favorable energetics for the Volmer–Heyrovsky or Volmer–Tafel mechanisms, and provide greater active-site densities, which result in reduced energy barriers [22]. Likewise, metal–nitride composites (e.g., Ni–MoN, Co–VN) have shown signs of improved catalytic kinetics as a result of favorable electron density and increased surface defect sites. These bimetallic systems can behave in a bifunctional manner and are often suitable for HER and OER applications in alkaline or neutral environments.

14.3.3 Stability and Corrosion Resistance in Harsh Electrolytes

The stabilization of the lattice by the covalent bond between the carbon or nitrogen and metal atoms allows carbides and nitrides to maintain stability in both acidic and alkaline conditions and resist surface reconstruction or corrosion [23]. For example, TiN and Mo_2C have shown little to no degradation over time while subjected to electrolysis, extending the field performance for years at a time [24]. Surface passivation also contributes to stability: thin oxide layers that can form on carbides or nitrides can serve as a barrier to corrosion while still allowing electrons and ions to transport [25]. Additionally, other less stable dopants like Fe, Cu, or P can enhance stability while still allowing some degree of catalytic activity through stabilizing the oxidation state of the surface to inhibit dissolution pathways.

14.3.4 Structure–Activity Relationship Principles

Comprehending the structure–activity relationships (SARs) is necessary for the targeted design of efficient carbide and nitride-based electrocatalysts. Key structural characteristics, which are important for the catalytic activity of carbide/nitride catalysts (such as their particle size, crystal orientation, defect density, and surface composition), guide the development of these candidates. As an example, nanosized carbides and nitrides have a greater surface-area-to-volume ratio, making them more effective catalysts because they have more active sites available for the electrocatalytic reaction. Defects and vacancies can also act as classes of catalysts because they enhance adsorption and activation of reac tants. The crystal facets themselves, especially (0001) stack planes in WC and (101) stacks in Mo_2C, are seen to have excellent catalytic activity—surpassing other facets possibly because of optimized interactions between surface atoms and the overall electronic configuration of the surface planes, respectively [26]. Newly developed computational approaches, including density functional theory (DFT), have been used to better understand reaction pathways and thermodynamics on carbide/nitride surfaces [27]. DFT predicts adsorbate–surface interactions, and these insights from theory, once correlated to experimental data, allow researchers to tailor material properties to specific electrocatalytic reactions in a systematic way while also maximizing activity, selectivity, and durability.

14.4 Synthesis and Strategies

The development of MCNs as future electrocatalysts has been attributed to their ability to modify electronic properties through a metal/nonmetal interaction in combination with the associated reaction intermediates. The strategies such as carbothermal reduction and nitridation methods, template-assisted synthesis, sol–gel and hydrothermal approaches, and plasma-enhanced and chemical vapor deposition (CVD) techniques are used for the synthesis of these electrocatalysts (Figure 14.4).

14.4.1 Carbothermal Reduction and Nitridation Methods

Carbothermal reduction and nitridation are the two most well-known routes to synthesize MCNs [28]. The starting materials are metal oxides or some precursors which are reduced with carbon materials at high temperatures (typically between 800 and 1,500 °C) under some inert or reactive atmosphere, such as argon, nitrogen, or ammonia [29, 30]. The reactions proceed according to the following general equations:

$$M_xO_y + yC = M_xC_y + yCO \text{ (Carbothermal reduction)}$$

$$M_xO_y + NH_3 = M_xN_y + H_2O + H_2 \text{ (Nitridation)}$$

The phase composition, particle size, and morphology of the materials are highly dependent on the precursor, given temperatures, and gas atmosphere used: for example, Mo_2C can be made from MoO_3 with carbon black at ~1,000 °C in an Ar/H_2 atmosphere [31], and TiN can be produced from TiO_2 when nitridized with NH_3 [32]. While these methods are simple, they can be problematic in terms of producing consistent particle size and phases. If the temperature is too high, sintering and grain growth may occur.

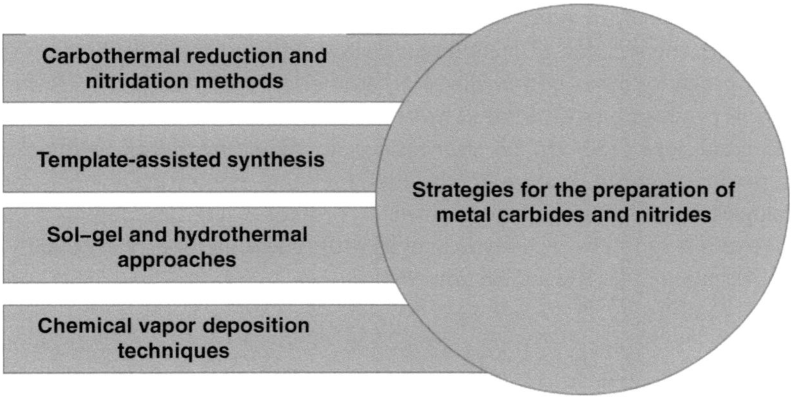

Figure 14.4 Strategies for the preparation of MCNs.

14.4.2 Template-assisted Synthesis

The use of template-assisted synthesis permits control to be exerted over morphology, porosity, and nanostructure during the synthesis of carbides and nitrides. Templates assist in the deposition or the formation of precursors of the desired phase. Hard templates (e.g., silica, alumina, and carbon) or soft templates (e.g., surfactants and polymers) will further allow the substrate material to be transformed into the desired phase via pyrolysis or chemical treatment [33]. An example of a mesoporous silica template is SBA-15, which can act as a template to make nanostructured materials (Mo_2C or WC) with ordered pores after carbonization and removal of the template [34, 35]. A similar approach can be made with polymeric templates (e.g., polyaniline [PANI] and polyacrylonitrile [PAN]), which can produce nitrogen-rich precursors of the metal nitrides of interest via annealing [36]. The presence of residual NH_3 groups is important because they contribute to the stabilization of the structure of the metal nitrides formed from polymeric precursor materials. The primary advantage of this approach is the opportunity to tune surface area and porosity, which are very significant to electrocatalytic performance. This approach involves long time scales, however, and template removal typically involves chemical etching or calcination, which can induce defects in the structure and/or collapse the sample altogether [37].

14.4.3 Sol–Gel and Hydrothermal Approaches

The sol–gel and hydrothermal synthesis techniques are versatile and low-temperature strategies to yield MCNs with controlled composition and nanostructure. The sol–gel method produces a precursor compound in low-temperature environment via hydrolysis, followed by condensation leads to the formation of metal alkoxides or salts. As condensation continues, these oxide chains grow and crosslink to form the gel, that is calcinated in carbon- or nitrogen-rich environments to produce the desired compound [38]. Arrangement of components in the sol–gel process allows for uniform and complete mixing at the molecular scale which can result in high levels of homogeneity of the precipitated product and phase purity. Hydrothermal synthesis involves the reaction of metal precursors in aqueous media at high temperatures and pressures in sealed autoclaves. Hydrothermal synthesis is also a favorable process to produce nanostructured precursors to metal nitrides and carbides under mild conditions (150–250 °C). Annealing can be performed to produce a crystalline product. Overall, both methods allow for some amount of fine control of morphologies such as nanorods, nanoparticles, or nanosheets, and doping with secondary elements to improve catalytic activity [10]. These methods can be ubiquitous to bulk scaling of electrocatalyst routes, thus they are extremely variable and scalable.

14.4.4 Plasma-enhanced and CVD Techniques

Plasma-enhanced and CVD techniques are sophisticated synthesis methods designed for producing highly uniform and crystalline carbide and nitride coatings.

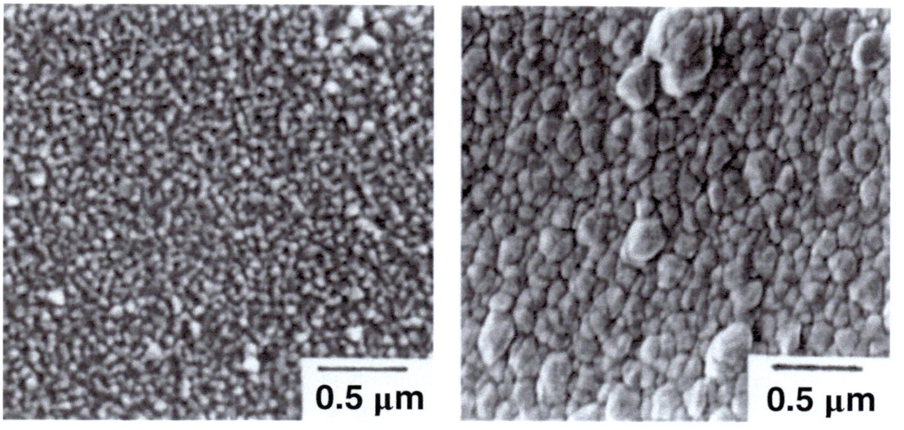

Figure 14.5 SEM images of TiN films fabricated by PECVD method. *Source*: [41]/with permission of Elsevier.

Volatile metal–organic precursors or halides are thermally activated and decomposed in reactive gases such as CH_4, NH_3, or N_2 at elevated temperatures or via energy deposition from the plasma. CVD provides precise control of film thickness, crystallinity, and composition, making it an ideal method for producing thin films and nanostructured coatings on diverse substrates. For example, CVD-prepared TiN and WC films have shown excellent adhesion, electrical conductivity, and, therefore, suitability for electrocatalytic and protective coatings [39, 40].

Plasma-enhanced chemical vapor deposition (PECVD) technology operates at lower-temperature conditions as it utilizes plasma, which drives the reaction kinetics and enables coating deposition onto temperature-sensitive substrates (Figure 14.5) [41, 42]. PECVD approaches have been used with great success, producing amorphous and nanocrystalline carbides/nitrides with excellent mechanical and catalytic properties [43].

14.4.5 Challenges in Phase Control, Morphology, and Scalability

Despite notable progress, there are still challenges in the synthesis of MCNs. One fundamental problem is phase control, with multiple polymorphs (e.g., α-Mo_2C, β-Mo_2C) existing depending on the synthesis conditions, which can have an impact on catalytic activity. Morphology control is equally important because agglomeration and uncontrolled growth can compromise active surface area and performance. In addition, developing scalable and economically feasible synthesis methods will continue to be a bottleneck for the development of materials for technological integration. Methods such as CVD and plasma-assisted deposition can provide high control, but are costly and resource-intensive. Conversely, methods such as carbothermal reduction are easy to use, but match less successful methods in producing heterogeneous products. Compared to other methods, hybrid synthesis processes that combine contributions from templating, solution-, and vapor-based schemes might strike a reasonable control/scalability balance. The next phase of synthesis research will involve developing machine learning and computational design techniques to better allow and predict synthesis optimization conditions and phase selectivity. There is also some interest in developing sustainable synthesis processes based on biomass-derived carbon materials and green precursors as an ecologically beneficial alternative.

14.5 MCNs for UOR

14.5.1 Molybdenum and Tungsten Carbides/Nitrides

Carbides and nitrides based on tungsten and molybdenum have emerged as active and durable electrocatalysts for the UOR. Carbides and nitrides have electronic structures that resemble noble metals, particularly platinum, due to the electronic states associated with their partially filled d-orbitals. This results in good electrical conductivity and high catalytic performance, both of which are necessary for favorable UOR kinetics. Molybdenum carbide is known to be highly active owing to its greater density of active sites and its ability to adsorb–desorb urea molecules. For molybdenum carbide, the reaction mechanism generally involves urea being adsorbed on to Mo during the first step and subsequent steps will dehydrogenate the urea to produce N_2 and CO_2. Molybdenum nitride (Mo_2N) exhibits good activity in alkaline conditions as a result of enhanced electron mobility and chemical stability [44]. Tungsten carbide and tungsten nitride (WN) have emerged as active and durable alternatives. Tungsten carbide shows enhanced activity toward oxygen evolution and has exhibited a Pt-like behavior during electrocatalysis and corrosion resistance [45]. Its catalytic activity can, however, be fundamentally limited by weathering or oxidation of the catalyst surface when the catalyst is under reversible conditions. Some strategies taken to mitigate the disadvantageous effects of oxidation of tungsten carbide surfaces are surface modification with Ni or Co as well as design of new active phases based on tungsten carbide/Ni heterostructures that endure the charge transfer and stability imposed by repeated UOR cycling [46].

14.5.2 Iron, Cobalt, and Nickel Carbides/Nitrides

Nickel, cobalt, and iron carbides and nitrides represent another important family of UOR electrocatalysts because they are low cost, abundant, and have significant catalytic activity with regard to the decomposition of urea. Among these, nickel-based systems have the highest intrinsic activity for UOR, primarily due to the redox couple of $Ni(OH)_2/NiOOH$, which serves as the primary active site for the oxidation of urea [47].

Nickel nitride (Ni_3N) and nickel carbide (Ni_3C) are good conductors of electricity and are resistant to CO_2 poisoning, a prevalent issue in UOR electrocatalysis. In addition, cobalt nitrides (Co_2N, Co_3N) and iron nitride (Fe_3N) have shown the ability to enhance oxidation of intermediates, which reduces the overpotential and increases reaction kinetics [48]. They also demonstrate synergistic effects when used in carbonaceous supports (e.g., graphene or carbon nanotubes) to improve dispersion and electron transport. As an example, Ni_3N nanoparticles on carbon nanofibers have been reported to show current density of 300 $mA \cdot cm^{-2}$ at 0.4 V vs. RHE, which makes TMNs attractive substitutes for noble metals in UOR electrocatalysis [49].

14.5.3 Mixed-metal Carbides and Nitrides

Mixed-metal carbides and nitrides (MMCs/MMNs) have gained increasing interest because they offer advantages of combining multiple metallic components that could yield synergistic electronic and structural effects. In many cases, these materials have displayed higher conductivity, tunable electronic properties, and better resistance to corrosion when compared to their monometallic counterparts. For example, bimetallic systems such as $Ni\text{–}Mo_2C$, $Co\text{–}Mo_2C$, and $Fe\text{–}Ni_3N$ exhibit higher electrocatalytic activity for UOR than monometallic counterparts due to their complementary interactions from the contribution of the metal sites [50, 51]. The Mo-based phases ultimately promote urea adsorption and activation, while the Ni or Co is responsible for the proper redox cycling for urea oxidation. Ternary nitrides, like $(Ni,Co)_3N$ or $(Fe,Mo)_2N$, have also exhibited tunable d-band centers and higher defect densities that improve charge transfer and intermediate desorption [52]. Their catalytic activity can be enhanced further with heteroatoms or dopants (e.g., P, S, or O) into the lattice, which promote lattice strain and enhanced reactivity on the surface [53].

14.5.4 Reaction Mechanism Insights and Performance Benchmarks

The enhanced catalytic performance of mixed-metal carbides and nitrides in UOR arises from their optimized electronic structures and synergistic active sites. The interaction between different metal components enables efficient adsorption, activation, and subsequent oxidation of urea molecules [50]. In particular, the presence of multiple active centers facilitates improved charge transfer kinetics and lowers the energy barriers of intermediate reaction steps. Defect-rich structures and tunable d-band centers further contribute to the efficient adsorption–desorption balance of reaction intermediates, which is critical for sustained catalytic activity. Additionally, strategies such as heteroatom doping and lattice strain engineering play a crucial role in modulating catalytic behavior by enhancing surface reactivity and optimizing binding energies. These modifications ultimately lead to improved catalytic efficiency, stability, and overall performance benchmarks in UOR systems.

14.6 MCNs for FOR

Furfural (a biomass-sourced platform molecule derived from hemicellulose) is an important feedstock that can be further converted into value-added chemicals via electrochemical oxidation. FOR consists of complex multistep pathways generating a variety of products due to catalyst composition, potential, and reaction environment.

The main oxidation pathways involve the conversion of furfural ($C_5H_4O_2$) to furoic acid ($C_5H_4O_3$) due to oxidation of the aldehyde function, and production of intermediates such as 2-furanone and maleic acid through C—C bond cleavage and ring-opening reactions via further oxidation [54, 55]. The selectivity of products is attributed to the electronic structure and surface chemistry of the used catalyst. Optimizing electrolyte pH, applied potential, and catalyst type are ways to control reaction routes/products.

Common oxidation pathways are as follows:

1. adsorption of furfural on the catalyst surface;
2. formation of electron transfer and a radical intermediate from furfural;
3. oxidation to furoic acid or opening of the 2-furan ring and conversion into open-chain products such as maleic acid or CO_2.

14.6.1 Ni-, Mo-, and Co-based Carbides/Nitrides for Furfural Valorization

TMCs and TMNs, especially those derived from nickel, molybdenum, and cobalt, have recently emerged as effective catalysts for the electrooxidation of furfural. These materials exhibit metallic conductivity, high corrosion resistance, and tunable electronic properties, making them favorable catalysts for biomass valorization reactions. Nickel-based carbides (Ni_3C) and nitrides (Ni_3N) exhibit strong adsorption for the aldehyde group of furfural, and enable the oxidation of furfural to furoic acid with high faradaic efficiency [56]. Reports have suggested that Ni_3N nanosheets on a conductive support (such as nickel foam) can achieve an impressive 90% or more selectivity to furoic acid at low overpotentials [57]. Molybdenum-based catalysts (Mo_2C and Mo_2N) also show excellent performance for the electrooxidation of furfural—this is partially attributed to their strong interactions with oxygen-containing intermediates in the reaction with 97.4% faradaic efficiency [58]. Mo sites activate C—H bonds and stabilize intermediates, and the carbides/nitrides provide these sites with electronic conductivity. Similarly, cobalt nitrides (Co_2N, Co_3N) have been shown to produce good catalytic activity for these types of reactions and are capable of maintaining stability [59]. The Co^{2+}/Co^{3+} redox couple is capable of driving oxidation reactions under alkaline conditions, and the presence of both Co and Mo species in the reaction has even enhanced the catalytic activity.

14.6.2 Selectivity Control and Product Distribution

The selectivity of the FOR is affected by multiple key factors, such as the electronic structure of the catalyst, binding energy of the intermediates, reaction potential, and electrolyte composition. MCNs allow for fine-tuning of these parameters by manipulation of the d-band and surface defect engineering [54]. For example, Ni_3N

and Mo_2C selectively promote the formation of furoic acid, while cobalt-based materials tend to promote deeper oxidation to maleic acid [60]. Reaction pathways also can be steered by tuning the applied potential; lower potentials will provide better selectivity toward oxidation to furoic acid, while higher potentials will lead to C—C bond scission and the formation of open-chain acids. Additionally, the incorporation of dopants, like Fe or Cu, into Ni- or Mo-based catalysts can facilitate further fine-tuning of the binding energy of the intermediates, providing for higher activity and selectivity in the reaction. Catalyst morphology tuning can also provide dimensional changes to allow greater access to active sites and transport for mass transfer, such as porous nanorods, nanosheets, and core–shell design. The selectivity is further affected by the selected electrolyte; alkaline media (KOH, NaOH) provide higher current densities and better selectivity relative to neutral or acidic electrolytes.

14.6.3 Coupling Furfural Oxidation with HER

A novel approach in electrocatalysis is the combination of the oxidation of furfural with the HER for the simultaneous production of high-value chemicals and hydrogen fuel (Figure 14.6) [61]. OER that limits traditional water electrolysis can be replaced by the FOR. This lowers the overall cell voltage due to the lower thermodynamic potential of FOR (~0.4 V vs. RHE) compared to OER (~1.23 V). This strategy not only preserves energy but also provides useful oxidation products.

Bifunctional catalysts based on metal-containing carbides and nitrides can be utilized to simultaneously drive FOR and HER. For example, Ni_3N/Ni_4P heterostructures promote charge transfer at the highly active interface which allows for the simultaneous oxidation of furfural at the anode and HER at the cathode. Other distinguishable performances can be seen with Mo_2C and Co_3N catalysts that have well-balanced binding energies for H* and furfural intermediates,

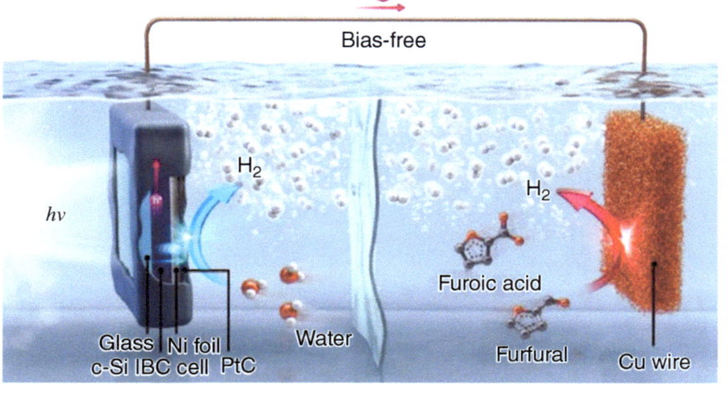

Figure 14.6 Furfural oxidation coupled with HER. *Source*: [61]/with permission of The Royal Society of Chemistry.

providing an excellent dual-function performance. This dual-reaction mode represents sustainable chemical synthesis, as it enables chemical production (via biomass valorization) while generating green hydrogen as a by-product. For optimization of this system, the catalyst structure, electrode configuration, as well as conditions should be tuned to maximize the stability and efficiency of the system.

14.7 Challenges and Future Perspectives

Realizing viable industrial-scale, waste-assisted water electrolysis with MCNs features several challenges, specifically in the aspects of stability, scalability, and control of the catalyst. Operationally, one of the most serious challenges is stability under prolonged, high-current-density operation, wherein MCNs are prone to oxidative dissolution and surface reconstruction, which necessitates advanced surface engineering and rational material design toward durable, highly crystalline structures based on in situ mechanistic studies. At the same time, the scalability issue prevents high-performance materials from proliferating beyond laboratory conditions to mass production, driven by the complexity and high costs associated with existing synthesis routes; future studies should directly focus on development of high-throughput "green" synthesis methods like molten salt or plasma processing and redundancy in standardization of mass-loading fabrication. Finally, in order to push performance to the next level, catalyst design would need to shift from bulk materials to new architectures, such as 2D MCN/MXene heterostructures or single-atom MCNs (SA-MCNs), exploiting control of doping and interfaces to tune electronic environments in active sites for selective reaction intermediates, which will require low-temperature synthesis routes to prevent aggregation of metal atoms.

Quickening this discovery process will significantly depend on machine learning and theoretical science for high-throughput virtual screening of thousands of compositions for MCNs, training predictive models to be used on reliable DFT data, and generating fully autonomous, closed-loop workflows for self-optimizing catalysts. Finally, commercial prospects will depend on integrated systems development and favorable economics; constructed electrolytic cells will need to efficiently treat relevant waste (real-world) streams and maximize product selectivity, validated through critical techno-economic and life cycle analyses to demonstrate MCNs-based electrolysis as a competitive and viable technology for coupled H_2 production and waste valorization.

14.8 Conclusion

MCNs have been established as proven, non-precious electrocatalysts that may transform waste-assisted water electrolysis by effectively producing H_2 while also upgrading waste streams such as urea and furfural. To move forward, it is vital to address the various long-term stability concerns along with synthesizing cost-effective and high-throughput new materials and scalability solutions via novel surface engineering. Furthermore, there is immense promise for the development of MCN electrocatalysts through atomic-scale catalyst control, such as the design of synergistic heterostructures and single-atom hybrids and also using machine learning and computational screening as a shortcut to predictive materials discovery. Successful resolution of materials and engineering challenges will make MCNs to seamlessly transition to a robust commercial application and usher in a better and more affordable future for green hydrogen production and waste valorization.

References

1 Dou, Y., Sun, L., Ren, J., & Dong, L. (2023). Opportunities and future challenges in hydrogen economy for sustainable development. In *Hydrogen Economy: Processes, Supply Chain, Life Cycle Analysis and Energy Transition for Sustainability* (eds. A. Scipioni, A. Manzardo, and J. Ren), 537–569. London: Academic Press.

2 Shiva Kumar, S. & Lim, H. (2022). An overview of water electrolysis technologies for green hydrogen production. *Energy Rep. 8*, 13793–13813.

3 S. Haghighat. (2017). Proton kinetics in electrochemistry: new directions and mechanistic analysis. University of Southern California.

4 Younus, H. A., Al Hajri, R., Ahmad, N., Al-Jammal, N., Verpoort, F., & Al Abri, M. (2025).Green hydrogen production and deployment: opportunities and challenges. *Discov. Electrochem. 2*, 32.

5 Arshad, F., Haq, T.-u., Hussain, I., & Sher, F. (2021). Recent advances in electrocatalysts toward alcohol-assisted, energy-saving hydrogen production. *ACS Appl. Energy Mater. 4*, 8685–8701.

6 Deng, C., Toe, C. Y., Li, X., Tan, J., Yang, H., Hu, Q., & He, C. (2022). Earth-abundant metal-based electrocatalysts promoted anodic reaction in hybrid water electrolysis for efficient hydrogen production: recent progress and perspectives. *Adv. Energy Mater. 12*, 2201047.

7 Rollinson, A. N., Jones, J., Dupont, V., & Twigg, M. V. (2011). Urea as a hydrogen carrier: a perspective on its potential for safe, sustainable and long-term energy supply. *Energy Environ. Sci. 4*, 1216–1224.

8 Wang, T., Tao, L., Zhu, X., Chen, C., Chen, W., Du, S., Zhou, Y., Zhou, B., Wang, D., Xie, C., Long, P., Li, W., Wang, Y., Chen, R., Zou, Y., Fu, X.-Z., Li, Y., Duan, X., & Wang, S. (2022). Combined anodic and cathodic hydrogen production from aldehyde oxidation and hydrogen evolution reaction. *Nat. Catal. 5*, 66–73.

9 Bhatt, M. D. & Lee, J. Y. (2020). Advancement of platinum (Pt)-free (non-Pt precious metals) and/or metal-free (non-precious-metals) electrocatalysts in energy applications: a review and perspectives. *Energy Fuels 34*, 6634–6695.

10 Zhong, Y., Xia, X., Shi, F., Zhan, J., Tu, J., & Fan, H. J. (2016). Transition metal carbides and nitrides in energy storage and conversion. *Adv. Sci. 3*, 1500286.

11 Zhan, G., Hu, L., Li, H., Dai, J., Zhao, L., Zheng, Q., Zou, X., Shi, Y., Wang, J., Hou, W., Yao, Y., & Zhang, L. (2024). Highly selective urea electrooxidation coupled with efficient hydrogen evolution. *Nat. Commun. 15*, 5918.

12 Xu, S., Ruan, X., Ganesan, M., Wu, J., Ravi, S. K., & Cui, X. (2024). Transition metal-based catalysts for urea oxidation reaction (uor): catalyst design strategies, applications, and future perspectives. *Adv. Funct. Mater. 34*, 2313309.

13 Climent, V., Rodes, A., Orts, J. M., Feliu, J. M., Perez, J. M., & Aldaz, A. (1997). On the electrochemical and in-situ Fourier transform infrared spectroscopy characterization of urea adlayers at Pt (100) electrodes. *Langmuir 13*, 2380–2389.

14 Tatarchuk, S. W., Medvedev, J. J., Li, F., Tobolovskaya, Y., & Klinkova, A. (2022). Nickel-catalyzed urea electrolysis: from nitrite and cyanate as major products to nitrogen evolution. *Angew. Chem. Int. Ed. 61*, e202209839.

15 Tian, J., Cheng, X., Fan, Y., Fang, W., Dong, P., Li, G., & Guo, Y. (2023). Progress of reactions between furfural and aliphatic alcohols via catalytic oxidation processes: reaction routes, catalysts, and perspectives. *Processes 11*, 640.

16 Ouyang, D., Gao, D., Hong, J., Jiang, Z., & Zhao, X. (2023). Promoting and controlling electron transfer of furfural oxidation efficiently harvest electricity, furoic acid, hydrogen gas and hydrogen peroxide. *J. Energy Chem. 79*, 135–147.

17 Boggs, B. K., King, R. L., & Botte, G. G. (2009). Urea electrolysis: direct hydrogen production from urine. *Chem. Commun. 2009*, 4859–4861.

18 Jiao, S., Fu, X., & Huang, H. (2022). Descriptors for the evaluation of electrocatalytic reactions: d-band theory and beyond. *Adv. Funct. Mater. 32*, 2107651.

19 Karuppasamy, K., Nichelson, A., Vikraman, D., Choi, J.-H., Hussain, S., Ambika, C., Bose, R., Alfantazi, A., & Kim, H.-S. (2022). Recent advancements in two-dimensional layered molybdenum and tungsten carbide-based materials for efficient hydrogen evolution reactions. *Nanomaterials 12*, 3884.

20 Wang, H., Li, J., Li, K., Lin, Y., Chen, J., Gao, L., Nicolosi, V., Xiao, X., & Lee, J.-M. (2021). Transition metal nitrides for electrochemical energy applications. *Chem. Soc. Rev. 50*, 1354–1390.

21 Rabi, O., Pervaiz, E., Ali, M., & Zahra, R. (2021). Bifunctional molybdenum carbide based hybrids for electrocatalytic water splitting. *Ceram. Int. 47*, 24949–24958.

22 Qiang, R., Du, Y., Chen, D., Ma, W., Wang, Y., Xu, P., Ma, J., Zhao, H., & Han, X. (2016). Electromagnetic functionalized Co/C composites by in situ pyrolysis of metal-organic frameworks (ZIF-67). *J. Alloys Compd. 681*, 384–393.

23 Chen, J. G. (1996). Carbide and nitride overlayers on early transition metal surfaces: preparation, characterization, and reactivities. *Chem. Rev. 96*, 1477–1498.

24 Liu, W., Wang, X., Wang, F., Du, K., Zhang, Z., Guo, Y., Yin, H., & Wang, D. (2021). A durable and pH-universal self-standing MoC–Mo2C heterojunction electrode for efficient hydrogen evolution reaction. *Nat. Commun. 12*, 6776.

25 Jena, S. & Swain, B. P. (2025). Surface passivation and surface treatment of coating materials. In *Advances in Mechanical Coating: Processing, Properties and Applications* (ed. B. P. Swain), 337–365. Singapore: Springer.

26 Lin, Z., Cai, L., Lu, W., & Chai, Y. (2017). Phase and facet control of molybdenum carbide nanosheet observed by in situ TEM. *Small 13*, 1700051.

27 Liu, L. M., Wang, S. Q., & Ye, H. Q. (2003). Adhesion of metal–carbide/nitride interfaces: Al/TiC and Al/TiN. *J. Phys.: Condens. Matter 15*, 8103.

28 Vaidhyanathan, B. & Rao, K. J. (1997). Synthesis of Ti, Ga, and V nitrides: microwave-assisted carbothermal reduction and nitridation. *Chem. Mater. 9*, 1196–1200.

29 Pavarajarn, V., Precharyutasin, R., & Praserthdam, P. (2010). Synthesis of Silicon Nitride Fibers by the Carbothermal Reduction and Nitridation of Rice Husk Ash. *J. Am. Ceram. Soc. 93*, 973–979.

30 Arik, H. (2003). Synthesis of Si3N4 by the carbo-thermal reduction and nitridation of diatomite. *J. Eur. Ceram. Soc. 23*, 2005–2014.

31 Sun, G.-D., Zhang, G.-H., Yan, B.-J., & Chou, K.-C. (2020). Study on the reduction of commercial MoO3 with carbon black to prepare MoO2 and Mo2C nanoparticles. *Int. J. Appl. Ceram. Technol. 17*, 917–931.

32 Kiyono, H., Miyake, Y., Nihei, Y., Tumura, T., & Shimada, S. (2012). Fabrication of Si3N4-based composite containing needle-like TiN synthesized using NH3 nitridation of TiO2 nanofiber. *J. Eur. Ceram. Soc. 32*, 1413–1417.

33 Pavlenko, V., Khosravi, H. S., Żółtowska, S., Haruna, A. B., Zahid, M., Mansurov, Z., Supiyeva, Z., Galal, A., Ozoemena, K. I., Abbas, Q., & Jesionowski, T. (2022). A comprehensive review of template-assisted porous carbons: Modern preparation methods and advanced applications. *Mater. Sci. Eng.: R: Reports 149*, 100682.

34 Hu, L., Ji, S., Jiang, Z., Song, H., Wu, P., & Liu, Q. (2007). Direct Synthesis and Structural Characteristics of Ordered SBA-15 Mesoporous Silica Containing Tungsten Oxides and Tungsten Carbides. *J. Phys. Chem. C 111*, 15173–15184.

35 Wu, P.-Y., Ji, S.-F., Hu, L.-H., Zhu, J.-Q., & Li, C.-Y. (2008). Preparation, characterization, and catalytic properties of the Mo2C/SBA-15 catalysts. *J. Porous Mater. 15*, 181–187.

36 He, T., Zhang, W., Manasa, P., & Ran, F. (2020). Quantum dots of molybdenum nitride embedded in continuously distributed polyaniline as novel electrode material for supercapacitor. *J. Alloys Compd. 812*, 152138.

37 Weers, M., von Seggern, A. R., Vocke, H., Taffa, D. H., & Wark, M. (2024). Two ways to more NH2-groups: formation of polymeric carbon nitride via melem tetramer nano sheets or supramolecular assembly of melamine and cyanuric acid for applications as photocatalyst. *ACS Appl. Nano Mater. 7*, 1402–1415.

38 Giordano, C., Erpen, C., Yao, W., Milke, B., & Antonietti, M. (2009). Metal nitride and metal carbide nanoparticles by a soft urea pathway. *Chem. Mater. 21*, 5136–5144.

39 Jeon, H., Lee, J.-W., Kim, Y.-D., Kim, D.-S., & Yi, K.-S. (2000). Study on the characteristics of TiN thin film deposited by the atomic layer chemical vapor deposition method. *J. Vac. Sci. Technol. A: Vac. Surf. Films 18*, 1595–1598.

40 Beadle, K. A., Gupta, R., Mathew, A., Chen, J. G., & Willis, B. G. (2008). Chemical vapor deposition of phase-rich WC thin films on silicon and carbon substrates. *Thin Solid Films 516*, 3847–3854.

41 Shimada, S., Yoshimatsu, M., Nagai, H., Suzuki, M., & Komaki, H. (2000). Preparation and properties of TiN and AlN films from alkoxide solution by thermal plasma CVD method. *Thin Solid Films 370*, 137–145.

42 Kafizas, A., Carmalt, C. J., & Parkin, I. P. (2013). CVD and precursor chemistry of transition metal nitrides. *Coord. Chem. Rev. 257*, 2073–2119.

43 Hamedani, Y., Macha, P., Bunning, T. J., Naik, R. R., & Vasudev, M. C. (2016). *Chemical Vapor Deposition: Recent Advances and Applications in Optical, Solar Cells and Solid State Devices*. Rijeka: IntechOpen.

44 Barakat, N. A. M. & Ali, M. A. (2022). Molybdenum carbide/Ni nanoparticles-incorporated carbon nanofibers as effective non-precious catalyst for urea electrooxidation reaction. *Sci. Rep. 12*, 22574.

45 Fan, J., Dou, Y., Jiang, R., Du, K., Deng, B., & Wang, D. (2021). Electro-synthesis of tungsten carbide containing catalysts in molten salt for efficiently electrolytic hydrogen generation assisted by urea oxidation. *Int. J. Hydrog. Energy 46*, 14932–14943.

46 Wang, L., Zhu, S., Marinkovic, N., Kattel, S., Shao, M., Yang, B., & Chen, J. G. (2018). Insight into the synergistic effect between nickel and tungsten carbide for catalyzing urea electrooxidation in alkaline electrolyte. *Appl. Catal. B: Environ. 232*, 365–370.

47 Ma, Y., Ma, C., Wang, Y., & Wang, K. (2022). Advanced nickel-based catalysts for urea oxidation reaction: challenges and developments. *Catalysts 12*, 337.

48 Chen, Y., Sun, P., & Xing, W. (2019). Cobalt nitride nanoflakes supported on Ni foam as a high-performance bifunctional catalyst for hydrogen production via urea electrolysis. *J. Chem. Sci. 131*, 101.

49 Barakat, N. A. M., El-Newehy, M. H., Yasin, A. S., Ghouri, Z. K., & Al-Deyab, S. S. (2016). Ni&Mn nanoparticles-decorated carbon nanofibers as effective electrocatalyst for urea oxidation. *Appl. Catal. A: Gen. 510*, 180–188.

50 Tiwari, D. K., Ghosh, T. K., Gopinathan, A. V., & Gangavarapu, R. R. (2025). Ball-milled Ni@Mo2C/C nanocomposites as efficient electrocatalysts for urea oxidation. *Environ. Sci. Pollut. Res.* https://doi.org/10.1007/s11356-025-36030-1

51 Zhang, Y., Zhen, X., Su, Z., Guo, J., Li, J., Liu, S., & Li, X. (2024). Ni/MoC@NC as bifunctional electrocatalyst coupled HER and urea oxidation for energy-efficient hydrogen production. *Int. J. Hydrog. Energy 60*, 46–54.

52 Deng, X., Imhanria, S., Sun, Y., Zhang, M., Cheng, Y., & Wang, W. (2022). Mo, Fe bimetallic carbide composite as high stability electrocatalyst for oxygen reduction reaction. *J. Environ. Chem. Eng. 10*, 108052.

53 Chen, P., Ye, J., Wang, H., Ouyang, L., & Zhu, M. (2021). Recent progress of transition metal carbides/nitrides for electrocatalytic water splitting. *J. Alloys Compd. 883*, 160833.

54 Zhang, X., Xu, S., Li, Q., Zhou, G., & Xia, H. (2021). Recent advances in the conversion of furfural into bio-chemicals through chemo-and bio-catalysis. *RSC Adv. 11*, 27042–27058.

55 Ayoub, N., Hamie, Z., Toufaily, J., Guenin, E., & Enderlin, G. (2024). Metal or metal-free catalysis for the oxidation of biosourced furfural. *ACS Sustain. Chem. Eng. 12*, 9568–9590.

56 Cheng, J., Zeng, Z., Huang, X., & Wei, Z. (2025). Electrooxidation of furfural on an electrodeposited NiMoP electrode to simultaneously produce hydrogen and 2-furancarboxylic acid at industrial current levels. *J. Power Sources 628*, 235919.

57 Zhang, N., Zou, Y., Tao, L., Chen, W., Zhou, L., Liu, Z., Zhou, B., Huang, G., Lin, H., & Wang, S. (2019). Electrochemical oxidation of 5-hydroxymethylfurfural on nickel nitride/carbon nanosheets: reaction pathway determined by in situ sum frequency generation vibrational spectroscopy,"*Angew. Chem. 131*, 16042–16050.

58 Ma, S., Wang, Z., Liu, B., Wang, X., Du, S., Song, Z., Bian, X., Hao, M., Huang, J., & Chen, Z. (2023). Interface-induced enhancement of paired electrolysis in 5-hydroxymethylfurfural-H2O system on symbiotic nickel/molybdenum nitride heterostructure. *Chem. Eng. J. 476*, 146490.

59 Zhang, D., Xing, M., Mou, X., Song, C., & Wang, D. (2023). Deep eutectic solvent induced ultrathin Co4N/N-doped carbon nanosheets self-supporting electrode for boosting hydrogen evolution integrated with biomass electrooxidation. *Appl. Surf. Sci. 608*, 155283.

60 Lin, Z. (2021). Catalytic transformation of biomass-derived oxygenates using transition metal carbide, nitride, and oxide surfaces. PhD thesis. Columbia University.

61 Pieczonka, N. P. W. & Aroca, R. F. (2008). Single molecule analysis by surfaced-enhanced Raman scattering. *Chem. Soc. Rev. 37*, 946–954.

15

Metal Sulfides and Phosphides Electrocatalysts for Furfural-assisted Water Electrolysis

Jessica J. P. Nascimento[1,2], *Paulo N. S. Casciano*[1], *Francisco W. P. Ribeiro*[3], *Thiago M. B. F. Oliveira*[4], *Adriana N. Correia*[1], *and Pedro de Lima-Neto*[1,*]

[1]*Department of Analytical and Physical Chemistry, Federal University of Ceará, Pici Campus, Fortaleza, Brazil*
[2]*Postgraduate Program in Engineering and Materials Science, Federal University of Ceará, Pici Campus, Fortaleza, Brazil*
[3]*Institute of Exact and Natural Sciences, University of International Integration of Afro-Brazilian Lusophony, Auroras Campus, Redenção, Brazil*
[4]*Center for Science and Technology, Federal University of Cariri, Juazeiro do Norte, Brazil*

*Corresponding author: pln@ufc.br

15.1 Introduction

Currently, energy and sustainability have formed a keyword pair that is on the minds and the lips of people concerned with the environmental problems associated with climate change. The origin of climate change is attributed to natural phenomena and human activities. For natural phenomena, it can be cited the volcanic activities and the occurrence of El Niño and La Niña, while for the human activities, it is possible to cite the use of fossil fuels, such as oil, coal, and natural gas, since their burning produces CO_2, one of the greenhouse gases (GHGs) which intensify global warming [1, 2].

Despite fossil fuels remaining as our primary energy sources, modern society has encouraged the search for energy solutions that are less harmful to the environment and are sustainable, aiming to reduce our dependence on fossil fuels and transition toward renewable energy sources such as solar, wind, hydropower, and biomass to mitigate climate changes. Indeed, this energy transition will require public policies that strengthen the research and development of greener technologies to accelerate our independence from fossil fuels and to diminish the impact of global warming, making it possible the decarbonization of air in the cities [3–6].

Among the renewable energy sources, hydrogen is emerging as a highly adaptable and multifunctional energy carrier that can support the decarbonization of multiple sectors. It also offers a solution to alleviate the growing strain on electricity grids, which are increasingly saturated by intermittent renewable energy sources. However, the widespread deployment of hydrogen technologies is currently hindered by several challenges, including inadequate infrastructure, limited market development, and excessive costs associated with production, storage, and distribution. These barriers are expected to be progressively overcome through the growth and expansion of hydrogen valleys [4].

Furthermore, the rapid increase in global population and industrial activity has significantly elevated the GHG emissions, with projections indicating a potential temperature rise of 3–6 °C by 2050. Immediate and coordinated action is required to limit global warming to 1.5 °C above preindustrial levels. Hydrogen, due to its high energy density and compatibility with renewable energy systems, presents a promising clean energy alternative for reducing GHG emissions. Nevertheless, its large-scale adoption is constrained by high production costs, insufficient infrastructure, and an immature value chain. Approximately 96% of global hydrogen is produced from fossil fuels, which contributes significantly to emissions, while only 4% is derived from water electrolysis. Green hydrogen, produced via electrolysis with efficiencies ranging from 55% to 80%, remains costly at \$2.28–7.39 per kg, compared to gray hydrogen at \$0.67–1.31 per kg, which emits approximately 8.5 kg of CO_2 per kg of hydrogen produced. Furthermore, hydrogen's low volumetric density complicates storage, and transportation remains risky and infrastructure-limited [7].

In addition, biomass has been proposed as a potential alternative to fossil fuels due to its carbon-neutral nature and renewable characteristics, as noted by Theerthagiri et al. [6]. Biomass is considered a renewable source of energy because these materials are replenished relatively quickly. The strategy involves utilizing biomass to replace nonrenewable carbon sources with sustainable carbon derived from diverse biomass feedstocks, including carbohydrates, triglycerides, lignin, and proteins [8]. Moreover, the chemical products produced from biomass can now be used for the electrosynthesis of compounds associated with the electrolysis of water. For instance, Nilges et al. [9] successfully reported an electrochemical route for producing *n*-octane from levulinic acid, a chemical derived from lignocellulosic biomass. These authors highlighted the advantages of the electrochemical route over the chemical route. According to them, the synthesis of octane from levulinic acid via chemical route is usually achieved via multistep processes, it requires temperatures between 250 and 400 °C and hydrogen pressures between 10 and 35 bar, while the electrochemical route allowed the synthesis be carried out at room temperature, in aqueous solutions; a hydrogen atmosphere or strong oxidizing or reducing chemicals are not required and the natural phase separation allows a simple separation of the final product [9]. Lastly, another advantage of the electrochemical route is that the electric energy required to promote the electrochemical reactions can be supplied by solar or wind energy. Another example is the report published by Lyu [10], in which the author investigated the electrochemical oxidation of furfural (FF), a commercial chemical produced from raw corncob, about hydrogen electrochemical production via water electrolysis [10]. His findings showed that hydrogen production from FF waste had a conversion rate of approximately 25% and a hydrogen yield of 0.48% by weight at 14.3 mA·cm^{-2}, with an energy consumption that was 50% lower than that of water electrolysis. The lower electric energy consumption required by Lyu's system to produce hydrogen from water electrolysis is explained by the fact that in this system, the oxygen evolution reaction (OER), that has a high standard electrode potential ($E^\circ = 1.229$ V vs. SHE), is replaced by oxidation of a chemical substance (like FF) at the anode, leading to lower energy consumption.

Nowadays, there is a growing interest in hybrid electrolysis systems, in which a chemical compound derived from biomass is oxidized at the anode. At the same time, electrochemical water splitting occurs at the cathode. Therefore, systems like these allow for the production of hydrogen with lower applied potential and, consequently, lower energy consumption. In addition, value-added chemical compounds can be produced at the anode compartment [11].

For a hybrid water electrolysis system, several electrode materials have been tested as anode and cathode. Helpful reviews highlighting some electrocatalyst materials that can be used as anodes and cathodes have been published in the literature [12–14]. Nowadays, metal sulfides and metal phosphides are non-noble materials that have gained attention as promising electrocatalyst materials for electrochemical water splitting [15–18]. Thus, in this chapter, an overview is provided for the use of metal sulfide and metal phosphide materials as electrocatalysts in FF-assisted water electrolysis systems. Finally, through an extensive literature review, the fundamental principles of using these electrocatalysts in FF-assisted water electrolysis systems are discussed.

15.2 Importance of Transition-metal-based Catalysts

Metal sulfide and phosphide-derived catalysts are gaining notable visibility as efficient and cost-effective replacements for noble metals in water electrolysis, given their favorable electronic structures, tunable surface chemistry, and robust kinetic activity under electrochemical conditions [19–22]. These features also make them relevant candidates for sustainable H_2 generation in integrated biomass platforms, such as FF-assisted water electrolysis, where the simultaneous conversion of organic substrates and reduced energy demand contribute to greater efficiency and environmental benefits [23–25].

Metal sulfides, metal phosphides, and their composites exhibit diverse crystal architectures, ranging from layered to non-layered structures, which directly influence their catalytic performance [21, 24, 26–29]. In metal sulfides, moderate metal–sulfur bonding energies balance the adsorption and desorption events of the intermediate reaction, allowing for the reactivation of their catalytic activity during the hydrogen evolution reaction (HER) or OER [21, 25, 27, 30]. As for phosphides, the incorporation of phosphorus alters the electron density around the metal centers, increasing electrical conductivity and modifying the d-band center to favor intermediate adsorption [31, 32]. These materials frequently exhibit defect-rich surfaces, high densities of active sites, and favorable charge-transfer characteristics [31]. Strong covalency in metal–phosphorus or metal–sulfur bonds and the semimetallic conductivity of these compounds are critically necessary for overpotential reduction and enhancement in electrocatalytic kinetics [20–22, 25, 27, 30].

This outstanding transfer rate is significantly enhanced in the context of FF-assisted water electrolysis [22]. FF has a highly oxygenated structure and serves as a valuable feedstock for electrochemical processes, enabling its selective transformation into high-value chemicals such as furfuralic acid and 2-furyl alcohol [19, 23, 31, 33]. Since metal sulfides and phosphides resist both alkaline and neutral electrolytes, they are ideal candidates for FF oxidation parallel to H_2 generation [21, 22]. This combined system also reduces reaction overpotentials, thereby enhancing the overall energy efficiency of water electrolysis and promoting the valorization of biomass-derived feedstocks for the generation of low-carbon hydrogen [34, 35].

Although traditional noble-metal-based catalysts (e.g., Pt, Pd, Ru, Rh, and Ir) and their oxides are considered reference catalytic materials in several redox reactions [19, 31, 33, 36], their high cost, scarcity, and limited long-term stability under variable pH conditions have resulted in increased search for more abundant alternatives [10, 37–39]. In contrast, transition-metal sulfides and phosphides offer a cost-effective and earth-abundant alternative [19–22]. Even though their electroactivity is lower than that of noble metals, their performance can be significantly improved through structural engineering, doping, or hybridization with conductive substrates (Cheng et al., 2024).

The catalytic performance of metal sulfides and phosphides is highly dependent on their synthesis route, which also influences morphology, crystallinity, porosity, surface area, and photochemical and electrochemical behaviors [40]. Figure 15.1 illustrates some of the major synthesis routes for metal sulfides and phosphides, along with the characteristic morphologies typically derived from them. Hydrothermal and solvothermal methods, for example, allow the growth of crystalline metallic

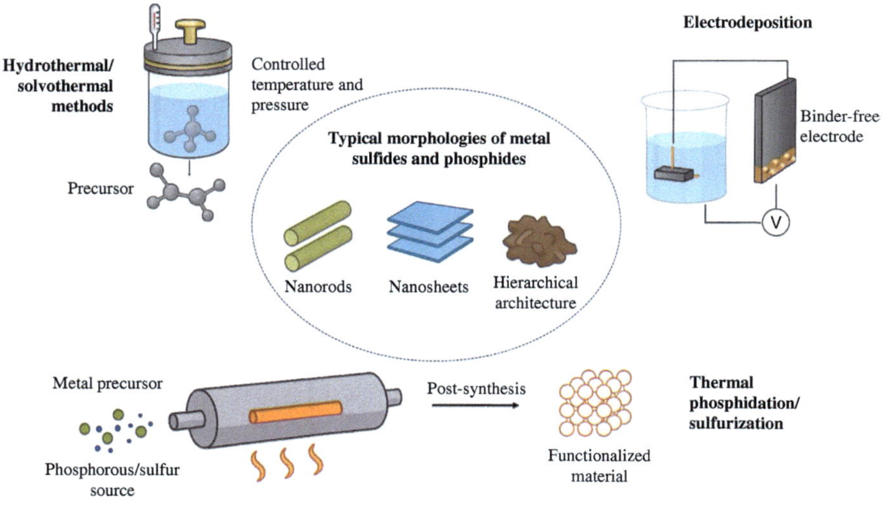

Figure 15.1 Schematic representation of typical synthesis routes and resulting morphologies for metal sulfides and phosphides.

nanostructures under controlled temperature and pressure conditions [41–43]. These wet-chemical approaches are widely used to produce nanorods, nanosheets, and hierarchical architectures [43]. Parameters such as precursor concentration, reaction time, and solvent type influence the nucleation and growth processes, affecting the number of available catalytic sites [42]. Electrodeposition of metal sulfides and phosphides is also a strategic approach for synthesizing thin layers of these catalysts on conductive substrates, allowing for precise control of thickness and composition [41, 43–45]. It is particularly advantageous for preparing binder-free electrodes with enhanced charge-transfer properties and mechanical stability [46]. There are also post-synthesis methods, such as thermal phosphidation/sulfurization, which involve treating metal precursors with sources of phosphorus or sulfur under inert atmospheres [47].

Each synthesis technique has implications for scalability, energy input, environmental impact, and material stability, critical factors for real-world implementation in electrochemical energy conversion systems [40, 43]. As different synthesis routes tailor the structural and electronic characteristics of these electrocatalysts, they are gaining wider acceptance for use in upcoming FF-assisted water electrolysis technologies.

15.3 Electrocatalytic Performance for FF Compounds Oxidation and HER

Electrocatalysis plays a crucial role in sustainable energy conversion by enhancing the rates of electrochemical reactions essential for several green technologies. Recently, increased attention has been given to processes such as water splitting coupled with oxidative biomass valorization, which enhance overall efficiency and are vital for producing hydrogen and value-added chemicals from renewable resources [48]. Water splitting is a promising strategy for sustainable hydrogen production; however, its practical application is limited by the slow kinetics of the OER, which necessitates high overpotentials and results in a reduction in energy conversion efficiency [49]. In this case, inorganic materials based on metal phosphides and sulfides can significantly reduce the overpotential required for OER, thereby improving the energy efficiency of water splitting [50, 51].

Furthermore, an additional strategy referenced in recent literature is based on the oxidation of biomass-derived compounds, which can be coupled with the HER, offering a sustainable route for energy generation and chemical synthesis [52–54]. Coupling biomass oxidation with the HER offers a versatile approach to sustainable energy production, significantly reducing the overpotential required for hydrogen production and improving the overall energy efficiency of the process. The high overpotential associated with the OER can be overcome by replacing it with the oxidation of biomass-derived compounds. Additionally, it also generates valuable chemicals from biomass, adding economic value to the process.

In recent studies, FF-based compounds (5-hydroxymethylfurfural [HMF], and furfuryl alcohol) have been identified as promising biomass-derived precursors for application in electrocatalytic systems [29, 31]. Developing efficient electrocatalysts for both FF compounds conversion and HER is essential for realizing this synergistic potential, requiring careful design and optimization of the catalyst material. The electrocatalyst should exhibit high activity and selectivity for both FF oxidation and HER, ensuring efficient conversion of FF into valuable products and high hydrogen production rates. The catalyst should also be stable and durable under the electrochemical conditions, maintaining its performance over extended periods of operation.

Currently, metal sulfides and phosphides are garnering growing interest as viable non-precious metal electrocatalysts, owing to their intrinsic abundance, cost-effectiveness, and environmentally benign properties [43]. Furthermore, these materials often exhibit comparable or even superior catalytic activity to noble metals under specific conditions, making them promising alternatives for various electrochemical reactions. A general scheme and advantages for simultaneous electrocatalytic HER and 5-HMF reactions are shown in Figure 15.2.

Transition-metal phosphides and sulfides have emerged as promising electrocatalysts due to their unique and intriguing electronic and catalytic properties,

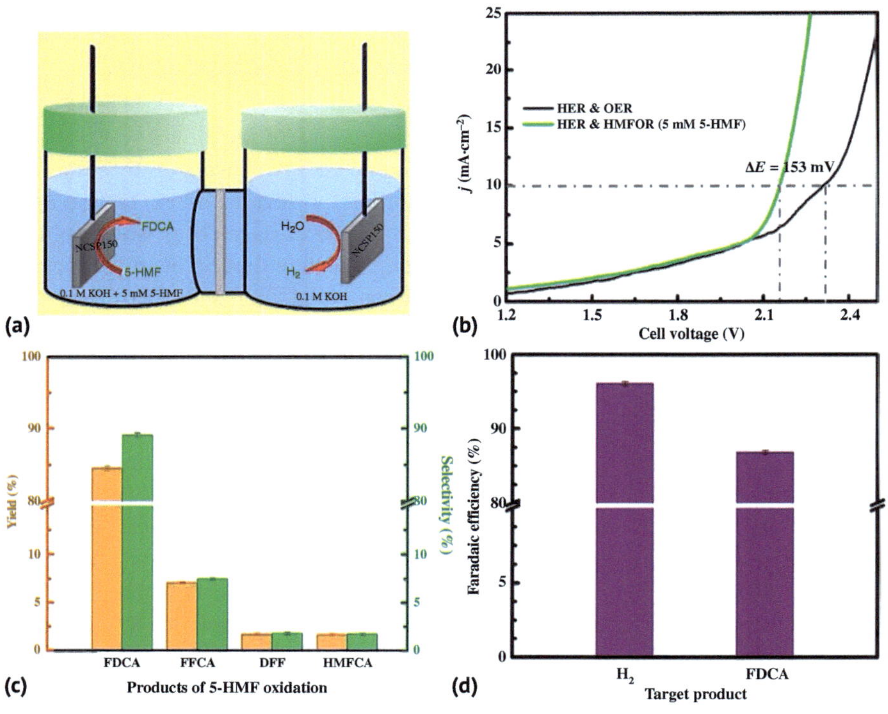

Figure 15.2 (a) General scheme of two-electrode electrolytic H-cell for simultaneous electrocatalytic HER and 5-HMF reactions. (b) Polarization curves of aqueous solution with and without 5-HMF. (c) Yields and selectivities of products from HMFOR. (d) FEs of H_2 and FDCA at a constant-cell-voltage KOH solution in two-electrode electrolytic H-cell. *Source*: Hung et al. (2024) [55]/with permission of Elsevier.

making them excellent candidates for various electrochemical applications [56]. These materials exhibit a combination of metallic and covalent bonding, which allows for tunable electronic structures and high catalytic activity. Their ability to form various stoichiometric compositions and crystal structures further enhances their versatility as electrocatalysts. The electronic interactions between the metal and sulfur or phosphorus atoms create active sites for electrochemical reactions, making these materials highly effective in catalyzing various energy conversion and storage processes [40]. In this context, the main results obtained in the development of metal sulfide and phosphide electrocatalysts for FF-assisted water electrolysis are presented in the following subsections.

15.3.1 Metal Phosphides Electrocatalysts for FF-assisted Water Electrolysis

You et al. [52] reported a noble-metal-free, bifunctional electrocatalyst consisting of three-dimensional nickel phosphide nanoparticle arrays on nickel foam (Ni_2P NPA/NF) that simultaneously catalyzes the HER and hydrogen peroxide (H_2O_2) oxidation. Significantly, replacing the OER with thermodynamically and kinetically more favorable 5-HMF oxidation reduced these voltages by over 200 mV. The partial surface oxidation of Ni_2P generates catalytically active oxidized nickel species, which are responsible for the high anodic activity.

Jiang et al. [57] demonstrated that electrodeposited cobalt phosphide (CoP) was an efficient bifunctional electrocatalyst for 5-HMF oxidation. It oxidizes 5-HMF at significantly lower potential (1.38 V vs. RHE at 20 $mA\cdot cm^{-2}$), making it advantageous for integrated electrocatalytic systems.

Zhang et al. [58] studied FF oxidation on Cu_3P and Ni_2P nanocrystals on carbon fiber cloth (CFC). Cu_3P/CFC achieved faradaic efficiencies (FEs) of 92–98% for FF hydrogenation to furfuryl alcohol due to favorable adsorption and high surface hydrogen coverage. This oxidation activity is linked to the formation of high-valent Ni species (oxyhydroxides) during operation, as confirmed by comprehensive surface analyses [58].

Wang et al. [59] demonstrated, through density functional theory (DFT) calculations, that electron transfer from CoOOH to CoP at the heterojunction interface creates hole accumulation on CoOOH, promoting 5-HMF adsorption and enhancing catalytic activity for the oxidation reaction. The synergistic effect of interfacial electronic modulation and advanced reactor design in enhancing biomass upgrading electrocatalysis is further demonstrated by the low operating voltage of 1.42 V required for the coupled hydrogen evolution and 5-HMF oxidation system, along with the high Faradaic efficiency (98.2%) and 2,5-furanedioic acid selectivity (99.4%) achieved at high 5-HMF concentrations in a flow reactor configuration [59].

Zhou et al. [60] developed coralline-like CoP nanorods anchored on Ni_2P–NiCoP nanosheets embedded in N-doped carbon (CoP/Ni_2P–NiCoP@NC), synthesized via a metal–organic frameworks-derived strategy. This architecture creates abundant heterogeneous interfaces, which play a crucial role in enhancing the electrocatalytic 5-HMF oxidation. Mechanistic insights supported by DFT calculations revealed that the heterogeneous interfaces modulate the d-band centers of Co and Ni, bringing them closer to the Fermi level. This electronic structure tuning enhances the adsorption of 5-HMF and reaction intermediates while reducing the energy barrier of the rate-determining step in the oxidation pathway [60].

Xu et al. [61] developed a deep eutectic solvent (DES)-induced in-situ etching and phosphorization approach to fabricate nickel phosphide, P and O co-doped carbon nanosheet composites (Ni_2P@POC) directly on nickel foam. The DES functioned dually as an etchant, extracting Ni^{2+} ions, and as a phosphorus source, facilitating the uniform formation of nickel phosphides within carbon films—the increase in electrochemically active surface area after activation was attributed to the formation of amorphous NiOOH. DFT calculations suggested that electron transfer from NiOOH to nickel phosphides and subsequently to the P and O co-doped carbon optimizes 5-HMF adsorption energetics, thereby promoting electrocatalytic activity [61].

Recent advances in the development of bifunctional electrocatalysts for biomass valorization and hydrogen evolution have highlighted Mn-doped $FePSe_3$ nanosheets (Mn–$FePSe_3$/NS) as an auspicious system. According to Zhang et al. [30], the Mn–$FePSe_3$/NS catalyst exhibits exceptional performance in the selective electro-oxidation (ECO) of 5-HMF to FDCA. The catalytic enhancements to Mn doping induced improvements in 5-HMF adsorption and reduced hydrogen adsorption free energy, which facilitate both anodic oxidation and cathodic reduction processes [30].

A bifunctional CoNiP nanosheet-integrated electrode (CoNiP-NIE) was developed by Song et al., demonstrating superior electrocatalytic performance for both HER and hydrogen peroxide (5-HMF) oxidation reaction. CoNiP-NIE achieved a high FE (>82%) for the oxidation of 5-HMF to FDCA. For HER, the catalyst exhibited a low overpotential of approximately 108 mV at 10 $mA \cdot cm^{-2}$, indicating high catalytic activity [62].

A recent study by Wang et al. demonstrated that Mo doping in cobalt phosphide (CoMoP) significantly enhances electrocatalytic performance for both HER and 5-HMF oxidation. Mo incorporation expands the CoP lattice, resulting in an upshift of the Co d-band center and a reduced hydrogen adsorption free energy, which together improve HER activity. Mo doping promotes adsorption and oxidation of 5-HMF on the catalyst surface by facilitating formation of active oxygen-containing species such as CoOOH and Mo–O [63].

Lin et al. [64] synthesized a nickel–phosphorus (Ni–P) catalyst via cathodic Ni plating on a Ni-deposited carbon paper substrate (NiP@Ni/C), which exhibits superior activity and selectivity for the 5-HMF oxidation reaction. Importantly, the unique bulk Ni–P component acts as a charge storage reservoir by stabilizing high oxidation states of nickel (Ni^{3+} and above), enabling continuous spontaneous oxidation of 5-HMF even under open-circuit conditions. The catalyst's combination of surface-active Ni^{3+} sites and bulk charge storage creates a synergistic core–shell effect, which drives efficient six-electron oxidation of 5-HMF with minimized intermediate accumulation [64].

Li et al. [65] developed a nickel phosphide electrocatalyst embedded in two-dimensional amorphous Al_2O_3 arrays on three-dimensional nickel foam (NiP–Al_2O_3/NF) via reduction of a layered double hydroxide (LDH) precursor. This unique

3D nano/micro-architecture features a multi-porous structure, interconnected electron-conductive networks, and highly dispersed Ni_2P active sites, yielding exceptional OER performance [65].

Bi et al. [66] reported porous Ce-doped CoP nanosheets with phosphorus vacancies using a DES. The optimized 5.2% Ce–CoP catalyst achieved complete conversion of 5-HMF. Electrochemical analyses revealed a high current density, reduced charge-transfer resistance, and increased surface area, indicating enhanced kinetics and active sites resulting from Ce doping and phosphorus vacancies [66].

A phosphorus-engineered perovskite-derived nickel phosphide catalyst, synthesized via a hydrothermal phosphorization method, was reported by Gong et al. [67]. Kinetic analysis revealed a first-order rate constant that was over four times higher than that of the non-phosphorus-engineered counterpart, indicating enhanced catalytic activity. The catalyst maintained structural integrity and catalytic efficiency after multiple cycles, demonstrating robust stability [67].

According to Xu et al. [68], the manganese-doped nickel phosphide catalyst (Mn-5Ni_2P) synthesized via a microwave-assisted DES approach, followed by phosphidation, achieved complete 5-HMF conversion. The catalyst demonstrated superior performance attributable to its unique microstructure, which integrates the high surface energy and active sites of the amorphous phase with the enhanced charge-transport properties of the crystalline phase [68].

The nickel–phytic acid (Ni–PA) hybrid catalyst demonstrates remarkable performance in the electrocatalytic oxidation of 5-HMF to FDCA. Under an applied potential of 1.6 V vs. RHE, Ni–PA achieves an FDCA yield of 99.1% with a FE of 90%, significantly outperforming Fe–PA and Cu–PA catalysts [69].

According to Xie et al., Ce-doped cobalt phosphide (Ce–Co_2P) nanoparticles embedded in nitrogen-doped carbon nanoarrays have demonstrated significant improvements in the electrochemical oxidation of 5-HMF coupled with HER. The enhanced performance is attributed to the Ce-induced modulation of electronic structures, which lowers the energy barrier of the rate-determining step via improved charge transfer and increased Co-site electron density. The formation of active $Co(OH)_2$ and CoOOH species during catalysis and durability tests confirms the robustness of the nanoarray architecture [70].

Dasgupta et al. [71] reported the molecular synthesis of amorphous nickel pnictide (NiP and NiAs) nanomaterials that undergo complete electrochemical reconstruction into nanocrystalline, potassium-intercalated γ-$NiOOH_x$ phases under alkaline conditions. Comparative studies revealed that the superior activity of NiP over NiAs is attributed to its smaller particle size, enhanced electrochemically active surface area, and a greater density of redox-active Ni sites [71].

15.3.2 Metal Sulfides Electrocatalysts for FF-assisted Water Electrolysis

You et al. [48] developed a bifunctional nickel sulfide-modified nickel foam (Ni_3S_2/NF) electrocatalyst that efficiently couples HER with oxidative upgrading of biomass-derived substrates, such as ethanol, benzyl alcohol, FF, furfuryl alcohol, and 5-HMF. This methodology replaced the OER with biomass oxidation, resulting in a considerable reduction of approximately 200 mV in cell voltage [48].

The study by Wang et al. [72] presented a Ni_3S_2/NF electrocatalyst synthesized via hydrothermal sulfidation for the efficient oxidation of 5-HMF to FDCA. Compared to bare nickel foam and commercial nickel-based catalysts (NiC, NiO, and NiS), Ni_3S_2/NF exhibits superior reaction rates, conversion, and stability, maintaining performance over six consecutive cycles. Mechanistic insights indicated the predominance of the pathway involving the oxidation of aldehyde groups to carboxyl groups [72].

A novel sulfur-modulated nickel/carbon composite catalyst (S–Ni@C) was synthesized using lignosulfonate as a renewable source of carbon and sulfur, via hydrothermal treatment followed by carbothermal reduction. The enhanced activity was attributed to sulfur-induced electronic modulation of nickel, promoting the in-situ formation of reactive nickel hydroxide species, combined with carbon shell coupling that improves electron transport [73].

Zhao et al. [74] developed the hierarchical NiCo-based sulfide (NiCo–S) nanosheets with atomic-level control of bimetallic sites and sulfur coordination for 5-HMF oxidation. The study elucidated that Co sites preferentially facilitate conversion of the aldehyde group to carboxyl, while Ni sites accelerate the subsequent oxidation steps toward FDCA formation. Sulfur coordination further promotes strong diatomic adsorption on the CHO group and improves electronic conductivity by increasing the density of states at the Fermi level [74].

Sun et al. [75] reported the development of a Co-doped Ni_3S_2 electrocatalyst supported on nickel foam ($Co_{0.4}NiS$@NF), which exhibits outstanding performance for 5-HMF oxidation under alkaline conditions. The Co doping significantly enhanced the electrochemical surface area, facilitating increased active-site exposure and accelerated electron/charge transfer during catalysis. In-situ Raman spectroscopy further identified rapid surface reconstruction of the catalyst into NiCo (oxy) hydroxides serving as active sites [75].

Wang et al. [76] demonstrated that controlled tungsten doping of nanostructured Ni_3S_2 significantly enhances its electrocatalytic performance for the ECO of 5-HMF to FDCA. The incorporation of high-valent W shifts the Ni_3S_2 d-band center upward, facilitating water adsorption and dissociation to generate abundant hydroxyl radical species, while simultaneously increasing the Ni valence state through electron withdrawal, thereby optimizing 5-HMF adsorption energy [76].

Li et al. [77] demonstrated the advances in ECO of 5-HMF to FDCA, showing that Mn-doped nickel sulfide ($Mn_{0.2}NiS$) nanosheets grown directly on 3D graphite felt substrates exhibit outstanding catalytic performance at industrially relevant current densities. Mn doping enhances the availability of active sites and facilitates electron-transfer kinetics [77, 78].

The electrocatalytic evaluation of bimetallic cobalt–nickel sulfide (Co_xNi_yS) catalysts derived from metal–organic frameworks for the 5-HMF oxidation was performed by Zeng et al. [79]; systematically varying the Co:Ni ratio, the authors demonstrated a volcano-type relationship between cobalt content and catalytic performance. Spectroscopy measurements identified the formation of high-valent

Ni^{3+} species (notably NiOOH) during the electrolysis process, which are implicated as the active sites driving 5-HMF oxidation. The presence of cobalt facilitates the generation of these Ni^{3+} species and synergistically enhances the catalytic efficiency [79, 80].

Chen et al. [81] developed a novel bio-derived electrocatalyst, Ni_3S_2 nanorods encapsulated in nitrogen-doped carbon directly grown on nickel foam (Ni_3S_2@NC/NF), synthesized via an in-situ sulfurization process using *Saccharomyces cerevisiae* as a sustainable source of carbon, nitrogen, and sulfur. The Ni_3S_2@NC/NF exhibited roughly double the current density of the nickel foam and NiO/NF benchmarks, attributing this enhancement to sulfide incorporation. Mechanistic studies indicated that oxidation follows a sequential pathway from 5-HMF through hydroxymethylfurancarboxylic acid intermediates to FDCA, with minimal generation of side products [81].

According to Xu et al. [82], the oxysulfide-coordinated amorphous nickel oxyhydroxide (NiOOH–SO_x) active phase on Ni_3S_2 precursors introduces numerous defects and unsaturated sites, which facilitate the anodic oxidation of 5-HMF to FDCA. Among several heteroatom dopants investigated, copper (Cu^{2+}) doping uniquely modulates the local coordination environment and electronic structure. The high-valent NiOOH–SO_x acts as the genuine active species, and Cu doping enhances electron transport and intermediate adsorption during 5-HMF oxidation. DFT calculations showed enhanced adsorption energies for reaction intermediates and improved electronic conductivity upon Cu incorporation [82].

Hung and Wu [55] used a nickel phosphide nanoparticle-decorated nickel cobalt sulfide nanosheet framework (NCSP150) fabricated on nickel foam, which exhibits exceptional bifunctional performance in alkaline media for simultaneous hydrogen generation and 5-HMF oxidation. Concurrently, during 5-HMF oxidation at 1.52 V vs. RHE, the catalyst achieves high FDCA selectivity and a FE of over 95% with over 95% 5-HMF conversion [55].

According to Pang et al. [83], ReS_2/Co_9S_8 heterostructure represents a significant advancement in bifunctional electrocatalysis, demonstrating remarkable activity and stability for both HER and FF oxidation. Synthesized via a hydrothermal method and subsequent calcination, this catalyst exhibits strong electronic interactions at the heterointerface, crucial for its enhanced performance. The Tafel slope of 37.8 $mV \cdot dec^{-1}$ confirmed rapid reaction kinetics via a Volmer–Tafel mechanism, while long-term electrochemical tests demonstrated negligible structural degradation [83].

Zhang et al. [28] developed a novel Ni_3S_2–Co_9S_8 heterojunction integrated within cobalt layered double hydroxide (Co LDH) nanosheets grown on nickel foam, forming a composite electrocatalyst (Ni_3S_2–Co_9S_8/Co LDH/NF) with remarkable activity toward OER and 5-HMF oxidation. The significantly reduced charge-transfer resistance was attributed to the synergistic effect at the Ni_3S_2–Co_9S_8 heterojunction in conjunction with the interconnected Co LDH nanosheet architecture. This combination enhances active-site exposure and facilitates electron transfer, while modulating the electronic structure to stabilize high-valent metal active species, which are essential for catalytic performance [28].

15.4 Pathways of FF Oxidation Reaction

Mechanistic insights obtained through electrochemical studies, in both coupled (FOR||HER) and half-cell (FOR [FF oxidation reaction]) systems, reveal the dependence of the reaction pathway, pH, and selectivity on the potential and the catalyst [84, 85]. This knowledge is essential for the optimized design of catalysts and the efficient control of FOR||HER electrocatalysis processes, aiming at the coproduction of valuable oxidized chemicals, reduced products, and hydrogen with lower energy demand [83–85]. The use of specific catalysts demonstrates the possibility of selective ECO for products such as furoic acid (FA), 5-hydroxyfuroic acid (HFA), 2(5*H*)-furanone or 2(3*H*)-furanone, 5-hydroxyfuran-2(5*H*)-one (HFN), and maleic acid (MA), with favorable kinetics evidenced by electrochemical analyses.

15.4.1 Pathways in Acidic Media

In Pt catalysts in an acidic electrolyte (0.25 M $HClO_4$), FF oxidation begins significantly above ~0.8 V vs. RHE. At the same time, the OER becomes relevant only above ~1.45 V vs. RHE [85]. At low potentials (0.9–1.1 V vs. RHE), the main products formed are FA and 5-HFA [84, 85], with FA formation being theoretically favored. Mechanistically, the ECO of FF to form FA begins with the dehydrogenation of the C—H bond of the aldehyde group, facilitated by catalysts, generating an FF radical (F—CO). This species subsequently reacts with adsorbed hydroxyls (OH*) to form FA, with this pathway being kinetically favored (Figure 15.3) [84–92]. At higher potentials (≥1.2 V vs. RHE), a change in selectivity toward 5-HFN is observed, with the appearance of MA [84, 85], and decomposition to CO_2 occurs via decarbonylated or decarboxylated intermediates, with CO being a potential "poison" for the catalyst surface at low potentials [85]. 5-HFA is generated by the oxidation (dehydrogenation) of F—CO at the five positions of FF. The formation of this compound in an alkaline medium is described as a four-electron process (Figure 15.3) [84, 85, 87]. The oxidation of FF to produce 2(3*H*)-furanone or 2(5*H*)-furanone, 5-HFN, and MA involves routes that begin with the oxidation of FF to FA. FA undergoes decarboxylation, generating a furyl intermediate (C_4) that, through hydration, forms 2-hydroxyfuran. This latter compound rapidly isomerizes into 2(5*H*)-furanone and subsequently undergoes hydration and decarboxylation, resulting in β-formylacrylic acid, a direct precursor of MA through hydration (Figure 15.3) [84, 85, 89, 91, 93].

15.4.2 Pathways in Alkaline/Neutral Media

Mechanistically, the formation of FA from FF in an alkaline medium involves the hydration of FF to a geminal diol, followed by deprotonation by hydroxyl ions (OH^-) to form the gem-diolate anion (GDA). Catalytic adsorption of GDA on the catalyst surface, followed by C—H cleavage coupled with electron transfer, results in the formation of FA at a low potential. Similar to the acidic medium, the

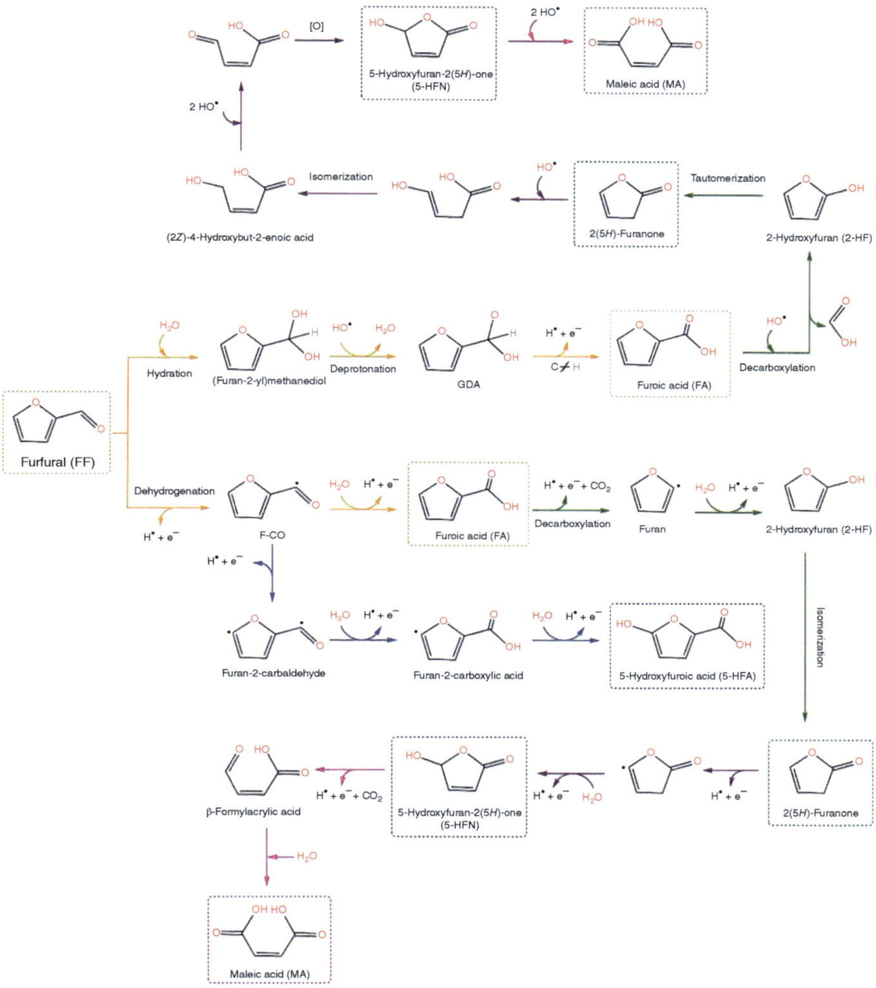

Figure 15.3 A scheme showing possible reaction mechanism for the electrochemical oxidation of FF under acidic and alkaline/neutral conditions, showing the formation of FA, 5-HFA, 2(5*H*)-furanone or 2(3*H*)-furanone, 5-HFN, and MA as main products.

formation of 5-HFA in an alkaline medium from FF occurs as a pathway parallel to the oxidation of the aldehyde group to form FA, which involves two electrons (Figure 15.3) [84, 85, 87]. The oxidation of FA in an alkaline/neutral medium leads to C—C cleavage (decarboxylation), forming 2-hydroxyfuran and formic acid (HCOOH), which decomposes to CO_2 under oxidation. 2-Hydroxyfuran tautomerizes to 2(5*H*)-furanone, whose subsequent oxidation, via the addition of hydroxyls in the reaction medium, forms 5-hydroxy-2(5*H*)-furanone, which is rapidly oxidized to MA (Figure 15.3) [84, 87, 89, 91, 93].

15.4.3 Electrochemical Synergy Between FOR and HER

The integration of FOR at the anode with the HER at the cathode establishes a promising electrochemical synergy. By replacing the energetically unfavorable OER, FOR valorizes biomass and simultaneously produces hydrogen, potentially reducing the total cell overpotential [83, 85, 87].

FOR||HER systems investigated with specific catalysts demonstrate this synergy. In systems catalyzed by rhenium disulfide and cobalt sulfide heterostructures (ReS_2/Co_9S_8 in 1 M KOH), FOR mainly produces FA with high FE of 86.3%, and coupling with HER significantly reduces cell voltage (1.43 V vs. RHE at 10 mA·cm^{-2}) compared to OER||HER (1.69 V). Half-cell studies confirmed the superior activity of the catalyst for FOR (160 mV at 10 mA·cm^{-2}) compared to OER (270 mV at 10 mA·cm^{-2}), with FOR operating effectively in the potential range of 1.33–1.53 V vs. RHE [83].

The use of symmetrical electrodes made of ruthenium and reduced graphene oxide nanocomposites (Ru/rGO in 1 M H_2SO_4) explores an ECO||ECH synergy, where the anodic ECO of FF generates H^+ ions and electrons that drive the cathodic electrohydrogenation (ECH) of FF to 2-methylfuran (2-MF). This integration yields a high FF conversion at the cathode (97%), primarily producing 2-MF (91% yield, 95% FE). Meanwhile, the anode produces FA (60%) and 5-HFA (14%), and other oxygenated products (MA [5%], 2(3*H*)-furanone [3%], and 5-HFN [4%]). The superiority of the ECO||ECH system over ECH in a half-cell (higher 2-MF yield and FF conversion) lies in the use of the anodic by-products (H^+ and e^-) as cathodic reagents, establishing an efficient internal cycle that optimizes energy use and eliminates the need for external sources for the reduction reaction [87].

An FOR||HER system with a nickel–molybdenum phosphide anode on a nickel foam substrate (PE-NiMoP/NF) and a ternary metal phosphide cathode (FeCoP) in 1 M NaOH, demonstrated that FOR is energetically more favorable than OER (150 mV reduction in cell overpotential at 500 mA·cm^{-2}). Linear sweep voltammetry (LSV) analyses confirmed higher catalytic activity for FOR||HER at potentials below 1.95 V vs. RHE. System stability was proven in a 35-hour test (500 mA·cm^{-2}, 50 °C, 20 mL·min^{-1}), with a stable cell voltage around 1.87 V. FOR produced FA with high selectivity (96.5%) and FE of 94.5%. The system's energy efficiency is attributed to the lower oxidation potential (1.43 V vs. RHE) and faster charge-transfer kinetics of FOR (R_{ct} = 0.13 Ω) compared to OER [36].

In contrast to other systems, the use of a copper sulfide catalyst on carbon paper (CuS/CP) in a ternary electrolyte ([Et_3NH]NO_3–MeCN–H_2O) demonstrated selective FF ECO to 5-hydroxyfuranone (HFO) under ambient conditions, achieving 83.6% selectivity and 70.2% conversion at 1.6 V vs. Ag/Ag^+. Additionally, the formation of MA was observed with a selectivity of 8.8%. The FE for the combined production of 5-HFO and MA was 77.1%. Electrochemical analyses (LSV and EIS) corroborated the effectiveness of CuS/CP in FF ECO, evidenced by increased current density and low charge-transfer resistance (34.51 Ω), indicating favorable reaction kinetics [92].

15.5 Challenges and Prospects

Hybrid water electrolysis aims to replace conventional water electrolysis by substituting the anodic OER with a chemical reaction that is more electrochemically favorable. Therefore, the main challenge faced by researchers who study and develop hybrid electrolysis systems, assisted with biomass derivatives, is to find novel electrocatalyst materials to overcome the difficulties related to the high overpotentials and poor selectivity and the reduction of the cost of electrical energy required for hydrogen production and conversion of biomass derivatives into compounds with added values [94]. In addition, the primary perspective of hybrid water electrolysis systems is to integrate them with renewable energy sources, such as solar energy and wind energy, which will promote an electrolysis process characterized by low environmental impact. On the other hand, the lack of global standards and investment uncertainty hinder the development of a robust hydrogen economy. The key findings underscore the need for coordinated efforts to enhance storage technologies, lower production costs, and establish favorable regulatory frameworks, thereby reinforcing the pivotal role of hydrogen in the global energy transition.

Although there are still challenges to overcome regarding the performance of electrocatalyst materials and the reduction of electrical energy consumption, hybrid water electrolysis assisted by biomass-derived compounds offers a promising path for more sustainable hydrogen production. Ultimately, further research is necessary to overcome these challenges and enable the practical application of hybrid water electrolysis for hydrogen production in industrial settings.

Acknowledgments

We express our gratitude to the following Brazilian agencies that funded this research: CNPq (Procs. 305103/2022-9, 302825/2022-3, and 303205/2023-7), FUNCAP (Procs. FPD-0213-00336.01.00/23, BP6-00241-00043.01.00/25, and Rede VERDES—07548003/2023), and CAPES (Finance Code 001).

References

1 Thomas, V. (2017). Climate Change and Natural Disasters: Transforming Economies and Policies for a Sustainable Future. London: Routledge. https://doi.org/10.4324/9781315081045

2 Emanuel, K. (2018). What We Know About Climate Change. Cambridge, MA: The MIT Press. https://doi.org/10.7551/mitpress/9610.001.0001

3 Araújo, K. (2014). The emerging field of energy transitions: Progress, challenges, and opportunities. Energy Res. Soc. Sci. 1, 112–121. https://doi.org/10.1016/j.erss.2014.03.002

4 Bampaou, M. & Panopoulos, K. D. (2025). An overview of hydrogen valleys: Current status, challenges and their role in increased renewable energy penetration. Renew. Sustain. Energy Rev. 207, 114923. https://doi.org/10.1016/j.rser.2024.114923

5 Saleh, H. M. & Hassam, A. I. (2024). The challenges of sustainable energy transition: A focus on renewable energy. Appl. Chem. Eng. 7(2), 2084. https://doi.org/10.59429/ace.v7i2.2084

6 Theerthagiri, J., Karuppasamy, K., Park, J., Rahamathulla, N., Kumari, M. L. A., Souza, M. K. R., Cardoso, E. S. F., Murthy, A. P., Maia, G., Kim, H. S., & Choil, M. Y. (2023). Electrochemical conversion of biomass-derived aldehydes into fine chemicals and hydrogen: A review. Environ. Chem. Lett. 21, 1555–1583. https://doi.org/10.1007/s10311-022-01543-5

7 Bhuiyan, M. M. H. & Siddique, Z. (2025). Hydrogen as an alternative fuel: A comprehensive review of challenges and opportunities in production, storage, and transportation. Int. J. Hydrog. Energy 102, 1026–1044. https://doi.org/10.1016/j.ijhydene.2025.01.033

8 Harnisch, F. & Urban, C. (2018). Electrobiorefineries: Unlocking the synergy of electrochemical and microbial conversions. Angew. Chem. Int. Ed. 57, 10016–10023. https://doi.org/10.1002/anie.201711727

9 Nilges, P., Santos, T. R., Harnisch, F., & Schröder, U. (2012). Electrochemistry for biofuel generation: Electrochemical conversion of levulinic acid to octane. Energy Environ. Sci. 5, 5231–5235. https://doi.org/10.1039/c1ee02685b

10 Lyu, X. (2021). Furfural and hydrogen production from corncob via tandem chemical and electrochemical approach. Bioresour. Technol. Rep. 15, 100790. https://doi.org/10.1016/j.biteb.2021.100790

11 Han, L., Chen, J., Chen, A., Gao, X., Wang, S., & Zhai, J. (2025). Energy scheduling for integrated electricity–hydrogen systems considering multiphysics dynamics of hybrid water and biomass electrolysis. Renew. Energy 244, 122635. https://doi.org/10.1016/j.renene.2025.122635

12 Wang, H.-Y., Sun, M.-L., Ren, J.-T., & Yuan, Z.-Y. (2023). Circumventing challenges: Design of anodic electrocatalysts for hybrid water electrolysis systems. Adv. Energy Mater. 13, 2203568. https://doi.org/10.1002/aenm.202203568

13 Veeramani, K., Janani, G., Kim, J., Surendran, S., Lim, J., Jesudass, S. C., Mahadik, S., Lee, H., Kim, T.-H., Kim, J. K., & Sim, U. (2023). Hydrogen and value-added products yield from hybrid water electrolysis: A critical review on recent developments. Renew. Sustain. Energy Rev. 177, 113227. https://doi.org/10.1016/j.rser.2023.113227

14 Gunerhan, A., Altuntas, O., & Açikkalp, E. (2025). A comprehensive analysis of the production of H_2 and value-added chemicals from the electrolysis of biomass and derived feedstocks. Biomass Bioenergy 192, 107510. https://doi.org/10.1016/j.biombioe.2024.107510

15 Vesborg, P. C. K., Seger, B., & Chorkendorff, I. (2015). Recent development in hydrogen evolution reaction catalysts and their practical implementation. J. Phys. Chem. Lett. 6, 951–957. https://doi.org/10.1021/acs.jpclett.5b00306

16 Ma, T. Y., Dai, S., & Qiao, S. Z. (2016). Self-supported electrocatalysts for advanced energy conversion processes. Mater. Today 9, 265–273. https://doi.org/10.1016/j.mattod.2015.10.012

17 Aldosari, O. F., Hussain, I., & Malaibari, Z. (2023). Emerging trends of electrocatalytic technologies for renewable hydrogen energy from seawater: Recent advances, challenges, and techno-feasible assessment. J. Energy Chem. 80, 658–688. https://doi.org/10.1016/j.jechem.2023.01.067

18 Hung, K. M. & Wu, J.-J. (2024). Bifunctional nickel phosphide nanoparticle/nickel cobalt sulfide nanosheet framework for electrocatalytic simultaneous hydrogen evolution and 2,5-furandicaroxylic acid production. Chem. Eng. J. 484, 149772. https://doi.org/10.1016/j.cej.2024.149772

19 Hasse, J. C., Ibánez, M. M., & Holenwiski, A. (2023). Impact of electrolyte composition on bulk electrolysis of furfural over platinum electrodes. ChemCatChem 15, e202300988. https://doi.org/10.1002/cctc.202300988

20 Sharma, P., Kumar, A., Zheng, G., Mashifana, T., Dhiman, P., Sharma, G., & Stadler, F. J. (2023). Current scenario in ternary metal indium sulphides-based heterojunctions for photocatalytic energy and environmental applications: A review. Mater. Today Commun. 36, 106741. https://doi.org/10.1016/j.mtcomm.2023.106741

21 Di, J. & Jiang, W. (2023). Recent progress of low-dimensional metal sulphides photocatalysts for energy and environmental applications. Mater. Today Catal. 1, 100001. https://doi.org/10.1016/j.mtcata.2023.100001

22 Li, M., Teng, Z., Liu, C., Meng, J., Sun, G., Liu, X., Lai, J., Chi, J., & Wang, L. (2025). Design principles of metal phosphides for the oxygen evolution reaction in seawater: Activity, selectivity and stability. Inorg. Chem. Front. 12, 2554–2586. https://doi.org/10.1039/d4qi03108c

23 Ge, M., Yin, H., Tian, W., Zhang, H., Li, S., Wang, S., & Chen, Z. (2024). Electrostatically induced furfural-derived carbon dots-CdS hybrid for solar light-driven hydrogen production. J. Colloid Interface Sci. 660, 147–156. https://doi.org/10.1016/j.jcis.2024.01.027

24 Selvam, R. & Gengan, S. (2025). Transition metal phosphides and nitrides: advanced electrodeposition strategies for efficient hydrogen evolution—Review. Top. Catal. https://doi.org/10.1007/s11244-025-02082-3

25 Yadav, N., Pan, K. K., Tripathi, K., Yadav, G., & Ahmaruzzaman, M. (2025). Harnessing metal sulphides for efficient hydrogen production employing photocatalytic water splitting: Current status and future direction. Int. J. Hydrog. Energy 101, 1221–1253. https://doi.org/10.1016/j.ijhydene.2024.12.516

26 Giannakoudakis, D. A., Colmenares, J. C., Tsiplakides, D., & Triantafyllidis, K. S. (2021). Nanoengineered electrodes for biomass-derived 5-hydroxymethylfurfural electrocatalytic oxidation to 2,5-furandicarboxylic acid. ACS Sustain. Chem. Eng. 9, 1970–1993. https://doi.org/10.1021/acssuschemeng.0c07480

27 Buravets, V., Miliutina, E., Burtsev, V., Severa, K., Shilenko, V., Rosenkranzova, J., Hönicke, P., Hrbek, T., Svorcik, V., & Lyutakov, O. (2025). High entropy 2D metals sulphides: Fast synthesis, exfoliation and electrochemical activity in overall water

splitting at alkaline pH. Appl. Surf. Sci. 681, 161600. https://doi.org/10.1016/j.apsusc.2024.161600

28 Zhang, T., Xiang, M., Yang, X., Fu, C., Yan, Z., & Xu, Z. (2025). Ni_3S_2–Co_9S_8 heterojunction integrated within cobalt layered double hydroxide nanosheets for 5-hydroxymethylfurfural oxidation and oxygen evolution reaction. ACS Appl. Nano Mater. 8, 7278–7289. https://doi.org/10.1021/acsanm.5c00659

29 Fu, Q., Xie, W., Yang, L., Yan, L., & Zhao, X. (2024). Recent advances in electrocatalytic upgrading of biomass-derived furfural. Chem. Eng. J. 485, 150083. https://doi.org/10.1016/j.cej.2024.150083

30 Zhang, H., Wang, Z., Zhang, J., & Dai, K. (2023). Metal-sulfide-based heterojunction photocatalysts: Principles, impact, applications, and in-situ characterization. Chin. J. Catal. 49, 42–67. https://doi.org/10.1016/S1872-2067(23)64444-4

31 Li, X., Cong, L., Lin, H., Liu, F., Han, F., & Lin, N. (2022). Electrochemical conversion of furfural to furoic acid: A more stable, efficient and energy-saving system. Sci. China Chem. 65, 2576–2587. https://doi.org/10.1007/s11426-022-1404-x.

32 Zhang, H., Qi, G., Liu, W., Zhang, S., Liu, Q., Luo, J., & Liu, X. (2023). Bimetallic phosphoselenide nanosheets as bifunctional catalysts for 5-hydroxymethylfurfural oxidation and hydrogen evolution. Inorg. Chem. Front. 10, 2423–2429. https://doi.org/10.1039/D3QI00013C

33 Azarian, G., Leili, M., Godini, K., Faradmal, J., & Bajalan, S. (2018). Furfural degradation using an electrochemical advanced oxidation process (EAOP): optimization of operating parameters using Taguchi approach. Desalin. Water Treat. 126, 287–295. https://doi.org/10.5004/dwt.2018.22895

34 Wang, X., Zhang, H., Liang, Y., Huang, L., & Wang, H. (2025). Enhanced electrooxidation of furfural to 2-furoic acid over $NiCo_2O_4$/NF: Optimization, in-situ monitoring and continuous AEM cell evaluation. Appl. Catal. A: Gen. 696, 120169. https://doi.org/10.1016/j.apcata.2025.120169

35 Kwon, Y., Schouten, K. J. P., van der Waal, J. C., de Jong, E., & Koper, M. T. M. (2016). Electrocatalytic conversion of furanic compounds. ACS Catal. 6, 6704–6714. https://doi.org/10.1021/acscatal.6b01861

36 Cheng, J., Zeng, Z., Huang, X., & Wei, Z. (2025). Electrooxidation of furfural on an electrodeposited NiMoP electrode to simultaneously produce hydrogen and 2-furancarboxylic acid at industrial current levels. J. Power Sources 628, 235919. https://doi.org/10.1016/j.jpowsour.2024.235919

37 Moura, J. V. B., de Souza, A. A. G., de Tarso Cavalcante Freire, P., da Luz Lima, C., & Oliveira, T. M. B. F. (2020). Blue-light-excited $NaCe(MoO_4)_2$ microcrystals for photoelectrochemical water splitting. Appl. Ceram. Technol. 18, 615–621. https://doi.org/10.1111/ijac.13697

38 Luo, H., Barrio, J., Sunny, N., Li, A., Steier, L., Shah, N., Stephens, I. E. L., & Titirici, M. M. (2021). Progress and perspectives in photo- and electrochemical-oxidation of biomass for sustainable chemicals and hydrogen production. Adv. Energy Mater. 11, 180–210. https://doi.org/10.1002/aenm.202101180

39 Rodrigues-Júnior, D. N., Sousa, N. G., Luna, F. M. T., Oliveira, T. M. B. F., Abreu, D. S., Schwarzacher, W., de Lima-Neto, P., & Correia, A. N. (2023). $Zn_xCo_{(1-x)}$ coatings from choline chloride-ethylene glycol deep eutectic solvent as electrocatalysts for hydrogen evolution reaction. J. Electroanal. Chem. 947, 117785. https://doi.org/10.1016/j.jelechem.2023.117785

40 Liu, Y., McCue, A. J., & Li, D. (2021). Metal phosphides and sulphides in heterogeneous catalysis: electronic and geometric effects. ACS Catal. 11, 9102–9127. https://doi.org/10.1021/acscatal.1c01718

41 Wang, J., Luo, J., Liu, D., Chen, S., & Peng, T. (2019). One-pot solvothermal synthesis of MoS_2-modified $Mn_{0.2}Cd_{0.8}S$/MnS heterojunction photocatalysts for highly efficient visible-light-driven H_2 production. Appl. Catal. B: Environ. 241, 130–140. https://doi.org/10.1016/j.apcatb.2018.09.033

42 Zeenat, Ahmad, Z., Maqbool, A., Hussain, M. A., Pashameah, R. A., Shahzadi, A., Nazar, N., Iqbal, S., Alanazi, A. K., Ashiq, M. N., & Abo-Dief, H. M. (2023). One-pot solvothermal synthesis of highly catalytic Janus transition metal phosphides (TMPs) for high performance OER. Fuel 331, 125913. https://doi.org/10.1016/j.fuel.2022.125913

43 Joo, J., Kim, T., Lee, J., Choi, S. I., & Lee, K. (2019). Morphology-controlled metal sulphides and phosphides for electrochemical water splitting. Adv. Mater. 31, 1806682. https://doi.org/10.1002/adma.201806682

44 Wang, C., Wu, Y., Zhou, Z., Wang, J., Pei, S., & Liu, S. (2022). Electrodeposited amorphous nickel-iron phosphide and sulfide derived films for electrocatalytic oxygen evolution. Int. J. Hydrog. Energy 47, 40849–40859. https://doi.org/10.1016/j.ijhydene.2022.09.200

45 Darband, G. B. (2024). Recent advances in phosphide-based nanostructures by electrodeposition for hydrogen evolution reaction. Curr. Opin. Electrochem. 46, 101507. https://doi.org/10.1016/j.jpowsour.2024.235919

46 Zhang, Y., Lin, L., Liu, J., Peng, J., & Chen, L. (2022). Hierarchical sulphide - phosphide NiSP/NF catalyst prepared by gradient electrodeposition for oxygen evolution reaction. J. Alloys Compd. 895, 162675. https://doi.org/10.1016/j.jallcom.2021.162675

47 Sepahdar, M. H., Masoudpanah, S. M., Bafghi, M. Sh., Mehri, M., & Aslibeiki, B. (2024). Effect of sulfidation and phosphidation on Fe doped Co LDH/nickel foam for efficient multifunctional water splitting electrodes, urea oxidation reaction, and seawater splitting: An experimental and computational study. Int. J. Hydrog. Energy 85, 733–743. https://doi.org/10.1016/j.ijhydene.2024.08.379

48 You, B., Liu, X., Jiang, N., & Sun, Y. (2016). A general strategy for decoupled hydrogen production from water splitting by integrating oxidative biomass valorization. J. Am. Chem. Soc. 138, 6707–6710. https://doi.org/10.1021/jacs.6b07127

49 Gidi, L., Amalraj, J., Tenreiro, C., & Ramírez, G. (2023). Recent progress, trends, and new challenges in the electrochemical production of green hydrogen coupled to selective electrooxidation of 5-hydroxymethylfurfural (HMF). RSC Adv. 13, 28307–28336. https://doi.org/10.1039/d3ra05623f

50 Peng, L., Shah, S. S. A., & Wei, Z. (2018). Recent developments in metal phosphide and sulfide electrocatalysts for oxygen evolution reaction. Chin. J. Catal. 39, 1575–1593. https://doi.org/10.1016/S1872-2067(18)63130-4

51 Shahroudi, A., Esfandiari, M., & Habibzadeh, S. (2022). Nickel sulfide and phosphide electrocatalysts for hydrogen evolution reaction: Challenges and future perspectives. RSC Adv. 12, 29440–29468. https://doi.org/10.1016/S1872-2067(18)63130-4

52 You, B., Jiang, N., Liu, X., & Sun, Y. (2016). Simultaneous H_2 generation and biomass upgrading in water by an efficient noble-metal-free bifunctional electrocatalyst. Angew. Chem. Int. Ed. 55, 9913–9937. https://doi.org/10.1002/anie.201603798

53 Xu, W., Zhou, B., Wang, Q., Xu, G., Li, N., Liu, W., & Zhang, Z. C. (2023). Energy-efficient electrochemical hydrogen production combined with biomass oxidation using polyoxometalate and metal salts. ChemCatChem 15, e202300522. https://doi.org/10.1002/cctc.202300522

54 Li, L., Lu, S., Fang, L., Wei, Y., & Yang, S. (2023). Review on biomass-derived carbon-based materials for electrocatalytic hydrogen production: state of the art and outlook. Energy Fuels 37, 18485–18501. https://doi.org/10.1021/acs.energyfuels.3c03231

55 Hung, K. M. & Wu, J. J. (2024). Bifunctional nickel phosphide nanoparticle/nickel cobalt sulfide nanosheet framework for electrocatalytic simultaneous hydrogen evolution and 2,5-furandicaroxylic acid production. Chem. Eng. J. 484, 149772. https://doi.org/10.1016/j.cej.2024.149772

56 Gong, Y., Lv, F., Lu, Y., Yu, Y., Niu, J., Lang, J., Deng, Y., Cao, X., & Gu, H. (2020). In situ surface-derivation of AgPdMo/MoS_2 nanowires for synergistic hydrogen evolution catalysis in alkaline solution. Nanoscale 12, 6472–6479. https://doi.org/10.1039/c9nr10331g

57 Jiang, N., You, B., Boonstra, R., Rodriguez, I. M. T., & Sun, Y. (2016). Integrating electrocatalytic 5-hydroxymethylfurfural oxidation and hydrogen production via Co–P-derived electrocatalysts. ACS Energy Lett. 1, 386–390. https://doi.org/10.1021/acsenergylett.6b00214

58 Zhang, X., Han, M., Liu, G., Wang, G., Zhang, Y., Zhang, H., & Zhao, H. (2019). Simultaneously high-rate furfural hydrogenation and oxidation upgrading on nanostructured transition metal phosphides through electrocatalytic conversion at ambient conditions. Appl. Catal. B: Environ. 244, 899–908. https://doi.org/10.1016/j.apcatb.2018.12.025

59 Wang, H., Zhou, Y., & Tao, S. (2022). CoP-CoOOH heterojunction with modulating interfacial electronic structure: A robust biomass-upgrading electrocatalyst. Appl. Catal. B: Environ. 315, 121588. https://doi.org/10.1016/j.apcatb.2022.121588

60 Zhou, M., Chen, J., & Li, Y. (2022). CoP nanorods anchored on Ni_2P-NiCoP nanosheets with abundant heterogeneous interfaces boosting the electrocatalytic oxidation of 5-hydroxymethyl-furfural. Catal. Sci. Technol. 12, 4288–4297. https://doi.org/10.1039/D2CY00683A

61 Xu, D., Yang, Y., Zhang, B., Yang, Z., Liu, S., & Mu, T. (2022). Deep eutectic solvent-induced in situ etching and phosphorization to form nickel phosphides for electrooxidation of 5-hydroxymethylfurfural. ChemSusChem 15, e202200822. https://doi.org/10.1002/cssc.202200822

62 Song, Y., Xie, W., Song, Y., Li, S., Jiang, S., Li, H., Lee, J. Y., & Shao, M. (2022). Bifunctional integrated electrode for high-efficient hydrogen production coupled with 5-hydroxymethylfurfural oxidation. Appl. Catal. B: Environ. 312, 121400. https://doi.org/10.1016/j.apcatb.2022.121400

63 Wang, H., Niu, C., Liu, W., & Tao, S. (2024). d-Electron tuned CoMoP for enhance 5-hydroxymethylfurfural oxidation and HER. Appl. Catal. B: Environ. 340, 123249. https://doi.org/10.1016/j.apcatb.2023.123249

64 Lin, R., Salehi, M., Guo, J., & Seifitokaldani, A. (2022). High oxidation state enabled by plated Ni-P achieves superior electrocatalytic performance for 5-hydroxymethylfurfural oxidation reaction. iScience 25, 104744. https://doi.org/10.1016/j.isci.2022.104744

65 Li, M., Chen, L., Ye, S., Fan, G., Yang, L., Zhang, X., & Feng, L. (2019). Dispersive non-noble metal phosphide embedded in alumina arrays derived from layered double hydroxide precursor toward efficient oxygen evolution reaction and biomass upgrading. J. Mater. Chem. A 7(24), 13695–13704. https://doi.org/10.1039/c9ta03580j

66 Bi, J., Ying, H., Xu, H., Zhao, X., Du, X., Hao, J., & Li, Z. (2022). Phosphorus vacancy-engineered Ce-doped CoP nanosheets for the electrocatalytic oxidation of 5-hydroxymethylfurfural. Chem. Commun. 58, 7817–7820. https://doi.org/10.1039/d2cc02451a

67 Gong, W., Li, J., Ma, J., Liu, D., Long, R., & Xiong, Y. (2023). Highly efficient electrocatalytic biomass valorization over a perovskite-derived nickel phosphide catalyst. Nanoscale Horiz. 8, 69–74. https://doi.org/10.1039/d2nh00391k

68 Xu, H., Bi, J., Sang, T., Wang, W., Hao, J., & Li, Z. (2023). Mn-doped Ni_2P: nanocrystal-decorated amorphous nanosheets for efficient electrooxidation of 5-hydroxymethylfurfural. Chem. Commun. 59, 8440–8443. https://doi.org/10.1039/d3cc02083e

69 Liu, S., Yuan, X., Huang, X., Huang, Y., Sun, C., Qian, K., & Zhang, W. (2023). Nickel-phytic acid hybrid for highly efficient electrocatalytic upgrading of HMF. Front. Chem. 11, 1199921. https://doi.org/10.3389/fchem.2023.1199921

70 Xie, S., Fu, H., Chen, L., Li, Y., & Shen, K. (2023). Carbon-based nanoarrays embedded with Ce-doped ultrasmall Co_2P nanoparticles enable efficient electrooxidation of 5-hydroxymethylfurfural coupled with hydrogen production. Sci. China Chem. 66, 2141–2152. https://doi.org/10.1007/s11426-023-1666-4

71 Dasgupta, B., Hausmann, J. N., Beltrán-Suito, R., Kalra, S., Laun, K., Zebger, I., Driess, M., & Menezes, P. W. (2023). A facile molecular approach to amorphous nickel pnictides and their reconstruction to crystalline potassium-intercalated γ-$NiOOH_x$ enabling high-performance electrocatalytic water oxidation and selective oxidation of 5-hydroxymethylfurfural. Nano Micro Small 19, 2301258. https://doi.org/10.1002/smll.202301258

72 Wang, W., Kong, F., Zhang, Z., Yang, L., & Wang, M. (2021). Sulfidation of nickel foam with enhanced electrocatalytic oxidation of 5-hydroxymethylfurfural to 2,5-furandicarboxylic acid. Dalton Trans. 50, 10922–10927. https://doi.org/10.1039/D1DT02025K

73 Kong, F. & Wang, M. (2021). Preparation of sulfur-modulated nickel/carbon composites from lignosulfonate for the electrocatalytic oxidation of 5-hydroxymethylfurfural to 2,5-furandicarboxylic acid. ACS Appl. Energy Mater. 4, 1182–1188. https://doi.org/10.1021/acsaem.0c02418

74 Zhao, Z., Guo, T., Luo, X., Qin, X., Zheng, L., Yu, L., Lv, Z., Ma, D., & Zheng, H. (2022). Bimetallic sites and coordination effects: electronic structure engineering of NiCo-based sulfide for 5-hydroxymethylfurfural electrooxidation. Catal. Sci. Technol. 12, 3817–3827. https://doi.org/10.1039/D2CY00281G

75 Sun, Y., Wang, J., Qi, Y., Li, W., & Wang, C. (2022). Efficient electrooxidation of 5-hydroxymethylfurfural using Co-doped Ni_3S_2 catalyst: Promising for H_2 production under industrial-level current density. Adv. Sci. 9, 2200957. https://doi.org/10.1002/advs.202200957

76 Wang, S., Yang, G., Jiao, Y., Liu, Y., Tian, C., Wu, A., & Yan, H. (2023). Tuning electronic structure of Ni_3S_2 with tungsten doping for high-performance electrooxidation of 5-hydroxymethylfurfural. Sci. China Chem. 66, 3636–3644. https://doi.org/10.1007/s11426-023-1818-8

77 Li, S., Wang, S., Wang, Y., He, J., Li, K., Xu, Y., Wang, M., Zhao, S., Li, X., Zhong, X., & Wang, J. (2023). Doped Mn enhanced NiS electrooxidation performance of HMF into FDCA at industrial-level current density. Adv. Funct. Mater. 33, 2214488. https://doi.org/10.1002/adfm.202214488

78 Li, X., Cong, L., Lin, N., & Tang, C. (2023). Efficient electrochemical upgradation strategies for the biomass derivative furfural. J. Mater. Chem. A 11, 23133–23147. https://doi.org/10.1039/D3TA03813K

79 Zeng, Y., Wang, K., Liu, Z., Peng, F., & Zhang, Q. (2023). MOF material-derived bimetallic sulfide Co_xNi_yS for electrocatalytic oxidation of 5-hydroxymethylfurfural. Nanomaterials 13, 2318. https://doi.org/10.3390/nano13162318

80 Chen, P., Wu, Y., Guo, X., Wang, M., Yu, C., Jiang, H., Zhou, W., Wu, G., & Yan, J. (2024). Rational design of FeCo-S/Ni_2P/NF heterojunction as a robust electrocatalyst for water splitting. Inorg. Chem. 63, 5520–5529. https://doi.org/10.1021/acs.inorgchem.3c04480

81 Chen, H., Chen, S. R., Yang, Q., You, L. X., Ye, Q. F., Wang, A. J., Lin, D. H., Jiang, Y. X., & Sun, S. G. (2024). Nickel sulfide nanorods encapsulated in nitrogen-doped carbon on nickel foam in situ derived from Saccharomycetes cerevisiae for the selective oxidation of biomass. Chem. Eng. J. 499, 155922. https://doi.org/10.1016/j.cej.2024.155922

82 Xu, P., Bao, Z., Zhao, Y., Zheng, L., Lv, Z., Shi, X., & Zheng, H. (2024). Anionic regulation and heteroatom doping of Ni-based electrocatalysts to boost HMF oxidation to FDCA coupled with hydrogen production. Adv. Energy Mater. 14, 230557. https://doi.org/10.1002/aenm.202303557

83 Pang, Q., Sun, K., Fan, X., Xiang, K., Li, B., Zhao, S., Kim, Y. D., Liu, Q., Liu, Z., & Peng, Z. (2024). Enhancement effect from ReS_2/Co_9S_8 heterostructure evolution for the highly effective furfural oxidation coupling with hydrogen production. Chem. Eng. J. 497, 154475. https://doi.org/10.1016/j.cej.2024.154475

84 Gong, L., Agrawal, N., Roman, A., Holewinski, A., & Janik, M. J. (2019). Density functional theory study of furfural electrochemical oxidation on the Pt (1 1 1) surface. J. Catal. 373, 322–335. https://doi.org/10.1016/j.jcat.2019.04.012

85 Román, A. M., Hasse, J. C., Medlin, J. W., & Holewinski, A. (2019). Elucidating acidic electro-oxidation pathways of furfural on platinum. ACS Catal. 9, 10305–10316. https://doi.org/10.1021/acscatal.9b02656

86 Kundu, A., Adak, M. K., Kumar, Y., & Chakraborty, B. (2022). Electrochemically derived crystalline CuO from covellite CuS nanoplates: A multifunctional anode material. Inorg. Chem. 61, 4995–5009. https://doi.org/10.1021/acs.inorgchem.1c03830

87 Bharath, G. & Banat, F. (2021). High-grade biofuel synthesis from paired electrohydrogenation and electrooxidation of furfural using symmetric Ru/reduced graphene oxide electrodes. ACS Appl. Mater. Interfaces 13(21), 24643–24653. https://doi.org/10.1021/acsami.1c02231

88 Jiang, N., Liu, X., Dong, J., You, B., Liu, X., & Sun, Y. (2017). Electrocatalysis of furfural oxidation coupled with H_2 evolution via nickel-based electrocatalysts in water. ChemNanoMat 3(7), 491–495. https://doi.org/10.1002/cnma.201700076

89 Li, X., Ho, B., Lim, D. S. W., & Zhang, Y. (2017). Highly efficient formic acid-mediated oxidation of renewable furfural to maleic acid with H_2O_2. Green Chem. 19(4), 914–918. https://doi.org/10.1039/c6gc03020c

90 Yuan, X., Huang, X., Hu, M., Liu, J., Guo, W., Hao, Y., Li, Z., Zhang, K., & Han, B. (2024). Highly efficient electrocatalytic oxidation of furfural to maleic acid over MOF-derived $MnCeO_x$. Green Chem. 26(9), 5365–5370. https://doi.org/10.1039/d4gc00408f

91 Zhang, X., Xu, S., Li, Q., Zhou, G., & Xia, H. (2021). Recent advances in the conversion of furfural into bio-chemicals through chemo- and bio-catalysis. RSC Adv. 11, 27042–27058. https://doi.org/10.1039/d1ra04633k

92 Zhu, P., Shi, M., Shen, Z., Liao, X., & Chen, Y. (2024). Electrocatalytic conversion of biomass-derived furan compounds: Mechanisms, catalysts and perspectives. Chem. Sci. 15, 4723–4756. https://doi.org/10.1039/d4sc00546e

93 Wu, H., Song, J., Liu, H., Xie, Z., Xie, C., Hu, Y., Huang, X., Hua, M., & Han, B. (2019). An electrocatalytic route for transformation of biomass-derived furfural into 5-hydroxy-2(5H)-furanone. Chem. Sci. 10, 4692–4698. https://doi.org/10.1039/c9sc00322c

94 Kashanian, M. & Ryan, S. M. (2024). Design of a supply chain network for chemicals from biomass using green electrochemistry. Clean. Logist. Supply Chain 10, 100132. https://doi.org/10.1016/j.clscn.2023.100132

16

Single-atom Electrocatalysts for Urea and Furfural-assisted Water Electrolysis

Xiangyu You[1], Julio C. Lourenço[2,], Tulio P. Porto[3], Dhananjai Pangotra[4], Luciana Vieira[4], Huize Wang[5], Marc Ledendecker[1,5], Robson S. Rocha[3], Alexsandro J. dos Santos[6], Marcos Roberto de Vasconcelos Lanza[2,*], Gilberto Maia[7], and Guilherme V. Fortunato[1,2,*]*

[1]*Sustainable Energy Materials, Technical University of Munich, TUM Campus Straubing for Biotechnology and Sustainability, Straubing, Germany*
[2]*São Carlos Institute of Chemistry, University of São Paulo, São Carlos, Brazil*
[3]*Lorena School of Engineering, University of São Paulo, Lorena, Brazil*
[4]*Department of Bio-Electro and Chemocatalysis (BioCat), Fraunhofer Institute for Interfacial Engineering and Biotechnology IGB, Straubing, Germany*
[5]*Forschungszentrum Jülich GmbH, Helmholtz Institute Erlangen-Nürnberg for Renewable Energy (IET-2), Erlangen, Germany*
[6]*Department of Chemistry and Environmental Sciences, São Paulo State University, São José do Rio Preto, Brazil*
[7]*Institute of Chemistry, Federal University of Mato Grosso do Sul, Campo Grande, Brazil*

*Corresponding authors: julio.lourenco@usp.br; marcoslanza@usp.br; g.fortunato@tum.de

16.1 Introduction

The transition toward sustainable chemical production requires decentralized solutions that simultaneously address environmental remediation and the valorization of biomass-derived molecules. Such approaches enable the local treatment of waste streams while generating valuable products, aligning closely with the principles of a circular economy and advancing global Sustainable Development Goals (SDGs), including clean water (SDG 6), affordable and clean energy (SDG 7), responsible production (SDG 12), and climate action (SDG 13).

Within this context, hybrid water electrolysis has emerged as a promising technology. Unlike conventional electrolysis, which relies on the oxygen evolution reaction (OER) at the anode, hybrid systems pair the hydrogen evolution reaction (HER) with alternative anodic processes. Replacing OER with thermodynamically more favorable oxidations reduces the overall energy input while simultaneously upgrading low-value or residual feedstocks into industrially relevant chemicals. This combination of improved energy efficiency and chemical valorization makes hybrid electrolysis a compelling approach for sustainable hydrogen production [1–5].

Among the various anodic reactions, the electrooxidation of urea and furfural stands out. Both exhibit lower theoretical redox potentials (<0.4 V vs. RHE compared to 1.23 V vs. RHE for OER in alkaline media), though in practice onset potentials typically exceed 0.9 V vs. RHE depending on the catalyst and electrolyte employed [4, 6–8]. Urea, which is common in wastewater and agricultural runoff, passes through a six-electron-oxidation pathway, whereas furfural, a biomass-derived platform molecule, provides pathways to furoic acid and maleic acid, among others. However, both processes require complex, multistep procedures and are hindered by a scarcity of selective, stable, and efficient electrocatalysts [1–5, 9, 10].

Single-atom catalysts (SACs) have recently emerged as a promising class of materials to address these challenges [11–15]. Composed of isolated metal atoms anchored to conductive supports, SACs maximize atomic utilization and provide highly tunable coordination environments. These atomically dispersed sites allow precise optimization of adsorption strength and reaction intermediates, thereby improving activity and selectivity [11–13, 15]. Moreover, engineered metal–support interactions (MSIs) stabilize the isolated atoms against migration, dissolution, or agglomeration under electrochemical conditions. Such features are particularly advantageous in hybrid electrolysis, where the simultaneous presence of reactive intermediates, oxidative environments, and dynamic potential cycling imposes severe stability demands that often degrade conventional catalysts [10, 11, 16].

This chapter examines the emerging role of SACs in urea and furfural-assisted water electrolysis. It begins by outlining the fundamental principles of SAC design and characterization, followed by a detailed discussion of the molecular pathways of urea and furfural oxidation. Recent advances in SACs for these processes are then reviewed, with emphasis on structure–property relationships, stability trends, and performance benchmarks. Finally, the chapter highlights key challenges, such as catalyst stability, product management, and device integration, and offers perspectives on future opportunities to implement SAC-based hybrid electrolyzers in decentralized systems.

16.2 Fundamentals of Single-atom Electrocatalysts

16.2.1 Definition and Features of SACs

SACs represent a rapidly evolving class of materials in which catalytically active metals are atomically dispersed and anchored onto diverse supports such as carbon, metal oxides, and metal carbides [17–19]. This concept has attracted attention across electrocatalysis and heterogeneous catalysis because it combines the advantages of homogeneous and heterogeneous systems into a single platform [17–19]. The most distinctive feature of SACs is their atomic dispersion, which endows them with several advantages compared to conventional nanoparticle-based catalysts:

Maximized atom utilization efficiency: in theory, every metal atom in SACs can serve as an active site, in stark contrast to nanoparticles where only surface atoms are catalytically available while a significant fraction remains buried and inactive [20, 21]. This property is especially appealing when dealing with scarce and costly noble metals such as Pt, Pd, and Au.

Well-defined and tunable coordination environments: in SACs, single metal atoms are stabilized through interactions with surrounding heteroatoms (e.g., N, S, O, or even other metals) [22, 23] or through anchoring at surface lattice sites, as shown in Figure 16.1. For metal-supported SACs, the distribution and stability of single atoms (SAs) are largely determined by the relative chemical potential of the constituent metals [24]. On metal oxide supports, active metal atoms may preferentially occupy coordinatively unsaturated sites, oxygen vacancies, or other surface defects [24, 25]. When carbon-based materials are employed as supports, SAs typically anchor at defect sites such as carbon vacancies or bind to functional groups embedded in the carbon matrix. In these cases, coordination structures such as $M{-}N_4$, $M{-}O_4$, $M{-}S_4$, or mixing or mixed motifs are frequently observed [20, 26]. Such atomically precise configurations enable rational tailoring of the electronic structure, adsorption energy of intermediates, and overall catalytic performance [20, 27]. Moreover, the enhanced, albeit not complete, uniformity of active sites enables more precise investigation of fundamental structure–property relationships that remain hidden in heterogeneous nanoparticle catalysts [21].

Unique catalytic behavior arising from a distinctive electronic structure: in SACs, the drastic reduction in coordination number relative to bulk or nanoparticle states, combined with a well-defined coordination environment provided by the support, gives rise to a unique electronic structure at the metal centers. This tailored electronic configuration modulates the adsorption modes and binding strength of reactant molecules and reaction intermediates, thereby altering activation barriers and overall reaction pathways [29, 30]. For instance, atomically dispersed precious metals can display several to tens of times higher intrinsic activity than their nanoparticle counterparts [24, 31, 32].

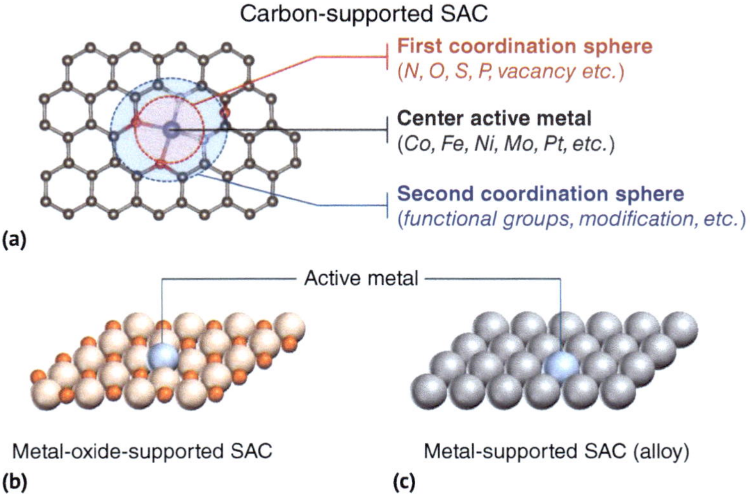

Figure 16.1 Schematic illustration of coordination environments in SACs using different support materials. Panel (a): *Source*: Tang et al. (2021) [28]/with permission of American Chemical Society.

Opportunities and challenges for complex reactions: while the strong adsorption capabilities and low coordination of isolated metal centers in SACs can be advantageous for activity, this same feature inherently precludes ensemble effects, thereby limiting their utility in reactions that necessitate the cooperation of adjacent metal atoms, as in C—C coupling [15, 33, 34]. As a result, SACs often deliver outstanding performance in single-site transformations (e.g., CO_2-to-CO conversion, HER, and selective oxidations) but face challenges in multi-electron or multi-atom processes [15, 33, 34]. In such cases, rational support design or dual atom configuration becomes essential to provide complementary active sites and facilitate interactions with reactants and intermediates throughout the reaction pathway [15, 30, 33, 34].

16.2.2 Common Synthesis Methods

The preparation of SACs is inherently challenging, particularly at metal loadings exceeding 1 wt.%, because isolated SAs tend to aggregate into clusters or nanoparticles due to their high surface energy [35]. To overcome this, advanced synthetic strategies are required to ensure that the metal precursors are discretely dispersed onto the support and that the desired local coordination environment is established to stabilize the isolated active centers. A variety of methods have been developed for SAC fabrication, including impregnation and coprecipitation, spatial confinement within porous matrices, coordination site construction, defect or vacancy engineering, electrochemical deposition, atomic-layer deposition (ALD), top-down synthesis from bulk metals or metal oxides, and even mechanochemical approaches such as ball milling [20, 36]. Each of these methods provides a pathway to control metal dispersion and anchoring, but they also face intrinsic limitations. The tendency for metal active centers to aggregate due to weak interactions with the support as well as the structural instability of the support are key issues that need to be addressed. The main challenges in SAC synthesis stem from the tendency of SAs to migrate and aggregate when their interaction with the support is insufficient, as well as from the structural fragility of some supports under reaction conditions [37–39]. To address these issues, effective strategies must ensure that metal species are well dispersed at low loadings, that the support is robust and rich in anchoring sites such as defects, vacancies, or functional groups, and that the MSIs are strong enough to stabilize isolated atoms throughout both synthesis and catalytic operation [40, 41].

Wet impregnation (Figure 16.2a) is one of the most straightforward and widely used methods for preparing supported catalysts. In this approach, a metal salt precursor is dissolved in a controlled volume of solvent. The solution is then mixed with the support material, and the resulting mixture is subjected to drying and thermal treatments. These steps remove precursor ligands and establish interactions between the metal species and the support [20, 23, 42]. Variants of this method involve directly impregnating complexes, such as metal acetylacetonates, onto the support, followed by drying to obtain the catalyst. In such cases, the active metal sites, retaining aspects of their original coordination, are anchored and stabilized on the support mainly through non-covalent interactions, including van der Waals

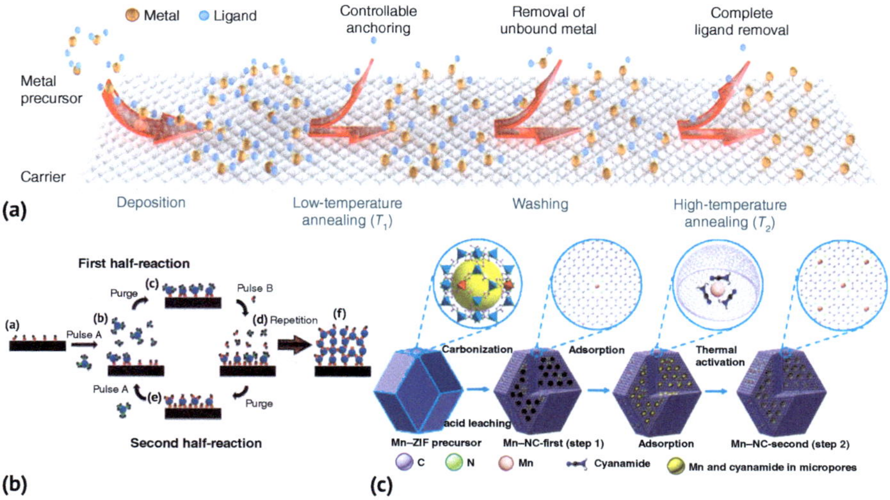

Figure 16.2 Common synthesis pathways for SACs. (a) Wet impregnation. *Source*: Hai et al. (2022) [50]/with permission of Springer Nature. (b) ALD. *Source*: Fonseca and Lu (2021) [46]/with permission of American Chemical Society. (c) MOF-derived method. *Source*: J. Li et al. (2018) [51]/with permission of Springer Nature.

forces, electrostatic effects, and soft donor–acceptor interactions [43, 44, 45]. ALD (Figure 16.2b) takes advantage of its ability to deposit uniform and conformal thin films with precise loading and different morphologies by simply adjusting the ALD cycles. In a typical ALD cycle, the sequence involves: first, introducing the initial precursor to the surface; second, flushing the chamber to remove excess species; third, delivering the second reactant; and finally, performing another purge to clear the by-products and unreacted molecules [46, 47]. Metal–organic frameworks (MOFs) (Figure 16.2c) are a kind of crystalline porous material composed of metal nodes and organic ligands in an ordered arrangement, enabling MOFs to be an ideal template to achieve uniform dispersion of metal sites on support [48]. In-situ construction of SACs by pyrolyzing the organic linker in MOFs with high-temperature posttreatment is possible. However, the migration and aggregation can be a challenge at high temperatures. To prevent possible sintering in the synthesis process, researchers have developed some strategies: mixing metal ions to separate the targeted metal, utilizing the functional groups, defects, and unsaturated ligands to immobilize SAs, and spatial confinement by encapsulating metal molecules within MOF pores [49].

16.2.3 Characterization Techniques

Compared to bulk materials or nanoparticles, SACs exhibit distinct morphological and electronic features arising from their atomic dispersion and diverse coordination environments. These differences necessitate advanced characterization methods to elucidate their structure–property relationships and catalytic behavior [52–54]. Electron microscopy provides direct evidence of atomic dispersion. In particular, aberration-corrected high-angle annular dark-field scanning transmission electron microscopy (AC-HAADF-STEM) enables visualization of isolated metal atoms, which appear as bright spots uniformly distributed on a darker support background, consistent with Rutherford scattering principles [52–54]. However, despite its powerful spatial resolution, AC-HAADF-STEM offers only local structural information from a limited region of the sample. This means that it does not provide statistically representative data on the overall dispersion or distribution of active sites. In addition, factors such as beam-induced atom migration or preferential imaging of specific surface areas can lead to misinterpretation. Therefore, microscopy results must always be complemented with ensemble-averaging techniques to verify atomic dispersion and quantify coordination environments.

Spectroscopic techniques are indispensable for probing the electronic and local coordination environments. X-ray photoelectron spectroscopy (XPS) provides insights into oxidation states and surface electronic structures, although its applicability can be limited by weak signals resulting from the low atomic content of SACs. In contrast, X-ray absorption spectroscopy (XAS) is a more powerful tool, providing detailed information on oxidation states, coordination numbers, and interatomic distances [52, 53]. Soft XAS probes light elements within the coordination sphere, while hard XAS is particularly effective for resolving the local environment of single metal centers and has become indispensable in SAC research [55–58]. A typical hard XAS spectrum consists of two regions: the X-ray absorption near edge structure (XANES), which reveals oxidation states and electronic configuration, and the extended X-ray absorption fine structure (EXAFS), where oscillations in absorption intensity arise from photoelectron scattering with neighboring atoms. Quantitative analyses of the EXAFS region, often using Fourier transform (FT) or wavelet transform (WT) methods, allow precise determination of interatomic distances and local coordination environments [55–57].

In addition to AC-HAADF-STEM and XAFS, solid-state NMR, often combined with dynamic nuclear polarization (DNP), has recently emerged as a powerful complementary tool to probe the local coordination environments and MSIs in SACs [59–61]. For example, ^{195}Pt solid-state NMR has been used to systematically characterize the coordination environments of Pt SAs on different supports and to translate their NMR spectra into quantitative fingerprints of site distribution and uniformity [60]. In another study, ^{17}O solid-state NMR in combination with DFT was employed to probe the interaction between Pt SAs and the oxygen lattice of CeO_2, thereby establishing a clear structure–activity relationship [61].

16.3 Advantages of Using SACs for Electrooxidation Reactions and the Stability Issues

The catalytic performance of SACs in electrocatalytic processes is closely related to the coordination environment of their atomic centers. The atoms surrounding the metal site, originating from the support, influence its electronic structure.

Intentionally modifying these coordination motifs, through substitution, heteroatom doping, or defect engineering, can markedly improve activity and selectivity [20, 26, 31, 34]. Robust MSIs further promote charge transfer, stabilize key intermediates, and suppress aggregation, thereby improving efficiency and stability. Such atomic-level control is particularly advantageous for complex, multi-electron processes, such as the urea oxidation reaction (UOR). For instance, it has been shown that Ni SAs incorporated into a Cu_2O surface exhibit synergistic behavior: Ni sites preferentially activate nitrogen in urea, while adjacent Cu atoms adsorb oxygen species, collectively lowering energy barriers for dehydrogenation and N–N coupling steps [62]. This illustrates how optimized coordination environments and cooperative interactions between SAs and supports can unlock high activity and selectivity in electrooxidation processes.

Despite these advantages, maintaining the SA configuration under operational conditions remains a key challenge, especially during long-term operation. In addition, the harsh electrochemical environment can induce support corrosion, oxidation, or other structural changes that compromise performance [37, 38]. Stability concerns extend beyond structural integrity to include recyclability, poisoning resistance, and minimal metal leaching under aqueous conditions [37, 63, 64]. Interestingly, reported instability is often more pronounced under reductive conditions than oxidative ones, though the extent strongly depends on the choice of metal and support [24, 37–39]. Addressing these issues requires stabilizing strategies centered on optimizing the MSIs. Robust supports with intrinsic corrosion resistance, or supports modified with defects and heteroatoms (N, O, S, and P), can enhance binding strength and tune the electronic structure of the active sites. Such modifications promote charge transfer between the metal center and the support, improving both activity and stability [24, 37–39, 45]. Considerable efforts have therefore focused on tailoring support chemistry and engineering metal–support interfaces to suppress aggregation and preserve atomic dispersion, which remain decisive factors for the practical deployment of SACs in electrooxidation reactions [26, 40, 41, 45].

16.4 UOR and SACs

16.4.1 Mechanisms of UOR, Kinetics, and Energy Advantages Compared to OER

Electrochemical UOR offers a dual benefit of wastewater remediation and energy-efficient hydrogen production [9, 10, 65]. Although UOR can, in principle, proceed in acidic, neutral, or alkaline electrolytes, alkaline conditions are generally preferred because they lower overpotentials, improve product selectivity, and enhance catalyst stability. The overall urea-assisted water-splitting process (UOR coupled with HER) is a six-electron transformation with a theoretical cell voltage of 0.37 V at standard conditions, following the pathway in Eq. (16.1) [9, 10]:

$$CO(NH_2)_2 + 6OH^- \longrightarrow N_2 + CO_2 + 5H_2O + 6e^- \tag{16.1}$$

In alkaline media, two mechanistic models are typically considered: the adsorbate evolution mechanism (AEM) and the lattice oxygen mechanism (LOM). AEM, particularly relevant to Ni-based catalysts, proceeds through sequential urea adsorption, progressive dehydrogenation of $-NH_2$ groups, C—N bond cleavage, N—N recombination to form N_2, and final desorption of CO_2 and N_2. Within this pathway, the dynamic formation of high-valent Ni species (e.g., Ni^{3+} in NiOOH; Figure 16.3a) is essential for driving the oxidative steps [66, 67], with *COO desorption often proposed as a kinetic bottleneck on Ni-based surfaces [66–69]. In such cases, tuning the electronic structure of Ni sites (e.g., via heteroatom doping or surface modification) to weaken *COO binding has been proposed as a route to enhance turnover rates.

In contrast, the LOM pathway incorporates the participation of lattice oxygen as a reactive species, enabling alternative reaction routes with lower energy barriers for CO_2 release [68–70]. This mechanism is often associated with materials

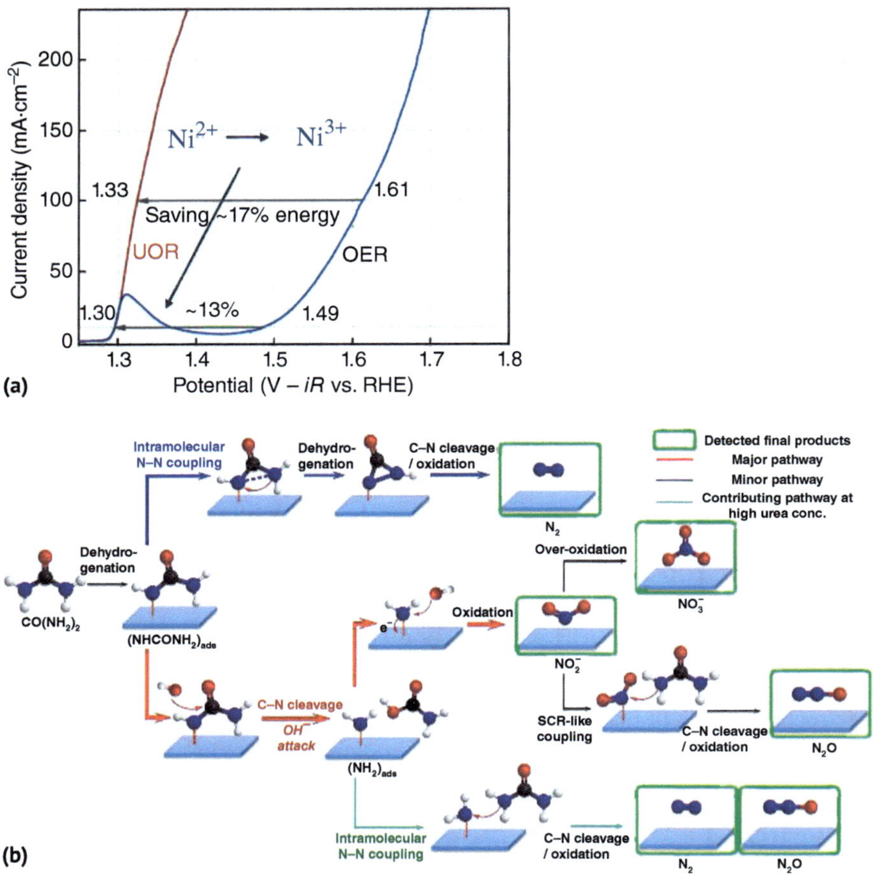

Figure 16.3 (a) Linear sweep voltammetry curves for UOR and OER using Ni—O—Ti active sites supported on Ti foam in 1 M KOH (with or without 0.33 M urea). *Source*: Zhan et al. (2024) [67]. Springer Nature. CC BY 4.0. (b) Schematic diagram of the possible pathways for urea oxidation. *Source*: Li et al. (2021) [72]/with permission of John Wiley & Sons.

capable of facile oxygen exchange, where oxygen vacancies generated during oxidation are subsequently replenished by OH^- from the electrolyte [66, 70, 71]. The direct involvement of lattice oxygen generally accelerates reaction kinetics and can overcome limitations inherent to the AEM pathway. Yet this enhanced reactivity comes at the cost of structural dynamics: repeated oxygen cycling can lead to surface reconstruction and catalyst degradation over extended operation [68, 69]. As a result, an emerging design challenge is to balance lattice oxygen mobility with structural robustness, for example, through defect engineering or the incorporation of stabilizing secondary cations.

However, practical UOR rarely yields only N_2 and CO_2 products. Instead, by-products such as NO_2^-, NO_3^-, N_2O, and CO are commonly detected, reflecting competing pathways and underscoring the complexity of the reaction (Figure 16.3b) [9, 10, 72–74]. Compared with HER and OER, UOR is kinetically sluggish because it involves multiple adsorption and desorption steps and site-blocking effects from intermediates such as *COO and $*NH_x$ [9, 10]. Moreover, partial N_2 selectivity further complicates the process, often promoting nitrate and nitrite formation under certain conditions [9, 10, 75].

Reaction selectivity is strongly influenced by electrolyte composition, applied potential, and catalyst surface chemistry [9, 10]. Recent studies have also emphasized the role of urea's functional groups in adsorption and cleavage [10, 76, 77]. For example, Wang et al. [77] showed that at a CoS_2/MoS_2 heterointerface, $—NH_2$ groups preferentially adsorb on CoS_2, whereas the C=O group binds to MoS_2, facilitating C—N bond scission through local electrophilic/nucleophilic interactions [76]. These findings highlight that UOR selectivity is governed not only by the stability of active metal species (e.g., Ni^{3+}) but also by how the catalyst surface engages with the electrophilic and nucleophilic sites of the urea molecule.

16.4.2 State-of-the-art Electrocatalysts and the Advantages of SACs for UOR

Ni is the benchmark catalyst for UOR, inspired by its similarity to urease enzymes, where hydroxide-bridged Ni^{2+} centers catalyze urea decomposition [78, 79]. Among non-noble metals, Ni exhibits superior UOR activity due to its optimal d-band position and reversible formation of NiOOH active species, which together enable efficient OH^- adsorption and C—N bond activation [70, 71, 74]. Gao et al. recently demonstrated that oxyanion-engineered Ni catalysts (Ni–SO_x, Ni–PO_x, and Ni–SeO_x) can effectively suppress competing OER. The optimized Ni–SO_x catalyst delivered current densities up to 323 $mA \cdot cm^{-2}$ at 1.65 V vs. RHE with 99.3% selectivity for nitrogen products [80]. In-situ XANES and operando Raman studies revealed dynamic transformations from Ni^{2+} to NiOOH-like species, which accelerate C—N bond scission while suppressing excess OH^- adsorption [80, 81]. The performance of Ni-based catalysts is generally governed by two factors: (i) nonmetallic dopants with distinct radii and electronegativities that modulate the Ni d-band center and optimize intermediate binding [82]; and (ii) the reversible Ni^{2+}/Ni^{3+} ($Ni(OH)_2$/NiOOH) redox cycle, which drives the six-electron UOR pathway [83, 84]. To exploit these features, diverse strategies such as crystallinity control, active-phase engineering, defect creation, heteroatom doping, and interface construction have been extensively investigated [82, 85].

When it comes to SACs, both Ni and other metals (Rh, Ir, Ru, Pt, Co, and Fe) have been proposed for UOR, with recent reviews summarizing their development [9, 13, 30, 86]. While many of these metals are intrinsically active for OER, which directly competes with UOR at high overpotentials, the atomic dispersion in SACs and the local coordination modulation enable precise tuning of adsorption sites, thereby suppressing OER and favoring UOR. Furthermore, because the operating potential of UOR is lower than that of OER, the use of functionalized carbon supports becomes especially advantageous. Carbon not only provides high surface area and conductivity but also allows manipulation of the local coordination of isolated metal centers, thereby optimizing activity and selectivity [13, 16]. As examples, Ni SACs coordinated with four nitrogen atoms (Ni_1—NC), synthesized via pyrolysis, delivered a 175-fold higher mass activity than Ni nanoparticles on carbon, attributed to the formation of a HOO—Ni—N_4 intermediate confirmed by in-situ Raman. Ni_1—NC also demonstrated remarkable stability over 10,000 cycles [16]. As another prominent example, Ni SACs anchored on N-doped carbon sheets proposed as a bifunctional catalyst for urea-assisted rechargeable zinc–air battery, showed UOR potential of 289 mV as compared to OER potential measured at 10 $mA{\cdot}cm^{-2}$, with Tafel slopes of ~41 $mV{\cdot}dec^{-1}$ and UOR current densities significantly superior than bulk NiO, NiOOH, Pt/C, and IrO_2 [87].

Beyond carbon, alternative supports have proven effective for stabilizing high loadings (>6 wt.%) of noble-metal-based SACs to achieve high-performance UOR catalysis. Kumar et al. employed a liquid N_2-quenching strategy to impose surface strain on cobalt oxide substrates, thereby anchoring substantial amounts of Pt, Ir, Ru, and Rh SAs while suppressing their migration tendency [88]. Among these, Rh SACs on strained Co_3O_4 (S—Co_3O_4) exhibited outstanding UOR activity, with an onset potential nearly 190 mV lower than OER at 100 $mA{\cdot}cm^{-2}$. DFT calculations attributed this enhancement to the strained interfacial environment, which simultaneously facilitates urea adsorption, intermediate stabilization, and electron transfer across the Rh—O—Co interface. However, these proof-of-concept studies remain limited by their restricted testing conditions—typically in three-electrode configurations and synthetic urea electrolytes rather than real wastewater matrices containing chloride, carbonate, phosphate, or organic contaminants that can poison active sites and accelerate corrosion. Moreover, the long-term stability of strain-engineered SACs under dynamic electrochemical operation is uncertain; surface reconstruction, strain relaxation, or partial aggregation could occur over extended cycling, diminishing the stabilizing effect.

16.4.3 Long-term Stability and Practical Perspectives of SACs for UOR

SACs combine high activity with nearly complete nitrogen selectivity (>99% in optimized systems), but their long-term stability remains the primary barrier to practical deployment. The atomically dispersed nature of the active sites, while key to their high performance, makes them intrinsically vulnerable to migration, agglomeration, and dissolution, particularly under high current densities (>100 $mA\cdot cm^{-2}$) and extended cycling [24, 38, 40, 55]. Support design plays a decisive role in mitigating these degradation pathways. Carbon offers high conductivity, large surface area, and versatile coordination environments, but is prone to oxidation at anodic potentials, especially above 1 V vs. RHE [37, 89, 90]. In contrast, oxides enhance MSIs that help stabilize isolated atoms but often have limited electronic conductivity. Hybrid supports that combine conductivity with anchoring strength are therefore emerging as promising solutions to balance these trade-offs. For rational design, advanced characterization is indispensable. Operando spectroscopy (XAS, Raman, and IR) enables tracking of dynamic oxidation states and coordination environments, while electrochemistry-coupled ICP–MS provides quantitative insights into metal and support dissolution at the ppb level [45, 91]. Together, these methods reveal degradation mechanisms and guide stabilization strategies. Given their exceptionally high mass activities and ultralow noble-metal loadings (<1 wt.% in many systems), SACs remain economically attractive for large-scale UOR. Yet translating laboratory advances into practical electrolyzer anodes will require demonstrating durable performance at high current densities and under realistic conditions, including complex wastewater matrices and integration into scalable architectures such as flow cells or membrane electrode assemblies.

16.5 Furfural Oxidation and SACs

16.5.1 Oxidation Pathways and Products of Furfural

Similar to the UOR discussed earlier, the furfural oxidation reaction (FOR) has emerged as a promising anodic alternative to the OER in hybrid electrolysis, enabling simultaneous hydrogen production and value-added biomass upgrading [3–5, 92–94]. Furfural is a key platform molecule derived from hemicellulose-rich agricultural residues and serves as a precursor to numerous higher-value chemicals and polymer precursors [5, 92, 95]. Its commercial production involves hemicellulose hydrolysis, xylose dehydration, furfural recovery, and purification, typically using batch or continuous reactors [96, 97]. Through ammoniation, hydrogenation, or oxidation, a wide range of derivatives can be obtained [98]. Among the oxidation products, furoic, succinic, and maleic acids, as well as lactones such as 2-(5*H*)-furanones, hold particular industrial significance [99, 100]. For instance, furoic acid is usually manufactured via the Cannizzaro reaction but can also be formed through a Baeyer–Villiger oxidation pathway yielding furoic and formic acids [98]. Maleic acid, generally obtained from butane or benzene [98, 101], can alternatively be produced from furfural through intermediates like 5-hydroxy-2(5*H*)-furanone [102, 103], while succinic acid, an important building block for pharmaceuticals and biopolymers, is mostly produced via catalytic hydrogenation of fossil-derived maleic anhydride [100, 104], though furfural oxidation also provides a renewable route [98].

Despite this promise, commercial furfural oxidation remains underdeveloped. Electrooxidation has recently emerged as a sustainable alternative, offering high

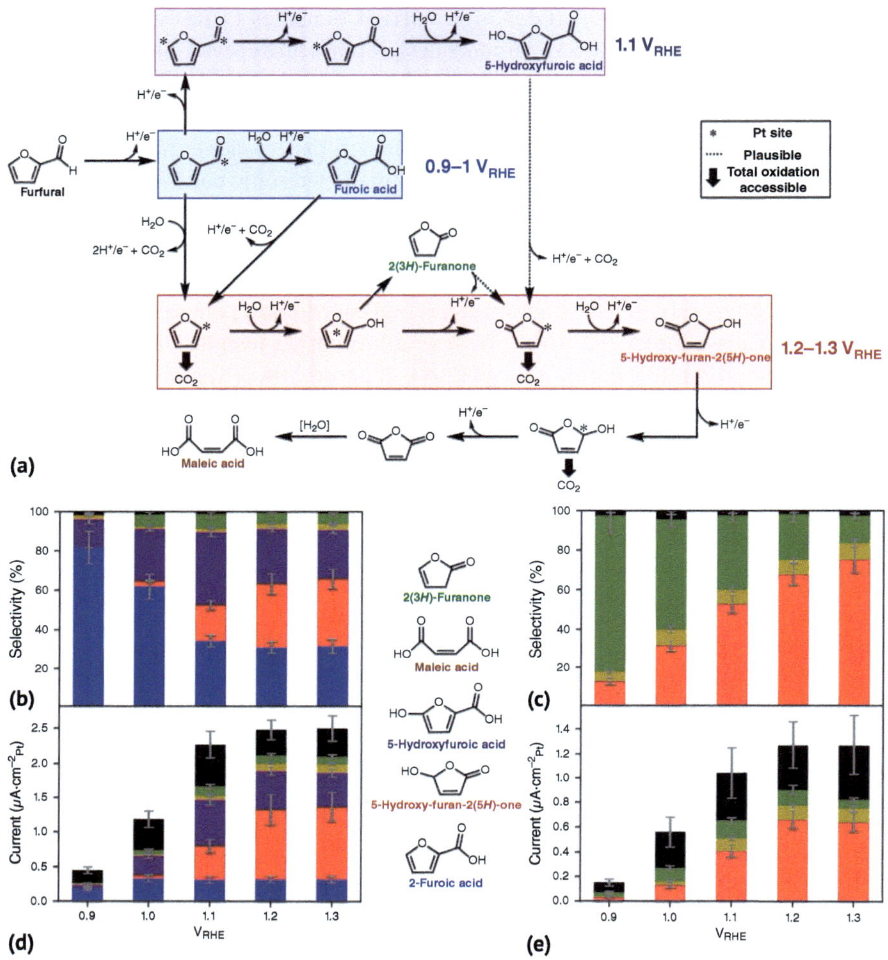

Figure 16.4 (a) Proposed reaction mechanism for the electrooxidation of furfural in acidic environment on Pt; selectivity and partial currents toward major products of the electrooxidation of (b, d) 100 mM furfural and (c, e) 100 mM furoic acid on Pt/C in 0.25 M $HClO_4$ at the steady state. Furoic acid (blue), 5-hydroxy-2(5*H*)-furanone (red), 5-hydroxyfuroic acid (purple), maleic acid (yellow), 2(3*H*)-furanone (green), and terminal CO_2 (black) (CO_2 current that is concomitant with C_4 products is allocated to those, respectively). *Source*: Román et al. (2019) [106]/with permission of American Chemical Society.

selectivity and environmental compatibility, although electrooxidation routes lack studies presenting the state-of-the-art catalysts for each pathway [105]. Understanding the possible reaction pathways is therefore crucial [106]. As shown in Figure 16.4a, 2-furoic acid results from a two-electron oxidation of furfural, occurring near 1.38 V vs. RHE in alkaline media and 0.9 V vs. RHE in acid media, using a Pt on carbon catalyst [106, 107]. A four-electron route instead yields 5-hydroxyfuroic acid within 0.9–1.1 V vs. RHE (acidic media) [106]. From 2-furoic acid, 2(3*H*)-furanone may be formed via a two-electron step, while 5-hydroxy-furan-2(5*H*)-one arises from a four-electron pathway; the latter can undergo further two-electron oxidation to maleic acid at 1.8–2 V vs. RHE in acidic media [94]. Figure 16.4b–e summarize activity and selectivity trends (Pt catalyst, acidic media), though these values strongly depend on pH, catalyst, reaction conditions, and the applied potential.

16.5.2 Electrocatalysts for FOR and Opportunities for SACs

The electrochemical oxidation of furfural can yield a variety of products, with furoic acid widely recognized as a key intermediate [94, 108, 109]. Recent mechanistic studies further suggested possible alternative pathways involving hydroxylated furan derivatives (e.g., 5-hydroxyfuroic acid) and even ring-opened species, highlighting the strong dependence of reaction selectivity on the catalyst and operating conditions [109, 110]. State-of-the-art electrocatalysts for FOR include Cu-based systems, doped or alloyed Ni oxides, Mn oxides, and noble metals such as Au, Pt, Pd, and Ru with the majority showing high selectivity toward 2-furoic acid [12, 92, 99, 106, 108, 111, 112]. The product distribution, however, can be tuned by incorporating secondary metals or heteroatoms (e.g., Ce, W, and Mn in mixed oxides), which alter the electronic structure and surface chemistry of the catalyst, thereby directing the reaction toward alternative products [113, 114]. A representative example is MOF-derived $MnCeO_x$, which in 0.5 M $KHCO_3$ selectively yields maleic acid. This behavior was attributed to abundant surface oxygen vacancies that strengthen furfural adsorption and promote activation at the interfacial region [114]. More broadly, the performance of FOR catalysts is strongly governed by the electronic configuration of the active sites, which must simultaneously support efficient charge transfer and the generation of reactive oxygen species (e.g., adsorbed OH and O) that mediate oxidation steps [114]. Surface defects and oxygen vacancies further enhance activity by facilitating water activation and increasing the supply of oxygen intermediates, thereby accelerating the overall reaction kinetics [113, 114].

Despite being more extensively studied in the thermochemical oxidation of furfural [115–117], SACs remain largely unexplored in electrooxidation. Their atomically defined active centers and tunable coordination environments, however, offer unique opportunities to precisely tailor product selectivity [5, 40, 115, 118]. However, efficient electrooxidation of furfural (and related substrates such as 5-hydroxymethylfurfural) requires the generation of reactive oxygen species at anodic potentials (typically requiring >1 V vs. RHE) [3, 92, 119, 120], where carbon supports become vulnerable to corrosion, raising concerns about the stability of conventional carbon-hosted SACs or molecular single-site complexes [14, 37, 45, 62]. The limited electrocatalytic studies to date already indicated a strong metal identity effect on furfural oxidation. For instance, while Cu nanoparticles readily oxidize furfural to 2-furoic acid, atomically dispersed $Cu–N_x$ sites on carbon are essentially inactive, reflecting their intrinsic preference for alcohol oxidation pathways [121]. Thermocatalytic studies on biomass conversion also hinted that the metal identity of SACs may impose divergent selectivity trends [5, 11, 115]. Taken together, these observations underscore the dual opportunity and challenge of incorporating SACs into furfural electrooxidation; their unique single-site nature could enable product selectivity engineering, but their coordination sensitivity and stability under oxidative electrochemical conditions remain critical bottlenecks that need to be addressed.

16.6 Hybrid Electrolyzer Configuration, Integration, and Challenges of Using SACs

Coupling FOR or UOR at the anode with HER at the cathode offers a sustainable route that simultaneously enables environmental remediation, biomass valorization, and green hydrogen production [122, 123]. This dual-functional electrolyzer strategy improves energy efficiency by replacing the sluggish and thermodynamically demanding OER with more favorable anodic alternatives such as FOR or UOR, which possess lower oxidation potentials [9, 107]. As illustrated in Figure 16.5, the substitution of anodic reaction reduces the cell voltage relative to the overall water splitting, thus lowering the electricity demand for hydrogen production.

16.6.1 System-level Considerations

From a systems perspective, OER in conventional electrolyzers is generally not limited by bulk water transport; limitations arise primarily at extremely high

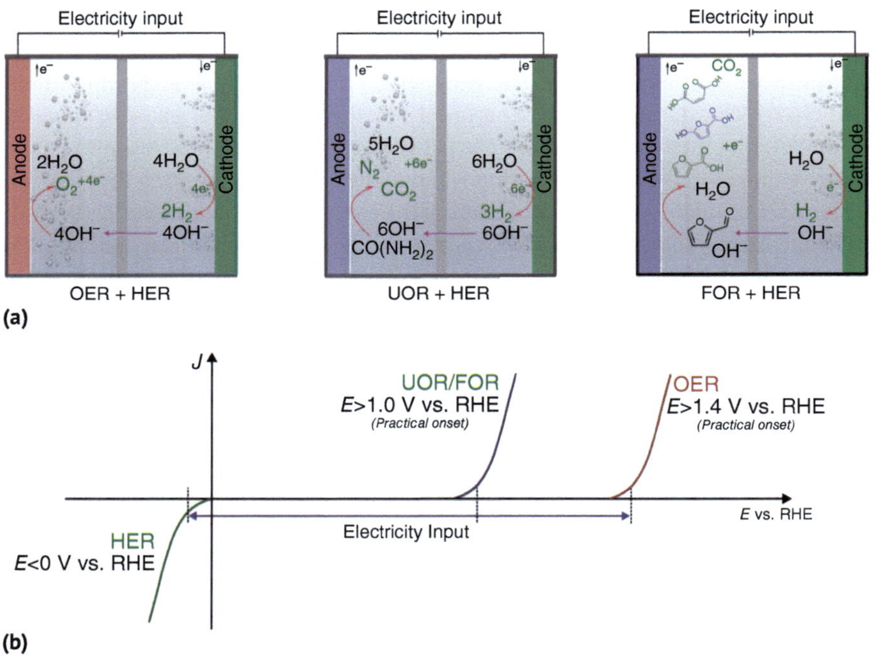

Figure 16.5 (a) Schematic illustrations of alkaline electrolyzer configurations for H_2 generation coupled with different anodic reactions (UOR, FOR, or OER). (b) Comparison of overall water splitting (OER + HER) with hybrid electrolyzers where OER is replaced by UOR or FOR, showing reduced cell voltage.

current densities, where issues such as water supply to the electrode surface and oxygen bubble management become critical for the system performance [1, 2]. In contrast, maintaining sufficient and homogeneous concentrations of furfural or urea in hybrid electrolyzers requires forced convection or flow-through operation to sustain reactant availability and solution homogeneity, which, if left unmanaged, can compromise both reaction selectivity and throughput. Paired-cell studies of FOR confirmed that circulating anolyte or similar flow configurations are necessary for sustained operation, but these approaches add complexity and cost while lowering overall device efficiency [107].

Electrolyzer design strongly influences system performance. Flow cells offer clear advantages over conventional H-type cells, providing enhanced mass transport, improved control over reaction conditions, and scalability [122]. For example, Cao et al. employed a continuous-flow reactor with a narrow interelectrode gap, achieving faster reaction rates and enhanced selectivity compared with batch cells [122]. More recently, Liu et al. demonstrated a membrane–electrode assembly (MEA) system for FOR–HER coupling, operating stably at 200 $mA{\cdot}cm^{-2}$ with a combined faradaic efficiency of 150% and low cell voltage (~2.7 V) over extended operation. Moreover, the Co-based catalyst material also exhibited excellent durability, with no performance degradation during a 36-hour stability test at current densities ranging up to 400 $mA{\cdot}cm^{-2}$ [107]. Figure 16.6a illustrates a typical MEA design for FOR–HER integration, while Figure 16.6b–d compare the performance of different catalysts, highlighting the influence of catalyst identity on product selectivity and current density.

Beyond architecture, the choice of electrolyte and membrane is equally critical. Alkaline conditions are especially advantageous, as they promote efficient and selective oxidation of furfural to 2-furoic acid while suppressing side reactions, resulting in both higher substrate conversion rates and product selectivity [124]. The use of anion-exchange membranes is a key strategy for enabling and sustaining high ionic conductivity in these systems, as they permit the transport of hydroxide ions while limiting the crossover of organic substrates and value-added products between the anode and cathode compartments [125]. This selectivity not only maintains overall cell efficiency but also protects downstream processes and the cathode from fouling, degradation, and catalyst poisoning, with the latter being critical for long-term operational stability [126]. Similar considerations apply to UOR, where alkaline electrolytes and anion-exchange membranes provide high activity and stability [127]. Moreover, the molecular composition and pore structure of anion-exchange membranes can be tailored to optimize conductivity, mechanical stability, and organic passage rates for specific electrochemical systems.

16.6.2 Catalyst-level Considerations and SAC Challenges

The use of SACs in these hybrid electrolyzers presents both unique opportunities and serious limitations. For OER, the high kinetic overpotentials, conductivity demand thresholds, and need for high active-site density make SACs generally unsuitable due to their low metal loadings and dependence on conductive supports.

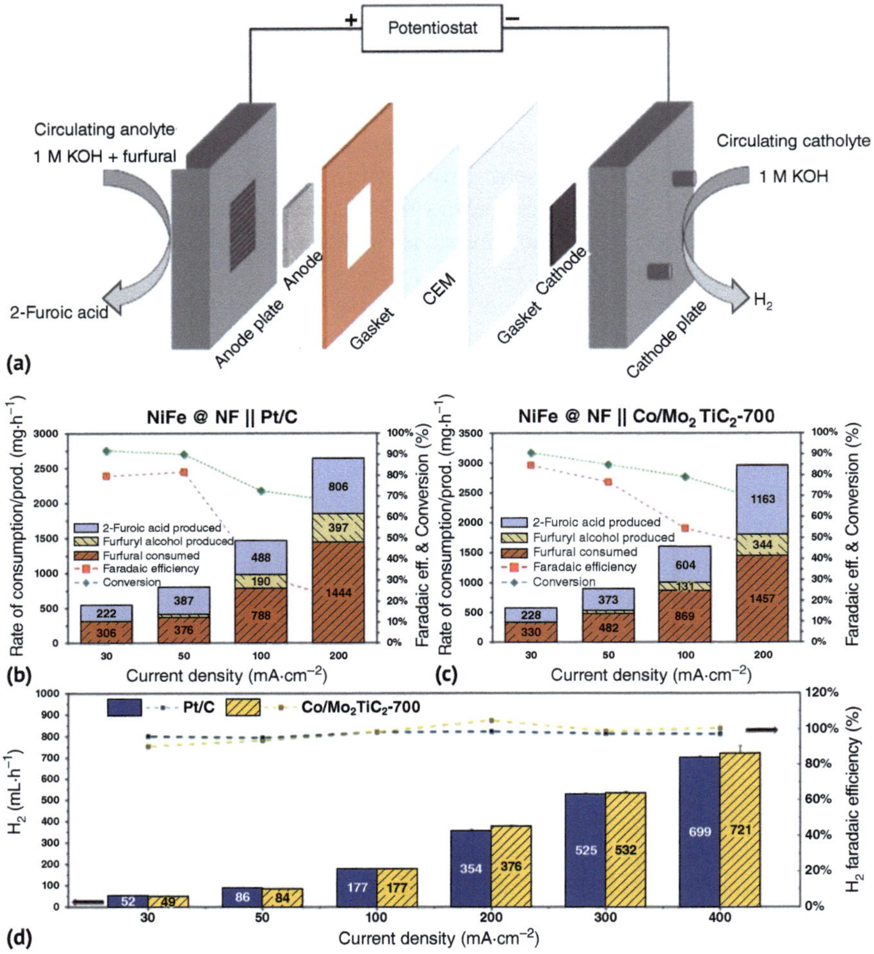

Figure 16.6 (a) Schematic of a typical MEA cell for paired FOR–HER operation. Production of oxidized furfural products in alkaline media using (b) Pt/C and (c) Co/Mo_2TiC_2 catalysts. (d) Comparison of selectivity and current density for different catalysts, illustrating the trade-off between activity and product distribution in FOR systems. *Source*: Liu et al. (2024) [107]/with permission of The Royal Society of Chemistry.

In contrast, SACs show stronger potential for UOR and FOR, where lower anodic barriers and specific substrate–catalyst interactions can be exploited. At industrially relevant current densities (>200–400 mA·cm^{-2}), SACs may undergo coordination restructuring, support corrosion, metal leaching, or poisoning by reaction intermediates/products [38, 39, 41, 45]. In UOR, real wastewater matrices exacerbate poisoning, while in FOR, the relatively higher potentials required for high current densities make support conductivity and carbon stability especially critical. Beyond intrinsic catalytic activity, several practical factors remain critical for integrating SACs into hybrid electrolyzers. Product management is a key challenge, since value-added products from FOR or UOR must be efficiently separated without crossover through membranes. Electrolyte compatibility is another bottleneck, as SAC supports optimized for stability in one medium may degrade under the alkaline conditions typically required for FOR and UOR.

Integration of SACs, therefore, must address additional hurdles such as robust product separation (to prevent crossover) and ensure support stability and electrolyte compatibility under typically alkaline conditions. Optimal support design, including defect engineering and the use of chemically inert supports, is crucial for stabilizing SAs and ensuring activity retention during prolonged operation in alkaline electrolytes [128, 129].

16.6.3 Selectivity–Current Density Trade-offs

Balancing high current density with product selectivity remains a persistent challenge in both FOR and UOR systems. At higher current densities, mass transport and reaction kinetics often push the system out of the optimal window for selective oxidation, resulting instead in increased side reactions and yield loss of desired products [106, 125]. For FOR, elevated currents promote undesired by-products, such as CO, formic acid, or polymeric "humin" species [130], resulting in side reactions that reduce yields of 2-furoic acid [75, 123]. Similarly, in UOR, NiOOH-based catalysts achieve high activities (~250 $mA \cdot cm^{-2}$) but often suffer from reduced selectivity at high overpotentials, where competing pathways such as the OER and partial urea oxidation dominate [123]. The interplay between mass transport, catalyst structure, and applied bias shapes product outcomes; for example, Zn-doped Ni-MOF catalysts have been engineered to enhance mass and charge transport, reduce the onset potential, and suppress OER, thereby extending the window for both high current density and good selectivity. Nevertheless, excessive currents still result in a decline of catalytic performance [131]. Optimizing flow cell parameters such as residence time and potential, represents a strategy to balance conversion efficiency with product selectivity [107, 122]. These dynamics are reflected in Figure 16.6b–d, which highlight the effect of catalyst variation on product distribution under alkaline conditions.

16.7 Conclusion

SACs are an innovative yet still evolving class of materials for hybrid hydrogen production. Their atomically dispersed active sites optimize metal use and allow precise control over adsorption and reaction pathways, offering potential to lower the energy required for hydrogen generation. Nonetheless, practical implementation faces hurdles. The stability of SACs under aggressive anodic and alkaline conditions, their vulnerability to metal leaching, and challenges in maintaining atomic dispersion at higher loadings hinder scalability. In hybrid electrolyzers, SACs serve as an effective platform to couple anodic reactions like urea or furfural oxidation with hydrogen evolution, enabling concurrent hydrogen production and waste or biomass valorization. However, attaining industrial current densities while maintaining activity and selectivity demands advances in support design, control over local coordination, and resistance to reconstruction during dynamic operation. Future research should prioritize creating resilient support structures, stabilizing active sites under real-world electrochemical conditions, and combining operando characterization with system-level testing. These strategies are essential for transforming SACs from promising laboratory prototypes into durable, scalable catalysts for energy-efficient and sustainable hydrogen production.

References

1 Abderrahmane, A. B., Tingry, S., Cornu, D., & Holade, Y. (2024). Progress in biomass electro-valorization for paired electrosynthesis of valuable chemicals and fuels. *Adv. Energy Sustain. Res. 5*(8), 2300302. https://doi.org/10.1002/aesr.202300302

2 Moges, E. A., Lakshmanan, K., Chang, C.-Y., Liao, W.-S., Angerasa, F. T., Dilebo, W. B., Edao, H. G., Tadele, K. T., Alemayehu, D. D., Bejena, B. D., Guta, C. B., Chang, C.-C., Tsai, M. C., Su, W.-N., & Hwang, B. J. (2024). Materials of value-added electrolysis for green hydrogen production. *ACS Mater. Lett. 6*(11), 4932–4954. https://doi.org/10.1021/acsmaterialslett.4c01173

3 Singh, B. & Verma, N. (2024). Exploring anodic oxidation reactions in hybrid water electrolysis: Special emphasis on substrate choice. *Catal. Sci. Technol. 14*(23), 6732–6760. https://doi.org/10.1039/D4CY00907J

4 Sun, H., Xu, X., Fei, L., Zhou, W., & Shao, Z. (2024). Electrochemical oxidation of small molecules for energy-saving hydrogen production. *Adv. Energy Mater. 14*(30), 2401242. https://doi.org/10.1002/aenm.202401242

5 Wang, H., Li, X., & Si, C. (2025). Electrocatalytic oxidation of biomass-derived chemicals using non-precious metal catalysts: Design strategies, performance characteristics and perspectives. *Green Chem. 27*(35), 10346–10371. https://doi.org/10.1039/D5GC00719D

6 Dabboussi, J., Eichel, R.-A., Kungl, H., Abdallah, R., & Loget, G. (2024). The rebirth of urea oxidation reaction for power-to-X and beyond. *Curr. Opin. Electrochem. 45*, 101468. https://doi.org/10.1016/j.coelec.2024.101468

7 Hasse, J. C., Manyé Ibáñez, M., & Holewinski, A. (2023). Impact of electrolyte composition on bulk electrolysis of furfural over platinum electrodes. *ChemCatChem 15*(23), e202300988. https://doi.org/10.1002/cctc.202300988

8 Protsenko, V. S. (2023). Thermodynamic aspects of urea oxidation reaction in the context of hydrogen production by electrolysis. *Int. J. Hydrog. Energy 48*(63), 24207–24211. https://doi.org/10.1016/j.ijhydene.2023.03.295

9 Ge, W., Lin, L., Wang, S.-Q., Wang, Y., Ma, X., An, Q., & Zhao, L. (2023). Electrocatalytic urea oxidation: Advances in mechanistic insights, nanocatalyst design, and applications. *J. Mater. Chem. A 11*(28), 15100–15121. https://doi.org/10.1039/D3TA02007J

10 Zhang, X., Feizpoor, S., Humayun, M., & Wang, C. (2024). Urea oxidation reaction electrocatalysts: Correlation of structure, activity, and selectivity. *Chem Catal. 4*(2), 100840. https://doi.org/10.1016/j.checat.2023.100840

11 Ma, J., Yan, H., Hou, X., Xu, M., & Cui, T. (2025). Single-atom catalysts for (photo)electrocatalytic biomass valorization to high-value-added chemicals. *Microstructures 5*(1), 2025002. https://doi.org/10.20517/microstructures.2024.38

12 Mukadam, Z., Liu, S., Pedersen, A., Barrio, J., Fearn, S., Sarma, S. C., Titirici, M.-M., Scott, S. B., Stephens, I. E. L., Chan, K., & Mezzavilla, S. (2023). Furfural electrovalorisation using single-atom molecular catalysts. *Energy Environ. Sci. 16*(7), 2934–2944. https://doi.org/10.1039/D3EE00551H

13 Urso, M., Ju, X., Nittoor-Veedu, R., Lee, H., Zaoralová, D., Otyepka, M., & Pumera, M. (2025). Single atom engineering for electrocatalysis: Fundamentals and applications. *ACS Catal. 15*(13), 11617–11663. https://doi.org/10.1021/acscatal.4c08027

14 Wang, Y.-c., Ma, X.-b., Ayeza, Wang, C.-x., Li, Y., Yang, C.-l., Wang, Z.-f., Wang, C., Hu, C., & Zhang, Y.-t. (2024). A review of carbon-supported single-atom catalysts for electrochemical reactions. *New Carbon Mater. 39*(3), 407–438. https://doi.org/10.1016/S1872-5805(24)60863-2

15 Wu, J., Shi, H., Li, K., & Guo, X. (2023). Advances and challenges of single-atom catalysts in environmental catalysis. *Curr. Opin. Chem. Eng. 40*, 100923. https://doi.org/10.1016/j.coche.2023.100923

16 Luo, F., Pan, S., Xie, Y., Li, C., Yu, Y., & Yang, Z. (2024). Atomically dispersed Ni electrocatalyst for superior urea-assisted water splitting. *J. Energy Chem. 90*, 1–6. https://doi.org/10.1016/j.jechem.2023.10.007

17 Guo, M., Meng, Q., Gao, M. L., Zheng, L., Li, Q., Jiao, L., & Jiang, H. L. (2025). Single-atom Pt loaded on MOF-derived porous TiO_2 with maximized Pt atom utilization for selective hydrogenation of halonitro-benzene. *Angew. Chem. Int. Ed. 64*(7), e202418964. https://doi.org/10.1002/anie.202418964

18 Liu, H., Cheng, M., Liu, Y., Wang, J., Zhang, G., Li, L., Du, L., Wang, G., Yang, S., & Wang, X. (2022). Single atoms meet metal–organic frameworks: collaborative efforts for efficient photocatalysis. *Energy Environ. Sci. 15*(9), 3722–3749. https://doi.org/10.1039/d2ee01037b

19 Qiao, B., Wang, A., Yang, X., Allard, L. F., Jiang, Z., Cui, Y., Liu, J., Li, J., & Zhang, T. (2011). Single-atom catalysis of CO oxidation using Pt_1/FeO_x. Nat. Chem. *3*(8), 634–641. https://doi.org/10.1038/nchem.1095

20 Wang, Y., Wang, D., & Li, Y. (2021). Rational design of single-atom site electrocatalysts: from theoretical understandings to practical applications. *Adv. Mater. 33*(34), 2008151. https://doi.org/10.1002/adma.202008151

21 Li, J., Yue, M.-F., Wei, Y.-M., & Li, J.-F. (2022). Synthetic strategies of single-atoms catalysts and applications in electrocatalysis. *Electrochim. Acta 409*, 139835. https://doi.org/10.1016/j.electacta.2022.139835

22 Zhang, M., Lu, X., Wu, Z., Sun, N., Liu, X., Wang, Y., Wang, L., & Yan, D. (2024). Advanced development of dual-atom catalysts: From synthesis methods to versatile electrocatalytic applications. *J. Power Sources 613*, 234923. https://doi.org/10.1016/j.jpowsour.2024.234923

23 Zhuang, J. & Wang, D. (2023). Recent advances of single-atom alloy catalyst: Properties, synthetic methods and electrocatalytic applications. *Mater. Today Catal. 2*, 100009. https://doi.org/10.1016/j.mtcata.2023.100009

24 Yang, X. F., Wang, A., Qiao, B., Li, J., Liu, J., & Zhang, T. (2013). Single-atom catalysts: a new frontier in heterogeneous catalysis. *Acc. Chem. Res. 46*(8), 1740–1748. https://doi.org/10.1021/ar300361m

25 Li, R., Luo, L., Ma, X., Wu, W., Wang, M., & Zeng, J. (2022). Single atoms supported on metal oxides for energy catalysis. *J. Mater. Chem. A 10*(11), 5717–5742. https://doi.org/10.1039/D1TA08016D

26 Song, W., Xiao, C., Ding, J., Huang, Z., Yang, X., Zhang, T., Mitlin, D., & Hu, W. (2024). Review of carbon support coordination environments for single metal atom electrocatalysts (SACS). *Adv. Mater. 36*(1), e2301477. https://doi.org/10.1002/adma.202301477

27 Zhuo, H. Y., Zhang, X., Liang, J. X., Yu, Q., Xiao, H., & Li, J. (2020). Theoretical understandings of graphene-based metal single-atom catalysts: Stability and catalytic performance. *Chem. Rev. 120*(21), 12315–12341. https://doi.org/10.1021/acs.chemrev.0c00818

28 Tang, C., Chen, L., Li, H., Li, L., Jiao, Y., Zheng, Y., Xu, H., Davey, K., & Qiao, S.-Z. (2021). Tailoring acidic oxygen reduction selectivity on single-atom catalysts via modification of first and second coordination spheres. *J. Am. Chem. Soc. 143*(20), 7819–7827. https://doi.org/10.1021/jacs.1c03135

29 Liu, L. & Corma, A. (2018). Metal catalysts for heterogeneous catalysis: from single atoms to nanoclusters and nanoparticles. *Chem. Rev. 118*(10), 4981–5079. https://doi.org/10.1021/acs.chemrev.7b00776

30 Xiao, J., Duan, C., Song, J., Sun, Y., & Pan, J. (2024). Research progress of single-atomic series catalysts for anodic coupled electrolysis of water and small molecules. *Chem. Eng. J. 500*, 157208. https://doi.org/10.1016/j.cej.2024.157208

31 Ou, H., Wang, D., & Li, Y. (2021). How to select effective electrocatalysts: Nano or single atom? *Nano Select. 2*(3), 492–511. https://doi.org/10.1002/nano.202000239

32 Wang, H., Liu, J.-X., Allard, L. F., Lee, S., Liu, J., Li, H., Wang, J., Wang, J., Oh, S. H., Li, W., Flytzani-Stephanopoulos, M., Shen, M., Goldsmith, B. R., & Yang, M. (2019). Surpassing the single-atom catalytic activity limit through paired Pt-O-Pt ensemble built from isolated Pt_1 atoms. *Nat. Commun. 10*(1), 3808. https://doi.org/10.1038/s41467-019-11856-9

33 Jeong, H., Shin, S., & Lee, H. (2020). Heterogeneous atomic catalysts overcoming the limitations of single-atom catalysts. *ACS Nano 14*(11), 14355–14374. https://doi.org/10.1021/acsnano.0c06610

34 Yu, Z., Xu, H., & Cheng, D. (2021). Design of single atom catalysts. *Adv. Phys. X 6*(1), 1905545. https://doi.org/10.1080/23746149.2021.1905545

35 Cheng, X., Shen, Z., Jiao, L., Yang, L., Wang, X., Wu, Q., & Hu, Z. (2021). Tuning metal catalysts via nitrogen-doped nanocarbons for energy chemistry: From metal nanoparticles to single metal sites. *EnergyChem 3*(6), 100066. https://doi.org/10.1016/j.enchem.2021.100066

36 Huang, Y., Xiong, J., Zou, Z., & Chen, Z. (2025). Emerging strategies for the synthesis of correlated single atom catalysts. *Adv. Mater. 37*(2), e2312182. https://doi.org/10.1002/adma.202312182

37 Kolle-Görgen, E., Fortunato, G., & Ledendecker, M. (2022). Catalyst stability in aqueous electrochemistry. *Chem. Mater. 34*(23), 10223–10236. https://doi.org/10.1021/acs.chemmater.2c02443

38 Lang, Z., Wang, X., Jabeen, S., Cheng, Y., Liu, N., Liu, Z., Gan, T., Zhuang, Z., Li, H., & Wang, D. (2025). Destabilization of single-atom catalysts: characterization, mechanisms, and regeneration strategies. *Adv. Mater. 37*(9), e2418942. https://doi.org/10.1002/adma.202418942

39 Rigby, K. & Kim, J.-H. (2023). Deciphering the issue of single-atom catalyst stability. *Curr. Opin. Chem. Eng. 40*, 100921. https://doi.org/10.1016/j.coche.2023.100921

40 Li, J., Chen, C., Xu, L., Zhang, Y., Wei, W., Zhao, E., Wu, Y., & Chen, C. (2023). Challenges and perspectives of single-atom-based catalysts for electrochemical reactions. *JACS Au 3*(3), 736–755. https://doi.org/10.1021/jacsau.3c00001

41 Tsipoaka, M. & Rownaghi, A. A. (2025). Toward understanding the stability of single-atom catalysts in oxygen evolution reactions. *Adv. Energy Mater. 15*(29), 2500635. https://doi.org/10.1002/aenm.202500635

42 Zhang, S., Wang, R., Zhang, X., & Zhao, H. (2024). Recent advances in single-atom alloys: preparation methods and applications in heterogeneous catalysis. *RSC Adv. 14*(6), 3936–3951. https://doi.org/10.1039/d3ra07029h

43 Sun, X., Dawson, S. R., Parmentier, T. E., Malta, G., Davies, T. E., He, Q., Lu, L., Morgan, D. J., Carthey, N., Johnston, P., Kondrat, S. A., Freakley, S. J., Kiely, C. J., & Hutchings, G. J. (2020). Facile synthesis of precious-metal single-site catalysts using organic solvents. *Nat. Chem. 12*(6), 560–567. https://doi.org/10.1038/s41557-020-0446-z

44 Yao, Y., Zhang, X., Cao, Y., Liu, L., Zhang, Y., & Zhang, S. (2024). Single-site Pd regulated by π–π stacking for high-selectivity cyclopropanation reaction. *J. Am. Chem. Soc. 146*(45), 31053–31061. https://doi.org/10.1021/jacs.4c10860

45 Fortunato, G. V., Gunnarson, A., Hosseini, H., You, X. Y., Bhuyan, P., Choi, J. S., Jeon, H. S., Zhao, X. S., Lourenço, J. C., Lim, S., Wang, H. Z., Buzanich, A. G., Radtke, M., Paciok, P., Lanza, M. R. V., Schüth, F., & Ledendecker, M. (2025). Stabilizing cationic palladium single-atom sites on heteroatom-doped carbon for selective hydrogen peroxide electrosynthesis. *Adv. Funct. Mater. 2025*, e16600. https://doi.org/10.1002/adfm.202516600

46 Fonseca, J. & Lu, J. (2021). Single-atom catalysts designed and prepared by the atomic layer deposition technique. *ACS Catal. 11*(12), 7018–7059. https://doi.org/10.1021/acscatal.1c01200

47 Sun, H., Tang, R., & Huang, J. (2022). Considering single-atom catalysts as photocatalysts from synthesis to application. *iScience 25*(5), 104232. https://doi.org/10.1016/j.isci.2022.104232

48 Chen, W., Ma, B., & Zou, R. (2025). Rational design and controlled synthesis of MOF-derived single-atom catalysts. *Acc. Mater. Res. 6*(2), 210–220. https://doi.org/10.1021/accountsmr.4c00330

49 Song, Z., Zhang, L., Doyle-Davis, K., Fu, X., Luo, J. L., & Sun, X. (2020). Recent advances in MOF-derived single atom catalysts for electrochemical applications. *Adv. Energy Mater. 10*(38), 2001561. https://doi.org/10.1002/aenm.202001561

50 Hai, X., Xi, S., Mitchell, S., Harrath, K., Xu, H., Akl, D. F., Kong, D., Li, J., Li, Z., Sun, T., Yang, H., Cui, Y., Su, C., Zhao, X., Li, J., Pérez-Ramírez, J., & Lu, J. (2022). Scalable two-step annealing method for preparing ultra-high-density single-atom catalyst libraries. *Nat. Nanotechnol. 17*(2), 174–181. https://doi.org/10.1038/s41565-021-01022-y

51 Li, J., Chen, M., Cullen, D. A., Hwang, S., Wang, M., Li, B., Liu, K., Karakalos, S., Lucero, M., Zhang, H., Lei, C., Xu, H., Sterbinsky, G. E., Feng, Z., Su, D., More, K. L., Wang, G., Wang, Z., & Wu, G. (2018). Atomically dispersed manganese catalysts for oxygen reduction in proton-exchange membrane fuel cells. *Nat. Catal. 1*(12), 935–945. https://doi.org/10.1038/s41929-018-0164-8

52 Kottwitz, M., Li, Y., Wang, H., Frenkel, A. I., & Nuzzo, R. G. (2021). Single atom catalysts: A review of characterization methods. *Chem. Methods 1*(6), 278–294. https://doi.org/10.1002/cmtd.202100020

53 Liu, Y., Su, X., Ding, J., Zhou, J., Liu, Z., Wei, X., Yang, H. B., & Liu, B. (2024). Progress and challenges in structural, in situ and operando characterization of single-atom catalysts by X-ray based synchrotron radiation techniques. *Chem. Soc. Rev. 53*(24), 11850–11887. https://doi.org/10.1039/d3cs00967j

54 Mitchell, S., Parés, F., Faust Akl, D., Collins, S. M., Kepaptsoglou, D. M., Ramasse, Q. M., Garcia-Gasulla, D., Pérez-Ramírez, J., & López, N. (2022). Automated image analysis for single-atom detection in catalytic materials by transmission electron microscopy. *J. Am. Chem. Soc. 144*(18), 8018–8029. https://doi.org/10.1021/jacs.1c12466

55 Bae, G., Han, S., Oh, H.-S., & Choi, C. H. (2023). Operando stability of single-atom electrocatalysts. *Angew. Chem. Int. Ed. 62*(19), e202219227. https://doi.org/10.1002/anie.202219227

56 Chen, Z., Walsh, A. G., & Zhang, P. (2024). Structural analysis of single-atom catalysts by X-ray absorption spectroscopy. *Acc. Chem. Res. 57*(4), 521–532. https://doi.org/10.1021/acs.accounts.3c00693

57 Creissen, C. E. & Fontecave, M. (2022). Keeping sight of copper in single-atom catalysts for electrochemical carbon dioxide reduction. *Nat. Commun. 13*(1), 2280. https://doi.org/10.1038/s41467-022-30027-x

58 Song, Z., Li, J., Davis, K. D., Li, X., Zhang, J., Zhang, L., & Sun, X. (2022). Emerging applications of synchrotron radiation X-ray techniques in single atomic catalysts. *Small Methods 6*(11), 2201078. https://doi.org/10.1002/smtd.202201078

59 Copéret, C., Liao, W.-C., Gordon, C. P., & Ong, T.-C. (2017). Active sites in supported single-site catalysts: An NMR perspective. *J. Am. Chem. Soc. 139*(31), 10588–10596. https://doi.org/10.1021/jacs.6b12981

60 Koppe, J., Yakimov, A. V., Gioffrè, D., Usteri, M.-E., Vosegaard, T., Pintacuda, G., Lesage, A., Pell, A. J., Mitchell, S., Pérez-Ramírez, J., & Copéret, C. (2025). Coordination environments of Pt single-atom catalysts from NMR signatures. *Nature 642*(8068), 613–619. https://doi.org/10.1038/s41586-025-09068-x

61 Wen, Y., Wang, F., Zhu, J., Wen, Q., Xia, X., Wen, J., Deng, C., Du, J.-H., Ke, X., Zhang, Z., Guan, H., Nie, L., Wang, M., Hou, W., Li, W., Tang, W., Ding, W., Chen, J., & Peng, L. (2025). Revealing the structure-activity relationship of Pt_1/CeO_2 with ^{17}O solid-state NMR spectroscopy and DFT calculations. *Nat. Commun. 16*(1), 3537. https://doi.org/10.1038/s41467-025-58709-2

62 Li, X., Lin, H., Huang, W., Xiong, S., & Huang, S. (2024). Enhanced urea oxidation catalysis through Ni single-atom doping on Cu_2O surfaces: a computational study. *J. Mater. Chem. A 12*(26), 15874–15884. https://doi.org/10.1039/D4TA02693D

63 Choi, J. S., Fortunato, G. V., Jung, D. C., Lourenco, J. C., Lanza, M. R. V., & Ledendecker, M. (2024). Catalyst durability in electrocatalytic H_2O_2 production: Key factors and challenges. *Nanoscale Horiz.* *9*(8), 1250–1261. https://doi.org/10.1039/d4nh00109e

64 Liu, J., Yang, J., Dou, Y., Liu, X., Chen, S., & Wang, D. (2025). Deactivation mechanism and mitigation strategies of single-atom site electrocatalysts. *Adv. Mater.* *37*(27), e2420383. https://doi.org/10.1002/adma.202420383

65 Yin, Y., Tan, Y., Wei, Q., Zhang, S., Wu, S., Huang, Q., Hu, F., & Mi, Y. (2021). Nanovilli electrode boosts hydrogen evolution: A surface with superaerophobicity and superhydrophilicity. *Nano Res.* *14*(4), 961–968. https://doi.org/10.1007/s12274-020-3133-x

66 Duan, N., Hou, T., Zheng, W., Qu, Y., Wang, P., Yang, J., Yang, Y., Wang, D., Chen, J., & Chen, Q. (2024). Insight into the structure of Mn-NiS_2 during urea oxidation using combined in situ X-ray absorption spectroscopy and attenuated total reflectance surface-enhanced infrared absorption spectroscopy. *ACS Catal.* *14*(3), 1384–1393. https://doi.org/10.1021/acscatal.3c05369

67 Zhan, G., Hu, L., Li, H., Dai, J., Zhao, L., Zheng, Q., Zou, X., Shi, Y., Wang, J., Hou, W., Yao, Y., & Zhang, L. (2024). Highly selective urea electrooxidation coupled with efficient hydrogen evolution. *Nat. Commun.* *15*(1), 5918. https://doi.org/10.1038/s41467-024-50343-8

68 Lu, S., Hummel, M., Gu, Z., Wang, Y., Wang, K., Pathak, R., Zhou, Y., Jia, H., Qi, X., Zhao, X., Xu, B. B., & Liu, X. (2021). Highly efficient urea oxidation via nesting nano-nickel oxide in eggshell membrane-derived carbon. *ACS Sustain. Chem. Eng.* *9*(4), 1703–1713. https://doi.org/10.1021/acssuschemeng.0c07614

69 Wang, L., Zhu, Y., Wen, Y., Li, S., Cui, C., Ni, F., Liu, Y., Lin, H., Li, Y., Peng, H., & Zhang, B. (2021). Regulating the local charge distribution of Ni active sites for the urea oxidation reaction. *Angew. Chem. Int. Ed.* *60*(19), 10577–10582. https://doi.org/10.1002/anie.202100610

70 Zhang, L., Wang, L., Lin, H., Liu, Y., Ye, J., Wen, Y., Chen, A., Wang, L., Ni, F., Zhou, Z., Sun, S., Li, Y., Zhang, B., & Peng, H. (2019). A lattice-oxygen-involved reaction pathway to boost urea oxidation. *Angew. Chem. Int. Ed.* *58*(47), 16820–16825. https://doi.org/10.1002/anie.201909832

71 Xu, S., Ruan, X., Ganesan, M., Wu, J., Ravi, S. K., & Cui, X. (2024). Transition metal-based catalysts for urea oxidation reaction (UOR): Catalyst design strategies, applications, and future perspectives. *Adv. Funct. Mater.* *34*(18), 2313309. https://doi.org/10.1002/adfm.202313309

72 Li, J., Li, J., Liu, T., Chen, L., Li, Y., Wang, H., Chen, X., Gong, M., Liu, Z.-P., & Yang, X. (2021). Deciphering and suppressing over-oxidized nitrogen in nickel-catalyzed urea electrolysis. *Angew. Chem. Int. Ed.* *60*(51), 26656–26662. https://doi.org/10.1002/anie.202107886

73 Schranck, A. & Doudrick, K. (2020). Effect of reactor configuration on the kinetics and nitrogen byproduct selectivity of urea electrolysis using a boron doped diamond electrode. *Water Res.* *168*, 115130. https://doi.org/10.1016/j.watres.2019.115130

74 Singh, R. K., Rajavelu, K., Montag, M., & Schechter, A. (2021). Advances in catalytic electrooxidation of urea: A review. *Energy Technol. 9*(8), 2100017. https://doi.org/10.1002/ente.202100017

75 Chen, W., Xu, L., Zhu, X., Huang, Y.-C., Zhou, W., Wang, D., Zhou, Y., Du, S., Li, Q., Xie, C., Tao, L., Dong, C.-L., Liu, J., Wang, Y., Chen, R., Su, H., Chen, C., Zou, Y., Li, Y., Liu, Q., & Wang, S. (2021). Unveiling the electrooxidation of urea: Intramolecular coupling of the N—N bond. *Angew. Chem. Int. Ed. 60*(13), 7297–7307. https://doi.org/10.1002/anie.202015773

76 Li, C., Liu, Y., Zhuo, Z., Ju, H., Li, D., Guo, Y., Wu, X., Li, H., & Zhai, T. (2018). Local charge distribution engineered by Schottky heterojunctions toward urea electrolysis. *Adv. Energy Mater. 8*(27), 1801775. https://doi.org/10.1002/aenm.201801775

77 Wang, T., Miao, L., Zheng, S., Qin, H., Cao, X., Yang, L., & Jiao, L. (2023). Interfacial engineering of Ni_3N/Mo_2N heterojunctions for urea-assisted hydrogen evolution reaction. *ACS Catal. 13*(7), 4091–4100. https://doi.org/10.1021/acscatal.3c00113

78 Barrios, A. M. & Lippard, S. J. (2000). Interaction of urea with a hydroxide-bridged dinuclear nickel center: An alternative model for the mechanism of urease. *J. Am. Chem. Soc. 122*(38), 9172–9177. https://doi.org/10.1021/ja000202v

79 Li, Z., Zheng, Y., Guo, H., Cheng, X., Huang, Y., Liu, C., Zang, J., & Dong, L. (2024). Ni-based electrocatalysts for urea oxidation reaction: Mechanistic insights and recent advancements. *J. Alloys Compd. 1008*, 176591. https://doi.org/10.1016/j.jallcom.2024.176591

80 Gao, X., Bai, X., Wang, P., Jiao, Y., Davey, K., Zheng, Y., & Qiao, S.-Z. (2023). Boosting urea electrooxidation on oxyanion-engineered nickel sites via inhibited water oxidation. *Nat. Commun. 14*(1), 5842. https://doi.org/10.1038/s41467-023-41588-w

81 Parvin, S., Aransiola, E., Ammar, M., Lee, S., Zhang, L., Weber, J., & Baltrusaitis, J. (2024). Tailored $Ni(OH)_2$/CuCo/$Ni(OH)_2$ composite interfaces for efficient and durable urea oxidation reaction. *ACS Appl. Mater. Interfaces 16*(49), 67715–67729. https://doi.org/10.1021/acsami.4c14041

82 Li, Q., Wang, J., Shi, Y., Li, H., Yang, H., Xiang, K., You, W., & Liang, J. (2025). Urea oxidation catalysts: A review on non-metallic enhancements in nickel-based electrocatalysts. *Mater. Horiz. 12*, 9952–9965. https://doi.org/10.1039/D5MH00919G

83 Cao, Q., Yuan, Y., Wang, K., Huang, W., Zhao, Y., Sun, X., Ding, R., Lin, W., Liu, E., & Gao, P. (2022). Phase and crystallinity regulations of $Ni(OH)_2$ by vanadium doping boost electrocatalytic urea oxidation reaction. *J. Colloid Interface Sci. 618*, 411–418. https://doi.org/10.1016/j.jcis.2022.03.054

84 Rao, R. R., Corby, S., Bucci, A., García-Tecedor, M., Mesa, C. A., Rossmeisl, J., Giménez, S., Lloret-Fillol, J., Stephens, I. E. L., & Durrant, J. R. (2022). Spectroelectrochemical analysis of the water oxidation mechanism on doped nickel oxides. *J. Am. Chem. Soc. 144*(17), 7622–7633. https://doi.org/10.1021/jacs.1c08152

85 Yang, H., Ge, L., Guan, J., Ouyang, B., Li, H., & Deng, Y. (2024). Synergistic engineering of heteroatom doping and heterointerface construction in V-doped $Ni(OH)_2$/FeOOH to boost both oxygen evolution and urea oxidation reactions. *J. Colloid Interface Sci. 653*, 721–729. https://doi.org/10.1016/j.jcis.2023.09.115

86 Zhao, Q., Gan, R., Ran, Y.-L., Ma, Q.-L., Chen, W.-W., Guo, Y.-X., Zhang, Y., & Wang, D.-S. (2024). Single-atom catalysts: Controlled synthesis and dynamic mechanism in electrochemical oxygen evolution substitution reactions. *Rare Met. 43*(10), 4903–4920. https://doi.org/10.1007/s12598-024-02786-7

87 Jiang, H., Xia, J., Jiao, L., Meng, X., Wang, P., Lee, C.-S., & Zhang, W. (2022). Ni single atoms anchored on N-doped carbon nanosheets as bifunctional electrocatalysts for urea-assisted rechargeable Zn-air batteries. *Appl. Catal. B: Environ. 310*, 121352. https://doi.org/10.1016/j.apcatb.2022.121352

88 Kumar, A., Liu, X., Lee, J., Debnath, B., Jadhav, A. R., Shao, X., Bui, V. Q., Hwang, Y., Liu, Y., Kim, M. G., & Lee, H. (2021). Discovering ultrahigh loading of single-metal-atoms via surface tensile-strain for unprecedented urea electrolysis. *Energy Environ. Sci. 14*(12), 6494–6505. https://doi.org/10.1039/D1EE02603H

89 Ashton, S. J. & Arenz, M. (2012). Comparative DEMS study on the electrochemical oxidation of carbon blacks. *J. Power Sources 217*, 392–399. https://doi.org/10.1016/j.jpowsour.2012.06.015

90 Fortunato, G. V., Kronka, M. S., Cardoso, E. S. F., dos Santos, A. J., Roveda, A. C., Jr., Lima, F. H. B., Ledendecker, M., Maia, G., & Lanza, M. R. V. (2022). A comprehensive comparison of oxygen and nitrogen functionalities in carbon and their implications for the oxygen reduction reaction. *J. Catal. 413*, 1034–1047. https://doi.org/10.1016/j.jcat.2022.08.001

91 Fortunato, G. V., Jung, D. C., Lourenço, J. C., Bhuyan, P., Choi, J. S., You, X., Lim, S., Melchionna, M., Sezen, H., Hofmann, J. P., Fornasiero, P., Lanza, M. R. V., & Ledendecker, M. (2025). Benchmarking the stability of state-of-the-art H_2O_2 electrocatalysts under acidic conditions. *ACS Catal. 15*(11), 8811–8821. https://doi.org/10.1021/acscatal.5c00868

92 Begildayeva, T., Theerthagiri, J., Lee, S. J., Min, A., Kim, G.-A., Manickam, S., & Choi, M. Y. (2024). Sustainable furfural biomass feedstocks electrooxidation toward value-added furoic acid with energy-saving H_2 fuel production using Pt-decorated Co_3O_4 nanospheres. *Energy Environ. Mater. 7*(2), e12563. https://doi.org/10.1002/eem2.12563

93 Fu, Q., Xie, W., Yang, L., Yan, L., & Zhao, X. (2024). Recent advances in electrocatalytic upgrading of biomass-derived furfural. *Chem. Eng. J. 485*, 150083. https://doi.org/10.1016/j.cej.2024.150083

94 Kubota, S. R. & Choi, K.-S. (2018). Electrochemical valorization of furfural to maleic acid. *ACS Sustain. Chem. Eng. 6*(8), 9596–9600. https://doi.org/10.1021/acssuschemeng.8b02698

95 Perez-Sena, W. Y., Tuppurainen, V., Gautier, A., Eränen, K., Ahola, J., Tanskanen, J., Salmi, T., Wärnå, J., & Murzin, D. Y. (2024). Biphasic furfural oxidation with hydrogen peroxide: process-level solvent screening and solvent effect on catalytic activity. *Ind. Eng. Chem. Res. 63*(36), 15809–15822. https://doi.org/10.1021/acs.iecr.4c02123

96 Bao, Y. Q., Du, Z. C., Liu, X. Y., Liu, H., Tang, J. S., Qin, C. R., Liang, C., Huang, C. X., & Yao, S. Q. (2024). Furfural production from lignocellulosic biomass: One-step and two-step strategies and techno-economic evaluation. *Green Chem. 26*(11), 6318–6338. https://doi.org/10.1039/d4gc00883a

97 Zhao, K., Wen, B., Tang, Q., Wang, F., Liu, X., Xu, Q., & Yin, D. (2024). Recent catalytic innovations in furfural transformation. *Green Chem. 26*(19), 9957–9992. https://doi.org/10.1039/D4GC01983K

98 Zhang, X., Xu, S., Li, Q., Zhou, G., & Xia, H. (2021). Recent advances in the conversion of furfural into bio-chemicals through chemo- and bio-catalysis. *RSC Adv. 11*(43), 27042–27058. https://doi.org/10.1039/D1RA04633K

99 Oyegoke, T., Dumeignil, F., Jibril, B. E. Y., Michel, C., & Wojcieszak, R. (2024). Exploring catalytic oxidation pathways of furfural and 5-hydroxymethyl furfural into carboxylic acids using Au, Pt, and Pd catalysts: A comprehensive review. *Catal. Sci. Technol. 14*(23), 6761–6774. https://doi.org/10.1039/D4CY00821A

100 Perez-Sena, W. Y., Paya, M., Eränen, K., Lassfolk, R., Lagerquist, L., Kumar, N., Aho, A., D'Angelo, A., Salmi, T., Wärnå, J., & Murzin, D. Y. (2024). Heterogeneous catalytic oxidation of furfural with hydrogen peroxide over a niobia catalyst. *Catal. Sci. Technol. 14*(7), 1942–1957. https://doi.org/10.1039/D3CY01686B

101 Bissot, T. C. & Benson, K. A. (1963). Oxidation of butane to maleic anhydride. *Ind. Eng. Chem. Prod. Res. Dev. 2*(1), 57–60. https://doi.org/10.1021/i360005a014

102 Thiyagarajan, S., Franciolus, D., Bisselink, R. J. M., Ewing, T. A., Boeriu, C. G., & van Haveren, J. (2020). Selective production of maleic acid from furfural via a cascade approach combining photochemistry and electro- or biochemistry. *ACS Sustain. Chem. Eng. 8*(29), 10626–10632. https://doi.org/10.1021/acssuschemeng.0c02833

103 Wu, H., Song, J., Liu, H., Xie, Z., Xie, C., Hu, Y., Huang, X., Hua, M., & Han, B. (2019). An electrocatalytic route for transformation of biomass-derived furfural into 5-hydroxy-2(5*H*)-furanone. *Chem. Sci. 10*(17), 4692–4698. https://doi.org/10.1039/c9sc00322c

104 Palai, Y. N., Shrotri, A., & Fukuoka, A. (2022). Selective oxidation of furfural to succinic acid over Lewis acidic Sn-beta. *ACS Catal. 12*(6), 3534–3542. https://doi.org/10.1021/acscatal.1c05348

105 Sohail, A. & Wattanakit, C. (2025). Electrocatalytic upgrading of furan derivatives. *Curr. Opin. Electrochem. 49*, 101628. https://doi.org/10.1016/j.coelec.2024.101628

106 Román, A. M., Hasse, J. C., Medlin, J. W., & Holewinski, A. (2019). Elucidating acidic electro-oxidation pathways of furfural on platinum. *ACS Catal. 9*(11), 10305–10316. https://doi.org/10.1021/acscatal.9b02656

107 Liu, X., Albloushi, M., Galvin, M., Schroeder, C. W., Wu, Y., & Li, W. (2024). A paired alkaline electrolyzer for furfural oxidation and hydrogen evolution over noble metal-free NiFe/Ni and Co/MXene catalysts. *Green Chem. 26*(22), 11351–11363. https://doi.org/10.1039/D4GC04447A

108 Román, A. M., Agrawal, N., Hasse, J. C., Janik, M. J., Medlin, J. W., & Holewinski, A. (2020). Electro-oxidation of furfural on gold is limited by furoate self-assembly. *J. Catal. 391*, 327–335. https://doi.org/10.1016/j.jcat.2020.08.034

109 Wu, Z., Liu, G., Song, Z., Hu, Y., Nie, T., & Song, Y.-F. (2025). Potential-dependent kinetics and reaction pathways of low-potential furfural electrooxidation with anodic H_2 production. *Small Sci. 5*(8), 2500132. https://doi.org/10.1002/smsc.202500132

110 Hasse, J. C., Agrawal, N., Janik, M. J., & Holewinski, A. (2022). ATR-SEIRAS investigation of the electro-oxidation mechanism of biomass-derived C_5 furanics on platinum electrodes. *J. Phys. Chem. C 126*(16), 7054–7065. https://doi.org/10.1021/acs.jpcc.2c01259

111 Thuriot-Roukos, J., Ferraz, C. P., Al Rawas, H. K., Heyte, S., Paul, S., Itabaiana, I., Jr, Pietrowski, M., Zieliński, M., Ghazzal, M. N., Dumeignil, F., & Wojcieszak, R. (2023). Supported gold catalysts for base-free furfural oxidation: the state of the art and machine-learning-enabled optimization. *Materials 16*(19), 6357. https://doi.org/10.3390/ma16196357

112 Wu, X., Guo, H., Jia, L., Xiao, Y., Hou, B., Gao, Y., & Li, D. (2023). Enhancement of Ru on furoic acid synthesis from furfural oxidation over $Ru_x/\delta\text{-}MnO_2$ catalysts. *Mol. Catal. 551*, 113604. https://doi.org/10.1016/j.mcat.2023.113604

113 Tran, V.-N., Ha, Q.-N., & Kuo, D.-H. (2025). Boosting furfural oxidation with CuWCe trimetal-doped Ni_3S_2 catalysts: Engineering in stabilizing the metastable phase for electrocatalysis. *J. Environ. Chem. Eng. 13*(6), 119372. https://doi.org/10.1016/j.jece.2025.119372

114 Yuan, X., Huang, X., Hu, M., Liu, J., Guo, W., Hao, Y., Li, Z., Zhang, K., & Han, B. (2024). Highly efficient electrocatalytic oxidation of furfural to maleic acid over MOF-derived $MnCeO_x$. Green Chem. *26*(9), 5365–5370. https://doi.org/10.1039/D4GC00408F

115 Chen, J. Y., Xiao, Y., Guo, F. S., Li, K.-M., Huang, Y. B., & Lu, Q. (2024). Single-atom metal catalysts for catalytic chemical conversion of biomass to chemicals and fuels. *ACS Catal. 14*(7), 5198–5226. https://doi.org/10.1021/acscatal.4c00208

116 De, S., Burange, A. S., & Luque, R. (2022). Conversion of biomass-derived feedstocks into value-added chemicals over single-atom catalysts. *Green Chem. 24*(6), 2267–2286. https://doi.org/10.1039/D1GC04285H

117 Vasconcelos, S. C., Marchini, L., Lima, C. G. S., Madriaga, V. G. C., Ribeiro, R. S. A., Rossa, V., Ferreira, L. E. M., de C. da Silva, F., Ferreira, V. F., Passos, F. B., Varma, R. S., Paixão, M. W., & Lima, T. M. (2022). Single-atom catalysts for the upgrading of biomass-derived molecules: An overview of their preparation, properties and applications. *Green Chem. 24*(7), 2722–2751. https://doi.org/10.1039/D1GC03809E

118 Liu, Z., Chen, X., Luo, M., Jiang, Q., Li, X., Yang, C., Zhang, Q., Ma, L., & Yan, L. (2025). Selective hydrogenation of furfural under mild conditions over single-atom $Pd_1/\alpha\text{-}MoC$ catalyst. *ChemSusChem 18*(7), e202401802. https://doi.org/10.1002/cssc.202401802

119 Gu, W., Pei, A., Zhang, S., Jiang, F., Jia, Y., Qin, Q., Du, R., Li, Z., Liu, R., Qiu, Y., Yan, K., Zhao, Y., Liang, C., & Chen, G. (2023). Atomic-interface effect of single-atom Ru/CoO_x for selective electrooxidation of 5-hydroxymethylfurfural. *ACS Appl. Mater. Interfaces 15*(23), 28036–28043. https://doi.org/10.1021/acsami.3c03430

120 Yang, S., Guo, Y., Zhao, P., Jiang, H., Shen, H., Chen, Z., Jiang, L., Xue, X., Zhang, Q., & Zhang, H. (2024). Unraveling the electrooxidation mechanism of 5-(hydroxymethyl)furfural at a molecular level via nickel-based two-dimensional metal–organic frameworks catalysts. *ACS Catal. 14*(1), 449–462. https://doi.org/10.1021/acscatal.3c04977

121 Zhou, Y., Slater, T. J. A., Luo, X., & Shen, Y. (2023). A versatile single-copper-atom electrocatalyst for biomass valorization. *Appl. Catal. B: Environ. 324*, 122218. https://doi.org/10.1016/j.apcatb.2022.122218

122 Cao, Y., Knijff, J., Delparish, A., d'Angelo, M. F. N., & Noël, T. (2021). A divergent paired electrochemical process for the conversion of furfural using a divided-cell flow microreactor. *ChemSusChem 14*(2), 590–594. https://doi.org/10.1002/cssc.202002833

123 Hu, X., Zhu, J., Li, J., & Wu, Q. (2020). Urea electrooxidation: Current development and understanding of Ni-based catalysts. *ChemElectroChem 7*(15), 3211–3228. https://doi.org/10.1002/celc.202000404

124 Chico-Mesa, L., Rodes, A., Arán-Ais, R. M., & Herrero, E. (2025). Insights into catalytic activity and selectivity of 5-hydroxymethylfurfural oxidation on gold single-crystal electrodes. *Nat. Commun. 16*(1), 3349. https://doi.org/10.1038/s41467-025-58696-4

125 Wang, H., Zheng, X., Fang, L., & Lu, S. (2023). Urea electrooxidation in alkaline environment: Fundamentals and applications. *ChemElectroChem 10*(13), e202300138. https://doi.org/10.1002/celc.202300138

126 Becker, H., Murawski, J., Shinde, D. V., Stephens, I. E. L., Hinds, G., & Smith, G. (2023). Impact of impurities on water electrolysis: A review. *Sustain. Energy Fuels 7*(7), 1565–1603. https://doi.org/10.1039/D2SE01517J

127 Sun, X. & Ding, R. (2020). Recent progress with electrocatalysts for urea electrolysis in alkaline media for energy-saving hydrogen production. *Catal. Sci. Technol. 10*(6), 1567–1581. https://doi.org/10.1039/C9CY02618E

128 Lee, S. A., Jun, S. E., Park, S. H., Kwon, K. C., Kang, J. H., Kwon, M. S., & Jang, H. W. (2024). Single atom catalysts for water electrolysis: From catalyst-coated substrate to catalyst-coated membrane. *EES Catal. 2*(1), 49–70. https://doi.org/10.1039/D3EY00165B

129 Li, W., Chen, Z., Jiang, X., Jiang, J., & Zhang, Y. (2022). Recent advances in the design of single-atom electrocatalysts by defect engineering. *Front. Chem. 10*, 1011597. https://doi.org/10.3389/fchem.2022.1011597

130 Calderón, J. C. V., Arora, J. S., & Mushrif, S. H. (2022). Mechanistic investigation into the formation of humins in acid-catalyzed biomass reactions. *ACS Omega 7*(49), 44786–44795. https://doi.org/10.1021/acsomega.2c04783

131 Shrestha, N. K., Patil, S. A., Salunke, A. S., Inamdar, A. I., & Im, H. (2023). Unprecedented urea oxidation on Zn@Ni-MOF with an ultra-high current density: understanding the competition between UOR and OER, catalytic activity limitation and reaction selectivity. *J. Mater. Chem. A 11*(27), 14870–14877. https://doi.org/10.1039/D3TA01962D

17

Non-oxide Electrocatalysts for Glucose-oxidation-coupled Water Electrolysis

P. H. Gangadharagowda[1], K. Chandana[2], S. K. Vinoth[3], K. Yogesh Kumar[4], and Arun Varghese[4,]*

[1]*Department of Chemistry, BGS Institute of Technology, Adichunchanagiri University, Mandya, India*
[2]*Department of Chemistry, Adichunchanagiri School of Natural Sciences, Adichunchanagiri University, Mandya, India*
[3]*Department of Chemistry, Jain University, Bengaluru, India*
[4]*Department of Sciences, Alliance School of Sciences, Alliance University, Central Campus, Bengaluru, India*

*Corresponding author: arunvarghese94@gmail.com

17.1 Introduction

The development of effective hydrogen production technology has accelerated due to the global search for clean and sustainable energy sources. With its high energy density and zero emissions, hydrogen has great potential to help solve the rising energy needs and environmental issues related to fossil fuels [1]. The simplicity of electrochemical water splitting and its potential for integration with renewable energy sources like solar and wind power make it stand out among other manufacturing techniques. However, there are a number of issues with traditional water electrolysis, mostly related to the anodic oxygen evolution reaction (OER). OER is a kinetically slow four-electron process that needs large overpotentials to reach respectable reaction rates. Furthermore, OER produces no value-added products, which reduces water electrolysis's economic appeal [2].

Using alternate oxidation processes that are thermodynamically more advantageous and able to generate value-added compounds in place of the anodic OER has become a popular tactic in recent years to get around these restrictions. The oxidation of glucose is one such potential method, as glucose, being a plentiful carbohydrate, can be easily produced from agricultural waste streams and biomass. Using glucose as anodic feedstock has two advantages: it lowers the total energy input needed for electrolysis while also yielding useful organic intermediates or products such as formic acid, gluconic acid, and glucaric acid. The initial oxidation of the aldehyde group to produce gluconic acid is one of the several probable mechanisms by which the glucose oxidation reaction (GOR) proceeds. Subsequent oxidation processes produce smaller carboxylic acids and other intermediates [3]. The general anodic half-reaction for glucose oxidation is represented in Eq. (17.1):

$$C_6H_6O_6 + H_2O \longrightarrow C_6H_6O_7 + 2H^+ + 2e^- \tag{17.1}$$

Compared to OER, which needs an overpotential to overcome the creation of strong oxygen–oxygen bonds, this first oxidation step is energetically more advantageous. The energy efficiency of hydrogen production is increased because the decreased energy barrier results in a lower total cell voltage. The creation of extremely stable and active electrocatalysts that can effectively support the GOR is an essential part of this coupled system [4]. Although oxide-based materials have historically dominated anodic oxidation catalysis, they frequently have poor selectivity, limited conductivity, and are vulnerable to deactivation because of structural deterioration or intermediate poisoning. As a result, non-oxide electrocatalysts are becoming more and more popular. Non-oxide electrocatalysts, such as carbides, phosphides, nitrides, and transition-metal chalcogenides, have shown exceptional catalytic qualities because of their high electrical conductivity, adjustable electronic structures, and large number of active sites. They are also appealing for glucose oxidation applications due to their strong chemical stability and resistance to surface passivation. Transition-metal phosphides, for instance, can speed up reaction kinetics by facilitating effective electron transfer and improving the adsorption of glucose and its intermediates. Likewise, metal carbides and nitrides frequently perform "platinum-like" functions, making them affordable substitutes for noble metals [5].

On non-oxide surfaces, the GOR usually entails a number of adsorption and dehydrogenation processes. Through the aldehyde or hydroxyl groups, the glucose molecule initially adsorbs onto the catalytic surface. Subsequent electron and proton exchanges ensue, resulting in the production of gluconic acid and other oxidation products. In order to stabilize chemical intermediates and reduce activation barriers at each stage, the catalyst surface is essential. Researchers want to increase the overall efficiency of the process as well as its selectivity for desired products by manipulating the surface structure and electrical characteristics of non-oxide materials. Moreover, decentralized hydrogen production employing renewable feedstocks is made possible by combining glucose oxidation with the cathodic hydrogen evolution process (HER [hydrogen evolution reaction]) [6]. By using glucose obtained from biomass, this co-electrolysis method not only addresses energy sustainability but also aids in waste valorization, which is consistent with the ideas of green chemistry and circular economy [7]. This chapter explores the latest developments in non-oxide electrocatalysts made especially for water electrolysis coupled with glucose oxidation. This chapter attempts to give a thorough grasp of how non-oxide electrocatalysts can open up new opportunities for clean and affordable hydrogen generation by combining recent developments and describing future approaches.

17.2 Importance of Glucose-oxidation-coupled Water Electrolysis

The pressing need to cut greenhouse gas emissions and switch to renewable energy sources has accelerated the global quest for clean and sustainable hydrogen generation technology in recent years. Water electrolysis is one of the most promising hydrogen generation techniques since it uses electricity to directly make ultrapure hydrogen from water [8]. However, the slow OER at the anode essentially limits this process. The energy consumption of the entire electrolytic process is greatly increased by the OER, a complicated, multistep reaction that necessitates a high overpotential. Furthermore, the anodic by-product of conventional water electrolysis, oxygen, has a low market value, which reduces the process's appeal from an economic standpoint [9]. A convincing remedy for these drawbacks is provided by the novel idea of combining water electrolysis and glucose oxidation. Researchers have shown the ability to lower energy usage while producing more value by substituting the GOR at the anode for the traditional OER. Thermodynamically, OER (1.23 V vs. RHE) produces substantially higher overall cell voltages than glucose oxidation to gluconic acid, which occurs at a much lower potential (around 0.05 V vs. RHE). One of the main financial obstacles to large-scale hydrogen generation is removed by this voltage reduction, which directly results in less electricity being used and increased energy efficiency [10].

In addition to the energy benefits, glucose oxidation produces useful organic compounds such as formic acid, gluconic acid, and glucaric acid. These substances are widely used in the chemical, textile, pharmaceutical, and food preservation industries, which boosts the process's economic feasibility and generates new revenue streams. Furthermore, cellulose hydrolysates and agricultural leftovers are two examples of renewable biomass sources that provide a plentiful supply of glucose. By converting waste or low-value biomass into high-value chemicals and green hydrogen, the use of these resources encourages biomass valorization and is consistent with the concepts of the circular economy. This method shows the potential of incorporating green chemistry techniques into energy production while also supporting the Sustainable Development Goals (SDGs) [11, 12].

The operational benefit to the anodic catalysts is another significant advantage of glucose-coupled electrolysis. The possibility of catalyst corrosion and degradation, which are frequent issues in traditional OER systems functioning under severe oxidative conditions, is decreased by the reduced anodic potential needed for glucose oxidation. Longer catalyst lifespans, reduced maintenance costs, and increased system durability are the results of this development. Additionally, the method provides flexibility for modular decentralized hydrogen production, especially in areas with abundant biomass. Local production of hydrogen and valuable chemicals is made possible by this decentralization, which lessens dependency on centralized infrastructures and improves energy accessibility for isolated or rural populations [13]. Glucose-coupled electrolysis makes a substantial contribution to mitigating climate change and promoting a sustainable future by utilizing renewable carbon sources and generating hydrogen without emitting CO_2 [14]. A proven technique for producing hydrogen is water electrolysis, which uses an electric current to divide water molecules into hydrogen and oxygen gases. The HER at the cathode and the OER at the anode are the two half-reactions that are usually involved in the process. Because of its intricate four-electron-transfer process and requirement for the formation of an O—O bond, the OER is inherently slow. To drive this reaction at feasible rates, a high overpotential, extra energy over the thermodynamic potential, is needed, greatly raising the electrolysis process's overall energy consumption [15].

Additionally, OER just produces oxygen gas, which is frequently vented and therefore not economically advantageous, without producing any valuable by-products. For the majority of electrode materials, the severe anodic conditions also present stability issues, which eventually cause corrosion and degradation. Together, these elements reduce the economic appeal of conventional water electrolysis, especially because it is fueled by pricey or sporadic renewable energy sources. In order to increase the overall energy efficiency and economic feasibility of electrolytic hydrogen production, researchers have been actively investigating alternate anodic reactions that are thermodynamically favorable and yield useful products. Among these, the oxidation of organics produced from biomass, such as glucose, has shown promise [16]. An attractive substitute for the OER is the glucose oxidation process. Glucose oxidation can effectively cut the total cell voltage required for hydrogen production because it requires a far lower potential thermodynamically than OER. For example, the typical potential for the oxidation of glucose to gluconic acid is approximately 0.05 V, which is significantly lower than 1.23 V for OER [17]. The process is more energy efficient overall because the decrease in anodic potential directly results in a lower energy input. Kinetically, there are several processes involved in the oxidation of glucose, and the catalyst must carefully facilitate each one. Usually, the reaction begins with glucose adhering to the catalyst surface, then its aldehyde group dehydrogenates to produce gluconolactone, which hydrolyzes to gluconic acid. Further oxidation can lead to deeper degradation products such as glucaric acid, formic acid, or even CO_2, depending on reaction conditions and catalyst properties [18]. The general representation of glucose oxidation coupled with water electrolysis is given in Figure 17.1. Controlling the reaction route to optimize selectivity toward desired value-added products while preserving high reaction rates is a significant problem in glucose oxidation kinetics. Because of their many active sites and adjustable electronic architectures, non-oxide electrocatalysts have shown excellent kinetic performance, allowing for better glucose conversion efficiencies at lower overpotentials. Transition-metal phosphides and nitrides are examples of non-oxide electrocatalysts that can alter the adsorption energy of glucose and intermediates, affecting the efficiency and selectivity of the reaction. To further refine these pathways, surface engineering techniques such as facet exposure, defect formation, and doping have been used [19].

Compared to conventional OER-based systems, the combination of glucose oxidation with water electrolysis offers significant benefits, particularly in terms of total efficiency and energy savings. Depending on catalyst design and system configuration, the necessary cell voltage for hydrogen production can be lowered by 0.4–0.6 V by replacing the slow OER with glucose oxidation. Because of this significant voltage drop, less electricity is used per kilogram of hydrogen produced,

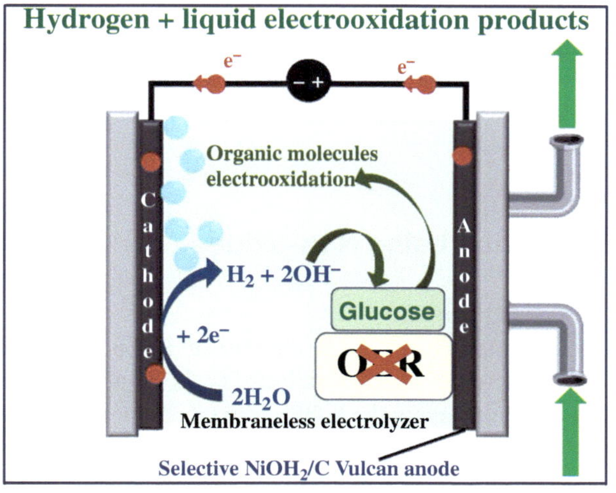

Figure 17.1 General representation of glucose oxidation coupled with water electrolysis. *Source*: [17] with permission of Elsevier.

increasing the process's economic appeal and environmental sustainability. This decreased energy requirement, when driven by renewable energy sources, significantly improves hydrogen production's environmental credentials and aids in the global push to cut carbon emissions and switch to clean energy technology [20]. In addition to being energy efficient, glucose oxidation makes it possible to produce useful organic compounds rather than low-value oxygen gas. Additional revenue streams are provided by goods like gluconic acid, which is utilized in the food and pharmaceutical industries, and glucaric acid, which is ranked among the best bio-based chemicals. Other substances, such as glycolic and formic acids, are used as chemical feedstocks and preservatives. Utilizing glucose obtained from biomass further advances the ideas of the circular economy by converting industrial or agricultural waste into clean hydrogen and usable compounds. The system stability is increased, maintenance expenses are decreased, and catalyst longevity is increased by the gentler anodic conditions. All things considered, this integrated strategy is a major step toward the generation of hydrogen and chemicals in a sustainable and profitable manner [21].

By turning waste into high-value products and green hydrogen, using glucose made from biomass promotes a circular economy and is consistent with SDGs. The economic and environmental viability of electrolytic hydrogen generation is greatly increased by this dual advantage. In addition to producing beneficial by-products and saving energy, glucose-coupled electrolysis increases catalyst stability by reducing corrosion and extending system lifetime under milder anodic conditions. Additionally, this method enables decentralized, modular hydrogen generation, which is especially helpful in areas with abundant biomass supplies and gives local groups the ability to produce their own energy [22]. Furthermore, managing agricultural waste and lowering greenhouse gas emissions are two further benefits of incorporating renewable carbon sources into hydrogen generation, which supports larger climate mitigation initiatives. All things considered, glucose-coupled water electrolysis is a paradigm shift that combines biomass valorization and sustainable energy conversion, opening the door to a more environmentally friendly, circular, and profitable hydrogen economy.

17.3 Performance Evaluation Using Non-oxide Electrocatalysts

Because of their distinct electronic structures, large number of active sites, and adjustable surface chemistries, non-oxide electrocatalysts, such as phosphides, borides, carbides, and metal–organic framework (MOF)-derived composites, have demonstrated remarkable promise in catalyzing GOR. These cutting-edge materials outperform conventional oxide-based catalysts by enabling better charge-transfer kinetics and increased selectivity for glucose oxidation pathways. The current developments in non-oxide electrocatalysts for glucose-oxidation-assisted water electrolysis, emphasizing important performance indicators, mechanistic understandings, and their revolutionary potential for biomass valorization and green hydrogen production, are discussed in detail.

Zhang et al. developed a novel method of producing hydrogen using a copper-based redox pair (Cu[I]/Cu[II]) with glucose assistance. In this configuration, the initial catalyst is a layer of copper hydroxide that is formed immediately on top of copper foam. Glucose is transformed into gluconic acid, and $Cu(OH)_2$ is reduced to Cu_2O when voltage is added. A continuous cycle between Cu(II) and Cu(I) is produced concurrently by the reoxidation of CuO back into $Cu(OH)_2$. Because of this continuous cycle, glucose can continue to be oxidized at the anode to gluconic acid. The rate-determining step is the initial stage of this reaction, according to theoretical calculations. The process is more efficient when the energy barrier is lowered, which is achieved by employing the $Cu(OH)_2$ electrode [23].

A sandwich-shaped CoFe-layered double hydroxide (CoFe-LDH) was built on ferric foam by Siwal et al. [24]. The homogeneous dispersion of Au is made possible by LDH's enormous surface area, increasing the number of active sites available to absorb OH^-. Higher intrinsic activity can result from the efficient adjustment of the electronic structure of Au by the CoFe-LDH. Additionally, intercalated Au can increase the electrocatalyst's conductivity [24]. The glucose-coupled water electrolysis was carried out by Liu et al., which makes use of the Co_3S_4/MnS p–p heterojunction catalyst. This technique demonstrates a reduction in the energy required for water splitting by replacing the high-energy OER at the anode with the GOR. For GOR, in an electrolyte with 0.3 M glucose and 1 M KOH, the Co_3S_4/MnS catalyst needs a potential of 1.40 V vs. RHE to reach a current density of 50 $mA{\cdot}cm^{-2}$. This is better than separate MnS (1.50 V) and Co_3S_4 (1.47 V) catalysts. The Co_3S_4/MnS catalyst requires just 1.63 V to generate a current density of 50 $mA{\cdot}cm^{-2}$ in 1 M KOH supplemented with 0.3 M glucose when used in a two-electrode system for total water splitting. The Co_3S_4/MnS catalyst outperforms other studied catalysts such as Co_3S_4, MnS, and NF in HER, exhibiting an overpotential of 281 mV at 50 mA·cm^{-2} in the glucose-containing electrolyte. Although electrolyte conductivity may be somewhat impacted by glucose, the Co_3S_4/MnS catalyst exhibits exceptional stability, continuing to function in glucose-containing solutions for up to 50 hours. This initial investigation of water splitting for hydrogen production aided by glucose oxidation provides a workable plan for effective bifunctional electrocatalysts [25].

When compared to OER, the designed porous Co–Ni alloy on carbon cloth (CC) shows considerably higher GOR activity (Figure 17.2). The thermodynamic potential of water oxidation (1.23 V vs. RHE) is even higher than the onset potentials for GOR (1.096 V vs. RHE for Co–Ni on CC). This indicates that, particularly in the low-potential zone, GOR requires substantially lower potentials than OER. The potential for Co–Ni/CC in GOR is 1.244 V vs. RHE at a high current density of 100 mA·cm^{-2}, which is significantly lower than its potential for OER at 1.586 V vs. RHE. Furthermore, the Co–Ni alloy maintains a very low Tafel slope (88 mV·dec^{-1}) for GOR, indicating improved GOR kinetics even in a high-viscosity glucose solution. The HER activity of porous Co–Ni alloy is essentially unaffected by the presence of glucose. Regardless of the presence or absence of glucose, the overpotential at 10 mA·cm^{-2} stays at 104 mV in alkaline conditions. This suggests that, although HER activity is generally unaffected, the anodic current density differential between GOR and OER determines the total efficiency improvement in glucose electrolysis. In a two-electrode system, the cell voltage to drive a current density of 10 mA·cm^{-2} is 1.39 V in 1 M KOH and 0.1 M glucose electrolyte, which is significantly lower than for water electrolysis (1.65 V). When combined with GOR, the Co–Ni alloy on CC exhibits superior overall water-splitting activity. It is determined that the glucose electrolysis efficiency for Co–Ni/CC at 1.5 V is 94.6%. In long-term chronopotentiometry testing, the Co–Ni alloy demonstrates exceptional activity and endurance as well. After 12 hours of operation, the cell voltage just slightly increases, remaining

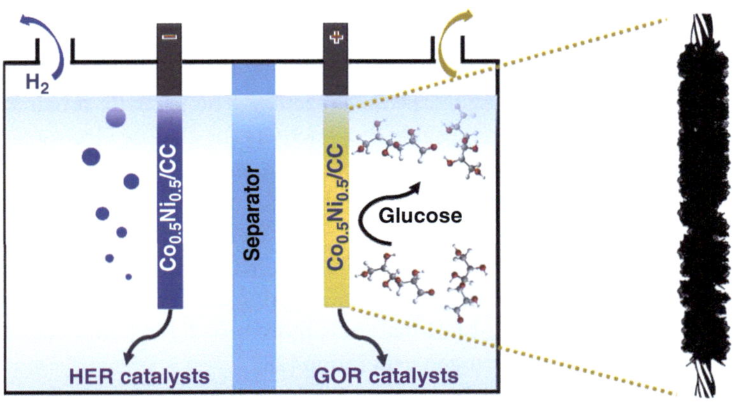

Figure 17.2 Co–Ni/CC catalyst used for glucose oxidation coupled with water electrolysis. *Source*: [26] with permission of Elsevier.

at 1.61 V. This study showed that employing an engineered porous Co–Ni alloy on CC as a bifunctional electrocatalyst to replace the energy-intensive OER with GOR considerably lowers the cell voltage needed for hydrogen production and presents a promising path toward energy-saving and value-added electrocatalysis [26].

Because of its exceptional electrochemical characteristics, the Co–Ni MOF offers an intriguing material platform for studying the influence of glucose oxidation in water electrolysis. Despite not being specifically used in this investigation to couple GOR and OER in a single integrated system, the Co–Ni MOF demonstrates remarkable performance in both reactions independently, highlighting its adaptability. In alkaline environments, the Co–Ni MOF exhibits improved kinetics, low charge-transfer resistance, and outstanding structural stability for water oxidation, achieving a low overpotential of 348 mV at 20 $mA{\cdot}cm^{-2}$. Its distinct layered structure with nano-cutting channels boosts active surface area and makes ion and electron transport more effective. High catalytic activity and long-term durability are guaranteed by the bimetallic synergy between nickel and cobalt, which makes it a promising option for OER applications in water splitting.

Concurrently, the same Co–Ni MOF has exceptional nonenzymatic glucose sensing capabilities through the GOR, exhibiting a broad linear range and a surprisingly low detection limit. The same structural and compositional characteristics that propel its OER performance, like layered morphology, accessible active sites, and the adaptable redox behavior of Ni and Co centers, are responsible for its glucose oxidation. These dual functions theoretically suggest that Co–Ni MOF may be used in future systems where glucose oxidation and water electrolysis are directly connected, providing simultaneous hydrogen production and glucose valorization. Even though this work showed these reactions separately, it established the foundation for creating integrated devices that take advantage of coupling, which is a creative step toward more multipurpose and sustainable electrochemical energy systems [27].

In another study, nickel/iron-layered double hydroxide (NiFe-LDH) has become a very appealing electrocatalyst for both OER and GOR. NiFe-LDH, the actual active phase that reduces thermodynamic barriers and enhances catalytic performance, is created by partially reconstructing the surface of the NiFe-LDH/NF-W electrode, which is created via an ultrasonic-assisted Fenton reaction. The partially repaired NiFe-LDH/NF-W electrode exhibits noticeably more activity for glucose electrooxidation than untreated counterparts, necessitating lower potentials for equivalent current densities. For instance, using glucose to achieve 10 $mA{\cdot}cm^{-2}$ requires only 1.365 V, 110 mV less than water oxidation, at 50 $mA{\cdot}cm^{-2}$. These potential decreases demonstrate that combining glucose electrooxidation with HER to produce hydrogen more efficiently is feasible. In addition to lowering energy expenses, this approach makes it easier to use NiFe-LDH materials more broadly in value-added organic synthesis and sustainable hydrogen generation [28].

Another study focused on synthesizing hollow RuV–$Co(OH)_2$ arrays as bifunctional electrocatalysts for both HER and GOR. ZIF-67, a sacrificial template co-doped with ruthenium and vanadium (V), was used to hydrothermally generate the hollow RuV–$Co(OH)_2$ arrays. Ru accelerated ZIF-67 etching, creating a distinctive hollow structure, while $NaVO_4$ was essential in the formation of the $Co(OH)_2$ phase. SEM, TEM, and XPS structural investigations verified the elements' uniform distribution and successful production. Excellent GOR performance was demonstrated by electrochemical tests, where RuV–$Co(OH)_2$ required just 1.24 V to reach a 10 $mA{\cdot}cm^{-2}$ current density, which is less than that of traditional catalysts. Its effectiveness was further validated by optimizing glucose content, and a hybrid electrolyzer built with this catalyst had a lower battery voltage of 1.42 V at 10 $mA{\cdot}cm^{-2}$, which is 240 mV lower than that of conventional water splitting (Figure 17.3). Strong mechanical and chemical stability were shown by the remarkable durability of RuV–$Co(OH)_2$, which included stable current densities for more than 30 hours without structural deterioration. Analysis following GOR verified that the crystalline integrity and hollow structure were retained. Ru and V doping increased the electronic structure by increasing the density of the active site and optimizing electron transport between Co and V, which improved catalytic efficiency, according to density functional theory (DFT) simulations. Following GOR, XPS examination revealed greater —OH/M—O ratios, indicating that glucose increases oxidation and OH adsorption. RuV–$Co(OH)_2$ is a highly effective bifunctional catalyst that provides a potential, energy-efficient pathway for the synthesis of hydrogen, along with useful by-products like gluconic and glucaric acids, thanks to the synergistic interaction of Ru and V [29].

A recently developed bifunctional electrocatalyst, hierarchical Fe-doped Ni_2P nanosheets coupled with carbon on nickel foam (Fe–Ni_2P@C/NF), is essential to this approach. This catalyst exhibits great efficiency in the monosaccharide oxidation reaction at the anode and the HER at the cathode. The technology significantly lowers the necessary cell voltages by using renewable biomass intermediates rather than water oxidation. For instance, better kinetics were demonstrated when 20 mM arabinose was present because only 1.55 V was required to reach 100 $mA{\cdot}cm^{-2}$, which is much lower than 1.66 V required for pure water splitting. The stability of the catalyst and efficacy were demonstrated by its remarkable monosaccharide conversion rates of over 99% and its outstanding faradaic efficiency of 98.2% for hydrogen generation. In situ surface oxidation, which produces active nickel oxides

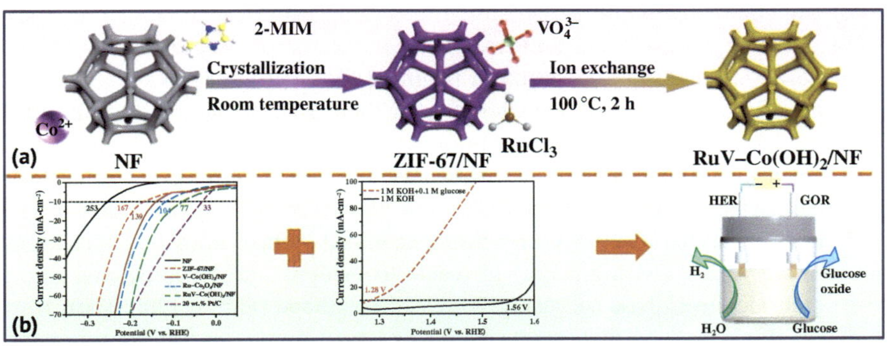

Figure 17.3 Results of glucose oxidation coupled with water electrolysis using RuV–$Co(OH)_2$ electrocatalyst. *Source*: [29] with permission of Elsevier.

or oxyhydroxides, is thought to be responsible for its endurance. Additionally, the carbon layer produced by the first synthesis, which involves soaking glucose, increases surface area and conductivity, thereby improving catalytic performance. This approach opens the door for the sustainable manufacture of important organic compounds in conjunction with energy-efficient hydrogen production [19].

The work by Li et al. focused on cobalt nanoparticles embedded in nitrogen-doped porous carbon (Co@NPC), a bifunctional electrocatalyst. This catalyst, which is made utilizing a biomass-based deep eutectic solvent (DES) and a straightforward, environmentally friendly pyrolysis technique, takes advantage of DES's dual functions as a structural template and a supply of nitrogen. The technique follows template removal procedures, which frequently result in material degradation and environmental problems. Co@NPC-800, the ideal catalyst synthesized at 800 °C, performed exceptionally well. It needed just 1.56 V to reach 10 mA·cm^{-2} in an integrated two-electrode glucose electrolytic cell, which is 180 mV less than traditional water splitting. The anodic potential was further lowered to 1.46 V by the presence of glucose, confirming the kinetic benefits of GOR over OER and improving system energy efficiency. Following 15 hours of glucose oxidation, product analysis showed that the main products were formic acid and lactic acid, with trace levels of gluconic and glucaric acids. The alkaline reaction media are responsible for the propensity for lactic acid generation. The synergistic effects of cobalt and nitrogen doping were demonstrated by DFT calculations, which revealed that pyridinic-N–Co sites promote HER by facilitating hydrogen adsorption, while graphitic-N–Co sites improve glucose adsorption for GOR. This study offered a low-cost, effective, and sustainable way to raise the energy conversion efficiency of electrolytic cells by combining the production of hydrogen with the value-added conversion of glucose [30].

Iron phosphide films (Fe_2P) produced in situ on stainless-steel mesh (SSM) were studied by Du et al. as effective electrocatalysts for the GOR in conjunction with hydrogen production. GOR lowers energy consumption by substituting the slow OER with the added advantage of generating valuable by-products like gluconolactone. A simple phosphidation procedure was used to prepare the Fe_2P/SSM catalyst, which produces strong nanorod structures embedded in the SSM that improve electrical conductivity and surface roughness. The successful generation and stability of the Fe_2P phase were confirmed by characterizations using XRD, SEM, and XPS. Fe_2P/SSM showed a lower onset potential for GOR (1.33 V vs. RHE) in the presence of glucose, suggesting better kinetics than OER and surpassing a number of previously published catalysts. Additionally, the Fe_2P/SSM catalyst demonstrated a greater electrocatalytically active surface area and demonstrated exceptional durability, functioning consistently at 10 mA·cm^{-2} for at least 24 hours. The configuration demonstrated a 300 mV improvement in cell voltage from 1.52 V (without glucose) to 1.22 V (with glucose) to reach 10 mA·cm^{-2} when integrated into a two-electrode electrolytic cell using Pt/C as the cathode (Figure 17.4). With 100% faradaic efficiency, this suggests increased hydrogen output and improved energy efficiency. Mechanical robustness was confirmed by post-electrolysis investigation, which revealed no catalyst detachment. All things considered, substituting GOR

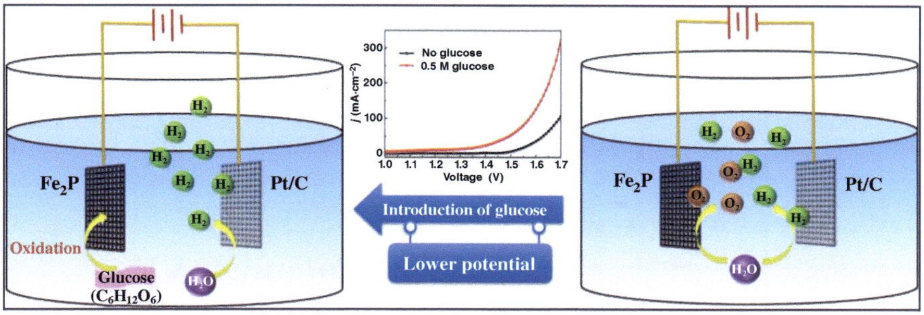

Figure 17.4 Representation of glucose oxidation coupled with water electrolysis using Fe_2P/SSM catalyst. *Source*: [31] with permission of Elsevier.

for OER using Fe_2P/SSM offers a viable, energy-efficient substitute for large-scale hydrogen production while repurposing waste streams high in glucose, hence facilitating both practical and industrial uses [31].

Another group of researchers created an ultrathin cobalt nickel hydroxide nanosheet ($CoNi(OH)_2$ NS) that was produced in situ on conductive CC as the electrocatalyst for GOR in order to implement glucose oxidation coupled with water electrolysis. The catalyst makes use of the superior electrocatalytic qualities of α-$Ni(OH)_2$, which are further improved by cobalt doping to increase stability and activity. The two-dimensional structure, meso-macro-porosity, and large surface area (406.6 $m^2 \cdot g^{-1}$) of the nanosheet maximize active-site exposure by facilitating effective ion and reactant transport. According to electrochemical studies, the cell voltage required to reach 100 $mA \cdot cm^{-2}$ decreased to 1.56 V when glucose was present, which is significantly less than 1.75 V required when glucose was absent. With glucose, the voltage was just 1.22 V at 10 $mA \cdot cm^{-2}$, suggesting a notable energy-saving benefit. The enhanced performance is ascribed to the synergistic effects of cobalt doping and micro–nano-structure engineering, which boost the production of Ni(III) active sites and increase electrochemical surface area. Higher Ni^{3+} concentration, the primary active species for GOR, was confirmed by XPS analysis, and cobalt doping stabilized the structure and stopped it from changing into the less active β-$Ni(OH)_2$ phase. Over a 12-hour period, the catalyst demonstrated constant voltage profiles and little metal dissolution, maintaining steady functioning. This study showed that employing tailored $CoNi(OH)_2$ NS to replace OER with GOR not only lowers energy input but also makes it possible to produce hydrogen in a safe, effective, and value-added manner, offering a viable path toward affordable and sustainable green hydrogen generation [32].

A bifunctional electrocatalyst called CoFe-MOF-74 integrated with multiwalled carbon nanotubes (CoFe-MOF-74@MWCNTs) was created in order to analyze the importance of glucose in water electrolysis. The overall charge-transfer efficiency and reactivity of the catalyst were improved by the synergistic interaction of CoFe-MOF's many active sites with MWCNTs' superior conductivity. The voltage required to reach 10 $mA \cdot cm^{-2}$ current density is 1.291 V (vs. RHE) (Figure 17.5) using glucose, according to experimental results. This is a remarkable 212 mV

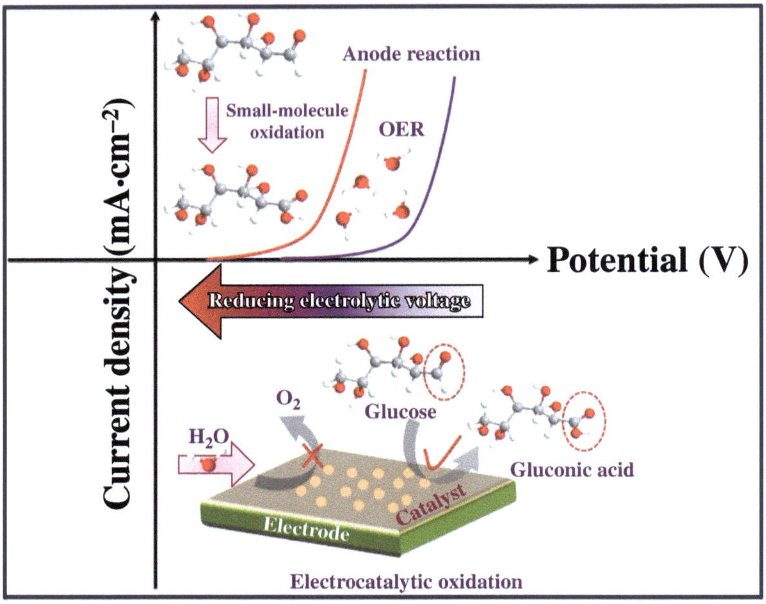

Figure 17.5 Representation of glucose oxidation coupled with water electrolysis using CoFe-MOF-74@MWCNTs catalyst. *Source*: [33] with permission of Elsevier.

decrease when compared to OER alone. This demonstrated the exclusive oxidation of glucose and demonstrated that GOR efficiently replaces OER, resulting in the CoFe-MOF-74@MWCNTs catalyst, significant energy savings, while avoiding gas bubble production at the anode. Strong glucose adsorption was revealed by DFT simulations, which also identified the dehydrogenation process as the rate-determining step. Together, the theoretical and experimental results demonstrated that CoFe-MOF-74@MWCNTs is a very promising multifunctional electrocatalyst for combining the oxidation of biomass molecules with the production of hydrogen, providing a workable plan to improve the sustainability and efficiency of electrolytic hydrogen generation [33].

In order to continuously monitor glucose levels in hypoxic conditions, another study presented an oxygen-independent, nonenzymatic glucose sensor that uses water-splitting-assisted electrocatalysis (WSE). The WSE sensor uses HER to eliminate the need for oxygen, in contrast to traditional enzyme-based sensors that rely on oxygen and produce hydrogen peroxide, which compromises accuracy and safety. In order to induce HER, a pretreatment potential of −2 V is provided to the Au electrode. This results in the production of hydroxide ions and the creation of localized alkaline conditions that are perfect for the oxidation of glucose. Even in tissues with low oxygen levels, dependable performance is ensured by subsequent detection at 0.20 V, which measures the anodic current without oxygen interference. The electrode is cleaned one last time at 1 V. The sensitivity, dependability, and biocompatibility of the sensor for both in-vitro and in-vivo applications were validated by demonstrations [34].

In order to stabilize glucose and guarantee precise readings, researchers utilized an Ni-based anodic catalyst in a low-concentration NaOH electrolyte (0.1 M) and an H-cell configuration to separate the anodic and cathodic compartments. The findings demonstrated that OER was gradually inhibited by increasing glucose concentration. At 40 mM glucose, GOR completely replaced OER, and no O_2 was observed, even at high applied potentials. This suggests a "membraneless electrolyzer region" for optimal performance, where high glucose levels promote GOR over OER. The study also looked at the onset potentials of O_2 production at different glucose concentrations, and it was shown that at 40 mM glucose, greater glucose levels completely suppress OER and raise onset potentials. After extended electrolysis, formic, lactic, and gluconic acids were identified by NMR analysis as the primary anodic result. The strategy's combined benefits of lower hydrogen production costs and the production of high-value by-products were proven by formic acid, which is acknowledged as a valuable chemical feedstock and an effective liquid hydrogen transporter. The study demonstrated a combination strategy that defines ideal conditions for membraneless electrolysis utilizing electrochemical data and in situ oxygen monitoring. These results provided a path for further research aimed at combining the oxidation of organic molecules with the synthesis of industrially significant compounds and green hydrogen [35].

In a different work by Wang et al., commercial cobalt–nickel foam (CNF) is modified using an ultrafast plasma treatment to control the high-valent Co^{3+} and Ni^{3+}, resulting in more active sites, quicker reaction kinetics, and an improved response to glucose. Through glucose electrooxidation and water splitting, the overpotential of the plasma-treated CNF at 10 $mA{\cdot}cm^{-2}$ was lowered to 133 mV in comparison to the OER [36]. Another work by Shrestha et al. demonstrated an approach to increase the efficiency of water electrolysis by substituting the slow OER with the GOR through the use of a Cu-doped MIL-88B(Fe) catalyst supported on NF. In addition to producing value-added chemicals like gluconic acid and enabling green hydrogen generation at reduced energy costs, this strategy makes use of glucose, a molecule obtained from renewable biomass, to get around the high energy barrier of OER. With NF acting only as a conductive substrate, the Cu-doped MIL-88B(Fe) catalyst was created as an ink and drop-cast over it to create thin, nanoporous films. The promise of glucose-assisted water splitting to save money and energy is confirmed by the improved GOR performance prior to OER at lower potentials. According to this study, the electronic structural modification of MOF is responsible for the remarkable catalytic performance. XPS examination revealed Fe(III) and Cu(II) oxidation states, indicating that Cu doping changes the valence state of the Fe nodes. Due to its greater electronegative nature, Cu partially oxidizes Fe by removing its electrons, which causes electron delocalization. Nyquist plots demonstrating decreased charge-transfer resistance confirmed the fact that this electron rearrangement enhances catalytic site activity and speeds up charge transfer. The Cu-doped MIL-88B/NF electrode demonstrated its promise for effective, sustainable hydrogen production by achieving a low GOR potential (1.35 V vs. RHE at 10 $mA{\cdot}cm^{-2}$), a high current density (350 $mA{\cdot}cm^{-2}$ at 1.48 V), and exceptional durability over 50 hours [37]. In the case of GOR at 1.83 V, Ni–P nanowire arrays

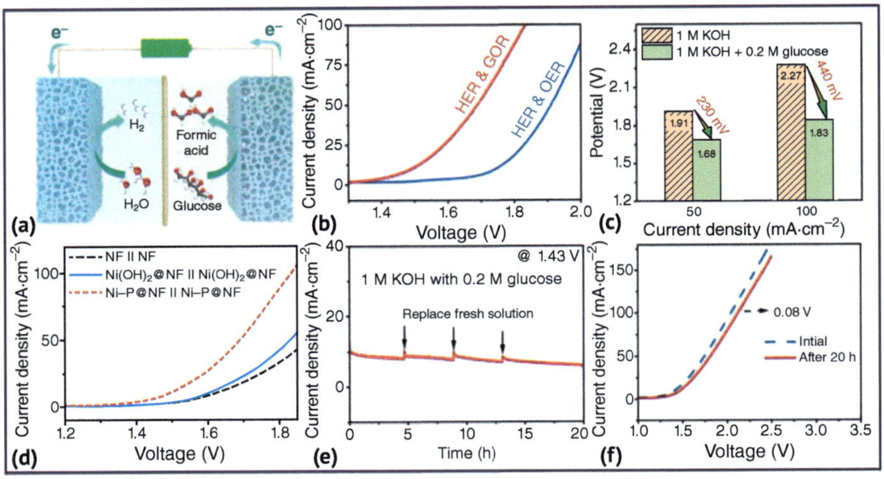

Figure 17.6 Results of glucose oxidation coupled with water electrolysis using Ni–P@NF catalyst. *Source*: [38] with permission of The Royal Society of Chemistry.

grown on NF (Ni–P@NF) attain a current density of 100 mA·cm^{-2}. The actual active species, NiOOH, is partially formed during operation by Ni–P. When combined with hydrogen evolution, this system shows long-term stability in alkaline conditions and operates at a significantly lower voltage than usual water-splitting voltages, 1.43 V for 10 mA·cm^{-2} (Figure 17.6) [38].

Due to the synergistic interactions between Ru and Ni borides, an Ru–Ni boride hybrid was synthesized, which performs exceptionally well in GOR, OER, and HER, requiring just 1.24 V for 10 mA·cm^{-2} glucose electrolysis (Figure 17.7) [39]. Another example is the remarkable HER activity of Ru–Co–Mn alloy nanoparticles on N-doped carbon (RuCoMn@NC), which outperforms commercial Pt/C catalysts in alkaline environments. With a cell voltage of 1.63 V for 10 mA·cm^{-2}, this method allows glucose electrolysis at a voltage far lower than that required for water splitting alone [40]. In the meantime, Cu/Cu_2O-decorated NF, which is produced using a hydrothermal process assisted by glycerol, exhibits outstanding GOR and HER performances. According to DFT simulations, the system supports high hydrogen production with a faradaic efficiency above 100% by favoring strong glucose intermediate adsorption and effective C—H cleavage [41]. Moreover, multilevel CuCoP/NF structures combine Cu(I) and CoP to provide extremely active surfaces that support the conversion of glucose to gluconic acid at low potentials and continue to function steadily even at very high current densities [42].

In addition to these electrocatalysts, photoelectrocatalytic glucose reformation under artificial sunlight has been investigated using a nanostructured WO_3 photoanode. With a faradaic yield of up to 64%, this device uses a "photocurrent doubling" effect to transform glucose into useful compounds, including erythrose, arabinose, and gluconic and glucaric acids, while achieving high photocurrents of 6.5 mA·cm^{-2} [43]. Furthermore, by using WSE to create a confined alkaline environment, CNTs incorporated in a Co-based zeolitic imidazolate framework (Co/

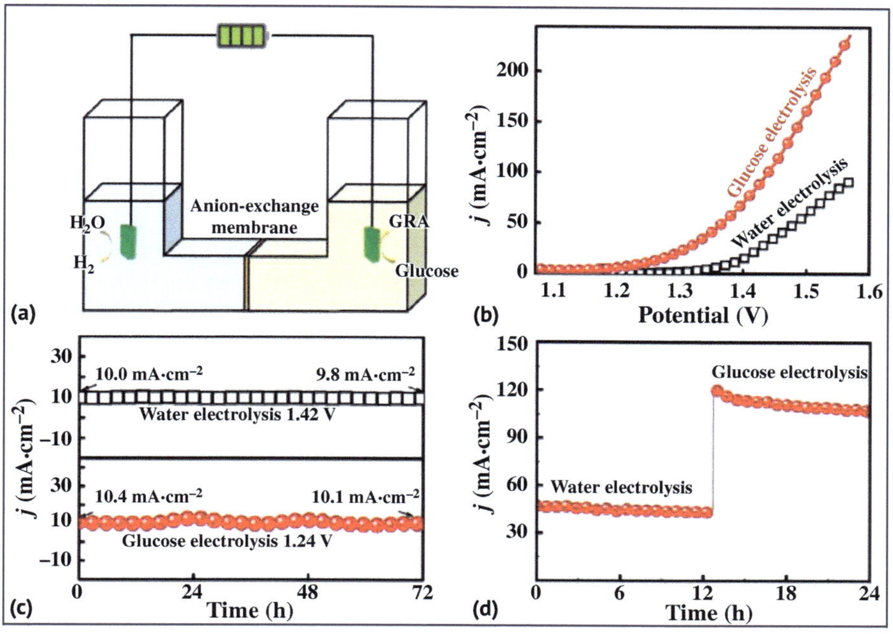

Figure 17.7 Results of glucose oxidation coupled with water electrolysis using Ru–Ni boride catalyst. *Source*: [39] with permission of John Wiley & Sons.

CoP) allow wearable, nonenzymatic electrochemical glucose detection under physiological pH. Together, these varied studies highlighted how combining GOR with water electrolysis lowers operating voltage, uses less energy, and allows for the coproduction of valuable organic compounds, providing a viable and profitable method of producing hydrogen [44].

The investigations included in this section highlight the enormous potential of non-oxide electrocatalysts in facilitating water electrolysis coupled with glucose oxidation. These materials retain high hydrogen evolution rates while achieving notable reductions in overpotential and energy input needs by the effective replacement of the energy-intensive OER with GOR. Additionally, the creation of important compounds like gluconic acid and formic acid provides an economic incentive, encouraging a value-added, circular approach to the production of renewable energy. Non-oxide catalysts have shown strong performance, improved selectivity, and long-term stability thanks to their customized compositions and electronic structures, which is a crucial step toward real-world use. There are still issues, nevertheless, such as managing intermediate products, integrating with actual biomass feedstocks, and catalyst deterioration with extended operation. It will take ongoing developments in reactor engineering, mechanistic knowledge, and materials design to overcome these obstacles. All things considered, the use of non-oxide electrocatalysts in the hydrogen production process aided by glucose oxidation offers a revolutionary route to sustainable energy systems. This strategy has a lot of potential to produce inexpensive, green hydrogen while promoting biomass valorization, which will ultimately help create a cleaner, more circular energy future.

17.4 Conclusions and Future Perspectives

In conclusion, a very promising substitute for the conventional oxygen evolution pathway is the incorporation of glucose oxidation into electrochemical water splitting. Significant decreases in overpotential and energy consumption have been shown by taking advantage of the kinetic benefits and thermodynamic favorability of glucose oxidation. This approach not only increases the efficiency of hydrogen production but also makes it possible to convert glucose into valuable chemical molecules. Catalytic activity, selectivity, and stability have been further enhanced by the introduction of sophisticated non-oxide electrocatalysts, such as metal phosphides, borides, and MOF-derived structures, providing strong platforms for real-world applications. Nevertheless, a number of difficulties still exist in spite of these noteworthy successes. To move this strategy from lab-scale experiments to practical implementation, problems with electrode fouling, catalyst durability during extended operation, and the difficulty of intermediate product separation must be resolved. Furthermore, to maximize catalyst design and reaction pathways, a deeper mechanistic knowledge through theoretical modeling and in situ characterization is also essential.

Future studies should concentrate on creating scalable and economical synthesis methods for high-performance non-oxide electrocatalysts specifically designed for the oxidation of glucose. Enhancing catalyst longevity, increasing selectivity toward desired oxidation products, and creating integrated systems that can effectively create hydrogen and valuable organic compounds at the same time should be the main priorities. Utilizing actual biomass-derived glucose streams instead of pure glucose offers a fascinating chance to value industrial and agricultural wastes, supporting the ideas of the circular economy and sustainable energy systems. To move closer to industrial-scale deployment, system-level research addressing reactor design, mass-transport constraints, and electrode architecture will also be essential. Overcoming existing barriers will need interdisciplinary collaborations that span materials science, electrochemistry, process engineering, and environmental science. In the end, glucose-oxidation-coupled water electrolysis has the potential to transform biomass usage and green hydrogen production, providing a combined benefit of environmental sustainability and energy efficiency. Future hydrogen generation technologies that are workable, profitable, and ecologically beneficial may be made possible by ongoing research in this area.

References

1 Zeradjanin, A. R., Topalov, A. A., Van Overmeere, Q., Cherevko, S., Chen, X., Ventosa, E., Schuhmann, W., & Mayrhofer, K. J. J. (2014). Rational design of the electrode morphology for oxygen evolution—Enhancing the performance for catalytic water oxidation. *RSC Adv.* *4*(19), 9579–9587. https://doi.org/10.1039/c3ra45998e

2 Lee, S. J., Theerthagiri, J., Nithyadharseni, P., Arunachalam, P., Balaji, D., Madan Kumar, A., Madhavan, J., Mittal, V., & Choi, M. Y. (2021). Heteroatom-doped graphene-based materials for sustainable energy applications: A review. *Renew. Sustain. Energy Rev. 143*, 110849. https://doi.org/10.1016/j.rser.2021.110849

3 Xue, D., Zheng, K., Tao, X., Feng, Z., Zhang, C., Su, R., Yang, L., Meng, X., & Xu, C. (2025). CuO@CoOOH heterojunction-induced built-in electric field for enhanced electrocatalytic glucose oxidation and hydrogen production. *Appl. Surf. Sci. 701*, 163267. https://doi.org/10.1016/j.apsusc.2025.163267

4 Ashoka, S., Sushmitha, M., Shirisha, M., Akshaya, H. M., Yogesh, K., & Girish Kumar, S. (2024). Glucose oxidation on monoclinic $Co_2V_2O_7$ nanorods: Intuitions on cell potential reduction in hybrid electrolyzer. *Mater. Sci. Semicond. Process. 184*, 108838. https://doi.org/10.1016/j.mssp.2024.108838

5 Li, Z., Zhang, X., Ou, C., Zhang, Y., Wang, W., Dong, S., & Dong, X. (2023). Transition metal-based self-supported anode for electrocatalytic water splitting at a large current density. *Coord. Chem. Rev. 495*, 215381. https://doi.org/10.1016/j.ccr.2023.215381

6 Varghese, A., Sunaja Devi, K. R., Pinheiro, D., & Jomy, J. (2024). Electrochemical investigations of chitosan/ZrO_2-Bi_2O_3 composite for advanced energy and environmental applications. *J. Environ. Chem. Eng. 12*(5), 113824. https://doi.org/10.1016/j.jece.2024.113824

7 Peng, X., Nie, X., Zhang, L., Liang, T., Liu, Y., Liu, P., Men, Y.-L., Niu, L., Zhou, J., Cui, D., & Pan, Y.-X. (2020). Carbon-coated tungsten oxide nanospheres triggering flexible electron transfer for efficient electrocatalytic oxidation of water and glucose. *ACS Appl. Mater. Interfaces 12*(51), 56943–56953. https://doi.org/10.1021/acsami.0c13547

8 Varghese, A., Devi, K. R. S., & Pinheiro, D. (2023). Rational design of PANI incorporated PEG-capped CuO/TiO_2 for electrocatalytic hydrogen evolution and supercapacitor applications. *Int. J. Hydrog. Energy 48*, 29552–29564. https://doi.org/10.1016/j.ijhydene.2023.04.114

9 Varghese, A., Sunajadevi, K. R. P., & Pinheiro, D. (2025). Composite revolution: unleashing the potential of polymers in sustainable energy and environmental applications. *Energy Adv. 4*, 743–762. https://doi.org/10.1039/D5YA00088B

10 Tayebi, M., Masoumi, Z., Lee, H., Hong, D., Seo, B., Lim, C., Kyung, D., & Kim, H. (2024). MOF-derived FeCoO/N-doped C bifunctional electrode for H_2 production through water and glucose electrolysis. *Adv. Sustain. Syst. 8*(11), 2400342. https://doi.org/10.1002/adsu.202400342

11 Patil, S. A., Khot, A. C., Kadam, K. D., Bui, H. T., Im, H., & Shrestha, N. K. (2023). Unlocking the catalytic potential of nickel sulfide for sugar electrolysis: green hydrogen generation from kitchen feedstock. *Inorg. Chem. Front. 10*(24), 7204–7211. https://doi.org/10.1039/D3QI01686B

12 Zhang, S., Duan, T., Yan, D., Yang, H., Cheng, L., Pei, Y., You, K., & Luo, H. (2024). The whole transfer layer of glucose-derived carbon enabled rapid charge transfer and hole storage for efficient water splitting. *Chem. Eng. J. 486*, 150176. https://doi.org/10.1016/j.cej.2024.150176

13 Ye, S., Feng, W., Li, J., Zhong, H., Weng, J., & Li, H. (2022). Assessing the role of sulfite in photoelectrocatalytic oxidation of glucose on Pt/TiO_2 for hydrogen production. *J. Electroanal. Chem. 927*, 116975. https://doi.org/10.1016/j.jelechem.2022.116975

14 Suraniti, E., Merzeau, P., Roche, J., Gounel, S., Mark, A. G., Fischer, P., Mano, N., & Kuhn, A. (2018). Uphill production of dihydrogen by enzymatic oxidation of glucose without an external energy source. *Nat. Commun. 9*(1), 3229. https://doi.org/10.1038/s41467-018-05704-5

15 Varghese, A., Sunajadevi, K. R. P., Pinheiro, D., Saravanakumar, B., & Selvaraj, M. (2025). Advancing energy production and storage: Polypyrrole/V_2O_5/MnO_2 composite as a high-performance electrocatalyst. *Int. J. Hydrog. Energy 123*, 238–246. https://doi.org/10.1016/j.ijhydene.2025.03.419

16 Theerthagiri, J., Senthil, R. A., Senthilkumar, B., Reddy, A., & Madhavan, J. (2017). Recent advances in MoS_2 nanostructured materials for energy and environmental applications—A review. *J. Solid State Chem. 252*, 43–71. https://doi.org/10.1016/j.jssc.2017.04.041

17 López-Fernández, E., Crisafulli, R., Dos Santos-García, A. J., Caravaca, A., & de Lucas-Consuegra, A. (2024). Can we completely suppress the oxygen evolution reaction in a glucose electrolyser? Three experimental evidences. *Int. J. Hydrog. Energy. 73*, 112–117. https://doi.org/10.1016/j.ijhydene.2024.06.020

18 Sayeed, M. A. & O'Mullane, A. P. (2017). A multifunctional gold-doped $Co(OH)_2$ electrocatalyst tailored for water oxidation, oxygen reduction, hydrogen evolution, and glucose detection. *J. Mater. Chem. A 5*(45), 23776–23784. https://doi.org/10.1039/C7TA08928G

19 Li, D., Li, Z., Zou, R., Shi, G., Huang, Y., Yang, W., Yang, W., Liu, C., & Peng, X. (2022). Coupling overall water splitting and biomass oxidation via Fe-doped Ni_2P@C nanosheets at large current density. *Appl. Catal. B: Environ. 307*, 121170. https://doi.org/10.1016/j.apcatb.2022.121170

20 Wei, M., Li, M., Gao, Q., Cai, X., Zhang, S., Fang, Y., Peng, F., & Yang, S. (2024). Bifunctional Ni foam supported TiO_2@Ni_3S_2 core@shell nanorod arrays for boosting electrocatalytic biomass upgrading and H_2 production reactions. *Small 20*(9), 2305906. https://doi.org/10.1002/smll.202305906

21 Zhu, X., Yuan, S., Ju, Y., Yang, J., Zhao, C., & Liu, H. (2019). Water splitting-assisted electrocatalytic oxidation of glucose with a metal–organic framework for wearable nonenzymatic perspiration sensing. *Anal. Chem. 91*(16), 10764–10771. https://doi.org/10.1021/acs.analchem.9b02328

22 Zhuang, S., Wang, L., Hu, H., Tang, Y., Chen, Y., Sun, Y., Mo, H., Yang, X., Wan, P., & Khan, Z. U. H. (2018). Ultrafast electrodeposition of Ni metal and NiFe hydroxide composites with heterogeneous nanostructures as high-performance multifunctional electrocatalysts. *ChemElectroChem 5*(18), 2577–2583. https://doi.org/10.1002/celc.201800819

23 Zhang, Y., Zhou, B., Wei, Z., Zhou, W., Wang, D., Tian, J., Wang, T., Zhao, S., Liu, J., Tao, L., & Wang, S. (2021). Coupling glucose-assisted Cu(I)/Cu(II) redox with electrochemical hydrogen production. *Adv. Mater. 33*(48), 2104791. https://doi.org/10.1002/adma.202104791

24 Siwal, S. S., Thakur, S., Zhang, Q. B., & Thakur, V. K. (2019). Electrocatalysts for electrooxidation of direct alcohol fuel cell: Chemistry and applications. *Mater. Today Chem. 14*, 100182. https://doi.org/10.1016/j.mtchem.2019.06.004

25 Liu, L., Li, X., Yang, R., Ma, H., Liu, E., Gao, T., & Sun, T. (2025). Co_3S_4/MnS p-p heterojunction as a highly efficient electrocatalyst for water splitting and electrochemical oxidation of organic molecules. *J. Colloid Interface Sci. 687*, 589–598. https://doi.org/10.1016/j.jcis.2025.02.115

26 Lin, C., Zhang, P., Wang, S., Zhou, Q., Na, B., Li, H., Tian, J., Zhang, Y., Deng, C., Meng, L., Wu, J., Liu, C., Hu, J., & Zhang, L. (2020). Engineered porous Co–Ni alloy on carbon cloth as an efficient bifunctional electrocatalyst for glucose electrolysis in an alkaline environment. *J. Alloys Compd. 823*, 153784. https://doi.org/10.1016/j.jallcom.2020.153784

27 Baby, N., Thangarasu, S., Murugan, N., Kim, Y. A., Lee, J. M., & Oh, T.-H. (2024). Versatile bimetallic metal-organic framework with nanocutting channels tailored for efficient electrocatalytic water oxidation and glucose detection. *J. Alloys Compd. 970*, 172601. https://doi.org/10.1016/j.jallcom.2023.172601

28 Sun, S., Wang, T., Liu, R., Sun, Z., Hao, X., Wang, Y., Cheng, P., Shi, L., Zhang, C., & Zhou, X. (2024). Ultrasonic-assisted Fenton reaction, inducing surface reconstruction, endows nickel/iron-layered double hydroxide with efficient water and organics electrooxidation. *Ultrason. Sonochem. 109*, 107027. https://doi.org/10.1016/j.ultsonch.2024.107027

29 Wu, Y., Li, M., Fang, W., Cao, Y., Xin, S., & Zhao, H. (2024). Hollow RuV-$Co(OH)_2$ arrays for efficiently electrocatalytic H_2 production assisted by glucose oxidation. *Appl. Surf. Sci. 657*, 159784. https://doi.org/10.1016/j.apsusc.2024.159784

30 Li, D., Huang, Y., Li, Z., Zhong, L., Liu, C., & Peng, X. (2022). Deep eutectic solvents derived carbon-based efficient electrocatalyst for boosting H_2 production coupled with glucose oxidation. *Chem. Eng. J. 430*, 132783. https://doi.org/10.1016/j.cej.2021.132783

31 Du, P., Zhang, J., Liu, Y., & Huang, M. (2017). Hydrogen generation from catalytic glucose oxidation by Fe-based electrocatalysts. *Electrochem. Commun. 83*, 11–15. https://doi.org/10.1016/j.elecom.2017.08.013

32 Lin, C., Li, H., Zhang, P., Deng, C., Meng, L., Zhou, Q., Wang, S., Wu, J., Liu, C., Tian, J., & Qian, Y. (2020). Boosting water electrolysis with anodic glucose oxidation reaction over engineered cobalt nickel hydroxide nanosheet on carbon cloth. *J. Electroanal. Chem. 861*, 113946. https://doi.org/10.1016/j.jelechem.2020.113946

33 Yuan, H., Weng, C., Zhang, X., Chen, L., Zhang, Q., Ma, L., & Liu, J. (2024). Construction of bifunctional MOF-based composite electrocatalysts promoting oxygen evolution reaction and glucose oxidation reaction, and their kinetic deciphering. *Mater. Today Phys. 49*, 101601. https://doi.org/10.1016/j.mtphys.2024.101601

34 Lei, L., Xu, C., Dong, X., Ma, B., Chen, Y., Hao, Q., Zhao, C., & Liu, H. (2023). Continuous glucose monitoring in hypoxic environments based on water splitting-assisted electrocatalysis. *Chemosensors 11*(2), 149. https://doi.org/10.3390/chemosensors11020149

35 Crisafulli, R., de la Hoz, A., de la Osa, A. R., Sánchez, P., & de Lucas-Consuegra, A. (2025). Studying the competition between glucose oxidation and oxygen evolution reaction: toward a membrane-free electrolyzer for the production of H_2 and added value products. *ACS Sustain. Chem. Eng.* *13*(13), 4963–4974. https://doi.org/10.1021/acssuschemeng.4c09249

36 Wang, Y., Yan, W., Ni, M., Zhu, C., & Du, H. (2023). Surface valence regulation of cobalt–nickel foams for glucose oxidation-assisted water electrolysis. *Chem. Commun.* *59*(17), 2485–2488. https://doi.org/10.1039/D2CC05270A

37 Shrestha, N. K., Patil, S. A., Salunke, A. S., Inamdar, A. I., & Im, H. (2023). Accelerating glucose electrolysis on Cu-doped MIL-88B for an energy-efficient anodic reaction in water splitting. *Dalton Trans.* *52*(31), 10933–10941. https://doi.org/10.1039/D3DT01773G

38 Lou, H., Yang, Y., Bu, X., Fan, H., Weng, D., Zhang, J., Gao, W., & Wen, D. (2025). Efficient electrocatalytic glucose oxidation coupled with water electrolysis driven by Ni-foam supported Ni–P nanowire arrays. *J. Mater. Chem. A* *13*(2), 1067–1073. https://doi.org/10.1039/D4TA06649A

39 Liu, H., Zhang, R., Chen, L., Wang, L., Guo, Y., & Yang, Y. (2021). Synergistic coupling of nickel boride with Ru cluster as a highly active multifunctional electrocatalyst for overall water splitting and glucose electrolysis. *Adv. Sustain. Syst.* *5*(8), 2000184. https://doi.org/10.1002/adsu.202000184

40 Liu, S., Zhang, E., Wan, X., Pan, R., Li, Y., Zhang, X., Su, M., Liu, J., & Zhang, J. (2022). Ru-Co-Mn trimetallic alloy nanocatalyst driving bifunctional redox electrocatalysis. *Sci. China Mater.* *65*(1), 131–138. https://doi.org/10.1007/s40843-021-1732-1

41 Nguyen, C. C., Nguyen, C. H. N., Nguyen, T. L., Le, T. H., Dinh, M. T. N., Van, D. T. T., Pham, H. T., Nguyen, H. N. T., Luu, T. A., Nguyen, N. L., Dao, M. U., Nguyen, Q. H., Le, H. K., & Le, Q. V. (2025). Cu/Cu_2O nanoparticle–decorated nickel foam as an electrocatalyst for hydrogen evolution and glucose oxidation. *ACS Appl. Nano Mater.* *8*(20), 10651–10662. https://doi.org/10.1021/acsanm.5c01579

42 Ma, X.-D., Liu, R., Yue, S., Xie, H.-J., & Xia, X.-H. (2025). Multilevel structured CuCoP with a synergistically catalytically active site designed for hydrogen evolution coupled to gluconic acid synthesis. *Rare Met.* *44*(5), 3141–3155. https://doi.org/10.1007/s12598-024-03103-y

43 Jakubow-Piotrowska, K., Witkowski, B., & Augustynski, J. (2022). Photoelectrocatalytic hydrogen generation coupled with the reforming of glucose into valuable chemicals using a nanostructured WO_3 photoanode. *Commun. Chem.* *5*(1), 125. https://doi.org/10.1038/s42004-022-00745-w

44 Zhang, Y., Qiu, Y., Ma, Z., Wang, Y., Zhang, Y., Ying, Y., Jiang, Y., Zhu, Y., & Liu, S. (2021). Core-corona Co/CoP clusters strung on carbon nanotubes as a Schottky catalyst for glucose oxidation assisted H_2 production. *J. Mater. Chem. A* *9*(17), 10893–10908. https://doi.org/10.1039/D0TA11850H

Part V

Future Prospects

18

Future Prospects and Commercialization of Energy-saving Hybrid Water Electrolysis

Juan A. Lopez-Ruiz and Talshyn Begildayeva*

Institute for Integrated Catalysis, Pacific Northwest National Laboratory, Richland, WA, USA

*Corresponding authors: juan.lopezruiz@pnnl.gov

18.1 Introduction

As we reviewed in previous chapters, hybrid electrolysis has emerged as a key technology in the global effort to electrify industry. However, there are several technical barriers that hinder widespread adoption, mainly a lack of demonstrated applications and performance for industrially relevant processes. Bridging the gap between academic research and industrial demonstrations requires concerted attention to (i) electrode and electrolyzer performance, (ii) energy utilization, and (iii) process compatibility, factors that tend to receive less emphasis in traditional academic studies (Figure 18.1). For example, why would a company consider utilizing hybrid electrolyzers if the performance, energy consumption, and compatibility with the existing process are worse than the incumbent, often thermocatalytic, system? Furthermore, hybrid electrolyzers are hardly operated under industrially relevant electrocatalytic conditions (e.g., >100 mA cm^{-2}, <3 V full-cell potential, >1,000-hour continuous operation, and real feedstocks) or reactor configurations (batch instead of flow electrolyzers). Thus, from an industrial perspective, it remains unclear if hybrid electrolyzers are a feasible technology that can be integrated with existing processes.

Key performance indicators used to benchmark electrolytic processes against incumbent thermocatalytic processes (e.g., normalized reaction rates, round-trip energy efficiency, and energy consumption) are rarely reported. Instead, conversions and current densities are the most commonly reported values, which cannot be used to compare against traditional technologies, thus hindering their consideration as an alternative to commercial processes. Likewise, the metrics needed to perform a basic energy and economic assessment are rarely reported even when the required process parameters (e.g., voltage and current) can be readily obtained from the potentiostat or power supplies.

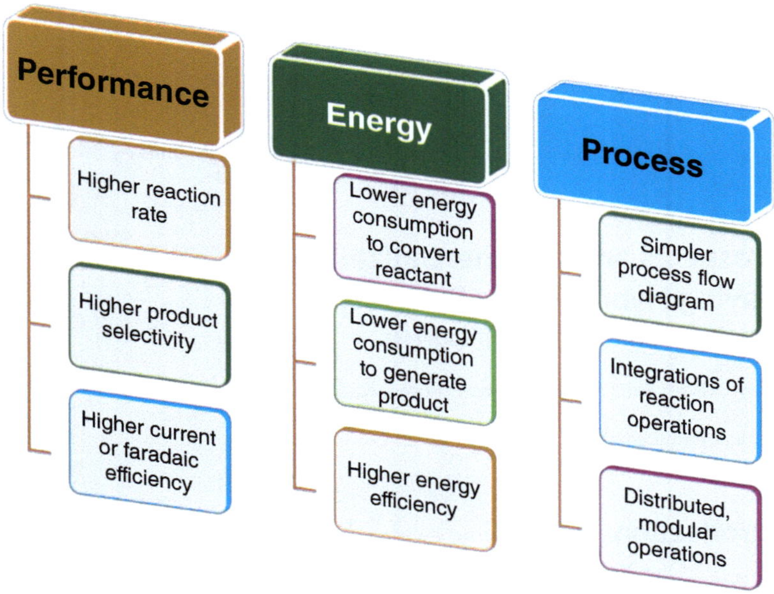

Figure 18.1 Key parameters needed to successfully integrate hybrid electrolyzers into industrial processes.

Finally, the main benefit of hybrid electrolyzers is process simplification through integration of reduction–oxidation reactions on the same device, which is rarely explored for industrially relevant applications beyond oxidation of organic compounds for H_2 production, thus limiting its applications to processes that need H_2. However, several applications exist in which electrocatalytic (reduction) reactions other than the hydrogen evolution reaction (HER) are more desirable, such as electrocatalytic reduction of organic and inorganic compounds into chemical intermediates. These applications make hybrid electrolyzers more attractive for industrial applications even if the performance and energy consumption are worse than incumbent conversion technologies.

This chapter reviews how we can advance the current state of research to accelerate commercialization of hybrid electrolyzers and provides guidelines for future research directions.

18.2 Bridging the Lab-to-plant Divide in Electrolyzers

Most academic studies continue to rely on batch H-cells or beaker setups, which are not necessarily well suited for scale-up. In contrast, industrial electrochemical operations such as water electrolysis, chlor-alkali processes, and adiponitrile synthesis employ flow electrolyzers that enable stable operation [1, 2]. While batch-type systems demonstrate proof-of-concept and help elucidate intrinsic kinetics and reaction mechanisms, continuous-flow processes are always preferred to demonstrate

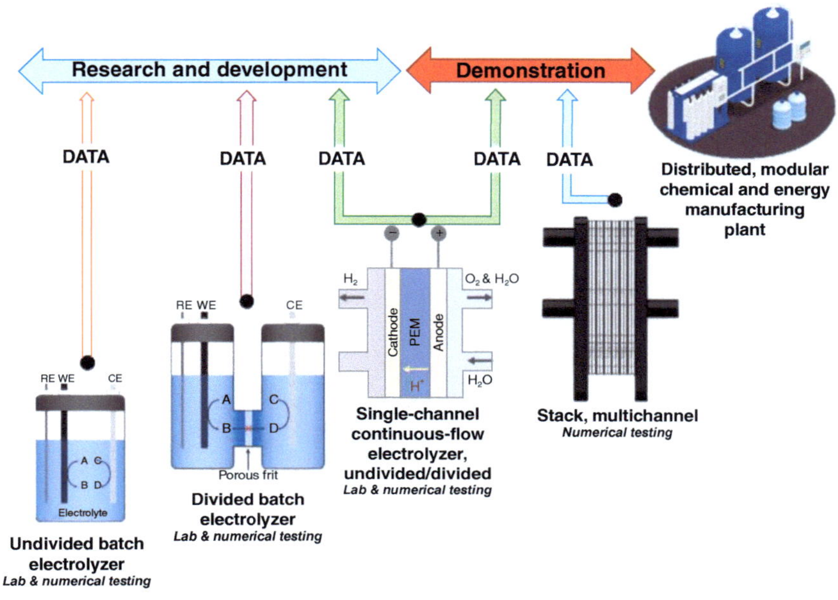

Figure 18.2 Typical electrolyzer scale-up steps from laboratory to plant scale. RE stands for reference electrode, WE stands for working electrode, and CE stands for counter-electrode. *Source*: Adapted from [3–5].

the scalability and stability of the electrocatalytic process (Figure 18.2). Table 18.1 provides a summary of different electrolyzer configurations traditionally used in different stages of research and their purposes.

Electrocatalytic reactions of model systems are typically performed in undivided batch reactors (10–1,500 mL glass beakers) in which the anodic and cathodic reactions are not usually separated by a membrane [7, 8], therefore affecting the overall current and faradaic efficiency (for anodic and cathodic reactions, respectively) in cases where anodic products can be reduced at the cathode. To improve production quantification and the efficiency of the electrocatalytic process, the field uses divided (two-compartment) batch reactors with membranes to independently quantify gas and liquid products from each chamber and simultaneously assess cathodic and anodic performances. Divided batch reactors are the simpler configuration for proof-of-concept hybrid electrolyzers; however, the anodic and cathode reaction pairs should be thoughtfully selected to align with real industrial applications. Divided batch electrolyzers require membranes (e.g., proton, anion exchange, and bipolar, depending on the pH of the electrolyte) and introduce additional complexity because they can interact with both the reactant and product, foul, degrade, or create pH imbalances between the anode and cathode, thus affecting the resistance of the electrolyzer and compensation of the electrode half-cell potential due to ohmic losses (i.e., *iR* compensation) [9, 10]. While batch electrolyzers might be appealing for research and development due to their simplicity (especially the

Table 18.1 Electrolyzer types with the main operational parameters.

Main parameter	Batch, undivided	Batch, divided	Flow, divided, single channel	Flow, divided, multichannel (stack)
Current density ($A \cdot cm^{-2}$)	0.01–0.1	0.01–0.1	0.1–1	0.1–1
Electrode cross-sectional area (cm^2)	1–5	1–5	5–100	100–500
Number of cells	1	1	1	>5
Operation conditions	Model	Model, real	Model, real	Real
Purpose	• Model studies • Single component • Proof-of-concept • High conversion • Buffered pH • Electrode screening	• Model studies • Proof-of-concept • High conversion • Separation of anode and cathode reactions • Buffered pH	• Separation of anode and cathode reactions • Kinetic studies at low and high conversions • Energy consumption • Long-term stability	• Energy consumption and efficiency • Stability under dynamic operations, scalability • Long-term stability

Pros	• Easy setup • Easy to keep pH constant • Does not require advanced controls • Easy *iR* compensation • Low-cost setup	• Separation of anodic and cathodic reactions • Independent measurement of anodic and cathodic liquid and gas products • Does not require advanced controls • Easy *iR* compensation • Low-cost setup	• Separation of anodic and cathodic reactions • Independent measurement of anodic and cathodic liquid and gas products • Kinetic experiments at different conversion levels • Easy change of electrolyte for kinetic studies without having to dismantle the cell	• Directly translatable to pilot systems • Independent measurement or resistance • Independent measurement of anodic and cathodic liquid and gas products
Cons	• Oxidation products can be reduced at the cathode • No separation of anodic from cathodic products • Difficult to deconvolute conversion effects from anode stability	• Membrane fouling • pH imbalances affect performance as a function of time • Difficult to deconvolute conversion effects from anode stability	• The *iR* compensation is affected by hydrodynamics • Complex setup • Pressure build-up due to gas evolution • System clogging due to precipitation of products • Membrane fouling	• Complex, multicomponent setup • Pressure build-up due to gas evolution • System clogging due to precipitation of products • High cost due to the high total surface area of electrodes, membrane needs, and equipment needed • Shunt currents

Source: Adapted from Begildayeva et al. [6].

undivided electrolyzers), they operate under differential conversion, and it is difficult to discern whether the changes in activity as a function of time are caused by changes in concentration of species (reactant or products) or electrode deactivation, which are critical to assess their feasibility for long-term operation.

In flow electrolyzers, the flow rate can be controlled (single-pass mode) to maintain low conversion of reagents and monitor electrode performance in the absence of conversion (i.e., mass transfer) effects and product inhibition. Furthermore, the flow rate serves as a carrier for the gases evolving at both the anode and cathode (e.g., O_2 or CO_2 at the anode, H_2 at the cathode), helping prevent a decrease in electrochemically active surface area (ECSA) of the electrode exposed to the electrolyte due to the accumulation of gases in each chamber. Additionally, continuous-flow electrolyzers are more industrially attractive [11, 12] for the ways in which they can (i) decrease the distance between the anode and cathode (i.e., interelectrode gap) and thereby decrease the electrolyzer resistance, (ii) improve mixing and mass transfer in the reaction chambers (i.e., laminar vs. turbulent flow) [13, 14], and (iii) improve heat transfer for reactions that take place above room temperature [15]. This is essential from the perspective of energy efficiency, which may not be a critical driver for academic research but is especially important for large-scale chemical and fuel production [16].

It is not just flow that is important, but also implementation of single-pass flow systems that go beyond recirculation [17], thus improving integration of electrolysis with separations (e.g., membrane, distillation) and other steps in the overall process [18]. For these reasons, flow is the standard for industrial electrosynthetic processes [16]. Different flow patterns can be established to handle multiphase flow, including segmented flows, where the flow pattern involves alternating bubbles and liquid electrolyte. Because flow processes have different characteristics than their batch counterparts, especially when working with multiphase flows (liquid and gas, aqueous and organic systems) [15], it is important for academic studies to increasingly implement flow studies to improve the translatability of results to real industrial processes. In cases where batch studies are preferred for academic investigation, there is increasing understanding of how to better translate these processes to operate under flow [16].

There can be competing desires when it comes to operating under a turbulent flow regime vs. a laminar flow regime. Turbulent flow assists mass transfer [14], which is favorable, but can also lead to undesired back-reduction or back-oxidation at the electrode opposite to where the product is generated. Thus, for certain undivided cell designs, laminar flow is favored and helps keep the products generated at the cathode and anode separate [13].

Flow electrolyzers are more complex systems due to the additional levels of automatization and control needed to operate them as well as the need for additional components (e.g., pumps, heaters) and unit operations (e.g., gas–liquid separators, heat exchangers), thus providing a more representative system for real chemical and fuel manufacturing processes. Therefore, it is essential to first confirm the electrochemical performance in flow systems before designing and building demonstration systems, which usually require stacks (100–500 cm^2 per cell and >5 cells per stack). Furthermore, unlike batch electrolyzers, flow electrolyzer performance is easier to automate and autonomize by including in-line detectors at the outlet of the electrolyzer and using the results to control the input variables to the system (e.g., flow rate, full- or half-cell potential, and current). Therefore, flow electrolyzers can be used to bridge the gap between laboratory and demonstration scales, enabling more accurate assessment of scalability and performance as well as energy and efficiency. Furthermore, the field needs to employ testing protocols through which electrode degradation can be observed (e.g., flow electrolyzers and long-term operations), thus allowing the efficacy of new deactivation mitigation strategies (e.g., better electrodes, different electrolyzer configurations, and cleaning protocols) to be evaluated.

18.3 Key Performance Indicators

A commercialization rule of thumb is that a new process needs to be at least 20% better than the existing process to be considered as a feasible alternative. Otherwise, it is not worthwhile for industry to consider changing their process due to the associated costs of dismantling the existing process and implementing new technology. The key performance indicators needed to assess the advancement of the electrocatalytic (conversion) process against an incumbent commercial (thermocatalytic conversion) process can be divided in two categories: (i) performance such as normalized reaction rate (Eq. 18.1) and current (Eq. 18.2) or faradaic efficiency (Eq. 18.3) for anodic and cathodic reactions, respectively, and product selectivity (Eq. 18.4); and (ii) energy, such as that required to convert the reactant or generate a product (Eq. 18.5), and energy efficiency (Eq. 18.6).

These key performance indicators need to be obtained under constant half-cell potential ($E_{\text{anode or cathode}}$, Eq. 18.7) as opposed to constant current (density), because the latter depends on electrocatalyst properties that can vary by orders of magnitude between different electrocatalytic systems such as ECSA (e.g., foils or wires vs. high-surface-area supports) or loading of the active phase (e.g., 0.1 against 50 wt.%). Comparing electrocatalyst performance at constant $E_{\text{anode or cathode}}$ may also be challenging because the ohmic (i.e., iR compensation) losses between different systems (batch or flow, divided or undivided) and electrolyte composition (nonconductive vs. conductive) can also vary by orders of magnitude. To avoid ohmic losses overcompensation, a compensation factor of 85% is usually selected [19, 20]. However, the compensation factor for iR losses is not usually accounted for or reported in academic studies. Thus, comparing electrocatalytic performance between different studies can be challenging.

Finally, other key performance indicators that might be more difficult to quantify without techno-economic analysis (TEA) and life cycle assessment (LCA) include simplifying process improvement, which could be associated with integrating different reaction unit operations (oxidation and reduction) into a single reactor (i.e., hybrid electrolyzer), modularization to enable distributed, small-scale manufacturing, or process simplification by removing the need for subsequent unit operations (e.g., separations if the product selectivity obtained in the electrolyzer is >90%).

In the following subsections, we will overview the most important, and often overlooked, factors that will enable commercialization of hybrid electrolyzers:

$$\text{Normalized reaction rate} = \frac{\text{Moles of reactant consumed or product generated}}{\text{Time} \times \text{Measurable electrocatalyst property}} \tag{18.1}$$

$$\text{Current efficiency (\%)} = \frac{\text{Moles of e}^- \text{ in the target product}}{\text{Total moles of e}^- \text{ generated at the anode}} \times 100 \tag{18.2}$$

$$\text{Faradaic efficiency (\%)} = \frac{\text{Moles of e}^- \text{ in the target product}}{\text{Total moles of e}^- \text{ consumed at the cathode}} \times 100 \tag{18.3}$$

$$\text{Product selectivity (\%)} = \frac{\text{Moles of product}}{\text{Total moles of reactant consumed}} \times 100 \tag{18.4}$$

$$\begin{gathered}\text{Energy consumption} \\ (\text{kWh} \cdot \text{kg}^{-1}_{\text{Reactant or Product}})\end{gathered} = \frac{E_{\text{electrolyzer}} \times i}{\begin{gathered}\text{Normalized} \\ \text{reaction rate}\end{gathered} \times \begin{gathered}\text{Measurable} \\ \text{electrocatalyst property}\end{gathered}} \tag{18.5}$$

$$\text{Energy efficiency (\%)} = \frac{\text{Energy in products}}{\text{Electrical energy input} + \text{Energy in reagents}} \times 100 \tag{18.6}$$

$$E_{\text{anode or cathode}}\ (\text{V}) = E_{0,\text{anode or cathode}} - i \times R \times \text{Compensation factor} \tag{18.7}$$

$$E_{\text{electrolyzer}} = E_{\text{anode}} - E_{\text{cathode}} \tag{18.8}$$

18.3.1 Normalized Reaction Rates

Although conversion and current density are the main performance indicators used to report electrocatalytic performance along with onset and overpotential, we want to emphasize that the normalized reaction rate is the best performance indicator needed for comparison with incumbent reaction technologies. For starters, conversion depends on the volume of solution used, concentration of reagent, and reaction time, which are rarely kept constant across different experiments due to the lack of standardization. Similarly, current density will depend on the measurable electrocatalyst property (e.g., ECSA, loading of active phase) and half-cell potential, which are also underreported. Comparing normalized reaction rates as a function of electrochemical reaction conditions is qualitatively useful to derive activity–structure–reaction condition relationships that can be used to improve electrocatalytic systems across various reaction environments. However, the comparison is meaningful only if the kinetic data are not biased by variations in the electrocatalyst properties (e.g., ECSA, mass of moles of active sites). As discussed in detail in recent

Table 18.2 Summary of performance indicators used to describe the performance of electrocatalysts.

	Pros	Cons
Product selectivity and conversion under constant current or charge conditions	• Useful for comparing the activity of a given catalyst for different molecules	• Does not give insight into the intrinsic activity of the catalyst because the half-cell potential is not specified • The intrinsic activity cannot be benchmarked against literature because the mass loading or electrochemical surface area is not specified
(Unnormalized) Reaction rates or selectivity at constant half-cell potential without a specified surface area	• Useful for understanding the effects of half-cell potential on reaction rate and selectivity	• The intrinsic activity of the electrocatalyst cannot be benchmarked against literature because the mass or ECSA of the electrocatalyst is not provided
Reaction rate, TOF, or current density at a given half-cell potential, where the rates are normalized to ECSA	• Useful for understanding the effects of electrocatalyst composition and surface properties on reaction rate and selectivity as a function of half-cell potential • Performance indicator needed for process scale-up and comparison with established processes	• The ECSA measurement might not be accurate enough and might not be possible for all electrocatalyst compositions

perspectives on benchmarking electrocatalysis [20, 21], it is important to normalize electrocatalytic rates (ideally to the number of active sites, but mass or surface area is also sufficient) to report normalized reaction rates.

In electrocatalyzed reactions, the reaction rates must be obtained under identical overpotential after being corrected for iR losses and pH (reported vs. RHE). In general, there are several types of reaction rates reported in the electrocatalytic literature that are summarized in Table 18.2, adapted from Akhadea et al. [11], but the reaction rate used ultimately depends on the ability of the researcher to obtain accurate quantitative measurements of the electrocatalyst active phase.

When possible, we recommend normalizing the reaction rate by the ECSA, which can be measured using various standardized techniques [22]. For unsupported metals, the surface area can be measured through capacitance measurements [20, 23, 24], where the area can be derived from the capacitance of the electrochemical double layer, using an assumption that the specific capacitance of metals is often between 10 and 40 $\mu C\ cm^{-2}$. In-situ coulometric stripping can also be used to measure the surface area of (i) platinum group metals (PGMs) using hydrogen underpotential deposition or CO stripping [21, 25–28], (ii) copper bimetal (e.g., Pb) underpotential deposition [29], or (iii) reducible metals by oxide reduction [30] or oxide formation [31]. The coulometric stripping method has the advantage of being usable on supported metals, as well as unsupported metals. As proposed by Boudart et al. in 1969 [32], rate normalization to the number of active sites available (i.e., turnover frequency [TOF]) determined using these surface-area quantification techniques is the best method for comparing the performance of different catalytic systems, which has been adopted by industry. When it is not possible to directly measure the active surface area, microscopy or X-ray diffraction can be used to approximate particle size to determine dispersion [33]. However, this method can be inaccurate, with up to two orders of magnitude difference in the theoretical surface area relative to stripping methods [29]. If quantifying the ECSA is difficult, we recommend normalization using the mass of the electrode active phase.

18.3.2 Energy Consumption and Efficiency

Industrial processes may consider new technologies if the energy consumption, capital costs, and operation costs are better (i.e., lower) than those of incumbent technologies. While a full TEA may be needed to completely address these points, especially for thermocatalytic processes, energy consumption (and costs) to convert a reactant or generate a product (Eq. 18.5) and improvement in overall energy efficiency (Eq. 18.6) can be readily calculated using process data from the potentiostat (i.e., full electrolyzer potential and current) and the reaction rate. Furthermore, these calculations are very beneficial to the researcher because they can easily identify how electrocatalytic processes can be improved (i.e., lowering anodic or cathodic half-cell potential, improving reaction rate or selectivity) to match the commercial process and assess the feasibility of doing so. For example, if improving the process requires lowering the full- or half-cell potential by an order of magnitude or increasing the reaction rate or selectivity by an order of magnitude, then the researcher needs to assess whether it is possible to do so and chart a path forward accordingly. If this is impossible, then the researcher must focus on different advantages beyond energy consumption such as modularity or potential improvements for upstream processing (e.g., higher selectivity to the target product would decrease separation costs), which is discussed in Section 18.4.

Another key aspect usually not considered is the overall energy efficiency of the process, such as energy inputs into the feedstocks including synthesizing, transporting, and conditioning. While a hybrid electrolyzer might be an appealing process for H_2 production because it replaces the anodic reaction from water oxidation to organic oxidation, synthesizing the organic compound (e.g., hydrazine, methanol) and conditioning it for electrolysis also requires energy, which needs to be accounted

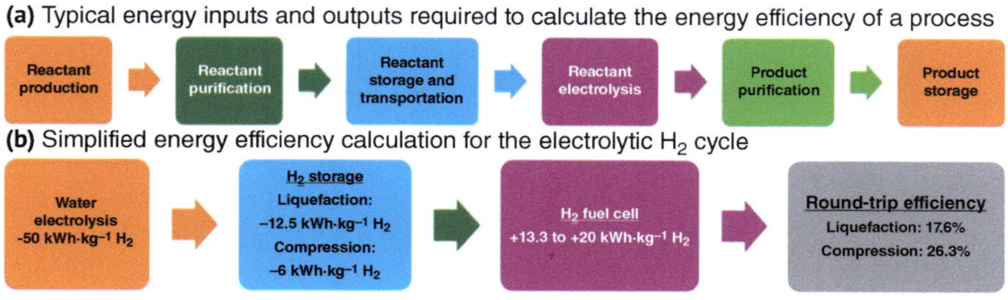

Figure 18.3 Summary of steps usually needed to determine the (round-trip) energy efficiency of an electrolytic process.

for in a full energy balance. Figure 18.3 highlights the main steps that need to be taken into consideration when calculating the energy efficiency of a hybrid electrolysis process.

While the academic researcher may be tempted to only account for the energy inputs starting at the electrolysis step, they must also account for the energy required to prepare the feedstock unless they are using waste feedstocks. For example, in the case of the electrolytic H_2 cycle, when factoring the energy required to produce and store (e.g., liquefaction or compression) H_2 to the energy recovered by reacting H_2 in a fuel cell, the electricity-to-electricity (i.e., round-trip efficiency) energy efficiency is typically between 17.6% and 26.3%. Transporting the stored H_2 over long distances or for extended periods will also require energy, which further decreases the overall (round-trip) energy efficiency of the process.

18.4 Process Consideration

In addition to key performance indicators and energy consumption, there are other equally important factors related to the process itself that can affect the commercialization of electrolyzers, which can only be quantified via TEA and LCA. These factors depend on the chemical manufacturing process, processing scale, and geographical location, which might ultimately depend on challenges regarding supply-chain issues. Compared to the traditional centralized thermocatalytic processing of chemical (including fertilizers) and fuel manufacturing [34, 35], hybrid electrolyzers enable distributed manufacturing in which processing can be colocated with feedstock generation or utilization, thus simplifying the process and enabling a more resilient and energy-efficient approach for chemical and fuel generation and energy utilization (Figure 18.4) [36–40]. One of the main issues regarding distributed manufacturing is the unfavorable economies of scale typically associated with thermocatalytic conversion routes (Figure 18.5). That is, the production cost increases as the processing scale decreases. However, electrocatalytic systems represent a paradigm shift because they are more cost-efficient for small, distributed operations due to their modular nature as compared to traditional thermocatalytic processes. As shown in Figure 18.5 adapted from Weber

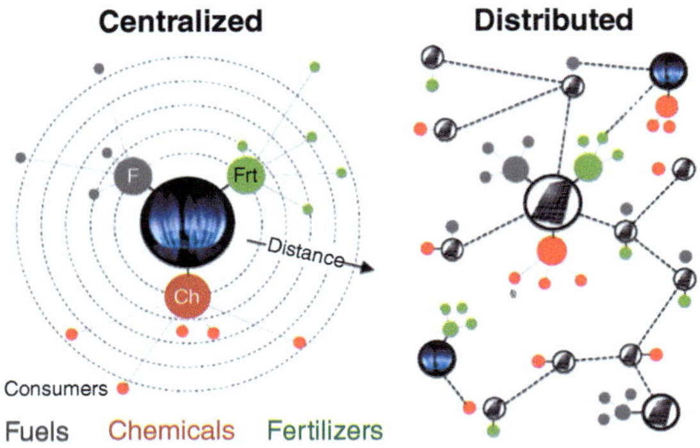

Figure 18.4 Depiction of an existing centralized scheme for production of fuels, chemicals, and fertilizers, where large plants dependent on fossil resources provide consumers with final products. On the right is shown a distributed process coupled to distributed production of electricity. *Source*: [39]/with permission of Elsevier.

et al. [41], for large operation scales above 1,000 barrels per day (kbpd), the traditional "stick-built" refinery of chemical plant construction used for centralized processes is similar to that of modular constructions. However, the production cost of modular construction is between one and two orders of magnitude cheaper than traditional, centralized "stick-built" constructions at 10 and 100 bpd. Furthermore, the cost of the modular facility decreases more steeply as more plants (modules) are constructed due to difference in the learning rate (i.e., a measurement of change in efficiency and reduction in cost or time as a task is repeated). A higher learning rate is expected with modular constructions because it is a repetitive, factory-controlled (i.e., automated) process, while a lower learning rate is expected for stick-built constructions because each project is unique and, therefore, less automated, and takes long to complete. Therefore, TEA/LCA studies for implementing hybrid electrolyzers for modular operation must be done at distributed scales for which thermocatalytic processes might not have been feasible before.

Furthermore, electrocatalytic systems have additional process advantages that could, on their own, accelerate their adoption in commercial distributed manufacturing processes, even though the key performance indicators are not better than those of incumbent (thermocatalytic) technologies. The main process differentiators are summarized in Table 18.3, which can be used to assess different scenarios using TEA and LCA, and to quantify capital and operational costs, as well as environmental impacts trade-offs of the different manufacturing configurations (i.e., centralized or distributed).

18.4.1 Integration of Oxidation and Reduction Processes into a Single Unit

Electrocatalytic processes require simultaneous (i) oxidation processes at the anode to generate electrons (e^-) and protons (H^+) and (ii) reduction processes at the cathode to consume e^- and H^+. A typical example is water electrolysis that oxidizes water ($H_2O \rightarrow 1/2O_2 + 2H^+ + 2e^-$) at the anode to generate H_2 ($2H^+ + 2e^- \rightarrow H_2$) at the

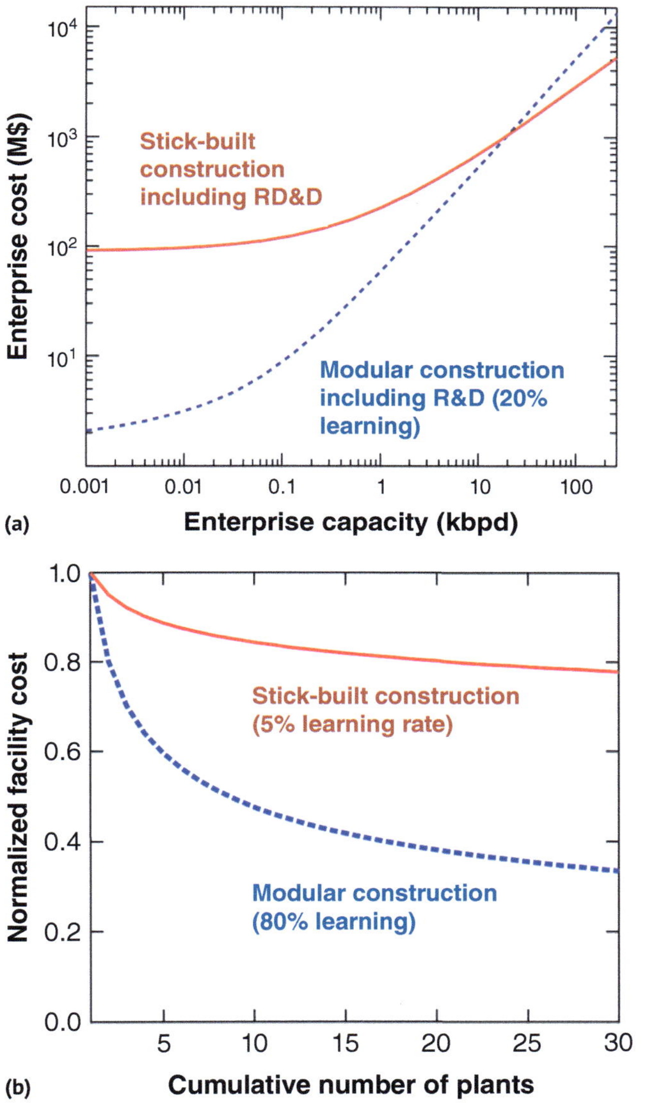

Figure 18.5 Comparison of (a) the cumulative enterprise costs of implementing stick-built (solid) and modular enterprises (broken) as a function of the size of the enterprise capacity in kbpd and (b) the change in facility cost as a function of the cumulative number of plants being built. R&D stands for research and development. RD&D stands for research, development, and demonstration. *Source*: [41]/John Wiley & Sons/CC BY 4.0.

cathode. The oxidation and reduction reactions have multiple applications of importance to chemical and fuel manufacturing and can be matched to perform different unit operations into a single step (Table 18.4). For example, as shown by Lopez-Ruiz and coworkers, electrochemical oxidation reactions can be used to treat wastewaters while simultaneously processing H_2 for bio-oil hydrotreating [19, 42, 43].

Table 18.3 Summary of main process differentiators between centralized and distributed manufacturing processes of chemicals and fuels.

Manufacturing configuration	Centralized (thermocatalytic)	Distributed (electrocatalytic)
Scales	Large	Small
Investment	Large investments, long lead times, leading to higher interest rates	Small investments, short lead times, leading to lower interest rates
Energy and resource utilization	Depends on energy distribution infrastructure (front of the meter); depends on transportation of resources	Couples production with decentralized power generation (behind the meter); uses locally available resources
Resilience and reliability	Single points of failure that can disrupt the supply chain; depends on transportation infrastructure	Point-of-use production; diversifies supply chain; does not depend on transportation infrastructure
Safety	Relies on high-pressure and -temperature process; requires large chemical storage	Lower temperature and pressure; requires smaller chemical storage (lower risk of accidents)
Process flexibility	Relies on energy-demanding high-pressure and -temperature operations; processes are optimized for steady, uniform feed	Relies on electricity-driven operations (instead of high-pressure and -temperature operations); optimized for variable feeds and dynamic operations

Table 18.4 Example of cathodic (reduction) and anodic (oxidation) reactions that can be integrated with hybrid electrolyzers for process simplification.

Cathodic reactions	Anodic reactions
Hydrogen generation	Water splitting/oxygen evolution
CO_2 reduction	Full organic oxidation to CO_2
Organic reduction	Organic decarboxylation to olefins and paraffins (Kolbe and non-Kolbe)
Coupling reactions	Partial organic oxidation to aldehydes and carboxylic acids
Oxygen reduction	Ammonia oxidation (to N_2)
Nitrate reduction to N_2 or NH_3	Chlorine oxidation (to hypochlorite)

Similarly, there are different electrocatalytic oxidation reactions of interest to the chemical sector, such as those related to olefins, paraffins, and alcohol production via (non-)Kolbe electrolysis [23, 24, 42, 44–49], partial electrocatalytic oxidation of alcohol to aldehydes and carboxylic acids [23, 24], and full electrocatalytic oxidation of waste nitrogen- and carbon-containing compounds into CO_2 and N_2 [19, 42, 43]. Similarly, the reduction reaction can be changed to directly reduce organic and inorganic compounds of interest to the commercial process such as electrocatalytic reduction of aldehydes, olefins, and unsaturated molecules [11, 50–56], CO_2 reduction [57–60], or nitrate reduction [61–64].

The flexibility of electrochemical systems and the variety of oxidation/reaction configurations that can be performed in a single unit make them a revolutionary technology to simplify processes and enable distributed manufacturing of fuels and chemicals.

18.4.2 Mild Operation Conditions

Unlike thermochemical oxidation and reduction reactions, electrochemical reactions can be performed at room temperature and atmospheric pressure without sacrificing reaction rate, which eliminates the need for heat integration and delivery units such as heat exchangers and steam plants, thereby simplifying the process. For example, Qiu et al. demonstrated that the electrochemical decarboxylation of carboxylic acids at room temperature and atmospheric pressure is one to two orders of magnitude higher than that of a thermocatalytic system at 300 °C and 57 bar [23, 24]. Therefore, the construction materials as well as the operation (no need to pressurize or heat up) are much simpler, making electrochemical systems more amenable for modularization.

18.4.3 Simpler and Safer Operation

Electrocatalytic systems can generate hazardous chemicals and intermediates on site (e.g., hydrogen peroxide, hypochlorite, and H_2) and on demand, thus eliminating the need for storage and transportation infrastructure and making the chemical process simpler and safer. These advantages can only be captured by TEA and LCA and can persuade a commercial process to adopt electrolysis technology even if there are no energy or performance advantages compared to the incumbent technologies.

An additional benefit of electrocatalytic systems is that the performance and stability of electrolyzers can be assessed by changes in current and voltage and through operando electrochemical characterization (e.g., cyclic voltammetry [CV], impedance spectroscopy, and linear sweep voltammetry [LSV]) without the need to shut down the electrolyzer or disassemble it. For example, if the current stays constant while operating at a constant potential, the process operator can infer that there is no electrocatalytic performance deactivation. Likewise, if the current or voltage starts to change, the process operator can identify which electrolyzer cell in a stack unit is failing and schedule its replacement without the need to shut down the entire process. Additionally, because these systems operate with electricity as the main driving force (instead of heat from a steam plant), they have simple start-ups and shutdowns and can be easily automated. Thus, the electrolyzer systems would need less supervision and fewer plant operators, resulting in lower maintenance costs.

18.4.4 Electrification With Diverse Sources

Because the main energy source of electrocatalytic systems is electrical power, they represent a unique approach to electrify distributed manufacturing of chemicals and fuels using local, renewable, or nonrenewable electricity sources. The hybrid electrolyzer can be sized up in accordance with the local electricity generation capacity. Therefore, as the electricity capacity increases, the size of the hybrid electrolyzer can be proportionally increased by connecting new electrolyzer modules (i.e., numbering up) instead of building larger electrolyzers, which simplifies the planning and expansion of chemical and fuel processing. Otherwise, the electrification of large, centralized fuel and chemical manufacturing would require dedicated power plants with power requirements beyond what could be locally provided with renewable sources.

18.5 Scale-up Considerations

Scaling electrolyzers from lab (watts) to demonstration (megawatts) and industrial scales (gigawatts) is a systems challenge that starts with clear and reproduceable lab performance and ends with disciplined, data-driven improvement in the field (Figure 18.6).

A rule of thumb is to demonstrate scalability by increasing the scale in staged steps (e.g., W → kW → tens of kW → MW → tens of MW → GW-class plants, measured as aggregate capacity), which can be done through a combination of increasing the electrode surface area per electrolyzer cell (e.g., 1 cm^2 → 10 cm^2 → 100 cm^2 → 1,000 cm^2

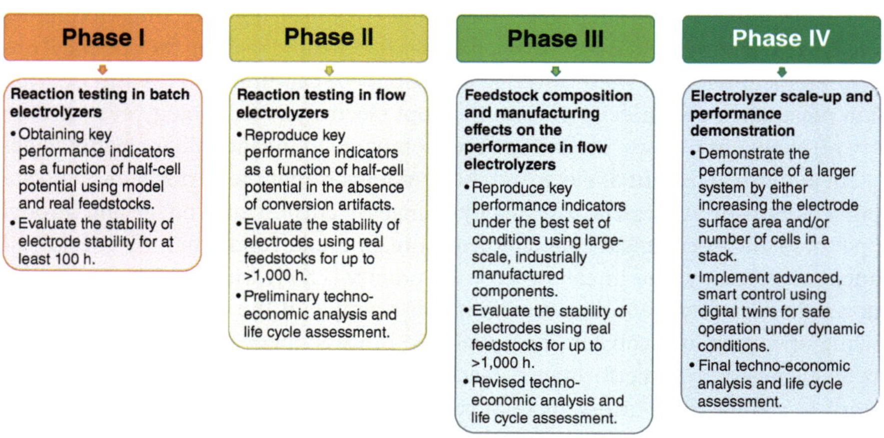

Figure 18.6 Different phases of electrolyzer scale-up.

but up to 1 m^2) and increasing the number of electrolyzer cells in the stack (usually starting at three electrolyzer cells per stack). After identifying key performance indicators in the W-scale system, an important consideration of the scale-up efforts is demonstrating that the key performance indicators are consistent across the different scales and that the quality of the manufactured components is consistent. TEA and LCA need to be continuously updated with real manufacturing and operating data as a function of the different scale-up stages to refine capital and operational cost estimates.

18.5.1 Electrolyzer Design

Designing electrolyzer cells and stacks for manufacturability, maintenance, and consistency is central to scale-up, which often falls outside the scope of academic research. As the active electrode area and the number of electrodes in the system increase, maintaining tight tolerances across the electrodes and over the active area and uniform compression under pressure and temperature excursions is essential to avoid local nonuniformities in current distribution (Figure 18.7) [65–67]. Furthermore, there is a trade-off between enlarging cell area and increasing the number of cells. On the one hand, larger areas simplify assembly and can decrease manufacturing costs per unit but exacerbate gas bubble coverage and

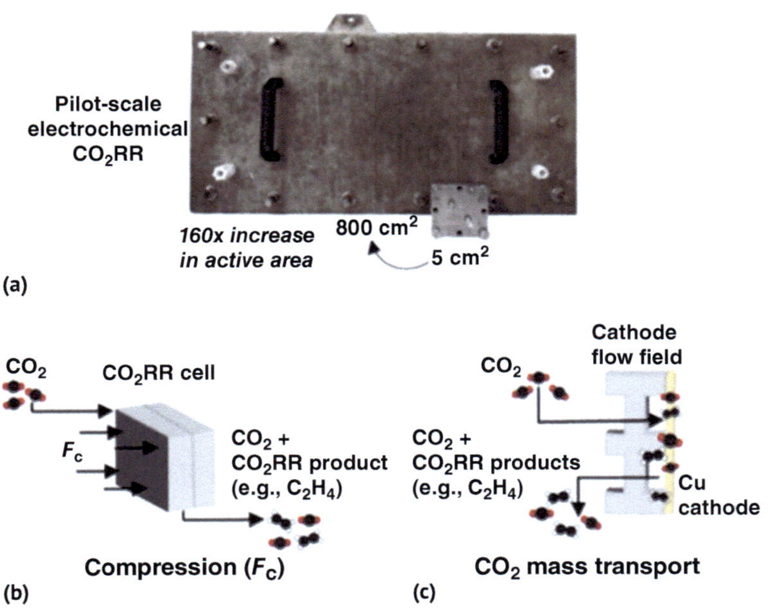

Figure 18.7 Size comparison schematic of MEA electrolyzers with 5 and 800 cm^2 active areas and key challenges in scaling up CO_2RR systems. (a) Comparison of bench (5 cm^2) and pilot (800 cm^2) MEA electrolyzer areas. (b) Compression is a key issue in pilot-scale CO_2RR. (c) CO_2 through the manifold to the catalyst surface is another key consideration in pilot-scale CO_2RR. *Source*: [66]/with permission of American Chemical Society.

thermal management. On the other hand, more cells allow smaller per-cell areas that can improve within-cell heat- and mass-transfer uniformity and bubble removal, but also increase header/manifold complexity and the risk/magnitude of shunt currents between cells [67, 68]. Thus, researchers need to validate the electrolyzer design with multiphysics modeling and iterate data-driven improvements [65].

The performance of individual electrolyzer cells can be tracked through electrochemical methods (e.g., polarization curves, electrochemical impedance spectroscopy) and modeled through multiphysics methods [68], which can help with operando identification of nonuniformities and component failures without the need to disassemble the electrolyzer. A modular "drag-and-drop" approach at the plant level allows for swapping underperforming modules without shutting down the entire electrolyzer system, but this only works if the electrolyzer designs are highly standardized with consistent mechanical, electrical, fluid, and control interfaces [69].

18.5.2 Component Manufacturing

Manufacturing key electrolyzer components, such as electrodes and membranes, requires scalable fabrication methods [69–71]. For example, direct coating methods must be tuned for coating uniformity (e.g., thickness, porosity), catalyst utilization (e.g., ionomer/catalyst ratio and distribution), and membrane integrity (e.g., compatibility with electrolyte, feedstock, and reaction products). Electrolyzer design and fabrication should aim for scalability, standardization, and quality assurance from the outset. Digital twins and factory simulation can be used to simulate layouts, material and people flows, recyclability, supply chain of materials of construction, and capacity scenarios, enabling the design of modular, reconfigurable systems without bespoke redesign at each scale, thereby accelerating and de-risking the scale-up process (Figure 18.8) [65, 72]. Economies of scale are realized

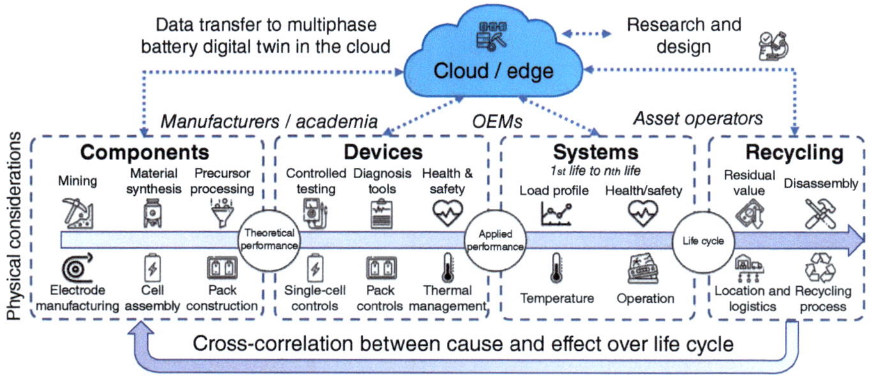

Figure 18.8 Vision for the ultimate digital twin for electrocatalytic systems in which lifetime performance is looped back to design and manufacturing choices. *Source:* [65]/with permission of Elsevier.

when components and processes are standardized and quality systems are robust [73]; this is especially important as manufacturers move toward cumulative multi-gigawatt-per-year (GW/yr) manufacturing capacities. Supplier engagement for critical component manufacture, qualification, and potential redesign is a core part of de-risking because field and factory performance data from scaled systems should feed back into the manufacturing process and drive continuous improvement.

18.5.3 Dynamic Operations and Advanced Controls

If hybrid electrolyzers are expected to operate under dynamic conditions, operability must be validated early in the scale-up process. Sensing, operando diagnostics, and control are the backbone of reliable scale-up and operation and are often underprioritized in academic laboratory studies [69, 71]. For example, the electrolyzer will need smart control strategies that enforce safe shutdown limits for pressure, temperature, flow rates, gas composition, and leak detection to protect the efficiency and integrity of the hybrid electrolyzer (Figure 18.9) [72]. Local, stack-level diagnostics (e.g., distributed temperature sensing, impedance spectroscopy) should be integrated. Programmable logic controllers that automate monitoring of flows, pressures, (normalized) reaction rates, current and faradaic efficiencies, and

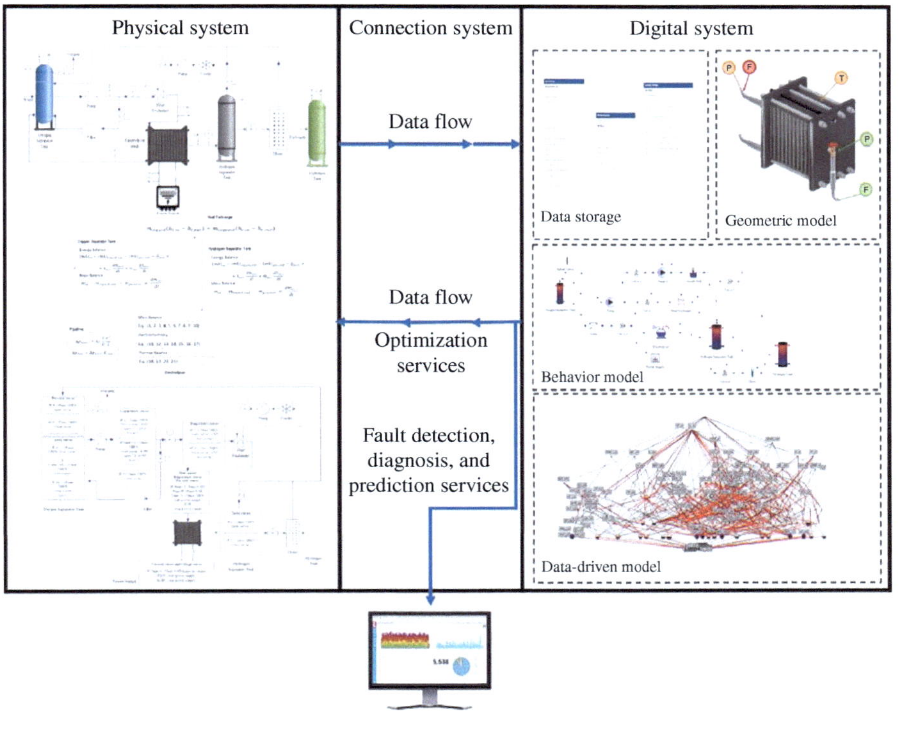

Figure 18.9 Example of data flow between physical and digital systems for advanced control and system diagnostic. *Source*: [72]/with permission of Elsevier.

product selectivity and overall energy consumption and efficiency can be combined with alarms and interlocks to ensure safe, efficient operation. In addition, AI and machine learning can augment control by (i) building soft sensors to estimate unmeasured states such as gas crossover risk or local membrane temperature, (ii) performing real-time anomaly detection on multivariate sensor streams to trigger early interventions, (iii) adapting set points via model predictive control to optimize ramp rates, turndown, and pressure/flow distribution under variable power, (iv) forecasting degradation to schedule maintenance and adjust operating envelopes, and (v) linking manufacturing fingerprints to control parameters for module-specific, feed-forward tuning.

18.6 Future Research Direction and Commercialization Strategies

Taken together, these future prospects and the commercialization of hybrid electrolyzers hinge on (i) reporting and optimizing appropriate key performance indicators, (ii) identifying appropriate use cases and integration scenarios, (iii) engineering and demonstrating scale-ups with standardized manufacturability and uniform performance, and (iv) operating with robust controls and diagnostics. Therefore, we urge the field to consider the following points so that new materials and reactions can be adequately benchmarked against existing, published works.

Proper electrocatalyst testing and performance reporting: the performance of any electrocatalyst is governed by half-cell potential (independent variable), which is often underreported or not properly *iR*-compensated, and mass-transport conditions. Therefore, it remains unclear whether apparent improvements in electrocatalyst performance arise from the new electrocatalyst design or simply from different testing protocols. Additionally, the dimensions, composition, or ECSA of new electrocatalysts may differ from previous formulations. Consequently, comparing electrocatalyst performance at equal current density rather than equal (*iR*-compensated) half-cell potential can be very misleading. We thus encourage the field to report reaction rates normalized by either (i) the mass of the electrode used (i.e., $kg_{COD} \cdot (h \cdot g_{electrode})^{-1}$), (ii) the cross-sectional area of the electrode (i.e., $kg_{COD} \cdot (h \cdot cm^2_{electrode})^{-1}$), or (iii) the ECSA (i.e., $kg_{COD} \cdot (h \cdot cm^2_{ECSA})^{-1}$).

Membranes and separators: while electrodes receive most of the attention in research and development, there are limited membrane choices (and separators) that remain effective under the diverse reaction conditions in which hybrid electrolyzers operate (e.g., acidic or basic, aqueous or nonaqueous). Thus, highly stable membranes are needed in a wide range of pH and in the presence of organic as well as aqueous systems that show no degradation. This highlights an emerging opportunity to codevelop new membrane materials for specific hybrid electrolyzer configurations that deliver high stability and conductivity under industrially relevant conditions, especially in the presence of contaminants and varying reactants and products.

Simulation: as high-throughput electrode synthesis and screening methods proliferate, larger datasets are produced, creating new challenges in data curation, processing, and analysis to identify meaningful trends. This creates a genuine need for artificial intelligence and machine learning approaches, which have already been applied to electrochemical reactor optimization. These approaches can be effective even in low-data regimes [74] and are applicable to a wider range of design-of-experiments problems beyond electrochemistry. Common challenges in the broader community include (i) the scarcity of negative/null results in published datasets, (ii) insufficient data and metadata standardization (formats, units, and test protocols), and (iii) unreported or poorly quantified uncertainties.

Proper demonstration and scale-up steps: electrode testing under model (idealized) conditions can be useful to quickly benchmark electrode performance under controlled conditions. However, electrodes must also be tested under realistic operating conditions because industrial feedstocks may contain inorganic or organic compounds that could inhibit performance or accelerate degradation. Furthermore, electrodes should be tested for significantly longer than 100 hours—preferably on the order of 1,000+ hours at relevant current density, temperature, and electrolyte composition—to demonstrate stability. Membrane electrode assembly (MEA)-based flow cells/flow electrolyzers are preferred for this testing. Once a robust performance baseline is established in flow electrolyzers, researchers should evaluate performance using commercially available, supplier-grade materials from potential suppliers (e.g., electrodes or electrode supports, membranes, gaskets, and other construction materials for the electrolyzer) and benchmark against in-house materials to support supplier qualification and de-risk scale-up.

Critical materials exposure and supply risk: due to metal scarcity, price volatility, and incomplete recycling infrastructure, researchers should focus on developing electrodes with low-to-no-PGM content that have validated durability. Furthermore, there should be an emphasis on the recyclability of these materials because they are often designed to be acid-resistant, which can hinder end-of-life separation and recovery. Similarly, using high-purity precursors may be infeasible at large scales (because of cost and supply-chain constraints), so researchers should systematically consider the effect of realistic impurities (e.g., metal cations, carbonates, and organics) on electrode and membrane performance and long-term stability. Furthermore, the correlation between device performance and manufacturing parameters for electrodes and electrolyzers (e.g., coating method and quality, compression/torque set points, tolerances, and construction materials) is underdeveloped and should be strengthened via designed experiments, statistical process control, and end-of-line/field performance traceability.

Durability under industrially relevant operations: there is limited validated performance data for cycling between low and high electrical loads, which can affect the overall efficiency and safety of the process due to, for example, shunt currents and gas crossover that are typically negligible in small laboratory-scale single-cell tests but can become significant in multi-cell stacks and common manifolds, especially at low loads. Additionally, feedstocks used in real applications may have variable compositions due to changes in upstream/downstream processing as well as seasonal (e.g., biomass) and geographical variations (e.g., biomass- and petroleum-derived streams). Furthermore, the release of metal ions (Ni, Fe, and Cr) from system materials and upstream/downstream unit operations can deposit on electrodes and membranes, creating hot spots and altered interfacial properties, leading to performance loss over time. Therefore, researchers need to systematically study how operation under simulated, industrially relevant duty cycles, including dynamic load profiles, feedstock compositions, and long-term electrolyte recirculation, might affect electrode and full electrolyzer performance. To that end, long-term studies exceeding ~1,000 hours at relevant current density and temperature should be performed, and electrodes and MEAs need to be physically (e.g., X-ray diffraction, X-ray photoelectron spectroscopy, and scanning electron microscopy/transmission electron microscopy) and electrochemically (e.g., CV, ECSA, electrochemical impedance spectroscopy, LSV, polarization curves, and gas-crossover measurements) characterized to understand degradation mechanisms and implement mitigation strategies.

Cost and energy realism: although energy consumption for converting reactants or generating target products can be calculated from electrolyzer performance data and is the basis for high-level TEA, it is rarely reported. Being able to compare energy-consumption values across different technologies, and by extension the operational electricity costs, will highlight strengths and identify potential areas for improvement for new hybrid processes. For example, a hybrid electrolyzer might generate a target product (e.g., H_2, CO) at higher efficiency; however, it might not be a suitable candidate because the overall electrical demand could far exceed existing grid capacity. Furthermore, commercial processes must understand the capital and operational costs of a new process to assess their feasibility and adoption. However, there is a lack of TEA/LCA grounded in performance data obtained under industrially relevant conditions and at scale. Therefore, researchers are encouraged to perform TEA (and LCA) for use cases grounded in real industrial needs with transparent system boundaries, clear normalization (e.g., kWh per kg product and inclusion/exclusion of balance-of-plant and dynamic operation), and uncertainty/sensitivity analyses. It is important to also consider potential changes to the entire process flow diagram and supply-chain infrastructure (e.g., avoiding H_2 transportation and storage when H_2 is produced on demand) in the TEA because process simplification and unit-operation integration might be a sufficient cost driver to adopt the technology even if the performance of the electrocatalytic process is similar to or worse than the incumbent process.

By addressing these gaps, the research community can help accelerate the commercial deployment of hybrid electrolyzers.

References

1 Botte, G. G. (2014). Electrochemical manufacturing in the chemical industry. *Electrochem. Soc. Interface 23*(3), 49. https://doi.org/10.1149/2.F04143if

2 Leech, M. C., Garcia, A. D., Petti, A., Dobbs, A. P., & Lam, K. (2020). Organic electrosynthesis: From academia to industry. *React. Chem. Eng. 5*(6), 977–990. https://doi.org/10.1039/D0RE00064G

3 Dunnill, C. W., Rearden, A., Mandale, S., Glover, K., & Phillips, R. (2020). Optimizing the design of an alkaline water splitting device test cell for renewable energy storage as hydrogen. *Arch. Chem. Chem. Eng. 2*(1), 1–9. https://doi.org/10.5281/zenodo.3964220

4 Göldner, V., Fangmeyer, J., & Karst, U. (2025). Mass spectrometric methods for the analysis of electrochemical transformation products. *Trends Anal. Chem. 185*, 118178. https://doi.org/10.1016/j.trac.2025.118178

5 Jin, H., Xu, J., Liu, H., Shen, H., Yu, H., Jaroniec, M., Zheng, Y., & Qiao, S.-Z. (2023). Emerging materials and technologies for electrocatalytic seawater splitting. *Sci. Adv. 9*(42), eadi7755. https://doi.org/10.1126/sciadv.adi7755

6 Begildayeva, T., Jiang, J., Ren, Z. J., & Lopez-Ruiz, J. A. (2025). Progress, gaps, and opportunities in electrocatalytic oxidation of wastewater: A review. *Water Res. 292*, 125164.

7 Mardani, S., Baghdadi, M., Torabian, A., & Goharrizi, B. A. (2023). Optimization of ammonia and COD removal from municipal wastewater effluent by electrochemical continuous flow reactor equipped with Ti/RuO_2 and Cu foam. *J. Water Process Eng. 55*, 104185. https://doi.org/10.1016/j.jwpe.2023.104185

8 Xiao, H., Xu, F., Chen, J., Hao, Y., Guo, Y., Zhu, C., Luo, S., & Jiang, B. (2023). Electrogenerated oxychlorides induced overlooked negative effects on electro-oxidation wastewater treatment in terms of over-evaluated COD removal efficiency and biotoxicity. *J. Hazard. Mater. 456*, 131667. https://doi.org/10.1016/j.jhazmat.2023.131667

9 Khalili, S., Afkhami, A., & Madrakian, T. (2023). Electrochemical simultaneous treatment approach: Electro-reduction of CO_2 at Pt/PANI@ZnO paired with wastewater electro-oxidation over PbO_2. *Appl. Catal. B: Environ. 328*, 122545. https://doi.org/10.1016/j.apcatb.2023.122545

10 Shao, D., Li, W., Wang, Z., Yang, C., Xu, H., Yan, W., Yang, L., Wang, G., Yang, J., Feng, L., Wang, S., Li, Y., Jia, X., & Song, H. (2022). Variable activity and selectivity for electrochemical oxidation wastewater treatment using a magnetically assembled electrode based on Ti/PbO_2 and carbon nanotubes. *Sep. Purif. Technol. 301*, 122008. https://doi.org/10.1016/j.seppur.2022.122008

11 Akhadea, S. A., Singh, N., Gutiérrez, O. Y., Lopez-Ruiz, J. A., Wang, H., Holladay, J. D., Liu, Y., Karkamkar, A., Weber, R. S., Padmaperuma, A. B., Lee, M.-S., Whyatt, G. A., Elliott, M., Holladay, J. E., Lercher, J. A., Rousseau, R., & Glezakou, V.-A. (2020). Electrocatalytic hydrogenation of biomass-derived organics: A review. *Chem. Rev. 120*(20), 11370–11419. https://doi.org/10.1021/acs.chemrev.0c00158

12 Atobe, M., Tateno, H., & Matsumura, Y. (2018). Applications of flow microreactors in electrosynthetic processes. *Chem. Rev. 118*(9), 4541–4572. https://doi.org/10.1021/acs.chemrev.7b00353

13 Horii, D., Fuchigami, T., & Atobe, M. (2007). A new approach to anodic substitution reaction using parallel laminar flow in a micro-flow reactor. *J. Am. Chem. Soc. 129*(38), 11692–11693. https://doi.org/10.1021/ja075180s

14 Ralph, T. R., Hitchman, M. L., Millington, J. P., & Walsh, F. C. (1996). Mass transport in an electrochemical laboratory filterpress reactor and its enhancement by turbulence promoters. *Electrochim. Acta 41*(4), 591–603. https://doi.org/10.1016/0013-4686(95)00346-0

15 Noël, T., Cao, Y., & Laudadio, G. (2019). The fundamentals behind the use of flow reactors in electrochemistry. *Acc. Chem. Res. 52*(10), 2858–2869. https://doi.org/10.1021/acs.accounts.9b00412

16 Tanbouza, N., Ollevier, T., & Lam, K. (2020). Bridging lab and industry with flow electrochemistry. *iScience 23*(11), 101720. https://doi.org/10.1016/j.isci.2020.101720

17 Maljuric, S., Jud, W., Kappe, C. O., & Cantillo, D. (2020). Translating batch electrochemistry to single-pass continuous flow conditions: an organic chemist's guide. *J. Flow Chem. 10*(1), 181–190. https://doi.org/10.1007/s41981-019-00050-z

18 Elsherbini, M. & Wirth, T. (2019). Electroorganic synthesis under flow conditions. *Acc. Chem. Res. 52*(12), 3287–3296. https://doi.org/10.1021/acs.accounts.9b00497

19 Bhatwadekar, S., Boruah, B., Riedel, N. W., Royer, N. D., Pedersen, T. H., & Lopez-Ruiz, J. A. (2025). Electrocatalytic oxidation of hydrothermal liquefaction-derived aqueous phase for on-site wastewater treatment and H_2 production. *Cell Rep. Phys. Sci. 6*(5), 102555. https://doi.org/10.1016/j.xcrp.2025.102555

20 Strange, L. E., Boruah, B., Riedel, N. W., Bowden, M. E., Engelhard, M. H., & Lopez-Ruiz, J. A. (2025). Electrocatalytic benchmarking of ruthenium-based bimetallic anodes for the electrocatalytic oxidation of biomass-derived wastewater. *Appl. Catal. O: Open 206*, 207055. https://doi.org/10.1016/j.apcato.2025.207055

21 Chen, J. G., Jones, C. W., Linic, S., & Stamenkovic, V. R. (2017). Best practices in pursuit of topics in heterogeneous electrocatalysis. *ACS Catal. 7*(9), 6392–6393.

22 Trasatti, S. & Petrii, O. (1991). Real surface area measurements in electrochemistry. *Pure Appl. Chem. 63*(5), 711–734.

23 Qiu, Y., Lopez-Ruiz, J. A., Sanyal, U., Andrews, E., Gutiérrez, O. Y., & Holladay, J. D. (2020). Anodic electrocatalytic conversion of carboxylic acids on thin films of RuO_2, IrO_2, and Pt. *Appl. Catal. B: Environ. 277*, 119277. https://doi.org/10.1016/j.apcatb.2020.119277

24 Qiu, Y., Lopez-Ruiz, J. A., Zhu, G., Engelhard, M. H., Gutiérrez, O. Y., & Holladay, J. D. (2022). Electrocatalytic decarboxylation of carboxylic acids over RuO_2 and Pt nanoparticles. *Appl. Catal. B: Environ. 305*, 121060. https://doi.org/10.1016/j.apcatb.2021.121060

25 Amouzegar, K. & Savadogo, O. (1998). Electrocatalytic hydrogenation of phenol on dispersed Pt: Reaction mechanism and support effect. *Electrochim. Acta 43*(5–6), 503–508.

26 Green, S. K., Lee, J., Kim, H. J., Tompsett, G. A., Kim, W. B., & Huber, G. W. (2013). The electrocatalytic hydrogenation of furanic compounds in a continuous electrocatalytic membrane reactor. *Green Chem. 15*(7), 1869–1879. https://doi.org/10.1039/C3GC00090G

27 Moniri, S., Van Cleve, T., & Linic, S. (2017). Pitfalls and best practices in measurements of the electrochemical surface area of platinum-based nanostructured electro-catalysts. *J. Catal. 345*, 1–10.

28 Zhao, B., Guo, Q., & Fu, Y. (2014). Electrocatalytic hydrogenation of lignin-derived phenol into alkanes by using platinum supported on graphite. *Electrochemistry 82*(11), 954–959. https://doi.org/10.5796/electrochemistry.82.954

29 Baturina, O. A., Lu, Q., Padilla, M. A., Xin, L., Li, W., Serov, A., Artyushkova, K., Atanassov, P., Xu, F., Epshteyn, A., Brintlinger, T., Schuette, M., & Collins, G. E. (2014). CO_2 electroreduction to hydrocarbons on carbon-supported Cu nanoparticles. *ACS Catal. 4*(10), 3682–3695. https://doi.org/10.1021/cs500537y

30 Fang, L.-l., Tao, Q., Li, M.-f., Liao, L.-w., Chen, D., & Chen, Y.-x. (2010). Determination of the real surface area of palladium electrode. *Chin. J. Chem. Phys. 23*(5), 543–548. https://doi.org/10.1088/1674-0068/23/05/543-548

31 Sheng, W., Myint, M., Chen, J. G., & Yan, Y. (2013). Correlating the hydrogen evolution reaction activity in alkaline electrolytes with the hydrogen binding energy on monometallic surfaces. *Energy Environ. Sci. 6*(5), 1509–1512. https://doi.org/10.1039/c3ee00045a

32 Boudart, M., Aldag, A., Ptak, L. D., & Benson, J. (1968). On the selectivity of platinum catalysts. *J. Catal. 11*(1), 35–45.

33 Carroll, K. J., Burger, T., Langenegger, L., Chavez, S., Hunt, S. T., Román-Leshkov, Y., & Brushett, F. R. (2016). Electrocatalytic hydrogenation of oxygenates using earth-abundant transition-metal nanoparticles under mild conditions. *ChemSusChem 9*(15), 1904–1910. https://doi.org/10.1002/cssc.201600290

34 Patrascu, M. (2023). Process intensification for decentralized production. *CEP:PI 184*, 109291. https://doi.org/10.1016/j.cep.2023.109291

35 Resasco, D. E., Wang, B., & Sabatini, D. (2018). Distributed processes for biomass conversion could aid UN Sustainable Development Goals. *Nat. Catal. 1*(10), 731–735. https://doi.org/10.1038/s41929-018-0166-6

36 Fernández, C. A., Chapman, O., Brown, M. A., Alvarez-Pugliese, C. E., & Hatzell, M. C. (2024). Achieving decentralized, electrified, and decarbonized ammonia production. *Environ. Sci. Technol. 58*(16), 6964–6977. https://doi.org/10.1021/acs.est.3c10751

37 Jewess, M. & Crabtree, R. H. (2016). Electrocatalytic nitrogen fixation for distributed fertilizer production? *ACS Sustain. Chem. Eng. 4*(11), 5855–5858. https://doi.org/10.1021/acssuschemeng.6b01473

38 Ji, Y. & Xia, C. (2025). Distributed production of ready-to-use liquid nitrogenous fertilizer. *Nat. Sustain. 8*, 992–993. https://doi.org/10.1038/s41893-025-01627-4

39 Martín, A. J. & Pérez-Ramírez, J. (2019). Heading to distributed electrocatalytic conversion of small abundant molecules into fuels, chemicals, and fertilizers. *Joule 3*(11), 2602–2621. https://doi.org/10.1016/j.joule.2019.09.007

40 Ramanujam, A., Constante-Flores, G. E., & Li, C. (2023). Distributed manufacturing for electrified chemical processes in a microgrid. *AIChE J. 69*(12), e18265. https://doi.org/10.1002/aic.18265

41 Weber, R. S., Holladay, J. E., Jenks, C., Panisko, E. A., Snowden-Swan, L. J., Ramirez-Corredores, M., Baynes, B., Angenent, L. T., & Boysen, D. (2018). Modularized production of fuels and other value-added products from distributed, wasted, or stranded feedstocks. *WIREs Energy Environ. 7*(6), e308. https://doi.org/10.1002/wene.308

42 Lopez-Ruiz, J. A., Qiu, Y., Andrews, E., Gutiérrez, O. Y., & Holladay, J. D. (2021). Electrocatalytic valorization into H_2 and hydrocarbons of an aqueous stream derived from hydrothermal liquefaction. *J. Appl. Electrochem. 51*(1), 107–118. https://doi.org/10.1007/s10800-020-01452-x

43 Lopez-Ruiz, J. A., Riedel, N. R., Boruah, B., Bhatwadekar, S., Strange, L. E., & Li, S. (2024). Low-temperature electrochemical wastewater oxidation. Report No. PNNL-35535. https://www.pnnl.gov/main/publications/external/technical_reports/PNNL-35535.pdf

44 Levy, P. F., Sanderson, J. E., & Cheng, L. K. (1984). Kolbe electrolysis of mixtures of aliphatic organic acids. *J. Electrochem. Soc. 131*(4), 773–777. https://doi.org/10.1149/1.2115697

45 Palkovits, S., & Palkovits, R. (2019). The role of electrochemistry in future dynamic bio-refineries: A focus on (non-)Kolbe electrolysis. *Chem. Ing. Tech. 91*(6), 699–706. https://doi.org/10.1002/cite.201800205

46 Sanderson, J. E., Levy, P. F., Cheng, L. K., & Barnard, G. W. (1983). The effect of pressure on the product distribution in Kolbe electrolysis. *J. Electrochem. Soc. 130*(9), 1844–1848. https://doi.org/10.1149/1.2120109

47 Teetz, N., Holtmann, D., Harnisch, F., & Stöckl, M. (2022). Upgrading Kolbe electrolysis—highly efficient production of green fuels and solvents by coupling biosynthesis and electrosynthesis. *Angew. Chem. Int. Ed. 61*(50), e202210596. https://doi.org/10.1002/anie.202210596

48 Wang, S., Ren, D., Du, Y., Zhang, M., Zhang, N., Sun, Y., & Huo, Z. (2023). A review of anodic catalysts and their application in (non-)Kolbe electrocatalytic decarboxylation of carboxylic acids. *Carbon Resour. Convers. 6*, 287–297. https://doi.org/10.1016/j.crcon.2023.04.004

49 Yuan, G., Wu, C., Zeng, G., Niu, X., Shen, G., Wang, L., Zhang, X., Luque, R., & Wang, Q. (2020). Kolbe electrolysis of biomass-derived fatty acids over Pt nanocrystals in an electrochemical cell. *ChemCatChem 12*, 642–648. https://doi.org/10.1002/cctc.201901443

50 Andrews, E., Lopez-Ruiz, J. A., Egbert, J., Koh, K., Sanyal, U., Miao, S., Li, D., Karkamkar, A., Derewinski, M. A., Holladay, J., Gutiérrez, O. Y., & Holladay, J. D. (2020). Performance of base and noble metals for electrocatalytic hydrogenation of bio-oil-derived oxygenated compounds. *ACS Sustain. Chem. Eng. 8*(11), 4407–4418. https://doi.org/10.1021/acssuschemeng.9b07041

51 Gutiérrez, O. Y., Grubel, K., Kothandaraman, J., Lopez-Ruiz, J. A., Brooks, K. P., Bowden, M. E., & Autrey, T. (2023). Using earth abundant materials for long duration energy storage: electro-chemical and thermo-chemical cycling of bicarbonate/formate. *Green Chem.* *25*(11), 4222–4233. https://doi.org/10.1039/D3GC00219E

52 Lopez-Ruiz, J. A., Andrews, E. A., Akhade, S. A., Lee, M., Koh, K., Sanyal, U., Yuk, S. F., Karkamkar, A. J., Derewinski, M. A., Holladay, J., Glezakou, V.-A., Rousseau, R., Gutiérrez, O. Y., & Holladay, J. D. (2019). Understanding the role of metal and molecular structure on the electrocatalytic hydrogenation of oxygenated organic compounds. *ACS Catal.* *9*(11), 9964–9972. https://doi.org/10.1021/acscatal.9b02921

53 Lopez-Ruiz, J. A., Sanyal, U., Egbert, J., Gutiérrez, O. Y., & Holladay, J. (2018). Kinetic investigation of the sustainable electrocatalytic hydrogenation of benzaldehyde on Pd/C: Effect of electrolyte composition and half-cell potentials. *ACS Sustain. Chem. Eng.* *6*(12), 16073–16085. https://doi.org/10.1021/acssuschemeng.8b02637

54 Page, J. R., Pophali, A., Kim, T., Lopez-Ruiz, J. A., Bliznakov, S., & Valla, J. A. (2024). Effect of Pt and Ru-based catalysts on the electrochemical hydrodeoxygenation of phenol to cyclohexane. *Catal. Sci. Technol.* *14*(19), 5559–5573. https://doi.org/10.1039/D4CY00634H

55 Sanyal, U., Lopez-Ruiz, J. A., Padmaperuma, A. B., Holladay, J., & Gutiérrez, O. Y. (2018). Electrocatalytic hydrogenation of oxygenated compounds in aqueous phase. *Org. Process Res. Dev.* *22*(12), 1590–1598. https://doi.org/10.1021/acs.oprd.8b00236

56 Sanyal, U., Yuk, S. F., Koh, K., Lee, M.-S., Stoerzinger, K., Zhang, D., Meyer, L. C., Lopez-Ruiz, J. A., Karkamkar, A., Holladay, J. D., Camaioni, D. M., Nguyen, M.-T., Glezakou, V.-A., Rousseau, R., Gutiérrez, O. Y., & Lercher, J. A. (2021). Hydrogen bonding enhances the electrochemical hydrogenation of benzaldehyde in the aqueous phase. *Angew. Chem. Int. Ed.* *60*(1), 290–296. https://doi.org/10.1002/anie.202008178

57 Gao, N., Quiroz-Arita, C., Diaz, L. A., & Lister, T. E. (2021). Intensified co-electrolysis process for syngas production from captured CO_2. *J. CO_2 Util.* *43*, 101365. https://doi.org/10.1016/j.jcou.2020.101365

58 Li, T., Lees, E. W., Goldman, M., Salvatore, D. A., Weekes, D. M., & Berlinguette, C. P. (2019). Electrolytic conversion of bicarbonate into CO in a flow cell. *Joule* *3*(6), 1487–1497. https://doi.org/10.1016/j.joule.2019.05.021

59 Möller, T., Ju, W., Bagger, A., Wang, X., Luo, F., Thanh, T. N., Varela, A. S., Rossmeisl, J., & Strasser, P. (2019). Efficient CO_2 to CO electrolysis on solid Ni–N–C catalysts at industrial current densities. *Energy Environ. Sci.* *12*(2), 640–647. https://doi.org/10.1039/C8EE02662A

60 Song, H., Fernández, C. A., Choi, H., Huang, P.-W., Oh, J., & Hatzell, M. C. (2024). Integrated carbon capture and CO production from bicarbonates through bipolar membrane electrolysis. *Energy Environ. Sci.* *17*(10), 3570–3579. https://doi.org/10.1039/D4EE00048J

61 Chauhan, R. & Srivastava, V. C. (2020). Electrochemical denitrification of highly contaminated actual nitrate wastewater by Ti/RuO_2 anode and iron cathode. *Chem. Eng. J.* *386*, 122065. https://doi.org/10.1016/j.cej.2019.122065

62 Sanjuán, I., García-Cruz, L., Solla-Gullón, J., Expósito, E., & Montiel, V. (2020). Bi–Sn nanoparticles for electrochemical denitrification: activity and selectivity towards N_2 formation. *Electrochim. Acta 340*, 135914. https://doi.org/10.1016/j.electacta.2020.135914

63 Zhang, K., Li, X. S., Duan, Y., King, D. L., Singh, P., & Li, L. (2013). Roles of double salt formation and $NaNO_3$ in Na_2CO_3-promoted MgO absorbent for intermediate temperature CO_2 removal. *Int. J. Greenh. Gas Control 12*, 351–358. https://doi.org/10.1016/j.ijggc.2012.11.013

64 Zhang, Z., Xu, Y., Shi, W., Wang, W., Zhang, R., Bao, X., Zhang, B., Li, L., & Cui, F. (2016). Electrochemical-catalytic reduction of nitrate over Pd–Cu/γAl_2O_3 catalyst in cathode chamber: Enhanced removal efficiency and N_2 selectivity. *Chem. Eng. J. 290*, 201–208. https://doi.org/10.1016/j.cej.2016.01.063

65 Dubarry, M., Howey, D., & Wu, B. (2023). Enabling battery digital twins at the industrial scale. *Joule 7*(6), 1134–1144. https://doi.org/10.1016/j.joule.2023.05.005

66 Nelson, V. E., O'Brien, C. P., Edwards, J. P., Liu, S., Gabardo, C. M., Sargent, E. H., & Sinton, D. (2024). Scaling CO_2 electrolyzer cell area from bench to pilot. *ACS Appl. Mater. Interfaces 16*(38), 50818–50825. https://doi.org/10.1021/acsami.4c11103

67 Richard, D., Jang, J., Çıtmacı, B., Luo, J., Canuso, V., Korambath, P., Morales-Leslie, O., Davis, J. F., Malkani, H., Christofides, P. D., & Morales-Guio, C. G. (2023). Smart manufacturing inspired approach to research, development, and scale-up of electrified chemical manufacturing systems. *iScience 26*(6), 106966. https://doi.org/10.1016/j.isci.2023.106966

68 Huang, D., Zhong, Z., Ai, X., Hu, K., Xiong, B., Wen, Q., Fang, J., & Cheng, S. (2024). Size design strategy for scaling up alkaline water electrolysis stack integrated with renewable energy source: A multiphysics modeling approach. *Energy Convers. Manag. 300*, 117955. https://doi.org/10.1016/j.enconman.2023.117955

69 Locci, C., Mertens, M., Höyng, S., Schmid, G., Bagus, T., & Lettenmeier, P. (2024). Scaling-up PEM electrolysis production: Challenges and perspectives. *Chem. Ing. Tech. 96*(1–2), 22–29. https://doi.org/10.1002/cite.202300111

70 Belsa, B., Xia, L., & García de Arquer, F. P. (2024). CO_2 electrolysis technologies: bridging the gap toward scale-up and commercialization. *ACS Energy Lett. 9*(9), 4293–4305. https://doi.org/10.1021/acsenergylett.4c00955

71 Riesener, M., Schukat, E., Schäfer, N., Lehde genannt Kettler, N., & Junglas, S. (2025). Ramping-up electrolyser manufacturing for the green hydrogen transition: a planning approach for scalable factory concepts. *Electrochem. Sci. Adv. 5*(3), e70005. https://doi.org/10.1002/elsa.70005

72 Monopoli, D., Semeraro, C., Abdelkareem, M. A., Alami, A. H., Olabi, A. G., & Dassisti, M. (2024). How to build a digital twin for operating PEM-electrolyser system—A reference approach. *Annu. Rev. Control 57*, 100943. https://doi.org/10.1016/j.arcontrol.2024.100943

73 Weber, R. S. & Holladay, J. E. (2018). Modularized production of value-added products and fuels from distributed, waste carbon-rich feedstocks. *Engineering 4*, 330–335. https://doi.org/10.1016/j.eng.2018.05.012

74 Blanco, D. E., Lee, B., & Modestino, M. A. (2019). Optimizing organic electrosynthesis through controlled voltage dosing and artificial intelligence. *Proc. Natl. Acad. Sci. USA 116*(36), 17683–17689. https://doi.org/10.1073/pnas.1909985116

Index

Note: Page numbers in *italics* and **bold** refers to figures and tables respectively.

f

g

n